Graduate Texts in Mathematics 134

Springer
New York
Berlin
Heidelberg
Barcelona
Budapest
Hong Kong
London
Milan
Paris
Santa Clara
Singapore
Tokyo

Graduate Texts in Mathematics

continued after index

Steven Roman

Coding and Information Theory

With 31 Illustrations

Springer

Steven Roman
Department of Mathematics
California State University at Fullerton
Fullerton, CA 92634
USA

Mathematics Subject Classifications (1991): 05-01, 15-01, 94-01, 94A15, 94A24, 94Bxx, 94B05, 94B15, 94B20, 94B35, 94B65

Library of Congress Cataloging-in-Publication Data
Roman, Steven.
Coding and information theory / Steven Roman.
p. cm. – (Graduate texts in mathematics)
Includes bibliographical references and index.
ISBN 0-387-97812-7
1. Coding theory. 2. Information theory. I. Title. II. Series.
QA268.R65 1992
003′.54–dc20 92-3828

Printed on acid-free paper.

Production managed by Henry Krell; manufacturing supervised by Robert Paella.
Camera-ready copy prepared by author.
Printed and bound by R.R. Donnelley & Sons Co., Harrisonburg, VA.
Printed in the United States of America.

9 8 7 6 5 4 3

ISBN 0-387-97812-7 Springer-Verlag New York Berlin Heidelberg
ISBN 3-540-97812-7 Springer-Verlag Berlin Heidelberg New York SPIN 10675954

To Donna

Preface

This book is an introduction to the subjects of information and coding theory at the graduate or advanced undergraduate level. Prerequisites include a basic knowledge of elementary probability, as well as a foundation in modern and linear algebra, both at the undergraduate level. A glance at the first section of the appendix will show the reader what is expected in the way of modern algebra. All material on finite fields is developed from scratch in the text.

I have tried in this book to provide a thorough but *basic* introduction to the subjects of coding and information theory. My intention is to describe as clearly as I can the fundamental issues involved in these two subjects, rather than trying to cover all aspects of the theory. There are a few places where I have included more than is necessary for this purpose. In such cases, the sections are marked with an asterisk and can be omitted without loss of continuity.

The first quarter of the book is devoted to information theory – enough to discuss the basic aspects of the subject and give a full statement of the Noisy Coding Theorem, as well as a complete proof, in the case of the binary symmetric channel. While the information theory portion of the book can be omitted, it does provide a solid foundation to help appreciate the issues involved in both subjects.

Chapter 1 covers the topic of entropy. In Chapter 2, we discuss noiseless coding, including the Noiseless Coding Theorem and the Huffman algorithm for efficient source encoding. Chapter 3 is devoted to noisy coding and culminates in the proof of the Noisy Coding Theorem (and its converses) for the binary symmetric channel, using

random encoding. The information theory portion of the book has a decidedly probabilistic flavor.

The remaining portion of the book is devoted to coding theory, and has a decidedly algebraic flavor. The approach is theoretical in nature. For example, we discuss encoding and decoding algorithms, but do not cover the shift-register circuits that might be used to implement these algorithms.

Chapter 4 begins with a brief review for those readers who have not read the information theory portion of the book, and continues with general remarks on codes, including a brief discussion of several families of codes. Chapter 5 covers linear codes in detail, and Chapter 6 covers the Hamming, Golay and Reed-Muller codes. The first three sections of Chapter 7 are devoted to a fairly thorough discussion of the theory of finite fields and are followed by a discussion of cyclic codes. In Chapter 8, we study several families of cyclic codes, and one family of non-cyclic codes.

The appendix, meant to serve as a reference, contains a review of topics from modern algebra, along with a discussion of Möbius inversion, and binomial inequalities, both of which are used in the text. There is also a discussion of some computational techniques for finite fields, such as Berlekamp's factoring algorithm that are not used in the text. Several tables are included, among which the finite field tables will prove very useful for the exercise sets.

For a course based completely on coding theory, the instructor may begin with Chapter 4. For a somewhat more balanced treatment, Sections 1.1, 1.2, and 3.1-3.3 could be covered rather lightly, skipping proofs, before beginning the coding theory portion of the text.

A few of the sections in the book are fairly technical and may be omitted by the instructor if desired without loss of continuity. These sections are:

1) Section 1.3 on entropy for countably infinite probability distributions and typical sequences,
2) Section 3.4 on the proof of the Noisy Coding Theorem,
3) Section 4.5 on the main coding theory problem (this section begins with a summary of results that can be covered quickly),
4) Section 5.2 on weight distributions of codes (one can state the MacWilliams identity for linear codes [Corollary 5.2.9], and cover the related examples in the text),
5) Section 5.4 on invariant theory and self-dual codes.

Irvine, December 1991 *Steven Roman*

Contents

Sections marked with an asterisk are optional.

Chapter 2

Chapter 3

Part 2 Coding Theory

Chapter 4

General Remarks on Codes 119

Chapter 5

Chapter 7
Finite Fields and Cyclic Codes 285

Introduction

The main problem of information and coding theory can be described in a simple way as follows. Imagine that a stream of source data, say in the form of bits (0's and 1's), is being transmitted over a communications channel, such as a telephone line. From time to time, disruptions take place along the channel, causing some of the 0's to be turned into 1's, and vice-versa. The question is "How can we tell when the original source data has been changed, and when it has, how can we recover the original data?"

The issue of accurate communication is an extremely important one and arises in a variety of situations. A particularly important area for error detection and correction is in communication from space vehicles. Data that is in storage is also subject to errors, due to imperfections in the storage medium, for instance, and is therefore a form of communications channel to which this question also applies.

Let us illustrate the issues involved in dealing with the question of how to detect and correct errors. One of the most fundamental models of a communications channel is the *binary symmetric channel*, pictured in Figure 1. This model describes a situation in which an error is made by the channel – that is, a 0 is turned into a 1 or vice-versa – with probability $p < 1/2$. Thus, regardless of the input, a *channel error*, or *bit error*, occurs with probability p.

Now, let us imagine that we must design a scheme for detecting, and hopefully correcting, bit errors. One possibility would be to do nothing. When a bit is received at the output of the channel, we simply decide that it was correct. In this case, the probability of making a *decision error* is p.

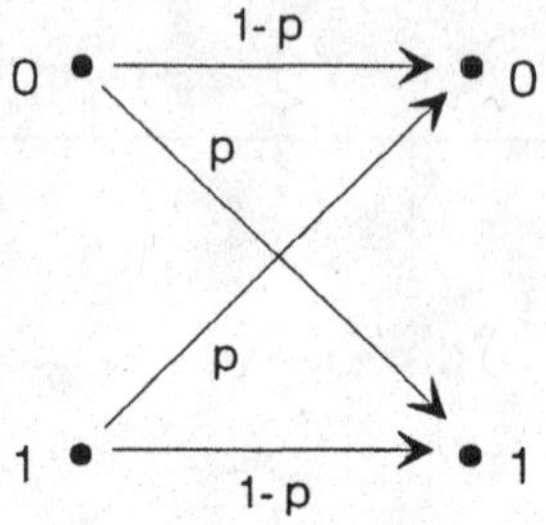

Figure 1

On the other hand, we can be a little more clever by instructing the sender to send each bit three times in succession. The receiver would then decide that the original source bit is the bit that appears a *majority* of the time at the output to the channel. For instance, if the output is 010, the receiver decides that the original bit was a 0.

Under the assumption that errors occur independently (not always a reasonable assumption, but certainly a convenient one), the probability of making a decision error is the probability that at least two bit errors have been made by the channel, which is

$$p_e = \binom{3}{2} p^2(1-p) + \binom{3}{3} p^3 = 3p^2 - 2p^3$$

Since this is less than p, for $p < 1/2$, we deduce that by *encoding* the original source data, we can reduce the probability of making a decision error. In this way, we are able to compensate in part for the loss that is inherent in the channel.

Taking this idea a step further, we can instruct the sender to send each bit $2n+1$ times in succession (we want an odd length so that majority decisions are always possible). As before, the receiver decides that the original source bit is the bit that appears a majority of the time in the output.

The probability of a decision error in this case is the probability that at least $n+1$ bit errors will be made by the channel. Under our independence assumption, the number of errors made by the channel has a binomial distribution with parameters $(2n+1,p)$, and so the expected number of errors is

$$(2n+1)p < n + \frac{1}{2}$$

for $p < 1/2$. Therefore, the weak law of large numbers tells us that the probability that at least $n+1$ channel errors are made tends to 0 as n tends to infinity. In other words, the probability that we make a

decision error tends to 0 as n gets large, and so we can compensate for channel errors to *any* desired degree by choosing n large enough.

However, we pay a heavy price for doing this, in terms of the efficiency of transmission of the source information. In particular, it takes a certain amount of time to send a bit through the channel, and if we send each bit 2n+1 times, we are spending 2n+1 units of channel time to send a *single* source bit. Thus, the *rate of source transmission* is $\frac{1}{2n+1}$ source bits per channel bit. For n large, this is likely to be an unacceptably low rate of transmission.

Nevertheless, we can see from this example that the basic idea is to *encode* source information, by adding additional information, sometimes referred to as *redundancy*, that can be used to detect, and perhaps correct, errors in transmission. The more redundancy that we add to the original source data, the more reliably we can detect and correct errors, but the less efficient we become at transmitting the source data, and so a compromise must be made.

Figure 2 shows the components that are involved in the communication process.

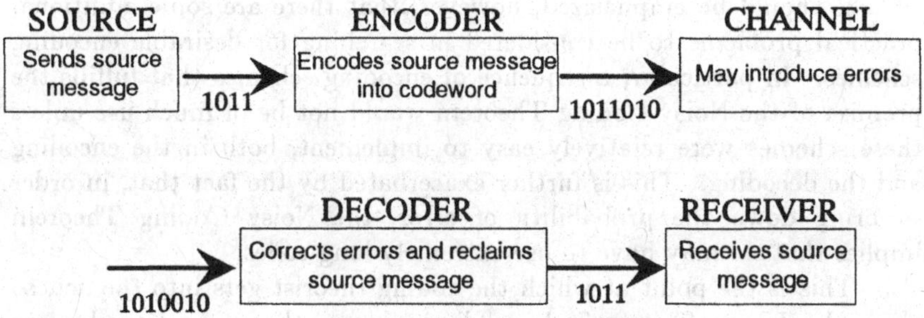

Figure 2

Information generated by the SOURCE is taken, either one bit at a time, or more often several bits at a time, and encoded by the ENCODER into a *codeword.* This is done by an *encoding scheme*, which consists of a *code* (set of codewords) and a function that assigns a codeword to each source message. The codeword is then sent over the CHANNEL, during which errors may be introduced. The DECODER accepts the output of the CHANNEL, attempts to correct any errors in transmission, and then recovers the original source message from the corrected codeword. Finally, the RECEIVER receives the (hopefully) correct source message.

Assuming that all source messages have the same length and all codewords have the same length, we can define the *rate of transmission*

of the communications scheme as the length of a source message divided by the length of the corresponding codeword. For instance, in the example shown in Figure 2, the rate of transmission is 4/7.

Now we come to a remarkable result known as the *Noisy Coding Theorem*. This theorem says that to every communications channel, there is a number $\mathfrak{C}$, called the *capacity* of the channel, with the following property. If we are willing to settle for a rate of transmission that is strictly below channel capacity, then there is an encoding scheme for the source data that will reduce the probability of a decision error to any desired level. In the case of the binary symmetric channel, the capacity is

$$\mathfrak{C}_{\text{bin sym}} = 1 + p \log_2 p + (1-p) \log_2(1-p)$$

For instance, if the probability of bit error is $p = 0.01$, then the capacity is $\mathfrak{C} \approx 0.91921$.

The Noisy Coding Theorem was first proved by Claude Shannon in 1948. Unfortunately, neither his proof, using a technique known as *random encoding*, nor any proof given since, is at all constructive, and at this point, no one has found a way to construct the encoding schemes promised by Shannon's theorem.

It should be emphasized, however, that there are some additional practical problems to be considered in searching for desirable encoding schemes. In particular, a sequence of encoding schemes that fulfills the promise of the Noisy Coding Theorem would not be of much use unless these schemes were relatively easy to implement, both in the encoding and the decoding. This is further exacerbated by the fact that, in order to bring down the probability of error, the Noisy Coding Theorem implies that we may have to use extremely long codes.

This is the point at which the coding theorist gets into the act, so to speak. In an effort to find encoding schemes that are relatively easy to implement, coding theorists have been led to search for codes that have considerable algebraic or geometric structure.

As a simple example, observe that the *source alphabet* $\mathbb{Z}_2 = \{0,1\}$ of the binary symmetric channel is in fact a group, under the operation of addition modulo 2. In symbols,

$$0 \oplus 0 = 0, \quad 0 \oplus 1 = 1 \oplus 0 = 1, \quad 1 \oplus 1 = 0$$

One way to use this algebraic structure is by taking source messages of length 7, say, and adjoining an *even parity check bit* to each message. That is, we add an 8th bit to the end of each source string in such a way that the total number of 1's in the resulting codeword is even. Put more formally, we encode the message string

$$a_1 \cdots a_7$$

as the codeword

$$a_1 \cdots a_7 a_8$$

where

$$a_1 \oplus \cdots a_7 \oplus a_8 = 0$$

Now, if a *single* error is made in transmission, the result will be a word that has *odd* parity, that is, an odd number of 1's. This tells the decoder that an error has occurred. Unfortunately, it does not allow for any error correction. Nevertheless, especially in situations where transmission can be repeated, this is a significant improvement over no error detection at all. In fact, this scheme is used by personal computers to detect errors in memory.

Let us consider a somewhat more sophisticated example. Observe that the set $\mathbb{Z}_2$ is actually a *field*, under addition and multiplication modulo 2. Furthermore, the set V(n,2) of all binary strings of length n is a vector space of dimension n over $\mathbb{Z}_2$, with componentwise addition modulo 2. For example, in V(4,2) we have

$$1101 \oplus 1001 = 0100$$

(Since the base field is $\mathbb{Z}_2$, scalar multiplication is trivial.) This implies that we may *add* source messages, as well as codewords.

Let us see how we might take advantage of this. Consider the matrix over $\mathbb{Z}_2$ defined by

$$G = \begin{bmatrix} 1 & 0 & 0 & 0 & 0 & 1 & 1 \\ 0 & 1 & 0 & 0 & 1 & 0 & 1 \\ 0 & 0 & 1 & 0 & 1 & 1 & 0 \\ 0 & 0 & 0 & 1 & 1 & 1 & 1 \end{bmatrix}$$

The rows of this matrix can be thought of as vectors in V(7,2), and since they are linearly independent, they span a 4-dimensional subspace of V(7,2). We denote this particular code by $\mathcal{H}$.

Now, let the message space be the vector space V(4,2) of all binary vectors of length 4. (We can only encode $2^4 = 16$ messages with this message space, but we are using it only for purposes of illustration.) If

$$\mathbf{a} = a_1 a_2 a_3 a_4$$

is a source message, we encode it by matrix multiplication, to produce the codeword **aG**. For instance, the source message $\mathbf{a} = 1011$ in Figure 2 becomes the codeword

$$\mathbf{c} = \mathbf{a}G = \begin{bmatrix} 1 & 0 & 1 & 1 \end{bmatrix} \begin{bmatrix} 1 & 0 & 0 & 0 & 0 & 1 & 1 \\ 0 & 1 & 0 & 0 & 1 & 0 & 1 \\ 0 & 0 & 1 & 0 & 1 & 1 & 0 \\ 0 & 0 & 0 & 1 & 1 & 1 & 1 \end{bmatrix} = \begin{bmatrix} 1 & 0 & 1 & 1 & 0 & 1 & 0 \end{bmatrix}$$

where all operations are performed modulo 2. Thus, $\mathbf{c} = 1011010$.

Notice that, since the leftmost 4×4 submatrix of G is an identity matrix, the first 4 bits of the codeword are identical to the source message. This makes recovering the source message from the codeword essentially trivial. The matrix G, whose rows form a basis for the code $\mathcal{H}$, is called a *generating matrix* for $\mathcal{H}$.

Now let us consider the problem of decoding with this code. Let H be the matrix

$$H = \begin{bmatrix} 0 & 0 & 0 & 1 & 1 & 1 & 1 \\ 0 & 1 & 1 & 0 & 0 & 1 & 1 \\ 1 & 0 & 1 & 0 & 1 & 0 & 1 \end{bmatrix}$$

It is not hard to show that each row of H is orthogonal to each row of G, where we use the ordinary dot product, except that all operations are performed modulo 2. For instance, taking the dot product of the first rows of each matrix gives

$$\begin{aligned} &\begin{bmatrix} 1 & 0 & 0 & 0 & 0 & 1 & 1 \end{bmatrix} \cdot \begin{bmatrix} 0 & 0 & 0 & 1 & 1 & 1 & 1 \end{bmatrix} \\ &= 1 \cdot 0 \oplus 0 \cdot 0 \oplus 0 \cdot 0 \oplus 0 \cdot 1 \oplus 0 \cdot 1 \oplus 1 \cdot 1 \oplus 1 \cdot 1 \\ &= 0 \oplus 0 \oplus 0 \oplus 0 \oplus 0 \oplus 1 \oplus 1 = 0 \end{aligned}$$

Thus, the rows of H, being linearly independent in V(7,2), form a basis for a 3-dimensional subspace $\mathcal{S}$ of V(7,2), with the property that $\mathcal{H} \subset \mathcal{S}^{\perp}$, where

$$\mathcal{S}^{\perp} = \{\mathbf{v} \in V(7,2) \mid \mathbf{v} \cdot \mathbf{s} = 0 \text{ for all } \mathbf{s} \in \mathcal{S}\}$$

is the orthogonal complement of $\mathcal{S}$. But it can be shown that $dim(\mathcal{S}^{\perp}) = dim(V(7,2)) - dim(\mathcal{S}) = 4$, and so $\mathcal{H} = \mathcal{S}^{\perp}$.

This gives us a very convenient description of the codewords in $\mathcal{H}$, for it tells us that a vector $\mathbf{c}$ is in $\mathcal{H}$ if and only if it is orthogonal to every row of the matrix H; that is,

$$\mathbf{c} \in \mathcal{H} \quad \text{if and only if} \quad \mathbf{c}H^{T} = \mathbf{0}$$

where $\mathbf{0}$ is the zero vector in V(7,2). For this reason, H is called a *parity check matrix* for the code $\mathcal{H}$.

Now, suppose that a codeword $\mathbf{c}$ is sent through the channel, but incurs a single error, say in the i-th position. Letting $\mathbf{e}_i$ be the vector in V(7,2) with a 1 in the i-th position, and 0's elsewhere, the output of the channel is the vector

$$\mathbf{x} = \mathbf{c} \oplus \mathbf{e}_i$$

Now let us compute

$$xH^T = (c \oplus e_i)H^T = cH^T \oplus e_iH^T = e_iH^T = \text{i-th column of } H$$

But notice that the matrix H has been cleverly designed so that its i-th column, read from the top down, is just the binary representation of the number i. In other words, the vector xH^T, known as the *syndrome* of the output vector **x**, when read as a binary number, tells us the precise location of the error. Thus, the code $\mathcal{H}$ can not only detect a single error, but can correct it as well!

For instance, a single error has been introduced into the codeword in Figure 2, and the output of the channel in this case is $\mathbf{x} = 1010010$. The syndrome of this vector is

$$xH^T = \begin{bmatrix} 1 & 0 & 1 & 0 & 0 & 1 & 0 \end{bmatrix} \cdot \begin{bmatrix} 0 & 0 & 1 \\ 0 & 1 & 0 \\ 0 & 1 & 1 \\ 1 & 0 & 0 \\ 1 & 0 & 1 \\ 1 & 1 & 0 \\ 1 & 1 & 1 \end{bmatrix} = \begin{bmatrix} 1 & 0 & 0 \end{bmatrix}$$

which is the binary number $100_2 = 4_{10}$. Hence, the error has occurred in the 4th position, as we can clearly see.

Thus, the code $\mathcal{H}$ is capable of correcting any single error in the transmission of a codeword. This code is one of the family of famous *Hamming codes*, which we will study at length in Chapter 6.

Let us examine the quality of the Hamming encoding scheme. Since each source symbol has length 4 and each codeword has length 7, the rate of transmission is 4/7. Furthermore, since any single error in a codeword will be corrected (but none others, as we will see), the probability of a decision error is the probability that at least 2 errors are made in the transmission of a 7-bit codeword. This probability is

$$p_e = 1 - 7p(1-p)^6 - p^7$$

where p is the probability of a channel error. (We are assuming a binary symmetric channel as before.) If $p = 0.01$, for instance, then $p_e \approx 0.00203$.

We could try to duplicate these numbers using a simple repeat-the-source strategy, by sending the same 3-bit source message three times in succession. This would give a rate of transmission equal to 1/3, which is less than 4/7. However, the probability of decision error using this scheme is the probability that more than one of the 3-bit messages is corrupted, and this is approximately 0.00259, which is larger than the corresponding number for the Hamming scheme.

As we mentioned in the preface, this book is divided into two parts. The first part of the book is devoted to the fundamentals of information theory, leading to a full statement of the Noisy Coding Theorem and its proof in the case of the binary symmetric channel. Then we proceed to the major portion of the book, which is devoted to a study of several families of codes that are used for error detection and correction.

Part 1

Information Theory

CHAPTER 1
Entropy

1.1 Entropy of a Source

Our goal in this section is to develop a satisfactory measure of the amount of information contained in an information source. Let us begin with a formal definition of the term source.

Definition A **source** is an ordered pair $\mathcal{S} = (S,P)$, where $S = \{x_1,\ldots,x_n\}$ is a finite set, known as a **source alphabet**, and P is a probability distribution on S. We denote the probability of x_i by p_i, or $p(x_i)$. ▯

Suppose that we *sample* a source $\mathcal{S} = (S,P)$, that is, we chose an element of S at random according to the probability distribution P. Thus, the probability that x_i is chosen is $p(x_i)$. Before the sampling takes place, there is a certain amount of *uncertainty* associated with the outcome, and after the sampling, we have gained a certain amount of *information* about the source. Thus, the concepts of uncertainty and information are related.

To illustrate this further, let us consider some extreme cases. If $p(x_1) = 1$, and $p(x_i) = 0$ for $i > 1$, then the element x_1 will always be chosen, and so in this case there is no uncertainty, that is, the uncertainty is zero. Put another way, we get no information from the sample, since there was nothing to learn about this source. Similarly, if only a "few" of the elements of S have nonzero probabilities of being chosen, then we can be reasonably certain about the outcome, and so the uncertainty is small and the amount of information in the source is

small. On the other hand, it seems clear that the uncertainty should be a maximum when each of the outcomes is equally likely, that is, when $p_i = \frac{1}{n}$ for all $i = 1,\ldots,n$. In this case, we get the maximum amount of information by sampling the source.

Since the concepts of uncertainty and information are in this sense equivalent, we will feel free to continue our discussion using either concept.

THE ENTROPY FUNCTION $H(p_1, \ldots, p_n)$

Now we wish to define a function $H(p_1,\ldots,p_n)$ to measure the uncertainty involved in sampling from a source. Notice that our notation alludes to the fact that the function H depends only on the probability distribution and not on the elements of the source alphabet. In order to define H, we need to look a bit more carefully at the notion of uncertainty. Of course, we want $H(p_1,\ldots,p_n)$ to be defined for all $p_1,\ldots,p_n$ satisfying $0 \leq p_i \leq 1$, $\sum p_i = 1$. Also, since a small change in probabilities should produce only a small change in uncertainty, we require that H be continuous. Next, when all outcomes are equally likely, it seems reasonable that the more outcomes there are, the greater should be the uncertainty. Thus, we require that

$$H(\tfrac{1}{n},\ldots,\tfrac{1}{n}) < H(\tfrac{1}{n+1},\ldots,\tfrac{1}{n+1})$$

Finally, suppose that the elements of $S = \{x_1,\ldots,x_n\}$ are partitioned into nonempty disjoint blocks $B_1,\ldots,B_k$, where $|B_i| = b_i$, and of course, $\sum b_i = n$. Consider the following experiment. We first pick a block B_i, with probability proportional to its size, that is, $P(B_i) = b_i/n$, and then we pick an element with equal probability, from the chosen block B_i. Now, if x_j is in block B_u, then since

$$P(x_j \mid B_i) = \begin{cases} 0 & \text{if } i \neq u \\ \frac{1}{b_u} & \text{if } i = u \end{cases}$$

we have

$$P(x_j) = \sum_{i=1}^{n} P(x_j \mid B_i)\, P(B_i) = \frac{1}{b_u}\,\frac{b_u}{n} = \frac{1}{n}$$

Hence, the probability of picking x_j is the same under these conditions as if we choose directly from S with equal probability. Therefore, the uncertainty in the outcomes should be the same.

Now, the uncertainty in choosing directly from S with equal probabilities is $H(\frac{1}{n},\ldots,\frac{1}{n})$. On the other hand, the uncertainty in choosing one of the blocks $B_1,\ldots,B_k$ is

$$H\left(\frac{b_1}{n},\ldots,\frac{b_k}{n}\right)$$

and, once a block has been chosen, we still have the uncertainty involved in choosing an element from that block. The average uncertainty in that process is

$$\sum_{i=1}^{k} P(B_i)\cdot(\text{Uncertainty in choosing from } B_i) = \sum_{i=1}^{k} \frac{b_i}{n} H\left(\frac{1}{b_i},\ldots,\frac{1}{b_i}\right)$$

Thus, we get

$$H\left(\tfrac{1}{n},\ldots,\tfrac{1}{n}\right) = H\left(\frac{b_1}{n},\ldots,\frac{b_k}{n}\right) + \sum_{i=1}^{k} \frac{b_i}{n} H\left(\frac{1}{b_i},\ldots,\frac{1}{b_i}\right)$$

In summary, we want our uncertainty function H to have the following properties. Let $\mathbb{Z}^+$ denote the positive integers.

1) $H(p_1,\ldots,p_n)$ is defined and continuous for all $p_1,\ldots,p_n$ satisfying $0 \le p_i \le 1$, $\sum p_i = 1$.

2) $H(\frac{1}{n},\ldots,\frac{1}{n}) < H(\frac{1}{n+1},\ldots,\frac{1}{n+1})$, for $n \in \mathbb{Z}^+$.

3) For $b_i \in \mathbb{Z}^+$, $\sum b_i = n$

$$H\left(\tfrac{1}{n},\ldots,\tfrac{1}{n}\right) = H\left(\frac{b_1}{n},\ldots,\frac{b_k}{n}\right) + \sum_{i=1}^{k} \frac{b_i}{n} H\left(\frac{1}{b_i},\ldots,\frac{1}{b_i}\right)$$

It turns out that properties 1)-3) uniquely define a function H.

Theorem 1.1.1 A function H satisfies properties 1)-3) if and only if it has the form

(1.1.1)
$$H_b(p_1,\ldots,p_n) = -\sum_{i=1}^{n} p_i \log_b p_i$$

where $b > 1$, and where we set $p \log_b p = 0$ for $p = 0$.

Proof. We will leave it as an exercise to show that the function defined by (1.1.1) satisfies properties 1)-3). For the converse, we proceed as follows.

Choose positive integers m and n for which $m \mid n$, and let $b_i = m$ for all $i = 1,\ldots,k$. Then, since $mk = \sum b_i = n$, we have $k = n/m$ and property 3 gives

$$\begin{aligned} H\left(\tfrac{1}{n},\ldots,\tfrac{1}{n}\right) &= H\left(\tfrac{m}{n},\ldots,\tfrac{m}{n}\right) + \sum_{i=1}^{k} \frac{m}{n} H\left(\tfrac{1}{m},\ldots,\tfrac{1}{m}\right) \\ &= H\left(\tfrac{m}{n},\ldots,\tfrac{m}{n}\right) + H\left(\tfrac{1}{m},\ldots,\tfrac{1}{m}\right) \end{aligned}$$

If $n = m^s$, where m and s are positive integers, this becomes

$$H\left(\frac{1}{m^s},\ldots,\frac{1}{m^s}\right) = H\left(\frac{1}{m^{s-1}},\ldots,\frac{1}{m^{s-1}}\right) + H\left(\frac{1}{m},\ldots,\frac{1}{m}\right)$$

Defining the function g by $g(n) = H\left(\frac{1}{n},\ldots,\frac{1}{n}\right)$, we get

$$g(m^s) = g(m^{s-1}) + g(m)$$

which implies that

$$g(m^s) = sg(m) \tag{1.1.2}$$

for all positive integers m and s. (Observe the appearance of logarithmlike properties.) Further, since property 2) is equivalent to saying that $g(n)$ is a strictly increasing function, we have

$$g(m^s) < g(m^{s+1})$$

and so

$$sg(m) < (s+1)g(m)$$

which implies that $g(m)$ must be positive.

Now, for positive integers r and t, let s be chosen so that

$$m^s \le r^t < m^{s+1} \tag{1.1.3}$$

Then since g is increasing, we have

$$g(m^s) \le g(r^t) < g(m^{s+1})$$

or

$$sg(m) \le tg(r) < (s+1)g(m)$$

or

$$\frac{s}{t} \le \frac{g(r)}{g(m)} < \frac{s+1}{t} \tag{1.1.4}$$

But from (1.1.3), we also have

$$s \log m \le t \log r < (s+1) \log m$$

and so

$$\frac{s}{t} \le \frac{\log r}{\log m} < \frac{s+1}{t}$$

Combining this with (1.1.4) gives

$$-\frac{1}{t} \le \frac{g(r)}{g(m)} - \frac{\log r}{\log m} < \frac{1}{t}$$

and since t was arbitrary, we conclude that

$$\frac{g(r)}{g(m)} = \frac{\log r}{\log m}$$

or

$$\frac{g(r)}{\log r} = \frac{g(m)}{\log m}$$

Hence, $g(r)/\log r$ is constant for all positive integers r, that is, $g(r) = C \log r$, where $C > 0$, since we have already established that $g(r) > 0$. By choosing the base of the logarithm appropriately, we may assume that $C = 1$, and so

$$(1.1.5) \qquad g(r) = \log_b r$$

for all positive integers r.

Now let us take another look at property 3), which in view of (1.1.5) can be written

$$H\left(\frac{b_1}{n},\ldots,\frac{b_k}{n}\right) = g(n) - \sum_{i=1}^{k} \frac{b_i}{n} g(b_i)$$

$$= \log_b n - \sum_{i=1}^{k} \frac{b_i}{n} \log_b b_i = - \sum_{i=1}^{k} \frac{b_i}{n} \log_b \frac{b_i}{n}$$

Since any positive rational numbers $p_1,,\ldots,p_k$ can be written in the form

$$\frac{b_1}{n},\ldots,\frac{b_k}{n}$$

simply by forming a common denominator, we get

$$(1.1.6) \qquad H(p_1,\ldots,p_k) = -\sum_{i=1}^{k} p_i \log_b p_i$$

for all positive rational numbers $p_1,\ldots,p_n$. Further, since the function H is assumed to be continuous, this must also hold for all positive real numbers $p_i,\ldots,p_k$. Finally, we observe that

$$\lim_{p\to 0+} p \log_b p = 0$$

and so (1.1.6) holds for all nonnegative real numbers $p_1,\ldots,p_k$. ∎

Definition Let $P = \{p_1,\ldots,p_n\}$ be a probability distribution. Then the quantity

$$H_b(p_1,\ldots,p_n) = -\sum_{i=1}^{n} p_i \log_b p_i = \sum_{i=1}^{n} p_i \log_b \frac{1}{p_i}$$

is called the b-ary **entropy** of the distribution P. If $\mathcal{S} = (S,P)$ is a source, with $P(x_i) = p_i$, then we refer to $H_b(\mathcal{S}) = H_b(p_1,\ldots,p_n)$ as the **entropy** of $\mathcal{S}$. □

The term *entropy* was first used by Clausius in 1864, and first introduced into information theory by Shannon in 1948. Note that entropy measures both the amount of uncertainty in a distribution before sampling, and the amount of information obtained by sampling.

We should emphasize that, while Theorem 1.1.1 tells us that there

is essentially only one function that satisfies the intuitive properties 1)-3) of uncertainty, this would not be useful if it were not for the fact that entropy does indeed play a key role in information theory, as we will see. In other words, the true justification for defining and studying the entropy function is its role in the upcoming theory and not that it happens to satisfy properties 1)-3).

A few remarks about the base b are in order here. For many results, the base does not need emphasis, and so we will adopt the "generic" notation

$$H(p_1,\ldots,p_n) = \sum_{i=1}^{n} p_i \log \frac{1}{p_i}$$

for entropy. Should this lead to any possibility of confusion, we will return to the subscripted notation. We should also mention that many books on information theory restrict attention to base 2, that is, to *binary* entropy, and use the notation $H(p_1,\ldots,p_n)$ for binary entropy.

THE UNITS OF ENTROPY

As to the matter of units, if we set $S = \{0,1,\ldots,k-1\}$, then it seems reasonable to say that sampling from S with equal probability gives an amount of information equal to one k-ary unit. For instance, if $S = \{0,1\}$, then sampling from S with equal probability gives one one binary unit of information, or one *bit* of information. Hence, since

$$H_k\left(\frac{1}{k},\ldots,\frac{1}{k}\right) = -\sum_{i=1}^{k} \frac{1}{k} \log_k \frac{1}{k} = 1$$

we see that H_k measures the number of k-ary units of information. In particular, the *binary entropy* $H_2(p_1,\ldots,p_n)$ measures information in *binary units,* or *bits*, and the *natural entropy* $H_e(p_1,\ldots,p_n)$ measures information in *natural units,* or *nats.*

Example 1.1.1 Sampling from the set $S = \{x_1,x_2,x_3\}$ with equal probabilities $p_i = \frac{1}{3}$ gives

$$H_2\left(\frac{1}{3},\frac{1}{3},\frac{1}{3}\right) = \frac{1}{3}\log_2 3 + \frac{1}{3}\log_2 3 + \frac{1}{3}\log_2 3 = \log_2 3 \approx 1.585 \text{ bits}$$

Sampling from $S=\{x_1,x_2,x_3\}$ with probabilities $p_1 = p_2 = \frac{1}{4}$ and $p_3 = \frac{1}{2}$ gives

$$H_2\left(\frac{1}{4},\frac{1}{4},\frac{1}{2}\right) = \frac{1}{4}\log_2 4 + \frac{1}{4}\log_2 4 + \frac{1}{2}\log_2 2 = 1.5 \text{ bits}$$

As expected, since we are more certain about the outcome in the second case, its uncertainty is smaller. ▯

Example 1.1.2 Table 1.1.1 shows the letters of the alphabet, along with their approximate probabilities of occurrence in the English language. (The letters are listed in decreasing order of frequency.) With the help of a computer, we see that the binary entropy is approximately 4.07991 bits. Thus, we get an average of 4.07991 bits of information by sampling a single letter from English text. ▯

TABLE 1.1.1

Letter	*Probability*	*Letter*	*Probability*
(Space)	0.1859	F	0.0208
E	0.1031	M	0.0198
T	0.0796	W	0.0175
A	0.0642	Y	0.0164
O	0.0632	P	0.0152
I	0.0575	G	0.0152
N	0.0574	B	0.0127
S	0.0514	V	0.0083
R	0.0484	K	0.0049
H	0.0467	X	0.0013
L	0.0321	Q	0.0008
D	0.0317	J	0.0008
U	0.0228	Z	0.0005
C	0.0218		

Example 1.1.3 The important entropy function

$$H_2(p,1-p) = p \log_2 \frac{1}{p} + (1-p) \log_2 \frac{1}{1-p}$$

(note the base 2) is often denoted by H(p) and called *the* entropy function. Its graph appears in Figure 1.1.1. ▯

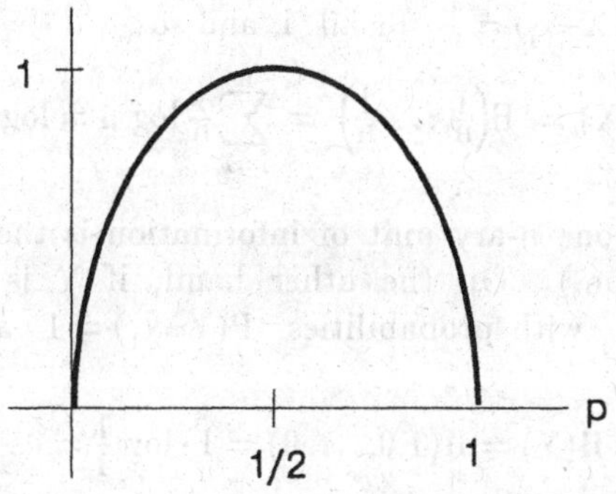

Figure 1.1.1 The entropy function H(p)

THE ENTROPY OF A RANDOM VARIABLE; JOINT ENTROPY

Most results concerning entropy are expressed in terms of the entropy of a random variable X or a random vector $\mathbf{X} = (X_1,\ldots,X_n)$. Accordingly, we make the following definitions. (These definitions will also help set our notation.)

Definition Let X be a random variable with range $S = \{x_1,\ldots,x_n\}$. If $P(X{=}x_i) = p(x_i)$, then the **entropy** of X is defined by

$$H(X) = \sum_{i=1}^{n} p(x_i) \log \frac{1}{p(x_i)} \qquad \square$$

Definition Let X and Y be random variables, where X has range $S_1 = \{x_1,\ldots,x_n\}$ and Y has range $S_2 = \{y_1,\ldots,y_m\}$. If $P(X{=}x_i,Y{=}y_j) = p(x_i,y_j)$, then the **joint entropy** of X and Y is defined by

$$H(X,Y) = \sum_{i,j} p(x_i,y_j) \log \frac{1}{p(x_i,y_j)}$$

The **entropy** of the random vector $\mathbf{X} = (X,Y)$ is defined by $H(\mathbf{X}) = H(X,Y)$. $\square$

Definition Let $X_1,\ldots,X_k$ be random variables, where X_i has range S_i. If $P(X_1{=}x_1,\ldots,X_n{=}x_n) = p(x_1,\ldots,x_n)$, then the **joint entropy** of $X_1,\ldots,X_k$ is defined by

$$H(X_1,\ldots,X_n) = \sum_{x_1 \in S_1,\ldots,x_n \in S_n} p(x_1,\ldots,x_n) \log \frac{1}{p(x_1,\ldots,x_n)}$$

The **entropy** of the random vector $\mathbf{X} = (X_1,\ldots,X_n)$ is defined by $H(\mathbf{X}) = H(X_1,\ldots,X_n)$. $\square$

Example 1.1.4 Suppose we sample with equal probability from the set $S = \{x_1,\ldots,x_n\}$. If the random variable X denotes the outcome of the sampling, then $P(X{=}x_i) = \frac{1}{n}$ for all i, and so

$$H(X) = H\left(\frac{1}{n},\ldots,\frac{1}{n}\right) = \sum_{i=1}^{n} \frac{1}{n} \log n = \log n$$

(This shows that one n-ary unit of information is the same as $\log_2 n$ bits of information.) On the other hand, if Y is the outcome of sampling from S with probabilities $P(Y{=}x_1) = 1$ and $P(Y{=}x_i) = 0$ for $i > 1$, then

$$H(Y) = H(1,0,\ldots,0) = 1 \cdot \log \frac{1}{1} = 0$$

We will prove in the next section that these are the extreme values for entropy. ☐

Example 1.1.5 If the random variables X and Y defined in Example 1.1.4 are independent, then

$$\begin{aligned} p(x_i,x_j) &= P(X=x_i,Y=x_j) \\ &= P(X=x_i)P(Y=x_j) = p(x_i)p(x_j) = p(x_i)\delta_{j,1} \end{aligned}$$

where $\delta_{j,1} = 1$ if $j = 1$, and 0 otherwise, and so the joint entropy of X and Y is

$$H(X,Y) = \sum_{i,j} p(x_i)\delta_{j,1} \log \frac{1}{p(x_i)\delta_{j,1}} = \sum_i p(x_i) \log \frac{1}{p(x_i)} = H(X)$$

This says that, under these circumstances, the information obtained by sampling both random variables is the same as the information obtained by sampling X alone. ☐

EXERCISES

1. Show that the function defined by (1.1.1) satisfies properties 1)-3).
2. Compute $H_2(1/8,1/8,3/4)$.
3. Compute $H_2(1/3,2/3)$.
4. Compute $H(1/a,\ldots,1/a,2/a,2/a)$.
5. Find a relationship between $H_b(X)$ and $H_c(X)$.
6. Compute $H'(p)$, where $H(p)$ is the entropy function of Example 1.1.3.
7. Prove that the entropy function $H(p)$ of Example 1.1.3 is symmetric about the line $x = 1/2$.
8. Suppose we toss a fair coin, and if the outcome is a heads, we toss it again. How much uncertainty is there in the outcome?
9. Suppose we toss a fair coin and roll a fair die. Do we get more information from this experiment or from the experiment of tossing three fair coins? four fair coins?
10. How much information do we get by sampling from a deck of cards if
    (a) each card is equally likely to be drawn?
    (b) the black cards are twice as likely to be drawn as the red cards?
11. Suppose that we roll a fair die that has two faces numbered 1, two faces numbered 2, and two faces numbered 3. Then we toss a fair coin the number of times indicated by the number on the die. How much information do we get by this procedure?

12. The accuracy of a certain radio station's weatherman at predicting rain is given by the following chart

	Actual rain	Actual no rain
Predicts rain	1/12	1/6
Predicts no rain	1/12	2/3

For instance, 1/12 of the time the weatherman predicts rain when in fact it does rain. Notice that the weatherman is correct 3/4 of the time. Now, an unemployed listener observes that he could be correct 5/6 of the time by simply always predicting no rain, and so he applies for the weatherman's job. However, the station manager declines to hire the listener. Why?

13. Let $S = \{0,1\}$ be a source with $P(0) = p$. Let X and Y be independent random samples from this source. Let Z be the number of 0's in the pair $\{X,Y\}$. Find the entropy of the random variable Z. Compute $H_2(X,Y) - H_2(Z)$ and interpret the result.

14. Let $S = \{0,1\}$ be a source, with $P(0) = p$. Sample this source independently twice, to get X_1 and X_2, and let $Y = 0$ if $X_1 = X_2$ and $Y = 1$ if $X_1 \neq X_2$.
(a) Find $H(Y)$.
(b) Show that $H(X_1,Y) = H(X_1,X_2)$. What does this say?
(c) Predict the value of $H(X_1,Y) - H(X_1)$, and then justify your prediction.

15. Let $\mathcal{S}_1 = (S_1,P_1)$ and $\mathcal{S}_2 = (S_2,P_2)$ be sources, with $S_1 = \{x_1,\dots,x_n\}$, $P_1(x_i) = p_i$ and $S_2 = \{y_1,\dots,y_m\}$, $P_2(y_i) = q_i$. Let $\lambda,\mu \geq 0$, $\lambda + \mu = 1$. Define the *mixed source* $\mathcal{S} = \lambda\mathcal{S}_1 + \mu\mathcal{S}_2$ to have alphabet $S_1 \cup S_2$ and probabilities $P(x_i) = \lambda p_i$, $P(y_i) = \mu q_i$.
(a) Calculate the entropy of $\mathcal{S}$.
(b) Determine the value of λ that maximizes this entropy.

Let E be an event with probability p. We define the ***information*** *obtained by an occurrence of E to be $I(p) = -\log_2 p$. Use this definition for Exercises 16-19.*

16. Show that $I(p)$ is characterized by the fact that it is the only continuous function on $(0,1]$ with the property that $I(pq) = I(p) + I(q)$ and $I(\frac{1}{2}) = 1$.

17. What is the relationship between $I(p)$ and $H(p_1,\dots,p_n)$?

18. (a) A personal computer monitor is capable of displaying pictures made up of pixels at a resolution of 640 columns by 480 rows. If each pixel can be in any one of 16 colors, estimate the amount of information in a random picture.

(b) Estimate the information obtained from a random speech of 1,000 words, assuming a 10,000 word vocabulary. (This shows that a picture is actually worth more than a thousand words!)

19. Using Table 1.1.1, compute I(E), I(R), I(Z).

1.2 Properties of Entropy

To establish the main properties of entropy, we begin with two lemmas.

Lemma 1.2.1 If $\ln x$ denotes the natural logarithm of x, then

$$\ln x \leq x-1$$

for all $x > 0$, with equality holding if and only if $x = 1$. □

Lemma 1.2.2 Let $P = \{p_1, p_2, \ldots, p_n\}$ be a probability distribution, that is, $0 \leq p_i \leq 1$ and $\sum p_i = 1$. Let $Q = \{q_1, q_2, \ldots, q_n\}$ have the property that $0 \leq q_i \leq 1$ and $\sum q_i \leq 1$ (note the *inequality* here). Then

$$\sum_{i=1}^{n} p_i \log \frac{1}{p_i} \leq \sum_{i=1}^{n} p_i \log \frac{1}{q_i}$$

where $0 \cdot \log \frac{1}{0} = 0$ and $p \cdot \log \frac{1}{0} = +\infty$ for $p > 0$. Furthermore, equality holds if and only if $q_i = p_i$ for all i.
Proof. By multiplying both sides of this inequality by an appropriate constant, we may assume that all logarithms are natural. Now let us show that

$$p_i \ln \frac{1}{p_i} \leq p_i \ln \frac{1}{q_i} + q_i - p_i \tag{1.2.1}$$

If $p_i = 0$, this becomes $0 \leq q_i$, which is certainly true. If $p_i \neq 0$ but $q_i = 0$, then (1.2.1) holds since the right side is $+\infty$. Finally, if p_i and q_i are positive, then we can write (1.2.1) in the form

$$p_i \ln \frac{q_i}{p_i} \leq q_i - p_i$$

or

$$\ln \frac{q_i}{p_i} \leq \frac{q_i}{p_i} - 1$$

which holds by Lemma 1.2.1. Summing (1.2.1) on i gives

$$\sum_i p_i \ln \frac{1}{p_i} \leq \sum_i p_i \ln \frac{1}{q_i} + \sum_i q_i - \sum_i p_i \leq \sum_i p_i \ln \frac{1}{q_i}$$

which proves the inequality. By looking at the proof of (1.2.1), and the conditions implying equality in Lemma 1.2.1, we see that equality holds if and only if $p_i = q_i$ for all i. ∎

THE RANGE OF THE ENTROPY FUNCTION

With Lemma 1.2.2 at our disposal, we can prove the following.

Theorem 1.2.3 Let X be a discrete random variable with range $\{x_1,\ldots,x_n\}$. Then

$$0 \le H(X) \le \log n$$

Furthermore, $H(X) = \log n$ if and only if $p(x_i) = \frac{1}{n}$ for all i, and $H(X) = 0$ if and only if $p(x_i) = 1$ for some i.

Proof. Applying Lemma 1.2.2 to the distribution of X and to the uniform distribution $Q = \{\frac{1}{n},\ldots,\frac{1}{n}\}$, we get

$$H(X) = \sum_{i=1}^{n} p(x_i) \log \frac{1}{p(x_i)} \le \sum_{i=1}^{n} p(x_i) \log \frac{1}{1/n}$$

$$= \sum_{i=1}^{n} p(x_i) \log n = (\log n) \sum_{i=1}^{n} p(x_i) = \log n$$

Thus, $H(X) \le \log n$. Furthermore, equality holds here precisely when equality holds in Lemma 1.2.2, that is, when $p(x_i) = \frac{1}{n}$, for all i. We will leave the proof of the remainder of the theorem as an exercise. ∎

Theorem 1.2.3 confirms the fact that the most information is obtained when sampling from a uniform distribution.

A GROUPING AXIOM FOR ENTROPY

Property 3 of Section 1.1 is an example of a *grouping axiom* for entropy. Here is another grouping axiom, whose proof we leave as an exercise.

Theorem 1.2.4 Let $\{p_1,\ldots,p_n,q_1,\ldots,q_m\}$ be a probability distribution. If $a = p_1 + \cdots + p_n$ then

$$H(p_1,\ldots,p_n,q_1,\ldots,q_m) = H(a,1-a) + aH\left(\frac{p_1}{a},\ldots,\frac{p_n}{a}\right) + (1-a)H\left(\frac{q_1}{1-a},\ldots,\frac{q_m}{1-a}\right)$$ □

PROPERTIES OF JOINT ENTROPY

Now let us consider some properties of joint entropy. It seems reasonable that the joint information obtained by sampling two random variables should be no greater than the sum of the information obtained by sampling each random variable separately, with equality holding precisely when the random variables are independent. Our next theorem confirms this.

Theorem 1.2.5 Let X and Y be discrete random variables. Then

$$H(X,Y) \leq H(X) + H(Y)$$

with equality holding if and only if X and Y are independent.
Proof. Since $\sum_j p(x_i,y_j) = p(x_i)$ and $\sum_i p(x_i,y_j) = p(y_j)$, we have

$$\begin{aligned} H(X) + H(Y) &= \sum_i p(x_i) \log \frac{1}{p(x_i)} + \sum_j p(y_j) \log \frac{1}{p(y_j)} \\ &= \sum_{i,j} p(x_i,y_j) \log \frac{1}{p(x_i)} + \sum_{i,j} p(x_i,y_j) \log \frac{1}{p(y_j)} \\ &= \sum_{i,j} p(x_i,y_j) \log \frac{1}{p(x_i)p(y_j)} \end{aligned}$$

Now, since $\sum_{i,j} p(x_i)p(y_j) = 1$, we may apply Lemma 1.2.2 to get

$$H(X) + H(Y) \geq \sum_{i,j} p(x_i,y_j) \log \frac{1}{p(x_i,y_j)} = H(X,Y)$$

Finally, according to Lemma 1.2.2, equality holds here if and only if

$$p(x_i)p(y_j) = p(x_i,y_j)$$

that is, if and only if X and Y are independent. ∎

Theorem 1.2.5 generalizes readily to more than two variables. We leave the proofs as exercises.

Corollary 1.2.6 Let $X_1,\ldots,X_n$ be discrete random variables. Then

$$H(X_1,\ldots,X_n) \leq H(X_1) + \cdots + H(X_n)$$

with equality holding if and only if the X_i are independent. □

Corollary 1.2.7 Let $X_1,\ldots,X_n$ and $Y_1,\ldots,Y_m$ be discrete random variables. Then

$$H(X_1,\ldots,X_n,Y_1,\ldots,Y_m) \leq H(X_1,\ldots,X_n) + H(Y_1,\ldots,Y_m)$$

with equality holding if and only if the random vectors $\mathbf{X} = (X_1,\ldots,X_n)$ and $\mathbf{Y} = (Y_1,\ldots,Y_m)$ are independent. □

THE ENTROPY FUNCTION H(p)

There is a very interesting relationship between the entropy

function

$$H(p) = p \log \frac{1}{p} + (1-p) \log \frac{1}{1-p}$$

($\log = \log_2$) first discussed in Example 1.1.3, and certain sums of binomial coefficients, which we will use in our proof of the Noisy Coding Theorem in Chapter 3 and again in later chapters.

Theorem 1.2.8 For $0 \leq \lambda \leq \frac{1}{2}$, we have

$$\sum_{k=0}^{\lfloor \lambda n \rfloor} \binom{n}{k} \leq 2^{nH(\lambda)}$$

where $H(\lambda) = -\lambda \log \lambda - (1-\lambda)\log(1-\lambda)$ is the entropy function.

Proof. If $\lambda = 0$, both sides of this inequality equal 1. If $\lambda = 1/2$, then since $H(1/2) = 1$, the right side is equal to 2^n, and so the inequality holds here as well. Let us assume that $0 < \lambda < 1/2$.

We begin with Chebyshev's inequality. If X is a random variable that takes on nonnegative values only, then Chebyshev's inequality says

$$P(X \geq a) \leq \frac{\mathcal{E}(X)}{a} \quad \text{for all } a > 0 \tag{1.2.2}$$

Suppose that X has the form $X = e^{tY}$, where Y is a random variable, and t is a real number. If we set $a = e^{tb}$, then (1.2.2) becomes

$$P(e^{tY} \geq e^{tb}) \leq \frac{\mathcal{E}(e^{tY})}{e^{tb}} \quad \text{for all } b \in \mathbb{R}$$

Now, if $t < 0$, then we have $e^{tY} \geq e^{tb}$ if and only if $tY \geq tb$ if and only if $Y \leq b$, and so this is equivalent to

$$P(Y \leq b) \leq \frac{\mathcal{E}(e^{tY})}{e^{tb}} \quad \text{for all } b \in \mathbb{R} \text{ and } t < 0 \tag{1.2.3}$$

If Y is a binomial random variable, with parameters (n,p), then

$$P(Y \leq b) = \sum_{k=0}^{b} \binom{n}{k} p^k q^{n-k}$$

where $q = 1 - p$, and where, if b is not an integer, the upper limit of summation is understood to be $\lfloor b \rfloor$. Furthermore, $\mathcal{E}(e^{tY})$ is the binomial moment generating function, which is well known to be

$$\mathcal{E}(e^{tY}) = (q + pe^t)^n$$

Thus, (1.2.3) becomes

$$\sum_{k=0}^{b} \binom{n}{k} p^k q^{n-k} \leq e^{-tb}(q + pe^t)^n$$

Setting $b = \lambda n$, where $0 < \lambda < 1$, we get

$$\sum_{k=0}^{\lambda n} \binom{n}{k} p^k q^{n-k} \le e^{-\lambda n t}(q + pe^t)^n \tag{1.2.4}$$

valid for $t < 0$.

By setting $x = e^t$ in the right hand side of (1.2.4) and minimizing over $0 < x < 1$ (which is equivalent to $t < 0$), we see that (1.2.4) is best when

$$e^t = \frac{\lambda q}{\mu p}$$

where $\mu = 1 - \lambda$, and $\lambda < p$. Substituting this value of e^t into the right hand side of (1.2.4) gives

$$\left(\frac{\lambda q}{\mu p}\right)^{-\lambda n}\left(q + p\frac{\lambda q}{\mu p}\right)^n = \left(\frac{\lambda q}{\mu p}\right)^{-\lambda n} q^n\left(1 + \frac{\lambda}{\mu}\right)^n$$

$$= \left(\frac{\lambda q}{\mu p}\right)^{-\lambda n}\left(\frac{q}{\mu}\right)^n = \lambda^{-\lambda n}\mu^{-\mu n}p^{\lambda n}q^{\mu n}$$

and so (1.2.4) becomes

$$\sum_{k=0}^{\lambda n} \binom{n}{k} p^k q^{n-k} \le \lambda^{-\lambda n}\mu^{-\mu n}p^{\lambda n}q^{\mu n} \tag{1.2.5}$$

for $\lambda < p$. (This is a useful result as well.) Setting $p = q = \frac{1}{2}$ gives

$$\sum_{k=0}^{\lambda n} \binom{n}{k} \le \lambda^{-\lambda n}\mu^{-\mu n}$$

for $\lambda < \frac{1}{2}$. But $\lambda^{-\lambda n}\mu^{-\mu n} = 2^{n[-\lambda \log \lambda - \mu \log \mu]} = 2^{nH(\lambda)}$, which gives the desired result. ∎

THE CONVEXITY OF THE ENTROPY FUNCTION

Let K be a subset of $\mathbb{R}^n$. We say that K is **convex** if $x,y \in K$ implies that

$$ax+(1-a)y$$

is also in K for $0 \le a \le 1$. Of course, the set

$$\{ax + (1-a)y \mid 0 \le a \le 1\}$$

is just the line segment connecting x and y, and so K is convex if and only if it contains all line segments connecting any pair of points of K. This idea is pictured in Figure 1.2.1.

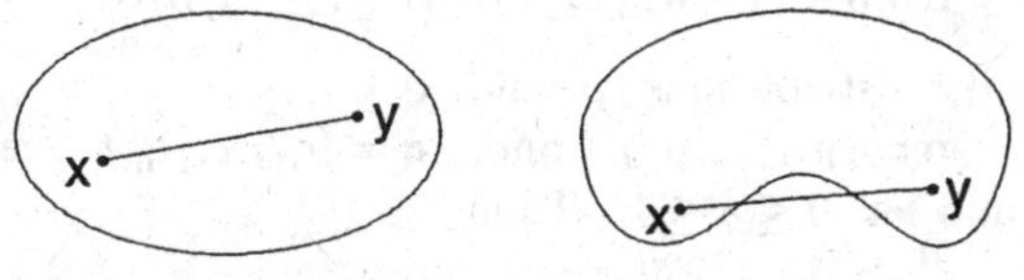

Figure 1.2.1

If K is a convex set, a real-valued function $f:K \to \mathbb{R}$ is **convex up** if

$$f(ax + (1-a)y) \leq af(x) + (1-a)f(y)$$

for every $x,y \in K$ and $0 \leq a \leq 1$. Similarly, a function $f:K \to \mathbb{R}$ is **convex down** (also called *concave*) if

$$f(ax + (1-a)y) \geq af(x) + (1-a)f(y)$$

for every $x,y \in K$ and $0 \leq a \leq 1$.

We may also characterize convex sets (and functions) in terms of convex combinations as follows. A **convex combination** of $x_1,\ldots,x_n \in K$ is an expression of the form $a_1x_1 + \cdots + a_nx_n$, where $0 \leq a_i \leq 1$ and $\sum a_i = 1$. A set K is convex if and only if every convex combination of elements of K is also in K. A function $f:K \to \mathbb{R}$ is convex up if and only if

$$f(a_1x_1 + \cdots + a_nx_n) \leq a_1f(x_1) + \cdots + a_nf(x_n)$$

for every convex combination $a_1x_1 + \cdots + a_nx_n$, and convex down if

$$f(a_1x_1 + \cdots + a_nx_n) \geq a_1f(x_1) + \cdots + a_nf(x_n)$$

for every convex combination $a_1x_1 + \cdots + a_nx_n$.

Now, the set $K = \{\mathbf{p} = (p_1,\ldots,p_n) \mid 0 \leq p_i \leq 1, \sum p_i = 1\}$ of all probability distributions is a convex subset of $\mathbb{R}^n$. For if $\mathbf{p} = (p_1,\ldots,p_n)$ and $\mathbf{q} = (q_1,\ldots,q_n)$ are in K, then

$$a\mathbf{p} + (1-a)\mathbf{q} = (ap_1 + (1-a)q_1,\ldots,ap_n + (1-a)q_n)$$

But $0 \leq a,p_i,q_i \leq 1$ implies that $0 \leq ap_i + (1-a)q_i \leq 1$ and further,

$$\sum_i [ap_i + (1-a)q_i] = a\sum_i p_i + (1-a)\sum_i q_i = a + (1-a) = 1$$

and so $a\mathbf{p} + (1-a)\mathbf{q} \in K$. Thus, K is convex. Our next result shows that the entropy function $H_b:K \to \mathbb{R}$ is convex down.

Theorem 1.2.9 The entropy function H is convex down on the set of probability distributions $\mathbf{p} = (p_1,\ldots,p_n)$, that is,

$$H(\mathbf{ap} + (1-a)\mathbf{q}) \geq aH(\mathbf{p}) + (1-a)H(\mathbf{q})$$

for all probability distributions $\mathbf{p}$ and $\mathbf{q}$.

Proof. Let $\mathbf{p} = (p_1,\ldots,p_n)$ and $\mathbf{q} = (q_1,\ldots,q_n)$ be probability distributions, and let $0 \leq a \leq 1$. Then

$$H(\mathbf{ap} + (1-a)\mathbf{q}) = \sum_i [ap_i + (1-a)q_i] \log \frac{1}{ap_i + (1-a)q_i}$$

$$= a\sum_i p_i \log \frac{1}{ap_i + (1-a)q_i}$$

$$+ (1-a)\sum_i q_i \log \frac{1}{ap_i + (1-a)q_i}$$

Now we use Lemma 1.2.2 to get

$$\geq a\sum_i p_i \log \frac{1}{p_i} + (1-a)\sum_i q_i \log \frac{1}{q_i}$$

$$= aH(\mathbf{p}) + (1-a)H(\mathbf{q})$$

which concludes the proof. ∎

ENTROPY AS AN EXPECTED VALUE

We conclude this section by mentioning that the entropy $H(X)$ can be thought of as the expected value of a certain random variable. By definition,

$$H(X) = \sum_i P(X{=}x_i) \log \frac{1}{P(X{=}x_i)}$$

Thus, if we define a random variable W whose value at x_i is $W(x_i) = \log \frac{1}{P(X=x_i)}$, then

$$H(X) = \mathcal{E}(W)$$

The suggestive notation $W = \log \frac{1}{P(X)}$ is often used for W.

EXERCISES

1. Finish the proof of Theorem 1.2.3.
2. Prove Theorem 1.2.4.
3. Prove Corollary 1.2.6.
4. Prove Corollary 1.2.7.
5. Show that $H(p_1,\ldots,p_n) = H(p_1,\ldots,p_n,0)$. Interpret this in words.

6. Let $X_1, X_2, \ldots, X_n$ be independent random variables with the same distribution as the random variable X. Prove that $H(X_1, \ldots, X_n) = nH(X)$.
7. Let $\{p_1, \ldots, p_n\}$ be a probability distribution, and let $q_m = p_{m+1} + \cdots + p_n$.
(a) Prove that
$$H(p_1, \ldots, p_n) \leq H(p_1, \ldots, p_m, q_m) + q_m \log(n-m)$$
(b) When does equality hold in this inequality?
8. Here is the most general grouping axiom. Let $P = \{p_1, \ldots, p_n\}$ be a probability distribution. Let $G_1, \ldots, G_k$ be disjoint, nonempty subsets of P whose union is P. Let $G_i = \{p_i(1), \ldots, p_i(g_i)\}$, where $|G_i| = g_i$. Prove that
$$H(p_1, \ldots, p_n) = H(g_1, \ldots, g_k) + \sum_i g_i H\left(\frac{p_i(1)}{g_i}, \ldots, \frac{p_i(g_i)}{g_i}\right)$$
9. Let X be a random variable, and let $Y = f(X)$. Prove that $H(Y) \leq H(X)$. Show that equality holds if and only if f is one-to-one on the set of all x such that $P(X = x) \neq 0$.
10. Let $P_1 = \{p_1, \ldots, p_n\}$ be a probability distribution, with $p_1 \geq p_2 \geq \cdots \geq p_n$. Suppose that $\epsilon > 0$ has the property that $p_1 - \epsilon \geq p_2 + \epsilon$. Show that
$$H(p_1, \ldots, p_n) \leq H(p_1 - \epsilon, p_2 + \epsilon, p_3, \ldots, p_n)$$
Interpret this in words?
11. Use Lemma 1.2.2 to prove that, if $\{p_1, \ldots, p_n\}$ is a probability distribution, then
$$x_1^{p_1} \cdots x_n^{p_n} \leq p_1 x_1 + \cdots + p_n x_n$$
where $x_1, \ldots, x_n$ are positive real numbers. This says that the geometric mean of the x_i is less than or equal to the arithmetic mean. Prove that equality holds if and only if the x_i are all equal. *Hint.* Consider the expressions $a_j x_j / \sum a_i x_i$.
12. Prove that a set K is convex if and only if every convex combination of elements of K is also in K.
13. Prove that a function is convex up if and only if $f(a_1 x_1 + \cdots + a_n x_n) \leq a_1 f(x_1) + \cdots + a_n f(x_n)$ for every convex combination $a_1 x_1 + \cdots + a_n x_n$. State and prove the analogous result for convex down functions.
14. If $W = \log(1/P(X))$, verify that $H(X) = \mathcal{E}(W)$.

1.3 Additional Properties of Entropy

In this section, we discuss two issues related to entropy – the entropy of countably infinite probability distributions and the matter of so-called *typical sequences.* This section can be omitted on first reading without loss of continuity.

THE ENTROPY OF COUNTABLY INFINITE DISTRIBUTIONS

Let us begin this section by defining entropy for countably infinite probability distributions.

Definition The **entropy** of a countably infinite probability distribution $\{p_1,p_2,\ldots\}$ is defined by

$$H(p_1,p_2,\ldots) = \sum_{i=1}^{\infty} p_i \log \frac{1}{p_i}$$ □

Notice that, since each term in the above sum is nonnegative, the sum either converges to a nonnegative real number or else diverges to $+\infty$.

Let us immediately prove the counterpart of Lemma 1.2.2 for infinite probability distributions.

Lemma 1.3.1 Let $\{p_1,p_2,\ldots\}$ be a countably infinite probability distribution, and let $\{q_1,q_2,\ldots\}$ have the property that $q_i \geq 0$ and $\sum q_i \leq 1$. Then, assuming that the sums converge, we have

$$\sum_{i=1}^{\infty} p_i \log \frac{1}{p_i} \leq \sum_{i=1}^{\infty} p_i \log \frac{1}{q_i}$$

with equality if and only if $p_i = q_i$.

Proof. We know from the proof of Lemma 1.2.2 that

$$p_i \log \frac{1}{p_i} \leq p_i \log \frac{1}{q_i} + q_i - p_i$$

Summing from 1 to n gives

$$\sum_{i=1}^{n} p_i \log \frac{1}{p_i} \leq \sum_{i=1}^{n} p_i \log \frac{1}{q_i} + \sum_{i=1}^{n} q_i - \sum_{i=1}^{n} p_i$$

Since each sum on the right side converges, we can take limits as follows

$$\lim_{n\to\infty} \sum_{i=1}^{n} p_i \log \frac{1}{p_i} \leq \lim_{n\to\infty} \left[\sum_{i=1}^{n} p_i \log \frac{1}{q_i} + \sum_{i=1}^{n} q_i - \sum_{i=1}^{n} p_i \right]$$

$$= \lim_{n\to\infty} \sum_{i=1}^{n} p_i \log \frac{1}{q_i} + \lim_{n\to\infty} \sum_{i=1}^{n} q_i - \lim_{n\to\infty} \sum_{i=1}^{n} p_i$$

$$= \sum_{i=1}^{\infty} p_i \log \frac{1}{q_i} + \sum_{i=1}^{\infty} q_i - \sum_{i=1}^{\infty} p_i$$

$$\leq \sum_{i=1}^{\infty} p_i \log \frac{1}{q_i}$$

from which the result follows. ∎

For our main result on entropy of countably infinite distributions, we need a result from the theory of infinite series. It is well known that if an infinite series $\sum r_i$ converges, then $r_i \to 0$ as $i \to \infty$. However, if the terms r_i are monotonically decreasing and nonnegative, then we can say much more.

Lemma 1.3.2 Let $r_1 \geq r_2 \geq \cdots$ be a monotonically decreasing sequence of positive real numbers. If the series $\sum r_i$ converges, then $ir_i \to 0$ as $i \to \infty$.

Proof. Since $\sum r_i$ converges and each r_i is positive, given any $\epsilon > 0$, there exists a number n such that

$$r_{n+1} + r_{n+2} + \cdots + r_m < \frac{\epsilon}{2} \tag{1.3.1}$$

for all $m \geq n$. In particular, taking $m = 2n$, we get

$$r_{n+1} + r_{n+2} + \cdots + r_{2n} < \frac{\epsilon}{2}$$

and since the sequence r_i is decreasing, this implies

$$nr_{2n} < \frac{\epsilon}{2}$$

or

$$2nr_{2n} < \epsilon$$

Hence, $ir_i \to 0$ as $n \to \infty$ for even i. A similar argument proves the result for odd i. ∎

Now we can prove the following result on the convergence of entropy.

Theorem 1.3.3 Let $\{p_1, p_2, \ldots\}$ be a countably infinite probability distribution.

1) If the sum $\sum p_i \log i$ converges, then so does $\sum p_i \log \frac{1}{p_i}$.

2) If $p_1 \geq p_2 \geq \cdots$ and if $\sum p_i \log \frac{1}{p_i}$ converges, then so does $\sum p_i \log i$.

Proof. Since the series $\sum (1/i^2)$ converges, we can set $S = \sum (1/i^2)$. (Actually, $S = \pi^2/6$, but we will not need this fact.) Then

$$\sum_{i \geq 1} \frac{1}{Si^2} = 1$$

and we may take $q_i = 1/Si^2$ in Lemma 1.3.1, which tells us that

$$\sum_i p_i \log \frac{1}{p_i} = \sum_i p_i \log \frac{1}{(1/Si^2)}$$

$$= \sum_i p_i \log Si^2 = \log S + 2 \sum_i p_i \log i$$

Hence, if $\sum p_i \log i$ converges, so does $\sum p_i \log(1/p_i)$, which proves part 1. As for part 2, we may assume that $p_i \neq 0$ for all i, for otherwise all sums are finite, and the result is obvious. Since the p_i form a monotonically decreasing sequence of positive real numbers for which $\sum p_i$ converges (to 1), we may apply Lemma 1.3.2, which tells us that $ip_i \to 0$ as $i \to \infty$. Hence, for any $\epsilon > 0$, there exists an N such that $i > N$ implies $ip_i < \epsilon$, that is, $i < \epsilon/p_i$. Thus,

$$\sum_{i>N} p_i \log i \leq \sum_{i>N} p_i \log \frac{\epsilon}{p_i} \leq \log \epsilon + \sum_{i>N} p_i \log \frac{1}{p_i}$$

which shows that if $\sum p_i \log \frac{1}{p_i}$ converges, so does $\sum p_i \log i$. ∎

We will leave it as an exercise to show that the condition of monotonicity is essential in part 2) of Theorem 1.3.3.

Example 1.3.1 Let us consider a situation where countably infinite sources arise naturally. Suppose we have a source $\mathcal{S}_1 = (S_1, P_1)$, where $S_1 = \{0,1\}$ and $P_1(0) = p$, that is outputting a stream of independent source symbols. Suppose further that a counting device intercepts this output and counts the number of 0's that occur before each 1. Thus, for example, if the output of the source is

001011000100000111...

the counter will output the sequence

2103500...

Now, the counter can also be thought of as a source $\mathcal{S}_2 = (S_2, P_2)$, where $S_2 = \{0,1,\ldots\}$ and P_2 is the geometric distribution, that is,

$$P_2(k) = P(k \text{ 0's followed immediately by a } 1) = p^k(1-p)$$

Let us compare the entropy of the two sources $\mathcal{S}_1$ and $\mathcal{S}_2$. Of course, $H(\mathcal{S}_1) = H(p, 1-p)$. On the other hand,

$$H(\mathcal{S}_2) = -\sum_{k\geq 0} p^k(1-p)\log p^k(1-p)$$

$$= -\sum_{k\geq 0} p^k(1-p)\log p^k - \sum_{k\geq 0} p^k(1-p)\log(1-p)$$

$$= -(1-p)\log p \sum_{k\geq 0} kp^k - (1-p)\log(1-p)\sum_{k\geq 0} p^k$$

Using the well-known formulas

$$\sum_{k\geq 0} p^k = \frac{1}{1-p} \quad \text{and} \quad \sum_{k\geq 0} kp^k = -\frac{p}{(1-p)^2}$$

we have

$$H(\mathcal{S}_2) = -(1-p)\left(-\frac{p}{(1-p)^2}\right)\log p - (1-p)\left(\frac{1}{1-p}\right)\log(1-p)$$

$$= \frac{p\log p - (1-p)\log(1-p)}{1-p} = \frac{H(\mathcal{S}_1)}{1-p}$$

Thus,

$$H(\mathcal{S}_2) = \frac{H(\mathcal{S}_1)}{1-p}$$

This tells us that, provided $p > 0$, there is more uncertainty in the number of consecutive 0's than in whether or not the next symbol will be a 0. Furthermore, as p gets close to 1, this discrepancy increases. □

TYPICAL SEQUENCES

Now let us return to finite probability distributions. We can get further insight into the concept of entropy by considering the idea of a *typical sequence.* Let $S = \{x_1, \ldots, x_m\}$ be a source, with distribution $P(x_i) = p_i$. Suppose that we repeatedly, and independently, sample from this source, obtaining a sequence $X_1, \ldots, X_n$ of independent random variables, each with the probability distribution P. Roughly how often should we expect a particular element x_i to occur in the samplings?

To answer this question, we consider each sampling as being a success if the outcome is x_i and a failure if the outcome is not x_i. Then each sampling is a Bernoulli trial, with probability of success equal to p_i. If we let $\mathbf{X} = (X_1, \ldots, X_n)$ and $S_i = S_i(\mathbf{X})$ be the number of successes in the n trials $X_1, \ldots, X_n$, then S_i has the binomial distribution

$$P(S_i{=}j) = \binom{n}{j} p_i^{\,j}(1-p_i)^{n-j}$$

with mean $\mu_i = np_i$ and variance $\sigma_i^{\,2} = np_i(1-p_i)$. Now, Chebyshev's inequality

$$P\left\{\left|\frac{S_i - \mu_i}{\sigma_i}\right| > k\right\} < \frac{1}{k^2} \tag{1.3.2}$$

tells us, in rough terms, that it is more likely that S_i is closer to $\mu_i = np_i$ than farther away from np_i. Loosely speaking, we expect the number of successes, that is, the number of x_i's, to be around $\mu_i = np_i$. This leads us to make the following definition.

Definition Let X be a random variable with range $\{x_1,\ldots,x_m\}$ and probability distribution $P(X{=}x_j) = p_j$. Let $X_1,\ldots,X_n$ be independent random variables with the same distribution as X. If $X_1 = \alpha_1,\ldots,X_n = \alpha_n$ is a particular sample of the X_i, we denote the number of x_i's among the sequence $\boldsymbol{\alpha} = (\alpha_1,\ldots,\alpha_n)$ by $S_i(\boldsymbol{\alpha})$. Then a sequence $\boldsymbol{\alpha}$ is a **k-typical sequence** if

$$\left|\frac{S_i(\boldsymbol{\alpha}) - \mu_i}{\sigma_i}\right| \le k \qquad \text{for all } i = 1,\ldots,n$$

that is, if

$$\left|\frac{S_i(\boldsymbol{\alpha}) - np_i}{\sqrt{np_i(1-p_i)}}\right| \le k \qquad \text{for all } i = 1,\ldots,n$$ □

Intuitively, $\boldsymbol{\alpha}$ is k-typical if the number of x_i's in $\boldsymbol{\alpha}$, for all i, is what Chebyshev's inequality says is "most likely."

Now we come to the connection between k-typical sequences and entropy. In particular, we show that for any k and for large n, of the m^n possible sample sequences, the number of k-typical sequences is approximately

$$m^{nH_m(X)}$$

where $H_m(X)$ is the m-ary entropy of X.

Theorem 1.3.4 For any $k > 0$,

1) The probability that a sampled sequence is not k-typical is at most m/k^2 and can therefore be made as small as desired by taking k sufficiently large.
2) If $\mathbf{X} = (X_1,\ldots,X_n)$ and if $\boldsymbol{\alpha} = (\alpha_1,\ldots,\alpha_n)$ is a k-typical sequence, then

$$m^{-nH_m(X) - \sqrt{n}C_k} \le P(\mathbf{X} = \boldsymbol{\alpha}) \le m^{-nH_m(X) + \sqrt{n}C_k}$$

where

$$C_k = \sum_{i=1}^{m} k\sqrt{p_i(1-p_i)}\,\log_m p_i$$

depends only on k and the probability distribution. Thus, for any k and for large n, the probability of getting a particular k-typical sequence is approximately

$$m^{-nH_m(X)}$$

3) The number $N_{k,n}$ of k-typical sequences of length n satisfies

$$\left(1-\frac{m}{k^2}\right) m^{nH_m(X) - \sqrt{n}C_k} \le N_{k,n} \le m^{nH_m(X) + \sqrt{n}C_k}$$

Thus, for any k and for large n, of the m^n possible sample sequences, approximately

$$m^{nH_m(X)}$$

are k-typical.

Proof. According to the definition of typical sequence, the probability that $\mathbf{X} = (X_1,\ldots,X_n)$ is not k-typical is

$$P(\mathbf{X}\text{ is not k-typical}) = P\left(\left|\frac{S_i - np_i}{\sigma_i}\right| > k, \quad \text{for some } i\right)$$

$$\le \sum_{i=1}^{m} P\left(\left|\frac{S_i - np_i}{\sigma_i}\right| > k\right)$$

Applying Chebyshev's inequality (1.3.2) gives

$$P(\mathbf{X}\text{ is not k-typical}) \le \sum_{i=1}^{m} \frac{1}{k^2} = \frac{m}{k^2}$$

This proves part 1).

Next, suppose that $\boldsymbol{\alpha} = (\alpha_1,\ldots,\alpha_n)$ is k-typical. Since the random variables X_i are independent, we have

$$P(\mathbf{X}=\boldsymbol{\alpha}) = P(X_1=\alpha_1,\ldots,X_n=\alpha_n) = \prod_{i=1}^{n} P(X_i=\alpha_i) = \prod_{i=1}^{n} p_i$$

Now, the factors p_i may not all be distinct, and we can collect like factors to get

$$P(\mathbf{X}=\boldsymbol{\alpha}) = \prod_{i=1}^{m} p_i^{S_i(\boldsymbol{\alpha})}$$

Taking logarithms gives

$$\log_m P(\mathbf{X}=\boldsymbol{\alpha}) = \sum_{i=1}^{m} S_i(\boldsymbol{\alpha}) \log_m p_i \tag{1.3.3}$$

Now, according to the definition of k-typical, we have

$$np_i - k\sigma_i \le S_i(\boldsymbol{\alpha}) \le np_i + k\sigma_i$$

and so (1.3.3) implies that

$$\sum_{i=1}^{m} [np_i - k\sigma_i] \log_m p_i \le \lg P(\mathbf{X}=\boldsymbol{\alpha}) \le \sum_{i=1}^{m} [np_i + k\sigma_i] \log_m p_i$$

or, since $\sigma_i = \sqrt{np_i(1-p_i)}$,

$$n \sum_{i=1}^{m} p_i \log_m p_i - \sqrt{n} \sum_{i=1}^{m} k\sqrt{p_i(1-p_i)} \log_m p_i \le \log_m P(\mathbf{X}=\boldsymbol{\alpha})$$

$$\le n \sum_{i=1}^{m} p_i \log_m p_i + \sqrt{n} \sum_{i=1}^{m} k\sqrt{p_i(1-p_i)} \log_m p_i$$

Setting

$$C_k = \sum_{i=1}^{m} k \sqrt{p_i(1-p_i)} \log_m p_i$$

we get

$$-nH(X) - \sqrt{n}\, C_k \le \log_m P(\mathbf{X}=\boldsymbol{\alpha}) \le -nH(X) + \sqrt{n}\, C_k$$

This gives part 2).

As for part 3), we first note that if E is a subset of a finite sample space, and if each element $e \in E$ has probability satisfying $r_1 \le P(e) \le r_2$, then

$$r_1 |E| \le \sum_{e \in E} P(e) \le r_2 |E|$$

and since $P(E) = \sum_{e \in E} P(e)$, we get

$$\frac{P(E)}{r_2} \le |E| \le \frac{P(E)}{r_1} \tag{1.3.4}$$

Now, in the case at hand, we let E be the set of k-typical sequences. Then from part 1) we get

$$(1.3.5) \qquad 1-\frac{m}{k^2} \le P(E) \le 1$$

and from part 2, we have for $\alpha \in E$,

$$(1.3.6) \quad m^{-nH_m(X)-\sqrt{n}C_k} \le P(\alpha) \le m^{-nH_m(X)+\sqrt{n}C_k}$$

Using (1.3.5) for lower and upper estimates on $P(E)$, and using (1.3.6) to obtain r_1 and r_2, (1.3.4) gives

$$\left(1-\frac{m}{k^2}\right)m^{nH_m(X)-\sqrt{n}C_k} \le |E| \le m^{nH_m(X)+\sqrt{n}C_k}$$

which is part 3). ∎

EXERCISES

1. Show that

$$\sum_{k\ge 0} p^k = \frac{1}{1-p} \quad \text{and} \quad \sum_{k\ge 0} kp^k = -\frac{p}{(1-p)^2}$$

2. Finish the proof of Lemma 1.3.2 by supplying the details to show that $ip_i \to 0$ for i odd.
3. Let $\{p_1,p_2,\ldots\}$ be a countably infinite probability distribution, and let X be a random variable with $P(X=n)=p_n$. Show that $H_2(X) \le 1+\mathcal{E}(X)$, where $\mathcal{E}(X)$ is the expected value of X. When does equality hold in this inequality? *Hint.* $\sum(1/2^{i+1})=1$.
4. Let $\{p_1,p_2,\ldots\}$ be a countably infinite probability distribution, whose entropy is finite. Let $\{p_1,p_2,\ldots\}$ be the disjoint union of the sets $\{q_i,q_2,\ldots\}$ and $\{r_1,r_2,\ldots\}$, where $\sum q_i = q$ and $\sum r_i = r$. (Thus $q+r=1$.) Prove that

$$H(p_1,p_2,\ldots) = H(q,r) + qH\left(\frac{q_1}{q},\frac{q_2}{q},\ldots\right) + rH\left(\frac{r_1}{r},\frac{r_2}{r},\ldots\right)$$

You will need a certain result about absolutely convergent series. What is this result?
5. Let $\{p_1,p_2,\ldots\}$ be a countably infinite probability distribution, whose entropy is finite. Suppose that $p_1 > p_2$ and that $p_1-\epsilon \ge p_2+\epsilon$ for some $\epsilon > 0$. Prove that

$$H(p_1,p_2,\ldots) \le H(p_1-\epsilon,p_2+\epsilon,\ldots) < \infty$$

6. (a) Show that the series $\sum \frac{1}{n \log^2 n}$ converges. Denote the sum by S.

   (b) Let $p_n = \frac{1}{Sn \log^2 n}$. Is $H(p_1, p_2, \ldots)$ finite or infinite?

7. (a) Show that the series $\sum \frac{1}{n^k}$ converges for $k \geq 2$. Denote the sum by S.

   (b) Let $p_n = \frac{1}{Sn^k}$. Show that $H(p_1, p_2, \ldots)$ is finite.

8. Let $S_1 = \{0,1,2\}$ be a source, with $P(0) = p$, $P(1) = q$, $P(2) = 1 - p - q$. Repeatedly performing the experiment of sampling this source until a 2 appears produces another source with alphabet $S_2 = \{a_1 \cdots a_k 2 \mid a_i \in S_1,\ a_i \neq 2\}$. Calculate the probability distribution and the entropy of this source.

9. Let $\{p_1, p_2, \ldots\}$ be a countably infinite probability distribution, whose entropy is finite. Can you approximate the entropy $H(p_1, p_2, \ldots)$ to any desired degree of accuracy by the entropy of a finite probability distribution? Explain.

10. Show by example that the requirement of being monotonically decreasing is essential to part 2) of Theorem 1.3.3. In other words, find an example of a probability distribution $\{p_1, p_2, \ldots\}$ for which the sequence $p_1, p_2, \ldots$ is not monotonically decreasing and for which $\sum p_i \log \frac{1}{p_i}$ converges, but $\sum p_i \log i$ does not converge.

CHAPTER 2

Noiseless Coding

2.1 Variable Length Encoding

Now we turn to a discussion of *source encoding for noiseless transmission.* When no errors can occur in the transmission of data, we may concentrate on the question of how to encode the data as efficiently as possible, in a sense we will make precise in a moment. First, let us set some basic terminology.

STRINGS AND CODES

Let $\mathcal{A} = \{a_1, \ldots, a_n\}$ be a finite set, which we refer to as an **alphabet.** A **string,** or **word,** over the alphabet $\mathcal{A}$ is any sequence of elements of $\mathcal{A}$. We will usually (but not always) write strings in the form

$$\mathbf{a} = a_{i_1} a_{i_2} \cdots a_{i_k}$$

using juxtaposition of symbols. Occasionally, for readability sake, we may include spaces, commas, parentheses, or other punctuation marks, between the symbols in a string. The **empty string** θ is the unique string with no symbols.

The **length** of a string **a**, denoted by *len*(**a**), is the number of alphabet symbols appearing in the string. The set of all strings over $\mathcal{A}$ will be denoted by $\mathcal{A}^*$.

Definition Let $\mathcal{A} = \{a_1, \ldots, a_r\}$ be a finite set, which we call a **code alphabet.** An **r-ary code** is a nonempty subset C of the set $\mathcal{A}^*$ of all

strings over $\mathcal{A}$. The size r of the code alphabet is called the **radix** of the code, and the elements of the code are called **codewords.** A code whose alphabet is $\{0,1\}$ is called a **binary code**, and a code whose alphabet is $\{0,1,2\}$ is called a **ternary code.** □

Definition Let $\mathcal{S} = (S,P)$ be a source. An **encoding scheme** for $\mathcal{S}$ is an ordered pair (C,f), where C is a code and $f:S \to C$ is an *injective* function, called an **encoding function.** □

Thus, an encoding function assigns a codeword from C to each source symbol in S.

AVERAGE CODEWORD LENGTH

For the purposes of noiseless encoding, the measure of efficiency of an encoding scheme is its *average codeword length.*

Definition The **average codeword length** of an encoding scheme (C,f) for a source $\mathcal{S} = (S,P)$, where $S = \{s_1,\ldots,s_n\}$, is defined by

$$AveLen(C,f) = \sum_{i=1}^{n} P(s_i)\,len(f(s_i))$$ □

Example 2.1.1 Consider the source $S = \{a,b,c,d\}$, with probabilities $P(a) = P(b) = 2/17$, $P(c) = 9/17$ and $P(d) = 4/17$. Consider also the encoding schemes $(C_1 f_1)$ and (C_2,f_2), where

$$C_1 = \{0,11,100,101\} \qquad C_2 = \{00,10,11,01010\}$$

$$\begin{array}{ll} f_1(a) = 11 & f_2(a) = 01010 \\ f_1(b) = 0 & f_2(b) = 00 \\ f_1(c) = 100 & f_2(c) = 10 \\ f_1(d) = 10 & f_2(d) = 11 \end{array}$$

Since

$$AveLen(C_1,f_1) = \frac{2}{17}\cdot 2 + \frac{2}{17}\cdot 1 + \frac{9}{17}\cdot 3 + \frac{4}{17}\cdot 2 = \frac{41}{17}$$

and

$$AveLen(C_2,f_2) = \frac{2}{17}\cdot 5 + \frac{2}{17}\cdot 2 + \frac{9}{17}\cdot 2 + \frac{4}{17}\cdot 2 = \frac{40}{17}$$

we see that (C_2,f_2) has a smaller average codeword length, and so is more efficient than (C_1,f_1), even though *the code* C_2 has longer codewords (on the average) than *the code* C_1. This emphasizes the fact that the average codeword length of an *encoding scheme* is not the same as the average codeword length of a *code*, since the former depends also on the probability distribution P. □

We should point out that it makes sense to compare the average codeword lengths of different encoding schemes only when the corresponding codes have the same radix. For in general, the larger the radix, the shorter we can make the average codeword length.

Our goal in this chapter is to determine the *minimum* average codeword length among all "good" encoding schemes (in a sense we will make precise soon), as well as to find a method for constructing such encoding schemes. As we will see, both goals are readily achieved.

FIXED AND VARIABLE LENGTH CODES

Definition If all the codewords in a code C have the same length, we say that C is a **fixed length code**, or **block code.** If C contains codewords of different lengths, we say that C is a **variable length code.**

Any encoding scheme that uses a fixed length code will be referred to as a **fixed length encoding scheme**, and similarly for **variable length encoding schemes.** □

When the probability distribution P is not uniform, variable length encoding is usually more efficient than fixed length encoding. As a simple example, consider a source with alphabet $S = \{s_1, \ldots, s_5\}$, whose probability distribution satisfies

$$P(s_1) = 1 - \epsilon \quad \text{and} \quad P(\{s_2, s_3, s_4, s_5\}) = \epsilon$$

Since a fixed length binary code must have codeword length at least 3, in order to encode 5 words, its average codeword length is also at least 3. On the other hand, using a variable length code, we may assign the codeword 0 to s_1 and the codewords 100, 101, 110, and 111 to the other source symbols, giving an average codeword length of $1 \cdot (1-\epsilon) + 3\epsilon = 1 + 2\epsilon$, which is less than 3 if $\epsilon < 1$.

UNIQUE DECIPHERABILITY

Even though variable length encoding schemes can be more efficient than fixed length schemes, there is a potential problem with variable length schemes, as illustrated by the following example.

$$S = \{a,b,c\}, \quad C = \{0,01,001\}$$
$$f(a) = 0, \quad f(b) = 01, \quad f(c) = 001$$

This encoding scheme is not *uniquely decipherable*, in the sense that the codeword string 001 could be decoded as *ab* or as *c*. In order to make this encoding scheme uniquely decipherable, we require a

codeword separator, such as /, which enables us to write the message *ab* as 0/01. Of course, the addition of a codeword separator adds to the overall length of encoded messages, which is contrary to the goal of efficient encoding. (Fixed length encoding schemes are automatically uniquely decipherable and need no codeword separator.)

The difficulty here can be traced to the fact that a string of code alphabet symbols may represent more than one string of codewords. This leads to the following definition.

Definition A code C is **uniquely decipherable** if whenever $\mathbf{c}_1,\ldots,\mathbf{c}_k$, $\mathbf{d}_1,\ldots,\mathbf{d}_j$ are codewords in C and

$$\mathbf{c}_1\cdots\mathbf{c}_k = \mathbf{d}_1\cdots\mathbf{d}_j$$

then $k = j$ and $\mathbf{c}_i = \mathbf{d}_i$ for all $i = 1,\ldots,k$. ▯

Clearly, the property of being uniquely decipherable is extremely desirable. Surprisingly, even a small change can make a code that is not uniquely decipherable into one that is.

Example 2.1.2 Let $S = \{a,b,c\}$ and consider the encoding scheme

$$C = \{1,01,001\}$$
$$f(a) = 1, \quad f(b) = 01, \quad f(c) = 001$$

This differs from the previous code only in that the codeword 0 is replaced by the codeword 1. However, this code is uniquely decipherable. To see this, observe that the symbol 1 acts as a kind of codeword separator, in the sense that the presence of a 1 indicates the end of a codeword. Thus, reading a codeword string from left to right, we *must* decode *when and only when* we encounter a 1. For instance, consider the string 1001011. Reading from left to right, we *must* decode 1 as *a*, 001 as *c*, 01 as *b*, and 1 as *a* to get the source string *acba*. No other decoding is possible. ▯

Although there are methods for showing that a particular code is uniquely decipherable, we shall not go into them here, since we will limit our discussion to a special type of uniquely decipherable code, without limiting our ability to be efficient.

To be more specific, one of the difficulties with unique decipherability is that, even though a code may have this property, it may be necessary to wait until the entire message has been received before we can begin to decode.

Example 2.1.3 Consider the code $C = \{0,01,001\}$ and the encoding function

$$f(a) = 0, \quad f(b) = 01, \quad f(c) = 011, \quad f(d) = 0111$$

It is not hard to see that this code is uniquely decipherable. Now suppose that the string 0111 is being transmitted. Just after receiving the first 0, we cannot tell whether it represents the source letter *a*, or the beginning of a different source letter. Similarly, when the first 01 is received, we cannot tell whether it represents a *b*, or the beginning of a *c* or *d*. In fact, we cannot decipher the source message 0111 until it has been completely received.

On the other hand, consider the code $D = \{0,10,110,1110\}$ and encoding function

$$g(a) = 0, \quad g(b) = 10, \quad g(c) = 110, \quad g(d) = 1110$$

In this case, individual codewords can be deciphered as soon as they are received, since the presence of a 0 indicates the end of a codeword. Thus, each source symbol can be decoded as soon as its codeword is received. □

INSTANTANEOUS CODES; THE PREFIX PROPERTY

The previous example prompts us to make the following definition.

Definition A code is said to be **instantaneous** if each codeword in any string of codewords can be decoded (reading from left to right) as soon as it is received. □

If a code is instantaneous, then it is also uniquely decipherable. However, as the code C of Example 2.1.3 illustrates, the converse is not true.

The property of being instantaneous is very desirable. Fortunately, there is a very simple way to tell when a code has this property. First we need a definition.

Definition A code is said to have the **prefix property** if no codeword is a prefix of any other codeword, that is, if whenever $\mathbf{c} = x_1x_2\cdots x_n$ is a codeword, then $x_1x_2\cdots x_k$ is not a codeword for $1 \leq k < n$. □

Given a code C, it is a simple matter to determine whether or not it has the prefix property. It is only necessary to compare each codeword with all codewords of equal or greater length to see if it is a prefix. For example, the code $\{1,01,001\}$ has the prefix property, since 1 is not a prefix of 01 or 001 and 01 is not a prefix of 001. However, the code $\{0,01,001\}$ does not have the prefix property, since

0 is a prefix of 01.

The importance of the prefix property comes from the following theorem, whose proof we leave as an exercise.

Theorem 2.1.1 A code C is instantaneous if and only if it has the prefix property. □

Example 2.1.4 Let n be a positive integer. A **comma code** is a code C with codewords

$$1,\ 01,\ 001,\ 0001, \ldots,\ \underbrace{0\cdots0}_{n-1}1,\ \underbrace{0\cdots0}_{n}$$

This terminology comes from the fact that the symbol 1 acts as a kind of comma, indicating the end of a codeword. (The last codeword is determined by the unique number of 0's that it contains.) Since comma codes have the prefix property, they are instantaneous. On the other hand, the code

$$1,\ 10,\ 100,\ 1000, \ldots,\ 1\underbrace{0\cdots0}_{n-1},\ \underbrace{0\cdots0}_{n}$$

does not have the prefix property, and so it is not instantaneous. However, it is uniquely decipherable, since we can decipher any string of codewords by reading from right to left, where a 1 indicates the *beginning* of a codeword. □

KRAFT'S THEOREM

The following remarkable theorem, published by L.G. Kraft in 1949, gives a simple criterion to determine whether or not there is an instantaneous code with given codeword lengths.

Theorem 2.1.2 (Kraft's Theorem)

1) If C is an r-ary instantaneous code with codeword lengths $\ell_1, \ldots, \ell_n$, then these lengths must satisfy **Kraft's inequality**

$$\sum_{k=1}^{n} \frac{1}{r^{\ell_k}} \le 1$$

2) If the numbers $\ell_1, \ell_2, \ldots, \ell_n$ and r satisfy **Kraft's inequality**, then there is an instantaneous r-ary code with codeword lengths $\ell_1, \ldots, \ell_n$.

Proof. Suppose first that $C = \{c_1, \ldots, c_n\}$ is an instantaneous r-ary code with codeword lengths $\ell_1, \ldots, \ell_n$. We will show that Kraft's inequality must hold. Let $L = \max\{\ell_i\}$. If $c_i = x_1x_2\cdots x_{\ell_i} \in C$, then

any word of the form

$$\mathbf{x} = x_1 x_2 \cdots x_{\ell_i} y_{\ell_i+1} \cdots y_L \tag{2.1.1}$$

where the y_j are any code symbols, cannot be in C, because c_i is a prefix of $\mathbf{x}$. But there are a total of $r^{L-\ell_i}$ words of the form (2.1.1). Summing on i, we see that there are

$$\sum_{i=1}^{n} r^{L-\ell_i} = r^L \sum_{i=1}^{n} \frac{1}{r^{\ell_i}}$$

words of length L that cannot be in C. However, the total number of words of length L over the code alphabet is r^L, and so we must have

$$r^L \sum_{i=1}^{n} \frac{1}{r^{\ell_i}} \leq r^L$$

or

$$\sum_{i=1}^{n} \frac{1}{r^{\ell_i}} \leq 1$$

which is Kraft's inequality.

Now suppose that $\ell_1, \ell_2, \ldots, \ell_n$ and r satisfy Kraft's inequality. We will show that there exists an instantaneous code C, over an alphabet $\mathcal{A} = \{a_1, a_2, \ldots, a_r\}$, with codeword lengths ℓ_i. Let α_j be the number of ℓ_i that are equal to j. Thus, α_1 is the number of desired codewords of length 1, α_2 is the number of desired codewords of length 2, and so on.

In order to construct the desired code, we want to select α_1 words of length 1, say the first α_1 code letters

$$a_1, a_2, \ldots, a_{\alpha_1} \tag{2.1.2}$$

This can be done as long as

$$\alpha_1 \leq r$$

Next, we want to select α_2 words of length 2. However, since our code must be instantaneous, we cannot allow any of the α_1 codewords in (2.1.2) to be prefixes of the new codewords. In other words, from among the r^2 possible words of length 2 over $\mathcal{A}$, we cannot select the $\alpha_1 r$ codewords that begin with any of the α_1 codewords in (2.1.2). This leaves $r^2 - \alpha_1 r$ codewords from which to choose α_2 codewords, and this can be done provided that

$$\alpha_2 \leq r^2 - \alpha_1 r$$

or

$$\alpha_1 r + \alpha_2 \leq r^2$$

The next step is to select α_3 codewords of length 3. There are r^3 such codewords, but as before, the requirement that C be instantaneous means that none of the $\alpha_1 r^2$ codewords of length 3 that begin with any of the α_1 elements in (2.1.2) can be used, nor can any of the $\alpha_2 r$ codewords of length 3 that begin with one of the α_2 previously chosen codewords of length 2. Thus, we are left with $r^3 - \alpha_1 r^2 - \alpha_2 r$ codewords of length 3, from which to pick α_3 codewords. This can be done provided that

$$\alpha_3 \leq r^3 - \alpha_1 r^2 - \alpha_2 r$$

or

$$\alpha_1 r^2 + \alpha_2 r + \alpha_3 \leq r^3$$

Continuing in this manner, we will get the system of inequalities

$$\begin{aligned} \alpha_1 &\leq r \\ \alpha_1 r + \alpha_2 &\leq r^2 \\ \alpha_1 r^2 + \alpha_2 r + \alpha_3 &\leq r^3 \\ &\vdots \\ \alpha_1 r^{n-1} + \alpha_2 r^{n-2} + \cdots + \alpha_n &\leq r^n \end{aligned} \tag{2.1.3}$$

Notice, however, that each inequality in (2.1.3) implies the previous one. Hence, as long as the last inequality is satisfied, we may construct the desired code. Dividing the last inequality in (2.1.3) by r^n gives

$$\frac{\alpha_1}{r} + \frac{\alpha_2}{r^2} + \cdots + \frac{\alpha_n}{r^n} \leq 1$$

which is equivalent to Kraft's inequality. ∎

It is important to note that Kraft's Theorem says that if the lengths $\ell_1, \ell_2, \ldots, \ell_n$ satisfy Kraft's Inequality, then there must exist *some* instantaneous code with these codeword lengths. It does *not* say that any code whose codeword lengths satisfy Kraft's inequality must be instantaneous. The next example shows that this is need not be the case.

Example 2.1.5 Consider the binary code $C = \{0,11,100,110\}$, with codeword lengths 1,2,3 and 3. Since $|\mathcal{A}| = 2$, the left side of Kraft's inequality is

$$\frac{1}{2} + \frac{1}{2^2} + \frac{1}{2^3} + \frac{1}{2^3} = 1$$

Hence, the lengths satisfy Kraft's inequality, but the code C is not instantaneous, since the second codeword is a prefix of the fourth. ▯

Let us give an example of the construction in the proof of Kraft's Theorem.

Example 2.1.6 Let $\mathcal{A} = \{0,1,2\}$ and $\ell_1 = \ell_2 = 1$, $\ell_3 = 2$, $\ell_4 = \ell_5 = 4$, $\ell_6 = 5$. Then since

$$\frac{1}{3} + \frac{1}{3} + \frac{1}{3^2} + \frac{1}{3^4} + \frac{1}{3^4} + \frac{1}{3^5} = \frac{3^4+3^4+3^3+3+3+1}{3^5} = \frac{196}{243} < 1$$

Kraft's Inequality is satisfied. Thus, according to Kraft's Theorem, we should be able to construct an instantaneous code over $\mathcal{A}$ with these codeword lengths.

The first step in constructing such a code is to choose the codewords of the smallest lengths $\ell_1 = \ell_2 = 1$. For this, we may as well choose

$$c_1 = 0 \quad \text{and} \quad c_2 = 1$$

Then we choose a codeword of length $\ell_3 = 2$. Since our code must be instantaneous, we cannot start this codeword with either 0 or 1, and so it must start with 2. Let us choose

$$c_3 = 20$$

Now we choose two codewords of length 4. These codewords cannot begin with either 0 or 1, and so they must begin with 2. However, they cannot begin with 20, since that is codeword c_3. Let us choose

$$c_4 = 2100 \quad \text{and} \quad c_4 = 2101$$

Finally, we choose a codeword of length 5 that begins with 211,

$$c_5 = 21100$$

Thus, $C = \{0,1,20,2100,2101,21100\}$. ▯

McMILLAN'S THEOREM

It is interesting to observe that Kraft's inequality is also necessary and sufficient for the existence of a *uniquely decipherable* code. Of course, Kraft's inequality is sufficient since any instantaneous code is also uniquely decipherable. The necessity of Kraft's inequality was proved by McMillan in 1956. (The proof given here is not McMillan's, however.)

Theorem 2.1.3 (McMillan's Theorem) If $C = \{c_1, c_2, \ldots, c_n\}$ is a uniquely decipherable r-ary code, then its codeword lengths $\ell_1, \ell_2, \ldots, \ell_n$ must satisfy Kraft's inequality

$$\sum_{k=1}^{n} \frac{1}{r^{\ell_k}} \leq 1$$

Proof. The following proof is the usual one given for this theorem, although it is not particularly intuitive. Suppose that α_k is the number of codewords in C of length k. Then we have

$$\sum_{k=1}^{n} \frac{1}{r^{\ell_k}} = \sum_{k=1}^{m} \frac{\alpha_k}{r^k}$$

Now let u be a positive integer, and consider the quantity

$$\left(\sum_{k=1}^{m} \frac{\alpha_k}{r^k}\right)^u = \left(\frac{\alpha_1}{r} + \frac{\alpha_2}{r^2} + \cdots + \frac{\alpha_m}{r^m}\right)^u$$

Multiplying this out gives

$$= \sum_{\substack{i_1, i_2, \ldots, i_u \\ 1 \leq i_j \leq m}} \frac{\alpha_{i_1}}{r^{i_1}} \cdots \frac{\alpha_{i_u}}{r^{i_u}} = \sum_{\substack{i_1, i_2, \ldots, i_u \\ 1 \leq i_j \leq m}} \frac{\alpha_{i_1} \alpha_{i_2} \cdots \alpha_{i_u}}{r^{i_1 + \cdots + i_u}}$$

Now, since $1 \leq i_j \leq m$, each sum $i_1 + \cdots + i_u$ is at least m and at most um. Collecting terms with a common sum $i_1 + \cdots + i_u$, we get

$$= \sum_{k=m}^{um} \left(\sum_{i_1 + \cdots + i_u = k} \alpha_{i_1} \alpha_{i_2} \cdots \alpha_{i_u} \right) \frac{1}{r^k} = \sum_{k=m}^{um} \frac{N_k}{r^k}$$

where

$$N_k = \sum_{i_1 + \cdots + i_u = k} \alpha_{i_1} \alpha_{i_2} \cdots \alpha_{i_u}$$

Now we are ready to use the fact that the code is uniquely decipherable. Recalling that α_i is the number of codewords in C of length i, we see that

$$\alpha_{i_1} \alpha_{i_2} \cdots \alpha_{i_u}$$

is the number of possible strings of length $k = i_1 + \cdots + i_u$ consisting of a codeword of length i_1, followed by a codeword of length i_2, and so on, ending with a codeword of length i_u.

Hence, the sum N_k is the total number of strings $c_1 \cdots c_u$ of

length k made up of exactly u codewords. Now, let

$$\mathcal{N} = \{c_1 \cdots c_u \mid c_i \in C, \ len(x) = k\}$$

be the set of all such strings of codewords. Hence, $N_k = |\mathcal{N}|$. Each $c_1 \cdots c_u \in \mathcal{N}$ can be though of simply as a string of length k over the r-ary alphabet $\mathcal{A}$, that is, as a member of $\mathcal{A}^*$, of length k. But there are r^k such strings in $\mathcal{A}^*$, and since C is uniquely decipherable, no two distinct elements of $\mathcal{N}$ represent the same string in $\mathcal{A}^*$. Hence,

$$|N_k| \leq r^k$$

and so

$$\left(\sum_{k=1}^{m} \frac{\alpha_k}{r^k}\right)^u \leq \sum_{k=m}^{um} \frac{N_k}{r^k} \leq \sum_{k=m}^{um} 1 \leq um$$

Taking u-th roots gives

$$\sum_{k=1}^{m} \frac{\alpha_k}{r^k} \leq u^{1/u} m^{1/u}$$

Since this holds for all positive integers u, we may let u approach ∞. But $u^{1/u} m^{1/u} \rightarrow 1$ as $u \rightarrow \infty$, and so we must have

$$\sum_{k=1}^{m} \frac{\alpha_k}{r^k} \leq 1$$ ∎

Again we should point out that if a code C has the property that its codeword lengths $\ell_1, \ldots, \ell_n$ satisfy Kraft's inequality, we cannot conclude that C must be uniquely decipherable.

Example 2.1.7 It is not possible to construct a uniquely decipherable code, over the alphabet $\{0,1,2,\ldots,9\}$, with 9 codewords of length one, 9 codewords of length two, 10 codewords of length three, and 10 codewords of length four. For if such a code existed, we would have (since $r = 10$)

$$\sum_{i=1}^{38} \frac{1}{r^{\ell_i}} = \frac{9}{10} + \frac{9}{10^2} + \frac{10}{10^3} + \frac{10}{10^4} = \frac{1001}{1000} > 1$$

Since this violates McMillan's Theorem, no such code can exist. □

Kraft's Theorem and McMillan's Theorem together imply the following results, whose proofs we leave as exercises.

Theorem 2.1.4 If a uniquely decipherable code exists with codeword lengths $\ell_1, \ell_2, \ldots, \ell_n$, then an instantaneous code must also exist with these same codeword lengths. □

Corollary 2.1.5 The minimum average codeword length, among all uniquely decipherable encoding schemes for a source $\mathcal{S}$, is equal to the minimum average codeword length among all instantaneous encoding schemes for $\mathcal{S}$. ☐

Hence, *in seeking to minimize the average codeword length over all uniquely decipherable encoding schemes, we may restrict attention to instantaneous codes.*

EXERCISES

In Exercises 1-7, determine whether or not there is an instantaneous code with given radix r *and codeword lengths. If so, construct such a code.*

1. $r = 2$, lengths 1,2,3,3
2. $r = 2$, lengths 1,2,2,3,3
3. $r = 2$, lengths 1,3,3,3,4,4
4. $r = 2$, lengths 2,2,3,3,4,4,5,5
5. $r = 3$, lengths 1,1,2,2,3,3,3
6. $r = 5$, lengths 1,1,1,1,1,8,9
7. $r = 5$, lengths 1,1,1,1,2,2,2,3,3,4
8. Is the code $C = \{0,10,1100,1101,1110,1111\}$ instantaneous? Is it uniquely decipherable?
9. Is the code $C = \{0,10,110,1110,1011,1101\}$ instantaneous? Is it uniquely decipherable?
10. Suppose that we want an instantaneous binary code that contains the codewords 0, 10 and 110. How many additional codewords of length 5 could be added to this code?
11. Prove that each inequality in (2.1.3) implies the previous one.
12. With reference to the proof of Kraft's Theorem, prove that

$$\frac{\alpha_1}{r} + \frac{\alpha_2}{r^2} + \cdots + \frac{\alpha_n}{r^n} \le 1$$

is equivalent to Kraft's inequality.

13. Prove Theorem 2.1.4.
14. Prove Corollary 2.1.5.
15. Prove that a code C is uniquely decipherable if and only if for any $x_1,x_2,\ldots,x_n,y_1,y_2,\ldots,y_n$ in C, we have $x_1x_2\cdots x_n = y_1y_2\cdots y_n$ implies $x_1 = y_1$, $x_2 = y_2,\ldots,x_n = y_n$.
16. Let C be instantaneous. Prove that the following are equivalent.
    (a) C is *maximal instantaneous* in the sense that no codeword can be added to C and still maintain the property of being instantaneous.
    (b) Every finite string of code symbols is the prefix of some string

of codewords in C.

(c) Equality holds in Kraft's inequality.

17. For a given binary code C, let N(k) be the total number of codeword strings that contain exactly k bits. For instance, if $C = \{c_1, c_2, c_3\}$, where $c_1 = 0$, $c_2 = 10$, $c_3 = 11$, then $N(3) = 5$, since the codeword strings $c_1c_1c_1, c_1c_2, c_1c_3, c_2c_1$ and c_3c_1 all contain exactly 3 bits, and no other codeword strings contain exactly 3 bits. For the code C, find a recurrence relation for N(k), and solve it.

2.2 Huffman Encoding

In 1952 D.A. Huffman published a method for constructing efficient instantaneous encoding schemes. This method is now known as *Huffman encoding.*

It is clear that the average codeword length of an encoding scheme is not affected by the nature of the source symbols themselves. Hence, for the purposes of measuring average codeword length, we may assume that the codewords are assigned directly to the probabilities. Accordingly, we may speak of *an encoding scheme* $(c_1,\ldots,c_n)$ for the probability distribution $(p_1,\ldots,p_n)$. When the probability distribution is understood, we may speak of *an encoding scheme* $(c_1,\ldots,c_n)$.

With this in mind, the average codeword length of an encoding scheme $(c_1,\ldots,c_n)$ is

$$AveLen(c_1,\ldots,c_n) = \sum_{i=1}^{n} p_i \, len(c_i)$$

We will use the notation $MinAveLen_r(p_1,\ldots,p_n)$ to denote the minimum average codeword length among all r-ary instantaneous encoding schemes for the probability distribution $(p_1,\ldots,p_n)$. By virtue of Corollary 2.1.5, this minimum is also over all uniquely decipherable encoding schemes.

Definition An **optimal** r-ary encoding scheme for a probability distribution $(p_1,\ldots,p_n)$ is an r-ary instantaneous encoding scheme $(c_1,\ldots,c_n)$ for which

$$AveLen(c_1,\ldots,c_n) = MinAveLen_r(p_1,\ldots,p_n)$$ □

Note that optimal encoding schemes are, by definition, instantaneous.

AN EXAMPLE OF HUFFMAN ENCODING

Before discussing Huffman encoding in general, we would do well to consider a specific example.

Example 2.2.1 Let us construct a 4-ary Huffman encoding scheme for the probability distribution

$$P = (0.24,0.20,0.18,0.13,0.10,0.06,0.05,0.03,0.01)$$

consisting of 9 probabilities.

With reference to Table 2.2.1, the first step is to arrange the probabilities in decreasing order of magnitude in the first column of a

table. Next, we replace the three smallest probabilities by their sum, rearrange the resulting probabilities in decreasing order, and place them in a new column. Notice that we have marked the sum 0.09 with an asterisk. Notice also that we have inserted a blank column labeled *code* between the two columns of probabilities. We will explain the purpose of this column momentarily. The process of combining the three smallest probabilities into a single probability is called a *Huffman reduction of size* 3. The next step is to perform a Huffman reduction of size 4, as shown in the fifth column of Table 2.2.1.

Thus, the probability columns in Table 2.2.1 are formed by the simple process of successive Huffman reductions and reorderings. As to the matter of the size of each reduction, when constructing an r-ary Huffman code, all reductions should have size r except possibly the first reduction, whose size is determined by the fact that we want the last probability column of the table to have exactly r entries. In this case $r = 4$. Thus, noting that a reduction of size s reduces the number of probabilities by $s - 1$, and since $9 - (3 - 1) - (4 - 1) = 4$, we see that the first reduction should have size 3. We will discuss the reduction size in more detail a bit later.

Once the probability columns in Table 2.2.1 are complete, we can construct a code for each probability column by working *from right to left*. For ease of readability, we will do this in Table 2.2.2, although the entire process can certainly be done in a single table. Since the last column contains only $r = 4$ entries (by design), we assign the code $0,1,\ldots,r-1$ to this column. To construct the next code, we "expand" the codeword associated with the probability marked with an asterisk, by concatenating that codeword with the symbols $0,1,\ldots,r-1$, as in the middle code column of Table 2.2.2. This expansion is repeated to obtain the first code column of Table 2.2.2, which is the desired Huffman encoding. ▯

TABLE 2.2.1

Probabilities	Code	Probabilities	Code	Probabilities	Code
0.24		0.24		0.38*	
0.20		0.20		0.24	
0.18		0.18		0.18	
0.13		0.13		0.13	
0.10		0.10			
0.06		0.09*			
0.05		0.06			
0.03					
0.01					

TABLE 2.2.2

Probabilities	Code	Probabilities	Code	Probabilities	Code
0.24	1	0.24	1	0.38*	0
0.20	2	0.20	2	0.24	1
0.18	3	0.18	3	0.18	2
0.13	00	0.13	00	0.13	3
0.10	01	0.10	01		
0.06	03	0.09*	02		
0.05	020	0.06	03		
0.03	021				
0.01	022				

MOTIVATION FOR THE GENERAL CASE

Since we are dealing with r-ary codes, we may as well assume that the code alphabet is $\{0,1,\ldots,r-1\}$. In view of the previous example, we make the following observations. Let

$$P = (p_1,\ldots,p_n)$$

be a probability distribution. Performing a Huffman reduction of size s gives the probability distribution

$$Q = (p_1,\ldots,p_{n-s},q)$$

where $q = p_{n-s+1} + \cdots + p_n$. Suppose that

$$D = (c_1,\ldots,c_{n-s},d)$$

is an optimal encoding scheme for this distribution. Then we can construct an encoding scheme for $(p_1,\ldots,p_n)$ by "expanding" the codeword d into s codewords

$$d0, d1, \ldots, d(s-1)$$

of length L, to get

$$C = (c_1,\ldots,c_{n-s},d0,d1,\ldots,d(s-1))$$

It is easy to see that this code has the prefix property, and so is instantaneous.

Since

$$AveLen(D) = \sum_{i=1}^{n-s} p_i \, len(c_i) + q \, len(d)$$

and

$$AveLen(C) = \sum_{i=1}^{n-s} p_i\ len(c_i) + \sum_{i=n-s+1}^{n} p_i\ [len(d)+1]$$

$$= \sum_{i=1}^{n-s} p_i\ len(c_i) + q[len(d)+1] = AveLen(D) + q$$

we have

(2.2.1) $$MinAveLen_r(p_1,\ldots,p_n) \leq AveLen(C) = AveLen(D) + q = MinAveLen_r(p_1,\ldots,p_{n-s},q) + q$$

From this we deduce that C will be optimal if and only if

(2.2.2) $$MinAveLen_r(p_1,\ldots,p_n) = MinAveLen_r(p_1,\ldots,p_{n-s},q) + q$$

Our goal then is to establish this result, by establishing the reverse inequality to (2.2.1)

(2.2.3) $$MinAveLen_r(p_1,\ldots,p_n) \geq MinAveLen_r(p_1,\ldots,p_{n-s},q) + q$$

Once we have done this, we will know that the process of successive "expansions" will produce optimal encoding schemes.

Thus, let $C = (c_1,\ldots,c_n)$ be an optimal encoding scheme for $(p_1,\ldots,p_n)$. Suppose that the maximum length of the codewords in C is L. If C happens to have s codewords of length L of the form

(2.2.4) $$d0, d1, \ldots, d(s-1)$$

for some d, then we can "contract" C into an encoding scheme

$$D = (c_1,\ldots,c_{n-s},d)$$

for $(p_1,\ldots,p_{n-s},q)$. Then

$$\begin{aligned} MinAveLen_r(p_1,\ldots,p_{n-s},q) &\leq AveLen(D) \\ &= \sum_{i=1}^{n-s} p_i\ell_i + (L-1)q \\ &= \sum_{i=1}^{n} p_i\ell_i - q \\ &= AveLen(C) - q \\ &= MinAveLen_r(p_1,\ldots,p_n) - q \end{aligned}$$

which shows that (2.2.3) holds.

Thus, we are left with the question of whether there is always an optimal encoding scheme $C = (c_1, \ldots, c_n)$ with codewords of the form (2.2.4).

THE GENERAL CASE

Before proceeding to this existence question, let us make a few remarks about the reduction size. As already mentioned, when constructing an r-ary Huffman code, we want each reduction to have size r, except possibly the first, whose size is determined by the requirement that the final column of probabilities should have size r. Noting that a reduction of size t reduces the number of probabilities by $t-1$, then if s is the size of the first reduction, and if u is the number of subsequent reductions of size r, we have

$$n - (s-1) - u(r-1) = r$$

or

$$s = n - (u+1)(r-1)$$

Since $2 \le s \le r$, we deduce that the first reduction size s is *uniquely* determined by the conditions

$$s \equiv n \mod (r-1), \quad 2 \le s \le r \tag{2.2.5}$$

It is useful to note that for binary Huffman encoding, where $r = 2$, condition (2.2.5) simply says that $s = r = 2$. Hence, in this case, the reduction size is always equal to 2.

Now we are ready for our existence result.

Theorem 2.2.1 Let $P = (p_1, \ldots, p_n)$ be a probability distribution, with $p_1 \ge p_2 \ge \cdots \ge p_n$. Then there exists an optimal r-ary encoding scheme $C = (c_1, \ldots, c_n)$ for P that has s codewords of maximum length L of the form

$$\mathbf{d}0, \mathbf{d}1, \ldots, \mathbf{d}(s-1)$$

where s is given by (2.2.5). As a result, for such probability distributions, we have

$$MinAveLen_r(p_1, \ldots, p_n) = MinAveLen_r(p_1, \ldots, p_{n-s}, q) + q$$

where $q = p_{n-s+1}, \ldots, p_n$.

Proof. Among all optimal r-ary encoding schemes for $(p_1, \ldots, p_n)$, choose one, say $C = (c_1, \ldots, c_n)$, that has smallest *total* codeword length

$$\sum_{i=1}^{n} \ell_i$$

where $\ell_i = len(\mathbf{c}_i)$. Since C is optimal, we have

$$AveLen(\mathrm{C}) = \sum_{i=1}^{n} p_i\, len(\mathbf{c}_i) = MinAveLen_r(p_1,\ldots,p_n)$$

We will show that if C does not have the desired property, then we can make certain changes in C in such a way that the resulting code will still be optimal, but will have the desired property.

Now, C must have the property that codewords with larger probability have smaller lengths, that is,

$$p_i > p_j \quad \text{implies} \quad len(\mathbf{c}_i) \le len(\mathbf{c}_j)$$

For if $len(\mathbf{c}_i) > len(\mathbf{c}_j)$, then we could interchange the codewords $\mathbf{c}_i$ and $\mathbf{c}_j$ and achieve a smaller average codeword length. By assumption, we have

$$p_1 \geq \cdots \geq p_n$$

and we may further assume, by rearranging if necessary, that

$$\ell_1 \le \cdots \le \ell_n \tag{2.2.6}$$

Now, if there are precisely k codewords in C of maximum length L, then (2.2.6) gives

$$\ell_{n-k} < \ell_{n-k+1} = \cdots = \ell_n = \mathrm{L}$$

We would like to show that $k \geq s$, where s is the length of the first reduction. This will tell us that C contains at least s codewords of maximum length L.

To do this, consider the Kraft sum

$$\mathrm{K} = \sum_{i=1}^{n} r^{-\ell_i}$$

We know that $\mathrm{K} \le 1$, since C is instantaneous. However, if

$$\mathrm{K} - r^{-\mathrm{L}} + r^{-(\mathrm{L}-1)} \le 1$$

then Kraft's inequality would hold with $\ell_n = \mathrm{L}$ replaced by $\ell_n - 1 = \mathrm{L} - 1$. This would mean that there was as instantaneous code with codeword lengths

$$\ell_1, \ldots, \ell_{n-1}, \ell_n - 1$$

which is a contradiction to the fact that C has minimum *total* codeword length among all optimal encoding schemes. (It may not be a contradiction to the fact that C is optimal, since p_n might be equal to 0.)

Thus, we have

$$1 + r^{-L} - r^{-(L-1)} < K \leq 1$$

Multiplying by r^L gives

$$r^L - r + 1 < r^L K \leq r^L$$

and so

$$\text{(2.2.7)} \qquad r^L K = r^L - r + \alpha, \text{ where } \alpha \in \{2,3,\ldots,r\}$$

Now let us make a few observations about $r^L K$.

1) Since $r-1 \mid r^L - r$, (2.2.7) implies that

$$r^L K \equiv \alpha \bmod (r-1)$$

2) Since

$$r^L K = \sum_{i=1}^{n} r^{L-\ell_i}$$

and since $r^u \equiv 1 \bmod (r-1)$ for any positive integer u, we have

$$r^L K \equiv n \bmod (r-1)$$

3) From 1 and 2, we get $\alpha \equiv n \bmod (r-1)$, and since $2 \leq \alpha \leq r$, we deduce from (2.2.5) that $\alpha = s$. Hence

$$r^L K = r^L - r + s$$

4) Since $r \mid r^L - r$, observation 3 implies that

$$r^L K \equiv s \bmod r$$

5) Since $L - \ell_i = 0$ if and only if $i > n-k$, we may write

$$r^L K = \sum_{i=1}^{n} r^{L-\ell_i} = \sum_{i=1}^{n-k} r^{L-\ell_i} + k$$

and so

$$r^L K \equiv k \bmod r$$

6) From 4) and 5) we get

$$s \equiv k \bmod r$$

But $2 \leq s \leq r$, and so we must have either $k \leq 0$, $k = s$ or $k > r$. Since the first inequality is not possible, and since $r \geq s$, we deduce that $k \geq s$, as desired.

Thus, we may assume that the last s codewords of C have maximum length L. Now suppose that $\mathbf{c}$ is one of these codewords of length L and that $\mathbf{c} = \mathbf{d}x$. Consider the words

(2.2.8) $$d0, d1, \ldots, d(s-1)$$

obtained from $\mathbf{c}$ be replacing the last symbol by $0,1,\ldots,s-1$.

If any of these words $\mathbf{x}$ is not in C, then since $k \geq s$, we know that there is another codeword in C of length L. Hence, we may replace that codeword by $\mathbf{x}$ and still have an optimal code. Finally, since the last k codewords in C have the same length, we can rearrange them, if necessary, to insure that the words (2.2.8) are the last s codewords. ∎

HUFFMAN'S ALGORITHM

Now we can present Huffman's algorithm.

Theorem 2.2.2 Consider the following algorithm $\mathcal{H}$ for producing r-ary encoding schemes C for probability distributions P.

1) If $P = (p_1, \ldots, p_n)$, where $n \leq r$, then let $C = (0, \ldots, n-1)$.
2) If $P = (p_1, \ldots, p_n)$, where $n > r$, then
   a) Reorder P if necessary so that $p_1 \geq p_2 \geq \cdots \geq p_n$.
   b) Let $Q = (p_1, \ldots, p_{n-s}, q)$, where $q = p_{n-s+1} + \cdots + p_n$, and where s is determined by (2.2.5).
   c) Perform the algorithm $\mathcal{H}$ on Q, obtaining an encoding scheme
   $$D = (\mathbf{c}_1, \ldots, \mathbf{c}_{n-s}, \mathbf{d})$$
   d) Let
   $$C = (\mathbf{c}_1, \ldots, \mathbf{c}_{n-s}, \mathbf{d}0, \mathbf{d}1, \ldots, \mathbf{d}(s-1))$$

Then the encoding scheme C is optimal.

Proof. The proof is by induction on the size n of the probability distribution. The result is clearly true for $n \leq r$. Suppose the result is true for all probability distributions of length less than n, and consider the encoding scheme $C = (\mathbf{c}_1, \ldots, \mathbf{c}_n)$ for $(p_1, \ldots, p_n)$, produced by $\mathcal{H}$.

Since $|Q| < n$ (where Q comes from part 2c), the inductive hypothesis implies that D (also from part 2c) is optimal. Thus, D is instantaneous, and by Theorem 2.2.1

$$AveLen(D) = MinAveLen_r(p_1, \ldots, p_{n-s}, q) = MinAveLen(p_1, \ldots, p_n) - q$$

Since D is instantaneous, so is C. Finally,

$$AveLen(C) = \text{AveLen}(D) + q = MinAveLen(p_1, \ldots, p_k)$$

and so C is optimal. ∎

We should remark that, because of step 2a, the Huffman algorithm may not always produce the same code when applied to the

same probabilities. In fact, as the next example demonstrates, we may get codes with different word lengths, even though the *average* word length must be the same.

Example 2.2.2 Table 2.2.3 shows the construction of a binary Huffman code, where the reduction size is 2. Table 2.24 shows the code that is obtained by a different ordering of the probabilities (note the different position of the asterisk in column 3). Even though these codes have different codeword lengths (the first has total codeword length 17, and the second has total codeword length 16), they are both Huffman codes, and they both have *average* codeword length 2.49, which is the minimum possible by Theorem 2.2.2. □

TABLE 2.2.3

Prob	Code	Prob	Code	Prob	Code	Prob	Code	Prob	Code
0.32	00	0.32	00	0.32	00	0.38*	1	0.62*	0
0.19	10	0.19	10	0.30*	01	0.32	00	0.30	1
0.19	11	0.19	11	0.19	10	0.30	01		
0.11	011	0.19*	010	0.19	11				
0.10	0100	0.11	011						
0.09	0101								

TABLE 2.2.4

Prob	Code	Prob	Code	Prob	Code	Prob	Code	Prob	Code
0.32	00	0.32	00	0.32	00	0.38*	1	0.62*	0
0.19	11	0.19*	10	0.30*	01	0.32	00	0.30	1
0.19	010	0.19	11	0.19	10	0.30	01		
0.11	011	0.19	010	0.19	11				
0.10	100	0.11	011						
0.09	101								

EXERCISES

In Exercises 1–6, find a Huffman encoding of the given probability distribution for the given radix.

1. $P = \{0.2,0.1,0.1,0.3,0.1,0.2\}$
   (a) $r = 2$ (b) $r = 3$ (c) $r = 4$ (d) $r = 5$

2. $P = \{0.9,0.02,0.02,0.02,0.02,0.02\}$
   (a) $r = 2$ (b) $r = 3$ (c) $r = 4$
3. $P = \{0.1,\ldots,0.1\}$
   (a) $r = 2$ (b) $r = 3$ (c) $r = 4$ (d) $r = 5$ (e) $r = 6$
4. $P = \{0.05,0.11,0.8,0.005,0.02,0.005,0.01\}$
   (a) $r = 2$ (b) $r = 3$ (c) $r = 4$
5. $P = \{0.3,0.05,0.03,0.02,0.3,0.1,0.15,0.05\}$
   (a) $r = 2$ (b) $r = 3$ (c) $r = 4$ (d) $r = 5$
6. $P = \{0.3,0.1,0.1,0.1,0.1,0.06,0.05,0.05,0.05,0.04,0.03,0.02\}$
   (a) $r = 2$ (b) $r = 3$ (c) $r = 4$ (d) $r = 5$
7. In the proof of Theorem 2.2.2, prove that C is instantaneous.
8. In the proof of Theorem 2.2.3, prove that $AveLen(C) = AveLen(D) + q$.
9. State a condition in terms of the sizes of the probabilities that will guarantee uniqueness (up to switching 0's and 1's) in Huffman encoding.
10. Determine the number of reductions in the Huffman encoding scheme in terms of the number n of probabilities and the radix r.
11. Explain how you would use the results of Theorem 2.2.2 to compute $MinAveLen_r(p_1,\ldots,p_n)$.
12. Show that the number of additions, concatenations and steps in reordering, in the Huffman construction process is at most Mn^2, where M does not depend on the number n of probabilities. (In other words, the number of operations is $O(n^2)$.) *Hint.* Try the binary case first.
13. Determine all probability distributions that have (00,01,10,11) and (0,10,110,111) as binary Huffman encodings.
14. Let C be a binary Huffman encoding for the uniform probability distribution $P = (1/n,\ldots,1/n)$, and suppose that the codeword lengths of C are ℓ_i. Let $n = a2^k$, where $1 \leq a < 2$.
    (a) Show that C has minimum *total* codeword length $T = \sum \ell_i$ among all instantaneous encodings for P.
    (b) Show that C has at least two codewords of maximum length $L = \max \ell_i$.
    (c) Show that the Kraft sum $\sum (1/r^{\ell_i})$ equals 1.
    (d) Show that $\ell_i = L$ or $\ell_i = L - 1$ for all i.
    (e) Let u be the number of codewords of length $L - 1$ and let v be the number of codewords of length L. Determine u, v and L in terms of a and k.
    (f) Find $MinAveLen_2(1/n,\ldots,1/n)$.

2.3 The Noiseless Coding Theorem

Now we are ready to discuss the main results on noiseless coding. As we know, the entropy $H(\mathcal{S})$ of a source $\mathcal{S}$ is the amount of information contained in the source. Further, since an instantaneous encoding scheme for $\mathcal{S}$ captures the information in the source, it is not unreasonable to believe that the average codeword length of such a code must be at least as large as the entropy $H(\mathcal{S})$. In fact, this is what the Noiseless Coding Theorem says. This theorem, first proved by Claude Shannon in 1948, also says that by clever encoding, we can make the average codeword length as close to the entropy as desired.

As in the previous section, the Noiseless Coding Theorem does not depend in any way on the nature of the source symbols, and so we will assume that codewords are assigned directly to the probabilities. Accordingly, we speak of *an encoding scheme* $(\mathbf{c}_1,\ldots,\mathbf{c}_n)$ for a probability distribution $(p_1,\ldots,p_n)$. Recall that the average codeword length of an encoding scheme $(\mathbf{c}_1,\ldots,\mathbf{c}_n)$ for $(p_1,\ldots,p_n)$ is

$$AveLen(\mathbf{c}_1,\ldots,\mathbf{c}_n) = \sum_{i=1}^{n} p_i\, len(\mathbf{c}_i)$$

We denote the length $len(\mathbf{c}_i)$ by ℓ_i.

Recall also that the r-ary entropy of a probability distribution (or of a source) is given by

$$H_r(p_1,\ldots,p_n) = \sum_{i=1}^{n} p_i \log_r \frac{1}{p_i}$$

Now we are ready for our first result.

Theorem 2.3.1 Let $C = (\mathbf{c}_1,\ldots,\mathbf{c}_n)$ be an instantaneous encoding scheme for $P = (p_1,\ldots,p_n)$. Then

$$H_r(p_1,\ldots,p_n) \le AveLen(\mathbf{c}_1,\ldots,\mathbf{c}_n)$$

with equality on the left side if and only if $len(\mathbf{c}_i) = \log_r(1/p_i)$.
Proof. Since C is instantaneous, Kraft's Theorem tells us that

$$\sum_{i=1}^{n} \frac{1}{r^{\ell_i}} \le 1$$

Thus, we may apply Lemma 1.2.2, with $q_i = 1/r^{\ell_i}$, to deduce that

$$H_r(p_1,\ldots,p_n) = \sum_{i=1}^{n} p_i \log_r \frac{1}{p_i} \le \sum_{i=1}^{n} p_i \log_r \frac{1}{q_i}$$

$$= \sum_{i=1}^{n} p_i \log_r r^{\ell_i} = \sum_{i=1}^{n} p_i \ell_i = AveLen(\mathbf{c}_1,\ldots,\mathbf{c}_n)$$

Furthermore, according to Lemma 1.2.2, equality holds here if and only if $q_i = p_i$, that is, if and only if $p_i = 1/r^{\ell_i}$ or $\ell_i = \log_r (1/p_i)$. ∎

We note that Theorem 2.3.1 holds with the word "instantaneous" replaced by "uniquely decipherable."

Example 2.3.1 The entropy of the probability distribution $P = (\frac{1}{2},\frac{1}{4},\frac{1}{4})$ is

$$H_2\left(\frac{1}{2},\frac{1}{4},\frac{1}{4}\right) = \frac{1}{2}\log_2 2 + \frac{1}{4}\log_2 4 + \frac{1}{4}\log_2 4 = \frac{3}{2}$$

Hence, any instantaneous *binary* encoding scheme for P must have average codeword length of at least $3/2$. The Huffman encoding scheme (assigned directly to the probabilities)

$$f\left(\frac{1}{2}\right) = 0, \quad f\left(\frac{1}{4}\right) = 10, \quad f\left(\frac{1}{4}\right) = 11$$

has length $\frac{1}{2}(1) + \frac{1}{4}(2) + \frac{1}{4}(2) = \frac{3}{2}$, which in this case is equal to the entropy. □

Example 2.3.2 The entropy of the probability distribution $P = (\frac{2}{3},\frac{1}{6},\frac{1}{6})$ is

$$H_2\left(\frac{2}{3},\frac{1}{6},\frac{1}{6}\right) = \frac{2}{3}\log_2 \frac{3}{2} + \frac{1}{6}\log_2 6 + \frac{1}{6}\log_2 6 \approx 1.2516$$

and so Theorem 2.3.1 says that any instantaneous binary encoding scheme for P must have average codeword length at least 1.2516. However, the Huffman encoding scheme

$$f\left(\frac{2}{3}\right) = 0, \quad f\left(\frac{1}{6}\right) = 10, \quad f\left(\frac{1}{6}\right) = 11$$

has average codeword length $4/3$. Thus, in this case, we cannot achieve an average codeword length as low as the entropy of the source. □

Example 2.3.3 In Example 1.1.2, we mentioned that the entropy of the source in Table 1.1.1 is approximately 4.07991. When the Huffman encoding scheme is applied to this source, the result is an encoding scheme with average codeword length approximately 4.1195, which is quite close to the entropy. □

Notice that the condition for equality in Theorem 2.3.1 is that

$$\ell_i = \log_r \frac{1}{p_i} = -\log_r p_i$$

which means that $\log_r p_i$ must be an integer. Since this is not often the case, we cannot often expect equality.

However, we can always find an instantaneous encoding scheme C for which the ℓ_i satisfy

$$\text{(2.3.1)} \qquad \log_r \frac{1}{p_i} \le \ell_i < \log_r \frac{1}{p_i} + 1$$

This follows from Kraft's inequality, since $\log_r \frac{1}{p_i} \le \ell_i$ is equivalent to

$$p_i \ge \frac{1}{r^{\ell_i}}$$

and so

$$\sum_{i=1}^{n} \frac{1}{r^{\ell_i}} \le \sum_{i=1}^{n} p_i = 1$$

Now, since

$$\ell_i < \log_r \frac{1}{p_i} + 1$$

the average codeword length of $C = (c_1, \ldots, c_n)$ satisfies

$$AveLen(C) = \sum_{i=1}^{n} p_i \ell_i \le \sum_{i=1}^{n} p_i \left(\log_r \frac{1}{p_i} + 1\right)$$

$$= \sum_{i=1}^{n} p_i \log_r \frac{1}{p_i} + \sum_{i=1}^{n} p_i = H_r(p_1, \ldots, p_n) + 1$$

Thus, we have found an instantaneous encoding scheme for $(p_1, \ldots, p_n)$ whose average codeword length is less than $H_r(p_1, \ldots, p_n) + 1$. Putting this together with Theorem 2.3.1, we get the Noiseless Coding Theorem.

Theorem 2.3.2 (The Noiseless Coding Theorem) For any probability distribution $P = (p_1, \ldots, p_n)$, we have

$$H_r(p_1, \ldots, p_n) \le MinAveLen_r(p_1, \ldots, p_n) < H_r(p_1, \ldots, p_n) + 1 \quad \square$$

We should point out that using (2.3.1) and the method of Example 2.1.6 to construct codes does not, in general, give the smallest possible average codeword length, and so is not as good as Huffman encoding.

EXTENSIONS OF A SOURCE

The Noiseless Coding Theorem determines $MinAveLen_r(p_1, \ldots, p_n)$ to within 1 r-ary unit, but this may still be too much for some purposes. Fortunately, there is a way to improve upon this, based on the following idea.

Definition Let $\mathcal{S} = (S,P)$ be a source. The **n-th extension** of $\mathcal{S}$ is $\mathcal{S}^n = (S^n,P^n)$, where S^n is the set of all words of length n over S, and P^n is the probability distribution defined for $\mathbf{x} = x_1 \cdots x_n$ by

$$P^n(\mathbf{x}) = P(x_1)\cdots P(x_n) \qquad \square$$

The probability distribution P^n is defined so that if $X_1,\ldots,X_n$ are *independent* random samplings of $\mathcal{S}$, where

$$P(X_i{=}s_j) = P(s_j)$$

for all i, then the distribution of the random vector $\mathbf{X} = (X_1,\ldots,X_n)$ is P^n. This allows for a simple proof, based on Corollary 1.2.6, of the fact that since we get n times as much information from an independently formed word of length n as we do from a word of length 1, the entropy of $\mathcal{S}^n$ should be n times the entropy of $\mathcal{S}$. We leave the details of this as an exercise.

Theorem 2.3.3 Let $\mathcal{S}$ be a source, and let $\mathcal{S}^n$ be its n-th extension. Then $H_r(\mathcal{S}^n) = nH_r(\mathcal{S})$. In terms of probability distributions, we have

$$H_r(P^n) = nH_r(P) \qquad \square$$

Applying the Noiseless Coding Theorem to the extension P^n, and using Theorem 2.3.3, gives the following.

Theorem 2.3.4 Let P be a probability distribution, and let P^n be its n-th extension. Then

$$H_r(P) \le \frac{\mathit{MinAveLen}(P^n)}{n} < H_r(P) + \frac{1}{n} \qquad \square$$

If $\mathcal{S} = (S,P)$ is a source with probability distribution P, and $\mathcal{S}^n = (S^n,P^n)$ is its n-th extension, then since each source symbol in $\mathcal{S}^n$ contains exactly n source symbols of $\mathcal{S}$, the quantity

$$\frac{\mathit{MinAveLen}(P^n)}{n}$$

is the minimum average codeword length *per source symbol of* $\mathcal{S}$. Theorem 2.3.4 says that, by encoding a sufficiently long extension of $\mathcal{S}$, we may make the minimum average codeword length per source symbol of $\mathcal{S}$ as close to the entropy $H_r(P)$ as desired.

Example 2.3.4 Consider the probability distribution $P = (\frac{1}{4},\frac{3}{4})$. A Huffman encoding for this distribution is

$$f\left(\frac{1}{4}\right) = 0, \quad f\left(\frac{3}{4}\right) = 1$$

whose average codeword length is $\frac{1}{4}+\frac{3}{4}=1$.

The second extension P^2 is

$$P^2 = \left(\frac{1}{4}\frac{1}{4},\frac{1}{4}\frac{3}{4},\frac{3}{4}\frac{1}{4},\frac{3}{4}\frac{3}{4}\right) = \left(\frac{1}{16},\frac{3}{16},\frac{3}{16},\frac{9}{16}\right)$$

A Huffman encoding scheme for P^2 is

$$f\left(\frac{1}{16}\right) = 010, \quad f\left(\frac{3}{16}\right) = 011, \quad f\left(\frac{3}{16}\right) = 00, \quad f\left(\frac{9}{16}\right) = 1$$

which has average codeword length

$$\frac{1}{16}\cdot 3 + \frac{3}{16}\cdot 3 + \frac{3}{16}\cdot 2 + \frac{9}{16}\cdot 1 = \frac{27}{16}$$

Hence, the average codeword length per source symbol is $27/32 = 0.84375$.

Similarly, we can show that the third extension P^3 has average codeword length per source symbol equal to 0.82292, and the average codeword length per source symbol for the fourth extension P^4 is 0.81836.

While it is not always true that the average codeword length per source symbol decreases monotonically as n increases, Theorem 2.3.4 guarantees that the limit must approach $H_r(P)$. ▯

EXERCISES

1. Provide the details of the proof of Theorem 2.3.2
2. Let $S = (a,b,c)$ and $P = (\frac{2}{3},\frac{2}{9},\frac{1}{9})$. Use (2.3.1), and the method of Example 2.1.6, to construct an encoding scheme, and compare its average codeword length with that of a Huffman encoding scheme.
3. Prove Theorem 2.3.3.
4. Prove Theorem 2.3.4.
5. Find a source $\mathcal{S}$ for which $MinAveLen_r(\mathcal{S}) < \frac{1}{2}\, MinAveLen_r(\mathcal{S}^2)$. What is the relevance of this to a theorem in this section?
6. Show that the Noiseless Coding Theorem is best possible by showing that, for any $\epsilon > 0$, there is a probability distribution for which $MinAveLen_r(p_1,\dots,p_n) - H_r(p_1,\dots,p_n) \geq 1-\epsilon$.
7. Let $D(p_1,\dots,p_n) = MinAveLen_r(p_1,\dots,p_n) - H_r(p_1,\dots,p_n)$. Show that a Huffman reduction cannot increase the value of $D(p_1,\dots,p_n)$. (In other words, $D(p_1,\dots,p_n) \geq D(p_1,\dots,p_{n-s},q)$, where s and q are as in the proof of Theorem 2.2.2.)
8. Consider the following method for constructing a binary encoding of a probability distribution $(p_1,\dots,p_n)$, where $p_1 \geq p_2 \geq \cdots \geq p_n$. Define q_j by

$$q_1 = 0, \quad q_i = p_1 + \cdots + p_{i-1} \quad (i > 1)$$

and let $m_i = \lceil -\log p_i \rceil$, where $\lceil x \rceil$ is the smallest integer at least as large as x. Finally, let $\mathbf{c}_i$ be the binary expansion of q_i, truncated after the first m_i bits to the right of the binary point. For example, if $q_i = 37/64$ and $p_i = 1/20$, then $m_i = \lceil -\log(1/20) \rceil = \lceil 4.32 \rceil = 5$, and $\mathbf{c}_i$ is the first 5 bits in the binary expansion of q_i, which is $\mathbf{c}_i = 10010$. (Note that $q_i = 100101$.)

(a) Show that the code $C = (\mathbf{c}_1, \ldots, \mathbf{c}_n)$ has the prefix property.

(b) Show that the average codeword length A of this code satisfies $H_2(p_1, \ldots, p_n) \le A \le H_2(p_1, \ldots, p_n) + 1$.

9. Consider a source that emits binary digits, with probabilities $P(0) = p$, $P(1) = 1 - p$. Suppose we encode the output of the source by counting 0's as follows

$$\begin{aligned} 1 &\to 1000 \\ 01 &\to 1001 \\ 001 &\to 1010 \\ &\vdots \\ 00000001 &\to 1111 \\ 00000000 &\to 0 \end{aligned}$$

In short, if fewer than 8 0's occur before a 1, we encode the string of 0's followed by the 1 with the codeword $1e_1e_2e_3$, where $e_1e_2e_3$ is the binary expression for the number of 0's. If eight 0's occur in a row, we encode this with a 0.

(a) Prove that this code is instantaneous.

(b) Define an *event* as the outputting of a codeword. Find the average number A_c of code bits per event.

(c) Find the average number A_s of source bits per event.

(d) For $p = .9$, compare the number A_c/A_s with the average codeword length per source symbol of the binary Huffman encoding of the fourth extension of the original source. What does this say about the optimality of Huffman encoding?

CHAPTER 3

Noisy Coding

3.1 The Discrete Memoryless Channel and Conditional Entropy

In the previous chapter, we discussed the question of how to most efficiently encode source information for transmission over a noiseless channel, where we did not need to be concerned about correcting errors. Now we are ready to consider the question of how to encode source data efficiently and, at the same time, minimize the probability of uncorrected errors when transmitting over a noisy channel.

The Noiseless Coding Theorem demonstrates the key role played by the concept of entropy in noiseless source encoding. As we will see, entropy also plays a key role in noisy channel encoding.

DISCRETE MEMORYLESS CHANNELS

We begin with the definition of a discrete memoryless channel.

Definition A **discrete memoryless channel** consists of an **input alphabet** $\mathcal{I} = \{x_1, \ldots, x_s\}$, an **output alphabet** $\mathcal{O} = \{y_1, \ldots, y_t\}$, and a set of **channel probabilities** $p(y_j \mid x_i)$, satisfying

$$\sum_{j=1}^{t} p(y_j \mid x_i) = 1$$

for all i. Intuitively, we think of $p(y_j \mid x_i)$ as the probability that y_j is received, given that x_i is sent through the channel.

Furthermore, if $\mathbf{c} = c_1 \cdots c_n$ and $\mathbf{d} = d_1 \cdots d_n$ are words of length n over $\mathcal{I}$ and $\mathcal{O}$, respectively, the probability $p(\mathbf{d} \mid \mathbf{c})$ that $\mathbf{d}$ is received, given that $\mathbf{c}$ is sent, is

$$p(\mathbf{d} \mid \mathbf{c}) = \prod_{i=1}^{n} p(d_i \mid c_i) \qquad \square$$

The term *discrete* refers to the fact that the input and output alphabets are finite, and we use the term *memoryless* since the probability that an output symbol d_i is received depends only on the current input c_i, and not on previous inputs. Notice also that a discrete memoryless channel has a certain time, or position, independence, expressed by the fact that the probability that an error occurs in a symbol does not depend on the position of that symbol in the word.

Our plan in this section is to discuss the concept of entropy, as it applies to a discrete memoryless channel. Since input to a channel is often probabilistic in nature, we think of the input to a channel as the values of a random variable X, with **input distribution** defined by $P(X=x_i) = p(x_i)$.

Each input X induces an output Y, with **output distribution** defined by

$$P(Y=y_j) = \sum_{i=1}^{s} p(y_j \mid x_i)p(x_i)$$

The **joint distribution** of X and Y is given by

$$P(X=x_i, Y=y_j) = p(y_j \mid x_i)p(x_i)$$

and the **backward channel probabilities** are

$$P(X=x_i \mid Y=y_j) = \frac{P(X=x_i, Y=y_j)}{P(Y=y_j)}$$

It is important to keep in mind that the output distribution, the joint distribution and the backward channel probabilities all depend on the input distribution, as well as the channel probabilities.

For convenience we will refer to a discrete memoryless channel simply as a *channel*. Also, when no confusion can arise, we will adopt the following notation

$$p(x_i) = P(X{=}x_i)$$
$$p(y_j) = P(Y{=}y_j)$$
$$p(x_i,y_j) = P(X{=}x_i,Y{=}y_j)$$
$$p(x_i \mid y_j) = P(X{=}x_i \mid Y{=}y_j)$$
$$p(y_j \mid x_i) = P(Y{=}y_j \mid X{=}x_i)$$

(Thus, for instance, the symbol $p(0 \mid 1)$ is ambiguous, but if we know that $u \in \mathcal{I}$ and $v \in \mathcal{O}$, then the symbols $p(u)$, $p(v)$, $p(u \mid v)$ and $p(v \mid u)$ are not ambiguous.)

A typical discrete memoryless channel is pictured in Figure 3.1.1.

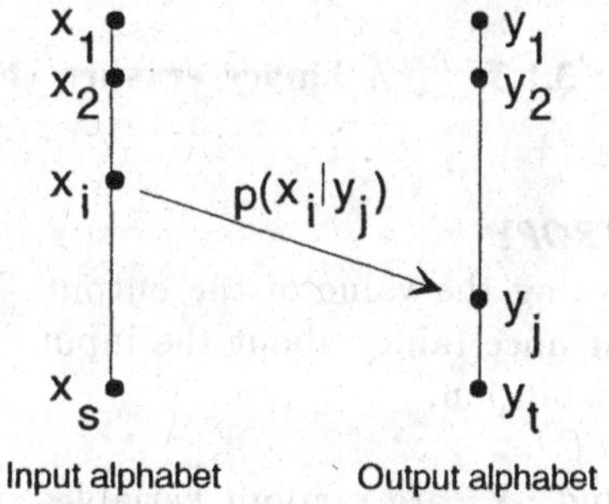

Figure 3.1.1 A typical discrete memoryless channel

Example 3.1.1 One of the most important discrete memoryless channels is the **binary symmetric channel**, which has input and output alphabets $\{0,1\}$, and channel probabilities

$$p(1 \mid 0) = p(0 \mid 1) = p \quad \text{and} \quad p(0 \mid 0) = p(1 \mid 1) = 1 - p$$

Thus, the probability of a bit error, or **crossover probability**, is p. This channel is pictured in Figure 3.1.2. □

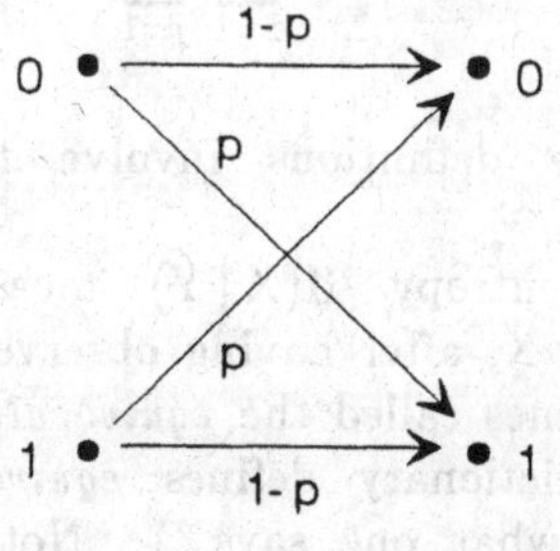

Figure 3.1.2 A binary symmetric channel

Example 3.1.2 The channel pictured in Figure 3.1.3 is called a **binary erasure channel**, since the output ? is interpreted as a loss, or erasure, of the input. □

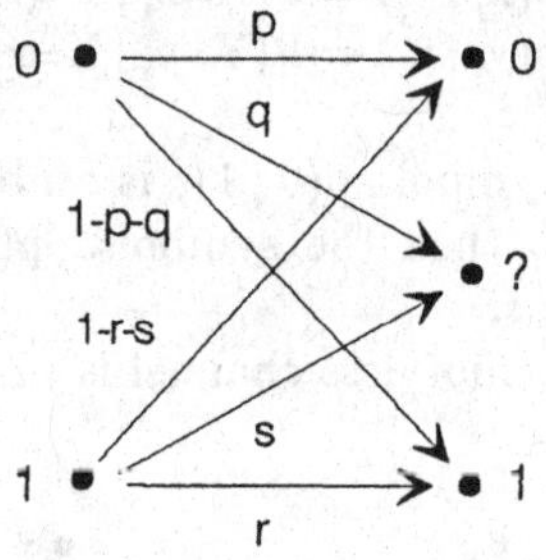

Figure 3.1.3 **A binary erasure channel**

CONDITIONAL ENTROPY

In general, knowing the value of the output Y of a channel will have an effect on our uncertainty about the input X. This leads us to make the following definition.

Definition If X and Y are random variables, then the **conditional entropy** of X, given $Y=y_j$, is defined by

$$H(X \mid Y=y_j) = \sum_{i=1}^{s} p(x_i \mid y_j) \log \frac{1}{p(x_i \mid y_j)}$$

Thus $H(X \mid Y = y_j)$ is the entropy of X, computed using the conditional probability distribution $p(\cdot \mid Y=y_j)$.

The **conditional entropy** of X, given Y, is the average conditional entropy of X, given $Y=y_j$, that is,

$$H(X \mid Y) = \sum_{j=1}^{t} H(X \mid Y=y_j) p(y_j) = \sum_{i=1}^{s} \sum_{j=1}^{t} p(x_i \mid y_j) p(y_j) \log \frac{1}{p(x_i \mid y_j)} \quad \square$$

Notice that these definitions involve the *backward* channel probabilities $p(x_i \mid y_j)$.

The conditional entropy $H(X \mid Y)$ measures the uncertainty remaining in the input X, after having observed the output Y. For this reason, it is sometimes called the *equivocation* of X with respect to Y. (Webster's dictionary defines *equivocation* as "To avoid committing oneself in what one says.") Note also that $H(X \mid Y)$ measures the amount of information remaining in X, after sampling Y, and so it can be interpreted as the *loss* of information about X

caused by the channel. (Put another way, $H(X \mid Y)$ is the amount of information about X that doesn't make it through the channel to Y.)

Example 3.1.3 Consider the channel pictured in Figure 3.1.4, which is a special case of the binary erasure channel.

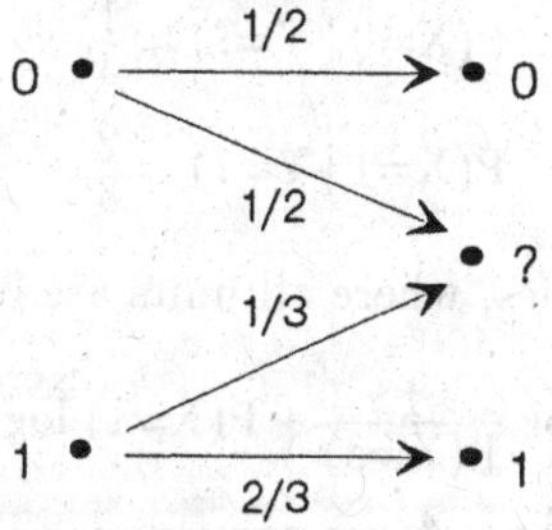

Figure 3.1.4

We are given the following input and channel probabilities

$$P(X=0) = \tfrac{1}{4}, \quad P(X=1) = \tfrac{3}{4}$$

$$P(Y=0 \mid X=0) = \tfrac{1}{2}, \quad P(Y=? \mid X=0) = \tfrac{1}{2}, \quad P(Y=1 \mid X=0) = 0$$

$$P(Y=0 \mid X=1) = 0, \quad P(Y=? \mid X=1) = \tfrac{1}{3}, \quad P(Y=1 \mid X=1) = \tfrac{2}{3}$$

The distribution of Y is computed as follows

$$P(Y = 0) = P(Y=0 \mid X=0)\, P(X=0) + P(Y=0 \mid X=1)\, P(X=1)$$

$$= \tfrac{1}{2} \cdot \tfrac{1}{4} + 0 \cdot \tfrac{3}{4} = \tfrac{1}{8}$$

and similarly for the others, giving

$$P(Y=0) = \tfrac{1}{8}, \quad (Y=?) = \tfrac{3}{8}, \quad P(Y=1) = \tfrac{1}{2}$$

The joint distribution is computed as follows

$$P(X=0,Y=0) = P(Y=0 \mid X=0)\, P(X=0) = \tfrac{1}{2} \cdot \tfrac{1}{4} = \tfrac{1}{8}$$

and similarly for the others, giving

$$P(X=0,Y=0) = \tfrac{1}{8}, \quad P(X=0,Y=?) = \tfrac{1}{8}, \quad P(X=0,Y=1) = 0$$

$$P(X=1,Y=0) = 0, \quad P(X=1,Y=?) = \tfrac{1}{4}, \quad P(X=1,Y=1) = \tfrac{1}{2}$$

The conditional distribution of X given Y is computed as follows

$$P(X=0 \mid Y=0) = \frac{P(X=0,Y=0)}{P(Y=0)} = \tfrac{1}{8} \cdot 8 = 1$$

as expected (see Figure 3.1.4). Similarly, we get

$$P(X=0 \mid Y=0) = 1, \quad P(X=0 \mid Y=?) = \tfrac{1}{3}, \quad P(X=0 \mid Y=1) = 0$$

$$P(X=1 \mid Y=0) = 0, \quad P(X=1 \mid Y=?) = \tfrac{2}{3}, \quad P(X=1 \mid Y=1) = 1$$

Next, we compute entropies, where all units are bits and the base is 2,

$$\begin{aligned} H(X) &= P(X=0) \log \frac{1}{P(X=0)} + P(X=1) \log \frac{1}{P(X=1)} \\ &= \tfrac{1}{4} \log 4 + \tfrac{3}{4} \log \tfrac{4}{3} \approx 0.811 \end{aligned}$$

$$\begin{aligned} H(Y) = P(Y=0) \log \frac{1}{P(Y=0)} &+ P(Y=?) \log \frac{1}{P(Y=?)} \\ &+ P(Y=1) \log \frac{1}{P(Y=1)} \approx 1.406 \end{aligned}$$

The conditional entropies are

$$\begin{aligned} H(X \mid Y=0) = P(X=0 \mid Y=0) \log \frac{1}{P(X=0 \mid Y=0)} \\ + P(X=1,Y=0) \log \frac{1}{P(X=1,Y=0)} = 0 \end{aligned}$$

and similarly,

$$H(X \mid Y=?) \approx 0.918, \quad H(X \mid Y=1) = 0$$

and so

$$\begin{aligned} H(X \mid Y) = H(X \mid Y=0)P(Y=0) + H(X \mid Y=?)P(Y=?) \\ + H(X \mid Y=1)P(Y=1) \approx 0.344 \end{aligned}$$

It is interesting to note that since $H(X \mid Y=?) > H(X)$, we actually have more uncertainty about the value of X *after* observing that $Y = ?$ than before observing Y at all! Put another way, there is more information in sampling X if we know that $Y = ?$ than if we do not know the value of Y. However, since $H(X \mid Y) < H(X)$, on the *average* we gain information by knowing the value of Y. □

Since the joint entropy $H(X,Y)$ represents the uncertainty in both X and Y, the quantity $H(X,Y) - H(Y)$ represents the uncertainty in both variables remaining after the uncertainty in Y has

been removed. It seems reasonable that this should be $H(X \mid Y)$, and the next theorem shows that this is indeed the case.

Theorem 3.1.1 If X and Y are random variables, then

$$H(X \mid Y) = H(X,Y) - H(Y)$$

Proof. We have

$$\begin{aligned} H(X \mid Y) &= \sum_{i,j} p(x_i \mid y_j)p(y_j) \log \frac{1}{p(x_i \mid y_j)} \\ &= \sum_{i,j} p(x_i,y_j) \log \frac{p(y_j)}{p(x_i,y_j)} \\ &= \sum_{i,j} p(x_i,y_j) \log \frac{1}{p(x_i,y_j)} - \sum_{i,j} p(x_i,y_j) \log \frac{1}{p(y_j)} \\ &= \sum_{i,j} p(x_i,y_j) \log \frac{1}{p(x_i,y_j)} - \sum_{j} p(y_j) \log \frac{1}{p(y_j)} \\ &= H(X,Y) - H(Y) \end{aligned}$$ ∎

We can also interpret the results of Theorem 3.1.1 by saying that the information obtained by sampling both X and Y is equal to the information obtained by sampling Y, plus the information obtained by sampling X given Y.

Theorem 3.1.1 has the following corollary, which says that sampling Y cannot increase the uncertainty in X.

Corollary 3.1.2 If X and Y are random variables, then

$$H(X \mid Y) \leq H(X)$$

with equality if and only if X and Y are independent. □

It is worth mentioning that, despite Corollary 3.1.2, it is possible that $H(X \mid Y{=}y) > H(X)$ for a given output y, as was the case in Example 3.1.3.

We will also have use for the conditional entropy $H(Y \mid X)$ of Y, given X, defined by

$$H(Y \mid X) = \sum_{i=1}^{s} H(Y \mid X{=}x_i)p(x_i) = \sum_{i=1}^{s} \sum_{j=1}^{t} p(y_j \mid x_i)p(x_i) \log \frac{1}{p(y_j \mid x_i)}$$

which, since it involves the (forward) channel probabilities, is a bit easier to compute than $H(X \mid Y)$.

Conditional entropy can also be defined for random vectors.

Definition If $X_1,\ldots,X_n$ and $Y_1,\ldots,Y_m$ are random variables, then the **conditional entropy** of $X_1,\ldots,X_n$, given that $Y_1 = y_1,\ldots,Y_m = y_m$, is defined by

$$H(X_1,\ldots,X_n \mid Y_1=y_1,\ldots,Y_m=y_m)$$

$$= \sum_{x_1,\ldots,x_n} p(x_1,\ldots,x_n \mid y_1,\ldots,y_m) \log \frac{1}{p(x_1,\ldots,x_n \mid y_1,\ldots,y_m)}$$

or, in vector notation,

$$H(\mathbf{X} \mid \mathbf{Y} = \mathbf{y}) = \sum_{\mathbf{x}} p(\mathbf{x} \mid \mathbf{y}) \log \frac{1}{p(\mathbf{x} \mid \mathbf{y})}$$

The **conditional entropy** of $\mathbf{X}$ given $\mathbf{Y}$ is defined by

$$H(\mathbf{X} \mid \mathbf{Y}) = \sum_{\mathbf{y}} H(\mathbf{X} \mid \mathbf{Y}=\mathbf{y}) p(\mathbf{y}) = \sum_{\mathbf{x},\mathbf{y}} p(\mathbf{x},\mathbf{y}) \log \frac{1}{p(\mathbf{x} \mid \mathbf{y})} \qquad \square$$

Conditional entropy can be thought of as the expectation of a certain random variable, just as can the (unconditional) entropy. Since

$$H(X \mid Y) = \sum_{i,j} p(x_i,y_j) \log \frac{1}{p(x_i \mid y_j)}$$

if we define a random variable U whose value at (x_i,y_j) is $\log \frac{1}{p(x_i \mid y_j)}$, then $H(X \mid Y) = \mathcal{E}(U)$. Thus,

$$H(X) = \mathcal{E}\left(\log \frac{1}{P(X)}\right) \quad \text{and} \quad H(X \mid Y) = \mathcal{E}\left(\log \frac{1}{P(X \mid Y)}\right)$$

SOME SPECIAL CHANNELS

The **channel matrix** of a channel is defined to be the matrix of transition probabilities

$$\begin{bmatrix} p(y_1 \mid x_1) & p(y_2 \mid x_1) & \cdots & p(y_t \mid x_1) \\ p(y_1 \mid x_2) & p(y_2 \mid x_2) & \cdots & p(y_t \mid x_2) \\ & & \vdots & \\ p(y_1 \mid x_s) & p(y_2 \mid x_s) & \cdots & p(y_t \mid x_s) \end{bmatrix}$$

Notice that each row of the channel matrix corresponds to a single input symbol, and each column corresponds to a single output symbol.

The channel matrix can be used to define certain special types of channels as follows. (We will leave verification of all equivalences as exercises.)

Definition

1) Intuitively, a channel is *lossless* if the input X is completely determined by the output Y. Specifically, a channel is **lossless** if any (and hence all) of the following equivalent conditions hold.
   a) There exist nonempty disjoint subsets $B_1,\ldots,B_s$ of the output alphabet with the property that $P(Y \in B_i \mid X = x_i) = 1$.
   b) For all input distributions, whenever $p(y_j) \neq 0$, there exists an x_i for which $p(x_i \mid y_j) = 1$.
   c) For all input distributions, the uncertainty in X, knowing Y, is zero, that is, $H(X \mid Y) = 0$.
2) Intuitively, a channel is *deterministic* if the output Y is completely determined by the input X. Specifically, a channel is **deterministic** if either (and hence both) of the following equivalent conditions hold.
   a) For all x_i, there exists a y_j for which $p(y_j \mid x_i) = 1$.
   b) For all input distributions, the uncertainty in Y knowing X is zero, that is, $H(Y \mid X) = 0$.
3) A channel is **noiseless** if it is both lossless and deterministic. Equivalently, a channel is noiseless if there exists an injection ϕ from the input alphabet $\{x_1,\ldots,x_s\}$ to the output alphabet $\{y_1,\ldots,y_t\}$ for which $p(\phi(x_i) \mid x_i) = 1$ for all i.
4) Intuitively, a channel is *useless* if knowledge about the input X tells us nothing about the output Y. Specifically, a channel is **useless** if any (and hence all) of the following equivalent conditions hold.
   a) The rows of the channel matrix are identical.
   b) For all input distributions, we have $H(X \mid Y) = H(X)$.
   c) The input X and the output Y are independent for all input distributions. ▯

Figures 3.1.5 and 3.1.6 illustrate these concepts.

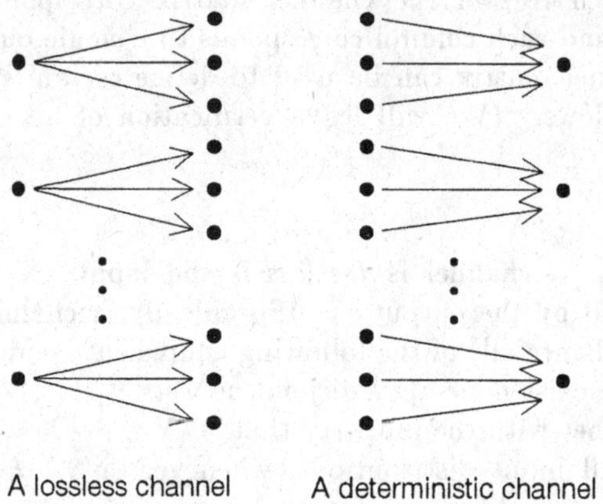

Figure 3.1.5

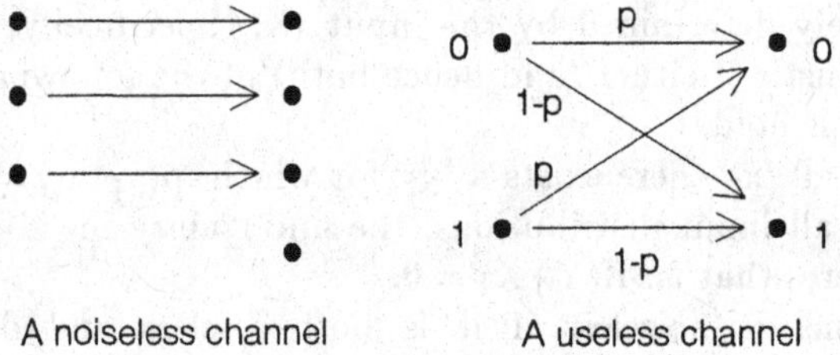

Figure 3.1.6

Definition

1) A channel is **row symmetric** if each row of the channel matrix consists of the same set of numbers, each occurring with the same frequency.
2) A channel is **column symmetric** if each column of the channel matrix consists of the same set of numbers, each occurring with the same frequency.
3) A channel is **symmetric** if it is both row symmetric and column symmetric. ▯

As an example, the channel with channel matrix

$$\begin{bmatrix} \frac{1}{3} & \frac{1}{3} & \frac{1}{6} & \frac{1}{6} \\ \frac{1}{6} & \frac{1}{6} & \frac{1}{3} & \frac{1}{3} \end{bmatrix}$$

is symmetric, as are the binary symmetric channels.

The following theorems describe the principle features of row and column symmetric channels. We leave proofs of these results as exercises.

Theorem 3.1.3 For a row symmetric channel, the uncertainty in Y, knowing X, is independent of the distribution of X, that is, $H(Y \mid X)$ is independent of the input distribution. In fact, we have

$$H(X \mid Y) = \sum_{j=1}^{t} p(y_j \mid x_i) \log \frac{1}{p(y_j \mid x_i)}$$

for *any* $i = 1, \ldots, s$. Put another way, a channel is row symmetric if knowledge about the output Y given by the input X does not depend on which input distribution is used. ▯

Theorem 3.1.4 For a column symmetric channel, a uniform input distribution produces a uniform output distribution. ▯

EXERCISES

1. Find the channel matrix for a binary symmetric channel.
2. Find the channel matrix for a binary erasure channel.
3. Give an example of a 3×3 channel matrix for a symmetric channel.
4. Consider the binary erasure channel with probabilities

$$P(X{=}0) = \tfrac{1}{3}, \quad P(X{=}1) = \tfrac{2}{3}$$

$$P(Y{=}0 \mid X{=}0) = \tfrac{1}{2}, \quad P(Y{=}? \mid X{=}0) = \tfrac{1}{2}, \quad P(Y{=}1 \mid X{=}0) = 0$$

$$P(Y{=}0 \mid X{=}1) = 0, \quad P(Y{=}? \mid X{=}1) = \tfrac{1}{3}, \quad P(Y{=}1 \mid X{=}1) = \tfrac{2}{3}$$

Compute $H_2(X)$, $H_2(Y)$, $H_2(X \mid Y)$ and $H_2(Y \mid X)$.

5. Consider the binary erasure channel with probabilities

$$P(X{=}0) = \tfrac{1}{3}, \quad P(X{=}1) = \tfrac{2}{3}$$

$$P(Y{=}0 \mid X{=}0) = \tfrac{2}{3}, \quad P(Y{=}? \mid X{=}0) = \tfrac{1}{6}, \quad P(Y{=}1 \mid X{=}0) = \tfrac{1}{6}$$

$$P(Y{=}0 \mid X{=}1) = \tfrac{1}{6}, \quad P(Y{=}? \mid X{=}1) = \tfrac{1}{6}, \quad P(Y{=}1 \mid X{=}1) = \tfrac{2}{3}$$

Compute $H_2(X)$, $H_2(Y)$, $H_2(X \mid Y)$ and $H_2(Y \mid X)$.

6. Consider a channel whose input is an integer from 0 to s, and whose output is the parity of the input. What kind of channel is this?

7. Consider a channel whose input alphabet is the set of all integers between $-n$ and n, and whose output is the square of the input. What type of channel is this? What type of channel do we get if we change the input alphabet to the set of all integers between 0 and n?
8. Give an example of a channel that is
   (a) lossless, but neither deterministic nor symmetric,
   (b) noiseless,
   (c) symmetric and lossless, but not deterministic,
   (d) useless and deterministic.
9. Verify the equivalences for the definition of a lossless channel.
10. Verify the equivalences for the definition of a deterministic channel.
11. Verify the equivalences for the definition of a noiseless channel.
12. Verify the equivalences for the definition of a useless channel.
13. Prove Theorem 3.1.3.
14. Prove Theorem 3.1.4.
15. Let $f:\{x_1,\ldots x_s\}\to\{y_1,\ldots y_s\}$ be a function from the input alphabet to the output alphabet of a channel.
   (a) Prove that $H(f(X)\mid X)=0$ for any input X.
   (b) Prove that $H(f(X))\leq H(X)$ for any input X. Under what conditions does equality hold in this inequality?
16. Prove that $H(X,Y\mid Z)\leq H(X\mid Z)+H(Y\mid Z)$, with equality if and only if $p(x_i,y_j\mid z_k) = p(x_i\mid z_k)p(y_j\mid z_k)$. What does this say in words?
17. Prove that $H(X,Y\mid Z)=H(X\mid Z)+H(Y\mid X,Z)$. What does this say in words?
18. Prove that $H(Z\mid X,Y)\leq H(Z\mid X)$, with equality if and only if $p(x_i,y_j\mid z_k) = p(x_i\mid z_k)p(y_j\mid z_k)$. What does this say in words?
19. (a) Prove that $H(X+Y\mid Y)=H(X\mid Y)$.
   (b) If X and Y are independent, prove that $H(X+Y)\leq \min\{H(X),H(Y)\}$.

3.2 Mutual Information and Channel Capacity

MUTUAL INFORMATION

Consider a channel with input X and output Y. The quantity

$$I(X;Y) = H(X) - H(X \mid Y)$$

is the amount of information in X, minus the amount of information *still* in X after knowing Y. In other words, I(X;Y) is the amount of information that we learn about X, by virtue of knowing Y, or, put yet another way, it is the amount of information about X that gets through the channel. Notice also that

$$\begin{aligned} I(X;Y) &= H(X) - H(X \mid Y) \\ &= H(X) - [H(X,Y) - H(Y)] = H(X) + H(Y) - H(X,Y) \end{aligned}$$

and so, by symmetry, we have $I(X;Y) = I(Y;X)$. This motivates the following definition.

Definition The **mutual information** of X and Y is defined by

$$I(X;Y) = H(X) - H(X \mid Y) = H(Y) - H(Y \mid X) \qquad \square$$

Notice that the quantity I(X;Y) depends upon the input distribution of X, as well as the channel probabilities $p(x_i \mid y_j)$.

Example 3.2.1 Referring to the channel in Example 3.1.3, we have

$$I(X;Y) = H(X) - H(X \mid Y) \approx 0.811 - 0.344 = 0.467 \text{ bits} \qquad \square$$

Example 3.2.2 Suppose that we toss a fair die, and if the outcome is a 1, 2, 3 or 4, we toss a fair coin once, but if the outcome is a 5 or 6, we toss a fair coin twice. How much information do we get about the outcome of the die from the number of heads?

This situation can be modeled by a discrete memoryless channel, as shown in Figure 3.2.1. We let $X = 0$ if the outcome of the die is a 1, 2, 3 or 4, and $X = 1$ if the outcome is a 5 or 6. (Clearly, we cannot distinguish between the outcomes 1, 2, 3 and 4, nor between 5 and 6.) We also let Y be the number of heads. Then

$$P(X{=}0) = \frac{2}{3} \quad \text{and} \quad P(X{=}1) = \frac{1}{3}$$

and

$$P(Y{=}0 \mid X{=}0) = \frac{1}{2}, \quad P(Y{=}1 \mid X{=}0) = \frac{1}{2}, \quad P(Y{=}2 \mid X{=}0) = 0$$

$$P(Y{=}0 \mid X{=}1) = \frac{1}{4}, \quad P(Y{=}1 \mid X{=}1) = \frac{1}{2}, \quad P(Y{=}2 \mid X{=}1) = \frac{1}{4}$$

Further, we have

$$H(Y \mid X{=}0) = H(\tfrac{1}{2},\tfrac{1}{2},0) = 1 \quad \text{and} \quad H(Y \mid X{=}1) = H(\tfrac{1}{4},\tfrac{1}{2},\tfrac{1}{4}) = \frac{3}{2}$$

and so

$$H(Y \mid X) = \frac{2}{3}\cdot 1 + \frac{1}{3}\cdot\frac{3}{2} \approx 1.166$$

We will leave the details of the following computations as an exercise.

$$P(Y{=}0) = \frac{5}{12}, \quad P(Y{=}1) = \frac{1}{2}, \quad P(Y{=}2) = \frac{1}{12}$$

$$H(Y) \approx 1.325$$

and so finally, we get

$$I(X;Y) = H(Y) - H(Y \mid X) \approx 1.325 - 1.166 = 0.159 \text{ bits} \qquad \square$$

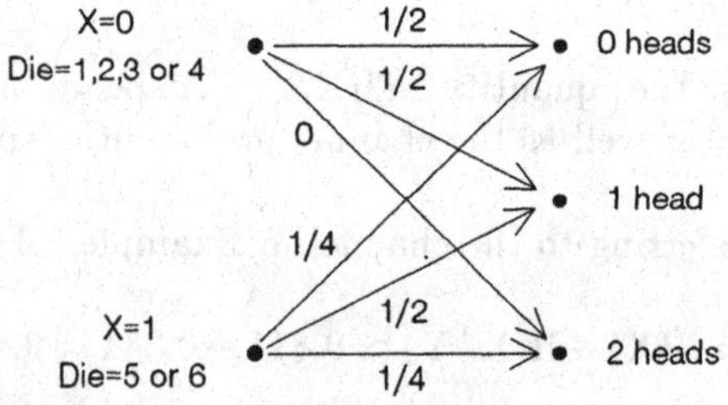

Figure 3.2.1

Mutual information can also be defined for random vectors.

Definition If **X** and **Y** are random vectors, then the **mutual information** of **X** and **Y** is defined by

$$I(\mathbf{X};\mathbf{Y}) = H(\mathbf{X}) - H(\mathbf{X} \mid \mathbf{Y}) \qquad \square$$

A SUMMARY OF PROPERTIES

For reference, let us list some of the properties of entropy and mutual information, along with a description of each property. We leave the proofs (where necessary) for the exercises.

Theorem 3.2.1 Let X and Y be random variables.

1) The information in both X and Y is less than or equal to the information in X plus the information in Y, with equality if and only if X and Y are independent

$$H(X,Y) \leq H(X) + H(Y)$$

2) The information in both X and Y is equal to the information remaining in X, given Y, plus the information in Y

$$H(X,Y) = H(X \mid Y) + H(Y)$$

3) The information about X obtained from Y is equal to the information in X minus the information remaining in X, given Y

$$I(X;Y) = H(X) - H(X \mid Y)$$

4) The information about X obtained from Y is equal to the information about Y obtained from X

$$I(Y;X) = I(X;Y)$$

5) The information about X obtained from Y is equal to the excess information obtained by sampling X and Y separately, rather than jointly

$$I(X;Y) = H(X) + H(Y) - H(X,Y)$$

6) The information about X obtained from Y is always nonnegative, and is equal to 0 if and only if X and Y are independent

$$I(X;Y) \geq 0$$

7) The information about X obtained from X is the information in X

$$I(X;X) = H(X)$$

□

Many of the relationships in Theorem 3.2.1 can be recalled from Figure 3.2.2.

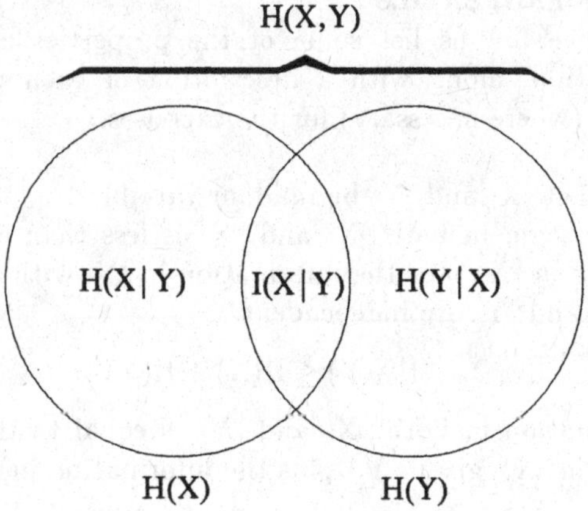

Figure 3.2.2 Relationships between entropy and information

THE CAPACITY OF A CHANNEL

We are now ready to define the concept of the capacity of a channel. This concept plays a key role in the main results of information theory.

Definition The **capacity** of a channel is the maximum mutual information $I(X;Y)$, taken over all input distributions $p(x_i)$ of X. In symbols,

$$\mathfrak{C} = \max_{p(x_i)} I(X;Y) \qquad \square$$

The problem of determining the capacity of an arbitrary channel seems to be very difficult. However, we can determine the capacities of some special types of channels, including the important symmetric channels. We leave the proof of the next theorem as an exercise.

Theorem 3.2.2

1) The capacity of a lossless channel is $\log s$, where s is the size of the input alphabet.
2) The capacity of a deterministic channel is $\log u$, where u is the size of the set $\{y_j \mid p(y_j \mid x_i) = 1 \text{ for some } x_i\}$.
3) The capacity of a noiseless channel is $\log s$, where s is the size of the input alphabet.
4) The capacity of a useless channel is zero. $\square$

Theorem 3.2.3 The capacity of a symmetric channel is

$$\mathcal{C}_{sym} = \log t - \sum_{j=1}^{t} p(y_j \mid x_i) \log \frac{1}{p(y_j \mid x_i)}$$

for any $i = 1, \ldots, s$. Furthermore, capacity is achieved by the uniform input distribution $p(x_i) = \frac{1}{s}$.
Proof. First, we observe that

$$H(Y \mid X) = \sum_i p(x_i) \left(\sum_j p(y_j \mid x_i) \log \frac{1}{p(y_j \mid x_i)} \right)$$

But since the channel is symmetric, the inside sum does not depend on i, and so we get

$$\text{(3.2.1)} \quad H(Y \mid X) = \left(\sum_j p(y_j \mid x_i) \log \frac{1}{p(y_j \mid x_i)} \right) \left(\sum_i p(x_i) \right)$$

$$= \sum_j p(y_j \mid x_i) \log \frac{1}{p(y_j \mid x_i)}$$

Now, since $H(Y \mid X)$ does not depend on the distribution of X, we have

$$\mathcal{C} = \max_{p(x_i)} I(Y;X)$$

$$= \max_{p(x_i)} [H(Y) - H(Y \mid X)]$$

$$= [\max_{p(x_i)} H(Y)] - H(Y \mid X)$$

But $H(Y)$ takes its maximum value of $\log t$ when Y has the uniform distribution, and this happens precisely when X has the uniform distribution $p(x_i) = 1/s$, for in this case,

$$P(Y{=}y_j) = \sum_{i=1}^{s} p(x_i) p(y_j \mid x_i) = \frac{1}{s} \sum_{i=1}^{s} p(y_j \mid x_i)$$

which is independent of j for a symmetric channel. Finally, using (3.2.1), we get

$$\mathcal{C} = \max_{p(x_i)} I(Y;X) = \log t - \sum_j p(y_j \mid x_i) \log \frac{1}{p(y_j \mid x_i)} \quad \blacksquare$$

Corollary 3.2.4 The capacity of the binary symmetric channel with crossover probability p is

$$\mathcal{C}_{\text{bin sym}} = 1 - p \log \frac{1}{p} - (1-p) \log \frac{1}{1-p} = 1 - H(p)$$

where H(p) is the entropy function. □

EXERCISES

1. Compute the remaining probabilities for Example 3.2.2.
2. Suppose you flip a fair coin twice, letting 0 represent heads and 1 represent tails. Thus, the outcome has the form (i,j), where i,j = 0 or 1. How much information about the outcomes of the tosses do you get from the sum i + j? How much information is there in the outcome of the two tosses?
3. An urn contains three balls, numbered 1, 2 and 3. Suppose you draw a ball at random. If the number on the ball is not 1, let Y be the number on the ball. If the number on the ball is 1, then draw another ball at random and let Y be the sum of the numbers on the two balls. Model this situation by a channel, and compute the information about the number drawn on the first draw, obtained from the value of Y. What information is obtained about the number drawn on the second draw from the value of Y?
4. Consider the special case of a binary erasure channel shown in Figure 3.2.3.

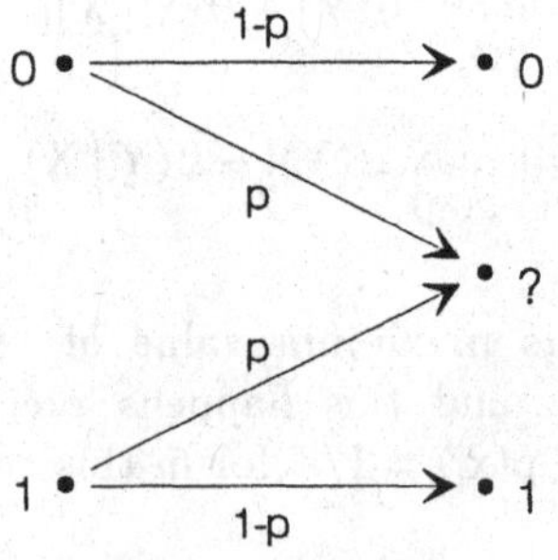

Figure 3.2.3

Calculate the mutual information I(X;Y) in terms of the input probability $p(0) = p_0$. Then determine the capacity of the channel, and an input distribution that achieves that capacity.

5. Consider the special case of a binary erasure channel shown in Figure 3.2.4.

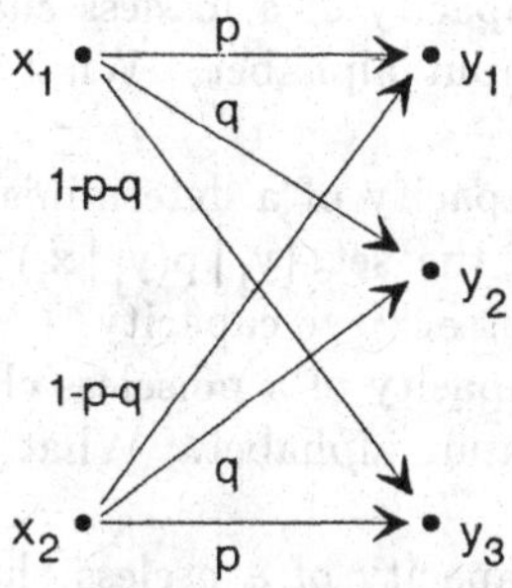

Figure 3.2.4

Calculate the mutual information $I(X;Y)$ in terms of the input probability $p(x_1) = p_1$. Then determine the capacity of the channel, and an input distribution that achieves that capacity.

6. Consider the channel shown in Figure 3.2.5.

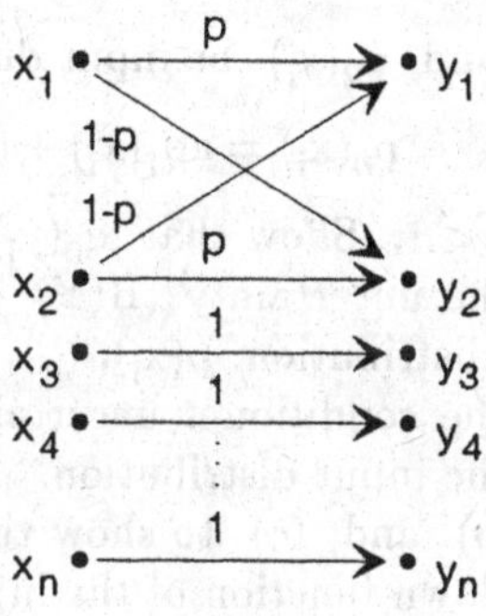

Figure 3.2.5

a) For $n = 3$, write the channel matrix of this channel. Find the capacity of the channel, and an input distribution that achieves capacity.

b) Repeat part a), for arbitrary n.

7. Show that the mutual information $I(X;Y)$ can be written in the form

$$I(X;Y) = \sum_{i=1}^{s} \sum_{j=1}^{t} p(x_i,y_j) \log \frac{p(x_i,y_j)}{p(x_i)p(y_j)}$$

8. Show that the capacity of a channel is always achieved by some input distribution. *Hint.* The mutual information function $I(X;Y)$ is a continuous function of the input distribution.

9. Can you express the mutual information $I(X;Y)$ as an expected value?
10. Show that the capacity of a lossless channel is $\log s$, where s is the size of the input alphabet. What input distribution achieves capacity?
11. Show that the capacity of a deterministic channel is $\log u$, where u is the size of the set $\{y_j \mid p(y_j \mid x_i) = 1 \text{ for some } x_i\}$. What input distributions achieve capacity?
12. Show that the capacity of a noiseless channel is $\log s$, where s is the size of the input alphabet. What input distributions achieve capacity?
13. Show that the capacity of a useless channel is zero. What input distributions achieve capacity?
14. Prove Corollary 3.2.4.
15. Supply proofs (where necessary) for Theorem 3.2.1.
16. Consider a channel, with channel probabilities $p(y_j \mid x_i)$. For each input distribution $p(x_i)$ for X, we get an output distribution $q(y_j) = \sum_i p(x_i)p(y_j \mid x_i)$ for Y, and hence also an uncertainty $H(Y)$.

    (a) Let $p_1(x_i)$ and $p_2(x_i)$ be input distributions, and let

    $$p_0(x_i) = ap_1(x_i) + (1-a)p_2(x_i)$$

    where $0 < a < 1$. Show that $q_0(y_j) = aq_1(y_i) + (1-a)q_2(y_i)$.

    (b) Show that the uncertainty $H(Y)$ is a convex down function of the *input* distribution $p(x_i)$.

    (c) Show that the conditional uncertainty $H(Y \mid X)$ is a *linear* function of the input distribution.

    (d) Use parts (b) and (c) to show that the information $I(X;Y)$ is a convex down function of the input distribution.

3.3 The Noisy Coding Theorem

The ultimate goal of accurate communication is to be able to reproduce the original source information, using the output information of the channel. This is done partly by means of a *decision scheme* that is used to guess the correct input to the channel, based on the output.

Unfortunately, most channels are not lossless, and so, in general, some information about the input to the channel is lost before reaching the output. As we have seen, the loss of information is given by $H(X \mid Y)$ and depends not only on the channel probabilities, but also on the input distribution.

Fortunately, however, there are ways to compensate for the loss of information in a given channel by *encoding* the source information before sending it through the channel. Let us consider a simple example of this. Suppose we wish to send information through a binary symmetric channel, with crossover probability $p < \frac{1}{2}$, as pictured in Figure 3.3.1.

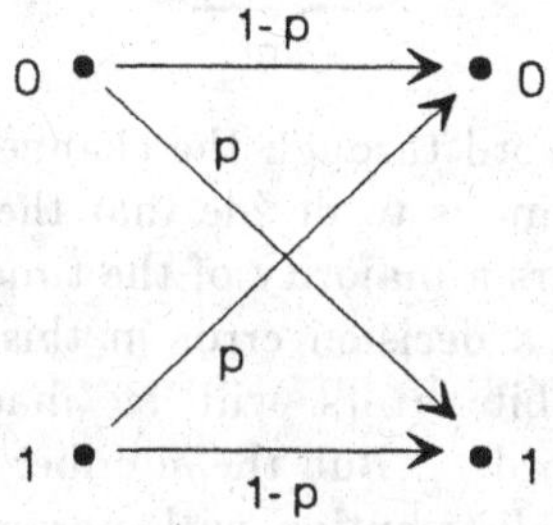

Figure 3.3.1

Here are some possible scenarios.

Scenario 1.
Each source symbol (0 or 1) is sent through the channel without encoding, and our decision scheme is simply to decide that the received symbol is the one that was sent. In this case, regardless of which source symbol is sent, the probability of making a *decision error* is p.

Scenario 2.
We encode each source symbol c_i by duplicating it two additional times, to get the *codeword* $c_ic_ic_i$. Then we send this codeword through the channel, one symbol at a time. Our decision scheme is to decide that the original source symbol is the symbol that appears a *majority* of the time in the output string. For instance, if the output string is 010, we

decide that 0 was sent.

Because we are considering only memoryless channels, the probability of making a *decision error*, that is, an error in deciding the original source symbol, is the probability that two or more bit errors are made by the channel in transmitting the codeword, and this probability is

$$\binom{3}{2}p^2(1-p)+\binom{3}{3}p^3=3p^2-2p^3$$

Since this is less than p, for $p<\frac{1}{2}$, we see that by encoding the original source symbol, we can reduce the probability of making a decision error. In this way, we are able to compensate in part for the loss that is inherent in the channel.

Scenario 3.
The previous scenario can be generalized as follows. We encode each source symbol c_i by writing it a total of $2n+1$ times (we want an odd string length so that majority decisions are always possible)

$$\underbrace{c_ic_i\cdots c_i}_{2n+1}$$

Then we send this codeword through the channel one bit at a time. As before, our decision scheme is to decide that the original source symbol is the symbol that appears a majority of the time in the output string.

The probability of a decision error in this case is the probability that at least $n+1$ bit errors will be made by the channel in transmitting the codeword. But the number of errors made by the channel has a binomial distribution with parameters $(2n+1,p)$, and so the expected number of errors is $(2n+1)p<n+\frac{1}{2}$ for $p<\frac{1}{2}$. Therefore, the weak law of large numbers tells us that the probability that at least $n+1$ channel errors are made tends to 0 as n tends to infinity. In other words, the probability that we make a decision error tends to 0 as n gets large. Thus, we can compensate for channel information loss to *any* desired degree by choosing n large enough.

However, we pay a heavy price for doing this, in terms of the efficiency of transmission of source information. In particular, it takes a certain amount of time to send an input symbol through the channel, and if we duplicate each source symbol $2n+1$ times, we are spending $2n+1$ units of channel time to send a *single* source symbol. Thus, the *rate of source transmission* is $\frac{1}{2n+1}$ *source bits* per channel bit. For n large, this is likely to be an unacceptably low rate of transmission.

We can see from the previous scenario that there are two somewhat contradictory goals at work in channel communication. On

the one hand, we want to minimize the probability of a decision error by encoding the source data, but on the other hand, we want to maximize the rate of transmission of that source data. The Noisy Coding Theorem, and its converse, tell us how we must compromise in dealing with these two opposing goals.

Recall that the capacity $\mathcal{C}$ of a channel is the maximum amount of information that can be sent through the channel, the maximum being taken over all input distributions. The Noisy Coding Theorem tells us that, as long as we are willing to settle for a source transmission rate that is *below* the capacity of the channel, that is, as long as we are willing to leave some room for judicious encoding, we can compensate for the loss of information to any desired degree of accuracy. To be more specific, for any number $R < \mathcal{C}$, and for any $\epsilon > 0$, there is an encoding scheme, and corresponding decision scheme, that transmits source data at rate R, with probability of error at most ϵ.

Let us now turn to the details.

THE CHANNEL

We begin with a discrete memoryless channel, with input and output alphabets

$$\mathcal{I} = \{x_1, \ldots, x_s\} \text{ and } \mathcal{O} = \{y_1, \ldots, y_t\}$$

and channel probabilities

$$p(y_j \mid x_i)$$

Our intention is to think of the channel as accepting codewords $\mathbf{c} = c_1 \cdots c_n$ from a code C over $\mathcal{I}$ of length n and size m, and outputting strings $\mathbf{d} = d_1 \cdots d_n$ of the same length over $\mathcal{O}$. (The **length** of a fixed length code is the length of any codeword in the code, and the **size** is the total number of codewords.)

Because the channel is memoryless, we have

$$p(\mathbf{d} \mid \mathbf{c}) = \prod_{i=1}^{n} p(d_i \mid c_i)$$

for all codewords $\mathbf{c}$.

Since our primary interest here is in codeword input, and not just symbol input, we will think of the input as a random vector $\mathbf{X}$, with **input distribution** defined by $P(\mathbf{X}=\mathbf{c}) = p(\mathbf{c})$. (For codes of length 1, codeword input is the same as symbol input.)

Each input $\mathbf{X}$ induces an output $\mathbf{Y}$ with **output distribution** defined by

$$P(\mathbf{Y}=\mathbf{d}) = \sum_{\mathbf{c}} p(\mathbf{d} \mid \mathbf{c}) P(\mathbf{X}=\mathbf{c})$$

The **joint distribution** of **X** and **Y** is given by

$$P(\mathbf{X}=\mathbf{c},\mathbf{Y}=\mathbf{d}) = p(\mathbf{d}\,|\,\mathbf{c})P(\mathbf{X}=\mathbf{c})$$

and the **backward channel probabilities** are

$$P(\mathbf{X}=\mathbf{c}\,|\,\mathbf{Y}=\mathbf{d}) = \frac{P(\mathbf{X}=\mathbf{c},\mathbf{Y}=\mathbf{d})}{P(\mathbf{Y}=\mathbf{d})}$$

When no confusion can arise, we will adopt the notation

$$\begin{aligned} p(\mathbf{c}) &= P(\mathbf{X}=\mathbf{c}) \\ p(\mathbf{d}) &= P(\mathbf{Y}=\mathbf{d}) \\ p(\mathbf{c},\mathbf{d}) &= P(\mathbf{X}=\mathbf{c},\mathbf{Y}=\mathbf{d}) \\ p(\mathbf{c}\,|\,\mathbf{d}) &= P(\mathbf{X}=\mathbf{c}\,|\,\mathbf{Y}=\mathbf{d}) \\ p(\mathbf{d}\,|\,\mathbf{c}) &= P(\mathbf{Y}=\mathbf{d}\,|\,\mathbf{X}=\mathbf{c}) \end{aligned}$$

THE DECISION SCHEME

A **decision scheme** is a partial function f from the set of output strings to the set of codewords. The word *partial* refers to the fact that f may not be defined for *all* output strings. The intention is that, if an output string **d** is received, and if f(**d**) is defined, then the decision scheme decides that f(**d**) is the codeword that was sent. If f(**d**) is not the codeword that was sent, we say that a **decision error**, or a **decoding error**, has been made.

By letting

$$B_{\mathbf{c}} = f^{-1}(\mathbf{c}) = \{\mathbf{d}\,|\,f(\mathbf{d})=\mathbf{c}\}$$

be the set of all outputs for which we decide that the correct input was **c**, we can also think of a decision scheme as a collection $\{B_{\mathbf{c}}\}$ of disjoint subsets of the set of output strings. This is pictured in Figure 3.3.2.

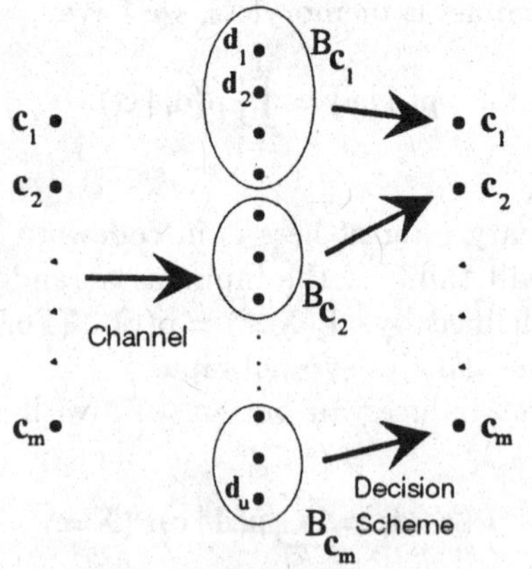

Figure 3.3.2

THE PROBABILITY OF A DECISION ERROR

For any decision scheme, if the codeword $\mathbf{c}$ is sent through the channel, then the probability of a decision error is given by

$$P(\text{error} \mid \mathbf{c}) = \sum_{\mathbf{d} \notin f^{-1}(\mathbf{c})} p(\mathbf{d} \mid \mathbf{c})$$

Hence, the probability p_e of a decision error is

$$p_e = \sum_{\mathbf{c}} P(\text{error} \mid \mathbf{c}) p(\mathbf{c}) = \sum_{\mathbf{c}} \sum_{\mathbf{d} \notin f^{-1}(\mathbf{c})} p(\mathbf{d} \mid \mathbf{c}) p(\mathbf{c}) \tag{3.3.1}$$

Notice that this probability depends on the input distribution $p(\mathbf{c})$, as well as on the decision scheme.

In order to see how we can determine a decision scheme that minimizes the probability of decision error, let us compute this error by conditioning on the output, rather than the input. If the output of the channel is $\mathbf{d}$, then the correct decision will be made if and only if $f(\mathbf{d})$ was the actual input. Hence,

$$P(\text{error} \mid \mathbf{d}) = 1 - p(f(\mathbf{d}) \mid \mathbf{d})$$

Averaging over all possible outputs, we have

$$p_e = \sum_{\mathbf{d}} P(\text{error} \mid \mathbf{d}) p(\mathbf{d}) = 1 - \sum_{\mathbf{d}} p(f(\mathbf{d}) \mid \mathbf{d}) p(\mathbf{d})$$

Now, this probability can be minimized by choosing a decision scheme that maximizes the sum on the far right. But since each term in this sum is nonnegative, and since the factors $p(\mathbf{d})$ do not depend on the decision scheme, we maximize this sum by choosing $f(\mathbf{d})$ so that $p(f(\mathbf{d}) \mid \mathbf{d})$ is as large as possible, for all $\mathbf{d}$. Let us summarize.

Definition For a given input distribution, any decision scheme f for which $f(\mathbf{d})$ has the property that

$$p(f(\mathbf{d}) \mid \mathbf{d}) = \max_{\mathbf{c}} p(\mathbf{c} \mid \mathbf{d})$$

for all output strings $\mathbf{d}$, is called an **ideal observer**. In words, an ideal observer is one for which $f(\mathbf{d})$ is a codeword most likely to have been sent, given that $\mathbf{d}$ was received. ▯

Notice that the probabilities $p(\mathbf{c} \mid \mathbf{d})$ are *backward* channel probabilities, and so they depend on the input distribution.

Theorem 3.3.1 For any given input distribution, an ideal observer decision scheme will minimize the probability p_e of a decision error, among all decision schemes. □

The ideal observer decision scheme has advantages as well as disadvantages. Perhaps its main disadvantage is that it depends on the input distribution. Thus, if the input distribution is changed, the ideal observer may also change. We can eliminate this dependency by considering the *maximum* probability of a decision error, defined by

$$p_e^{max} = \max_{\mathbf{c}} P(\text{error} \mid \mathbf{c})$$

This probability depends only on the decision scheme (and the channel, of course), and has the virtue that if $p_e^{max} < \epsilon$, then we have a *uniform* bound on the error probability, in the sense that the probability of error is small no matter what the input is, and so for *all* input distributions, we have

$$p_e = \sum_{\mathbf{c}} P(\text{error} \mid \mathbf{c}) p(\mathbf{c}) \le p_e^{max} < \epsilon$$

The difficulty here is that, unfortunately, we do not have a general method for finding decision schemes that make p_e^{max} small.

Another way to remove the dependency on the input distribution is to consider a uniform input distribution $p(\mathbf{c}) = \frac{1}{m}$, where m is the size of the code. Then, according to (3.3.1), the probability of a decision error is given by

$$p_e^{av} = \frac{1}{m} \sum_{\mathbf{c}} P(\text{error} \mid \mathbf{c})$$

This is referred to as the **average probability of error.** (This terminology is a bit misleading, since (3.3.1) is also an average probability of error, the average being taken over the input distribution $p(\mathbf{c})$. Perhaps a better term would have been *uniform probability of error.*)

Since, in the uniform case

$$p(\mathbf{c} \mid \mathbf{d}) = \frac{p(\mathbf{d} \mid \mathbf{c})\, p(\mathbf{c})}{p(\mathbf{d})} = \frac{1}{m p(\mathbf{d})} p(\mathbf{d} \mid \mathbf{c})$$

we have

$$\max_{\mathbf{c}} p(\mathbf{c} \mid \mathbf{d}) = \max_{\mathbf{c}} \frac{1}{m p(\mathbf{d})} p(\mathbf{d} \mid \mathbf{c}) = \frac{1}{m p(\mathbf{d})} \max_{\mathbf{c}} p(\mathbf{d} \mid \mathbf{c})$$

In other words, for the uniform input distribution, maximizing the backward channel probabilities $p(\mathbf{c} \mid \mathbf{d})$ is equivalent to maximizing the (forward) channel probabilities $p(\mathbf{d} \mid \mathbf{c})$. Again we summarize.

Definition Any decision scheme f for which $f(\mathbf{d})$ has the property that

$$p(\mathbf{d} \mid f(\mathbf{d})) = \max_{\mathbf{c}} p(\mathbf{d} \mid \mathbf{c})$$

for all output strings $\mathbf{d}$, is called a **maximum likelihood decision scheme.** In words, $f(\mathbf{d})$ is an input string with the property that, for no other input string would it be more likely that the output $\mathbf{d}$ was received. ▯

Theorem 3.3.2 For the uniform input distribution, an ideal observer (which minimizes the *average* probability of error), is the same as a maximum likelihood decision scheme. ▯

The maximum likelihood decision scheme has the advantage that it is often easy to implement. We will discuss this in considerable detail when we turn to coding theory in the second part of this book. On the other hand, it may seem to have the disadvantage that minimizing the *average* probability of error is not as good as minimizing the *maximum* probability of error. However, as we will see in the next section, for the asymptotic results of the Noisy Coding Theorem, it turns out to be just as good.

THE RATE OF A CODE

Before we can state the Noisy Coding Theorem formally, we need to discuss in more detail the notion of rate of transmission. Let us suppose that the source information is in the form of strings of length k, over the input alphabet $\mathcal{I}$ of the channel, and that the code C consists of codewords of length n over $\mathcal{I}$.

If we denote the number of codewords by $|C|$, then it is customary to say that the code C has **length** n and **size** $|C|$, or is an $(n, |C|)$-code. (The notation $(n, |C|)$-code is used by coding theorists, but the components are often reversed in books on information theory.)

Now, since the channel must transmit n code symbols in order to send k source symbols, the rate of transmission is $R = \frac{k}{n}$ source symbols per code symbol. Further, since there are s^k possible source strings, the code must have size at least s^k in order to accommodate all of these strings. Assuming that $|C| = s^k$, we have

$$k = \log_s |C|$$

and so

$$R = \frac{\log_s |C|}{n}$$

Thus we have the following.

Theorem 3.3.3 An s-ary $(n, |C|)$-code C transmits at a rate equal to

$$R = \frac{\log_s |C|}{n}$$

source (s-ary) digits per code (s-ary) digit. ▯

Definition The number

$$R(C) = \frac{\log_s |C|}{n}$$

is called the **rate** of the code C. ▯

THE NOISY CODING THEOREM

Now we can state the Noisy Coding Theorem. Let $\lceil x \rceil$ denote the smallest integer greater than or equal to x.

Theorem 3.3.4 (The Noisy Coding Theorem) Consider a discrete memoryless channel with capacity $\mathfrak{C}$. For any positive number $R < \mathfrak{C}$, there exists a sequence C_n of s-ary codes, and corresponding decision schemes f_n, with the following properties.

1) C_n is an $(n, \lceil s^{nR} \rceil)$-code, that is, C_n has length n and rate at least R.
2) The maximum probability of error of f_n approaches 0 as $n \to \infty$, that is,

$$p_e^{max}(n) \to 0$$ ▯

We will give a formal proof of the Noisy Coding Theorem, for the binary symmetric channel, in the next section. Let us instead look at the idea behind the proof, since that will shed considerable light on the current state of affairs in information and coding theory.

In the next section, we will prove a result that says that replacing the words *maximum probability* in the Noisy Coding Theorem by *average probability* results in an equivalent theorem, and so we may prove the Noisy Coding Theorem in this equivalent form.

The first step in the proof is to describe a conceptually simple decision scheme that applies to any code $C_n = \{c_1, \ldots, c_m\}$ of length n and size m. The idea is that we decide in favor of the *unique* codeword that is "closest" to the received word, in a sense we will make precise in

the next section. If a unique closest codeword does not exist, we simply admit an error.

The next step is to find an upper bound on the *average* probability of error based on this decision scheme, say of the form

$$p_e^{av}(c_1,\ldots,c_m) \le U_n(c_1,\ldots,c_m)$$

The notation $p_e^{av}(c_1,\ldots,c_m)$ emphasizes the dependence of the average probability of error on the choice of code.

Next, we take the viewpoint that we can pick the codewords $\mathbf{c}$ independently and at random, with each binary string equally likely to occur. This enables us to think of the $\mathbf{c}$ as *random variables* ω_i, and so we have

$$p_e^{av}(\omega_1,\ldots,\omega_m) \le U_n(\omega_1,\ldots,\omega_m)$$

Taking expected values gives

$$\mathcal{E}(p_e^{av}(\omega_1,\ldots,\omega_m)) \le \mathcal{E}(U_n(\omega_1,\ldots,\omega_m))$$

But, for any random variable X, it must be true that some value of X is no greater than the expected value $\mathcal{E}(X)$ of X. Hence, there must exist some code $c_1,\ldots,c_m$ for which

$$p_e^{av}(c_1,\ldots,c_m) \le \mathcal{E}(p_e^{av}(\omega_1,\ldots,\omega_m)) \le \mathcal{E}(U_n(\omega_1,\ldots,\omega_m))$$

Finally, we show that, provided $m \le \lceil s^{nR} \rceil$, where $R < \mathcal{C}$, the right-hand side $\mathcal{E}(U_n(\omega_1,\ldots,\omega_m))$ tends to 0 as $n \to \infty$, and hence so does the left hand side. This will complete the proof.

The type of proof just outlined is known as *proof by random coding*, and was used by Claude Shannon in his original paper. In effect, it says that not only do codes exist with the desired properties, but that a *randomly chosen* code is reasonably likely to have these properties!

Unfortunately, proof by random coding is not constructive, that is, the proof does not tell us how to construct the codes promised in the theorem. What is most remarkable about the present state of affairs is that, despite the fact that the proof says that there are a lot of codes that perform as promised in the theorem, no one has yet been able to actually construct such codes.

There are some additional practical problems to be considered in searching for desirable codes. In particular, a sequence of codes that fulfills the promise of the Noisy Coding Theorem would not be of much use unless the corresponding decision schemes were relatively easy to implement. Also, in order to bring the probability of error down to a desired level, the Noisy Coding Theorem implies that we may have to increase the length of the code to a perhaps unworkable size. Thus,

simply finding codes that satisfy the Noisy Coding Theorem does not solve the problem of reliable communication. As we will see in the sequel, in an effort to find decision schemes that are relatively easy to implement, coding theorists have been led to search for codes that have considerable algebraic or geometric structure.

THE WEAK CONVERSE OF THE NOISY CODING THEOREM

The converse of the Noisy Coding Theorem takes two common forms – the *strong converse* and the *weak converse.* This terminology is a bit misleading, since the strong converse is an asymptotic result, whereas the weak converse is not and, in fact, the strong converse does not imply the weak converse.

Our plan is to state and prove the weak converse now, and state the strong converse later in the section. We will prove the strong converse, for the case of the binary symmetric channel, in the next section.

We begin with an important result relating the information loss $H(\mathbf{X}\mid\mathbf{Y})$ to the probability of error. Consider a discrete memoryless channel, a code C, with corresponding decision scheme f, and an input distribution $p(\mathbf{c})$. Suppose that a word $\mathbf{d}$ is received. Then it seems reasonable that the uncertainty $H(\mathbf{X}\mid\mathbf{Y}{=}\mathbf{d})$ in $\mathbf{X}$, given that $\mathbf{d}$ was sent, should be no larger than

the uncertainty in deciding correctly or incorrectly,
given $\mathbf{d}$ was sent
+
P(being correct) · uncertainty in being correct,
given that we are correct
+
P(being incorrect) · uncertainty in being incorrect,
given that we are incorrect

But if we are correct, then the uncertainty in being correct is zero, and so we have

the uncertainty in deciding correctly or incorrectly,
given $\mathbf{d}$ was sent
+
P(being incorrect) · uncertainty in being incorrect,
given that we are incorrect

This can be made more precise by using a grouping axiom for entropy. Let $C = \{\mathbf{c}_1,\ldots,\mathbf{c}_m\}$ be a code of length n and size m. Let

$\mathbf{X}$ and $\mathbf{Y}$ be the input and output random vectors associated with the input distribution $p(\mathbf{c})$. Thus,

$$P(\mathbf{X}=\mathbf{c}) = p(\mathbf{c})$$

and

$$P(\mathbf{Y}=\mathbf{d}) = p(\mathbf{d}) = \sum_{\mathbf{c}} p(\mathbf{d} \mid \mathbf{c})p(\mathbf{c})$$

Let us temporarily fix the output at $\mathbf{d}$ and consider the conditional probability distribution of $\mathbf{X}$ given that $\mathbf{Y} = \mathbf{d}$, defined by

(3.3.2) $$p'(\mathbf{c}) = p(\mathbf{c} \mid \mathbf{d}) = P(\mathbf{X}=\mathbf{c} \mid \mathbf{Y}=\mathbf{d})$$

If we assume for concreteness that $f(\mathbf{d}) = \mathbf{c}_1$, then

(3.3.3) $$p'_e = P(\text{error} \mid \mathbf{d}) = 1 - p(\mathbf{c}_1 \mid \mathbf{d}) = 1 - p'(\mathbf{c}_1)$$

Now we wish to employ a special case of the grouping axiom of Theorem 1.2.4, whose proof is left as an exercise.

(3.3.4) $$H(p_1,\ldots,p_m) = H(1-p_1) + (1-p_1)H\left(\frac{p_2}{1-p_1},\ldots,\frac{p_m}{1-p_1}\right)$$

Letting $p_1 = p'(\mathbf{c}_1)$, $p_2 = p'(\mathbf{c}_2),\ldots,p_m = p'(\mathbf{c}_m)$, we have, in view of (3.3.2) and (3.3.3),

$$H(\mathbf{X} \mid \mathbf{Y}=\mathbf{d}) = H(p'_e) + p'_e H\left(\frac{p'(\mathbf{c}_2)}{1-p'(\mathbf{c}_1)},\ldots,\frac{p'(\mathbf{c}_m)}{1-p'(\mathbf{c}_1)}\right)$$

This equation is the precise formulation of our previous discussion.

Since $H(q_2,\ldots,q_m) \le \log(m-1)$, for all probability distributions $q_2,\ldots,q_m$, we can write

$$H(\mathbf{X} \mid \mathbf{Y}=\mathbf{d}) \le H(p'_e) + p'_e \log(m-1)$$

Averaging over all outputs $\mathbf{d}$ gives

$$\begin{aligned} H(\mathbf{X} \mid \mathbf{Y}) &= \sum_{\mathbf{d}} p(\mathbf{d})H(\mathbf{X} \mid \mathbf{Y}=\mathbf{d}) \\ &\le \sum_{\mathbf{d}} p(\mathbf{d})[H(p'_e) + p'_e \log(m-1)] \\ &= \sum_{\mathbf{d}} p(\mathbf{d})H(p'_e) + \log(m-1) \sum_{\mathbf{d}} p(\mathbf{d})p'_e \end{aligned}$$

Since the entropy function $H(p)$ is convex down, this implies

$$H(\mathbf{X} \mid \mathbf{Y}) \le H\Big(\sum_{\mathbf{d}} p(\mathbf{d})p'_e\Big) + \log(m-1) \sum_{\mathbf{d}} p(\mathbf{d})p'_e$$

Finally, we observe that

$$\sum_{\mathbf{d}} p(\mathbf{d})p'_e = \sum_{\mathbf{d}} p(\mathbf{d})[1 - p(\mathbf{c}_1 \mid \mathbf{d})] = 1 - p(\mathbf{c}_1) = 1 - p(f(\mathbf{d})) = p_e$$

is the probability of a decision error. Thus,

$$H(\mathbf{X} \mid \mathbf{Y}) \le H(p_e) + p_e \log(m-1)$$

This important result is known as *Fano's inequality.*

Theorem 3.3.5 (Fano's Inequality) For any decision scheme (C,f) with $|C| = m$, and any input distribution, if p_e denotes the probability of a decision error, then

$$H(\mathbf{X} \mid \mathbf{Y}) \le H(p_e) + p_e \log(m-1) \qquad \square$$

Now we can turn to the weak converse of the Noisy Coding Theorem, which says that, for any sequence of codes whose rate R exceeds the capacity of the channel, and for any corresponding decision schemes, the average probability of error must be bounded away from 0.

Theorem 3.3.6 (Weak Converse to the Noisy Coding Theorem) Consider a discrete memoryless channel, with capacity $\mathfrak{C}$. Suppose that C_n is a sequence of $(n, \lceil s^{nR} \rceil)$-codes, with corresponding decision schemes f_n, and that the average probability of error of f_n is $p_e^{av}(n)$. Then if $R > \mathfrak{C}$, there exists a constant $\delta > 0$ for which

$$p_e^{av}(n) \ge \delta$$

for *all* n.

Proof. Let $m = \lceil s^{nR} \rceil$. Also, let $\mathbf{X} = (X_1, \ldots, X_n)$ be the input random vector, where X_i is the i-th input to the channel. For the uniform input distribution $p(\mathbf{c}) = \frac{1}{m}$, we have $H(\mathbf{X}) = \log m$. Hence,

$$I(\mathbf{X} \mid \mathbf{Y}) = H(\mathbf{X}) - H(\mathbf{X} \mid \mathbf{Y}) = \log m - H(\mathbf{X} \mid \mathbf{Y})$$

Now, we leave it as an exercise to show that

$$I(\mathbf{X} \mid \mathbf{Y}) \le \sum_{i=1}^{n} I(X_i \mid Y_i) \le n\mathfrak{C} \tag{3.3.5}$$

and so

$$\log m - H(\mathbf{X} \mid \mathbf{Y}) \le n\mathfrak{C}$$

Hence, Fano's inequality gives

$$\text{(3.3.6)} \qquad \log m - n\mathcal{C} \le H(\mathbf{X} \mid \mathbf{Y})$$
$$\le H(p_e^{av}(n)) + p_e^{av}(n) \log(m-1)$$
$$\le 1 + p_e^{av}(n) \log m$$

and so

$$\frac{\log m - n\mathcal{C} - 1}{\log m} \le p_e^{av}(n)$$

or

$$1 - \frac{n\mathcal{C}+1}{\log m} \le p_e^{av}(n)$$

Now, if $R = \mathcal{C}+\epsilon$, with $\epsilon > 0$, then $m = \lceil 2^{nR} \rceil \ge 2^{n(\mathcal{C}+\epsilon)}$, and so we we get

$$1 - \frac{n\mathcal{C}+1}{n(\mathcal{C}+\epsilon)} \le p_e^{av}(n)$$

or, after rearranging,

$$\frac{\epsilon - \frac{1}{n}}{\epsilon + \mathcal{C}} \le p_e^{av}(n)$$

Since the left side of this inequality approaches $\frac{\epsilon}{\epsilon + \mathcal{C}} > 0$, we deduce the existence of a $\delta_1 > 0$ and an $N \ge 0$ such that

$$p_e^{av}(n) \ge \delta_1 > 0, \qquad \text{for all } n > N$$

Now, for a fixed n, if $p_e^{av}(n) = 0$, then the second inequality in (3.3.6) tells us that

$$\log m - n\mathcal{C} \le 0$$

which is equivalent to $\log m \le n\mathcal{C}$, or $R \le \mathcal{C}$. Since this is not the case, we deduce that $p_e^{av}(n) > 0$ for each n. Taking

$$\delta = \min\{p_e^{av}(1), \ldots, p_e^{av}(N), \delta_1\}$$

proves the theorem. ∎

THE STRONG CONVERSE OF THE NOISY CODING THEOREM

We conclude this section by stating the strong converse of the Noisy Coding Theorem, which we will prove in the next section.

Theorem 3.3.7 **(Strong Converse to the Noisy Coding Theorem)** Consider a discrete memoryless channel, with capacity $\mathfrak{C}$. Suppose that C_n is a sequence of $(n,\lceil s^{nR}\rceil)$–codes, with corresponding decision schemes f_n, and that the average probability of error of f_n is $p_e^{av}(n)$. Then if $R > \mathfrak{C}$, we must have

$$p_e^{av}(n) \to 1$$

as $n \to \infty$. □

EXERCISES

1. Consider the channel with channel matrix

$$\begin{bmatrix} \frac{1}{6} & \frac{1}{3} & \frac{1}{2} \\ \frac{1}{3} & \frac{1}{2} & \frac{1}{6} \\ \frac{1}{2} & \frac{1}{6} & \frac{1}{3} \end{bmatrix}$$

For the input distribution $p(x_1) = \frac{1}{2}$, $p(x_2) = p(x_3) = \frac{1}{4}$, find the best decision procedure and the associated average and maximum probabilities of error.

2. Consider the channel shown in Figure 3.3.3.

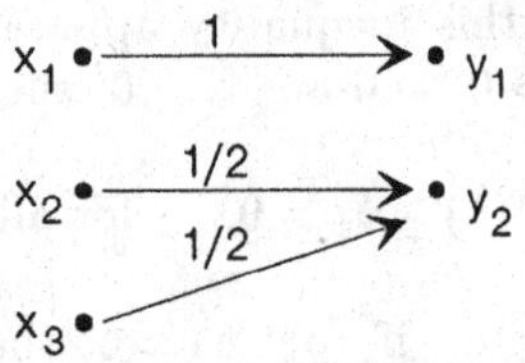

Figure 3.3.3

Without doing any calculations, characterize the condition that the channel loss is zero, in terms of the input distribution. Then verify your thoughts algebraically.

3. Verify equation (3.3.4).
4. Verify inequality (3.3.5).
5. Explain why the strong converse of the Noisy Coding Theorem does not imply the weak converse.
6. Let $\{p_1,\ldots,p_n\}$ be a probability distribution. Let $p_{max} = \max\{p_i\}$.

   (a) Show that $H(p_1,\ldots,p_n) \geq H(p_i)$ for any i.

   (b) Show that $H(p_1,\ldots,p_n) \geq \log \frac{1}{p_{max}}$.

   (c) Show that

$$H(p_1,\ldots,p_n) \geq 2(1-p_{max})$$

Hint. Use part (a) for $p_{max} \geq \frac{1}{2}$, and part (b) for $p_{max} \leq \frac{1}{2}$.

(d) For the ideal observer, show that

$$H(\mathbf{X} \mid \mathbf{d}) \geq 2\ P(\text{error} \mid \mathbf{d})$$

(e) For the ideal observer, show that the average probability of error satisfies

$$p_e^{av} \leq \frac{1}{2} H(\mathbf{X} \mid \mathbf{Y})$$

7. Suppose we choose a code $\mathbf{c}_1,\ldots,\mathbf{c}_m$ at random. Use Chebyshev's inequality

$$P(X \geq \delta) \leq \frac{\mathcal{E}(X)}{\delta}$$

to show that, for any ϵ, we can ensure that the probability that $p_e^{av}(\mathbf{c}_1,\ldots,\mathbf{c}_m)$ is greater than or equal to δ is at most ϵ, if the length n of the code is sufficiently large.

8. Let CH1 and CH2 be channels with

$$\mathcal{I}_1 = \{x_{11},\ldots,x_{s_1 1}\},\quad \mathcal{O}_1 = \{y_{11},\ldots,y_{t_1 1}\}$$

$$\text{Channel Matrix} = \left(p(y_{j1} \mid x_{i1})\right)$$

and

$$\mathcal{I}_2 = \{x_{12},\ldots,x_{s_2 2}\},\quad \mathcal{O}_2 = \{y_{12},\ldots,y_{t_2 2}\}$$

$$\text{Channel Matrix} = \left(p(y_{j2} \mid x_{i2})\right)$$

Define a new channel, called the **product** of the two channels, with

$$\mathcal{I} = \{x_{i1}x_{j2}\}_{i=1,\ldots,s_1;\ j=1,\ldots,s_2}$$

$$\mathcal{O} = \{y_{i1}y_{j2}\}_{i=1,\ldots,t_1;\ j=1,\ldots,t_2}$$

and

$$\text{Channel Matrix} = \left(p(x_{i1}x_{j2} \mid y_{i1}y_{j2})\right)$$

where

$$p(x_{i1}x_{j2} \mid y_{i1}y_{j2}) = p(y_{j1} \mid x_{i1}) \cdot p(y_{j2} \mid x_{i2})$$

In words, the product channel is used, for input $x_{i1}x_{j2}$, by sending the first component x_{i1} through the first channel, and the second component x_{j2} through the second channel, simultaneously. Show that the capacity $\mathcal{C}$ of the product is equal to $\mathcal{C} = \mathcal{C}_1 + \mathcal{C}_2$, where $\mathcal{C}_i$ is the capacity of the i–th channel.

9. Let BSC be a binary symmetric channel, with crossover probability p. Consider the following three channels, each with input and output alphabets consisting of the set 2^3 of all binary words of length 3. Let $i_1i_2i_3$ be the input and $o_1o_2o_3$ be the output.
   (a) Output o_1 is determined by sending i_1 through BSC. Outputs o_2 and o_3 are chosen randomly, with each possibility being equally likely.
   (b) Output o_1 is determined by sending the *majority* bit from among i_1, i_2 and i_3 through BSC. Then $o_2 = o_3 = o_1$.
   (c) Output o_k is determined by sending i_k through BSC.

   For each channel, assume a uniform input distribution, and a decision scheme that simply declares that the input is the same as the output. Find the probability of a decision error in each case.
10. The *Hamming codes*, which we will study in detail later in the book, have the following properties.
   (i) For $r = 1,2,\ldots$, the Hamming code $\mathcal{H}_2(r)$ has length $n = 2^r - 1$ and size

   $$|\mathcal{H}_2(r)| = 2^{2^r - r - 1}$$

   (The codeword alphabet is binary.)
   (ii) The Hamming distance between any pair of codewords is at least 3.
   (iii) Any binary word of length n is either a codeword, or else it has distance 1 from exactly one codeword.

   (a) Calculate the rate of the Hamming code $\mathcal{H}_2(r)$. What happens to the rate as $n \to \infty$?
   (b) What is the ideal observer decision scheme for this code, assuming a uniform input distribution into a binary symmetric channel with crossover probability $p < \frac{1}{2}$?
   (c) Compute $P(\text{error} \mid \mathbf{d})$, under the ideal observer decision scheme.
   (d) Compute p_e, under the ideal observer decision scheme.

3.4 Proof of the Noisy Coding Theorem and Its Strong Converse

In this section, we prove the Noisy Coding Theorem and its strong converse for the case of the binary symmetric channel. The proofs in this case include the major ideas embodied in proofs of fuller versions of these theorems, and have the advantage of bringing these ideas closer to the surface.

Before beginning, we need a few preliminary definitions.

Definition The set of all binary words of length n will be denoted by $\mathbf{2}^n$. Thus, $|\mathbf{2}^n| = 2^n$. If $\mathbf{x}$ and $\mathbf{y}$ are binary words of the same length, then the **Hamming distance** $d(\mathbf{x},\mathbf{y})$ from $\mathbf{x}$ to $\mathbf{y}$ is the number of positions in which the two words differ. □

For instance, if $\mathbf{x} = 11001$ and $\mathbf{y} = 01011$, then $d(\mathbf{x},\mathbf{y}) = 2$, since the two words differ in two positions (the first and fourth positions).

Definition If $\rho \geq 0$, then the **closed ball** with center $\mathbf{x}$ and radius ρ is the set

$$\overline{B}(\mathbf{x},\rho) = \{\mathbf{y} \mid d(\mathbf{x},\mathbf{y}) \leq \rho\}$$ □

Let us dispose of a technical matter in a separate lemma, rather than interrupt the upcoming proofs.

Lemma 3.4.1 Let $s \geq 2$ be an integer. Let $0 < R < \mathcal{C}$. Then there exists an R' satisfying

$$R + \frac{\log_s 2}{n} + \frac{1}{n} \leq R' < \mathcal{C} \tag{3.4.1}$$

for all n sufficiently large, and for this R', we have

$$\frac{1}{2}\lceil s^{nR'}\rceil \geq \lceil s^{nR}\rceil \tag{3.4.2}$$

for n sufficiently large.

Proof. The existence of R' is clear. Since $nR' \geq nR + \log_s 2 + 1$, we have

$$\frac{1}{2}\lceil s^{nR'}\rceil \geq \frac{1}{2} s^{nR'} = s^{nR' - \log_s 2} \geq s^{nR+1} = s(s^{nR}) \geq s(\lceil s^{nR}\rceil - 1)$$
$$= s\lceil s^{nR}\rceil - s \geq \lceil s^{nR}\rceil$$

the last inequality holding provided that $\lceil s^{nR}\rceil \geq \frac{s}{s-1}$, which is certainly true for n sufficiently large. ∎

MORE ON THE PROBABILITY OF ERROR

Our plan is to prove a version of the Noisy Coding Theorem with the words *maximum probability* replaced by *average probability.* As promised in the previous section, we have the following result, which shows that the two versions are equivalent.

Theorem 3.4.2 Consider a discrete memoryless channel, and let $\mathfrak{C}$ be any positive real number. The following two statements are equivalent.

1) For any $R < \mathfrak{C}$, there is a sequence of $(n,\lceil s^{nR}\rceil)$-codes C_n, and corresponding decision schemes f_n, for which the *maximum* probability of error $p_e^{max}(n)$ approaches 0 as $n \rightarrow \infty$.
2) For any $R < \mathfrak{C}$, there is a sequence of $(n,\lceil s^{nR}\rceil)$-codes E_n, and corresponding decision schemes g_n, for which the *average* probability of error $p_e^{av}(n)$ approaches 0 as $n \rightarrow \infty$.

Proof. As we have already noted, if the maximum probability of error is less than ϵ, then so is the average probability of error. Hence, statement 1) implies statement 2).

As for the converse, suppose statement 2) holds and we are given an $R < \mathfrak{C}$. First, choose an R' satisfying (3.4.1). Then, according to statement 2, there is a sequence of $(n,\lceil s^{nR'}\rceil)$–codes E_n, and decision schemes g_n, for which the *average* probability of error tends to 0 as $n \rightarrow \infty$. Hence, for any $\epsilon > 0$, if n is sufficiently large, we may find a code

$$E_n = \{e_1, \ldots, e_u\}$$

where $u = \lceil s^{nR'}\rceil$, and a decision scheme g_n, for which the average probability of error is less than $\epsilon/2$, that is,

$$\frac{1}{u}\sum_{i=1}^{u} p(\text{error} \mid e_i) < \frac{\epsilon}{2}$$

Now, we leave it as an exercise to show that if the average of a set of nonnegative numbers is equal to a, then at least half of the numbers must be less than or equal to $2a$. In the present context, this tells us that at least half of the codewords $e_i \in E_n$ must satisfy $P(\text{error} \mid e_i) < 2(\epsilon/2) = \epsilon$. Using *only* these codewords, we get a code

$$C_n = \{c_1, \ldots, c_v\}$$

of length n and size $v \geq \frac{1}{2}\lceil s^{nR'}\rceil$. Furthermore, by restricting the decision scheme g_n to those output strings d_j for which $g_n(d_j) \in C_n$, we get a maximum probability of error at most ϵ. Finally, since (3.4.1) holds, so does (3.4.2), and so

$$|C_n| \geq \frac{1}{2}\lceil s^{nR'}\rceil \geq \lceil s^{nR}\rceil$$

Then we can, if necessary, take a subset of C_n (and further restrict the decision scheme) to obtain the code described in statement 1). This proves the theorem. ∎

PROOF OF THE NOISY CODING THEOREM

In order to prove the Noisy Coding Theorem, we will use Theorem 1.2.8. For convenience, let us restate it here as a lemma.

Lemma 3.4.3 If $0 \leq \lambda \leq \frac{1}{2}$, then

$$\sum_{k=0}^{\lambda n} \binom{n}{k} \leq 2^{nH(\lambda)}$$

where $H(\lambda) = \lambda \log \frac{1}{\lambda} + (1-\lambda) \log \frac{1}{1-\lambda}$ is the entropy function. □

We are now ready for the proof of the Noisy Coding Theorem.

Theorem 3.4.4 (The Noisy Coding Theorem) Consider a discrete memoryless channel with capacity $\mathfrak{C}$. For any positive number $R < \mathfrak{C}$, there exists a sequence C_n of s-ary codes, and corresponding decision schemes f_n, with the following properties.

1) C_n is an $(n, \lceil s^{nR} \rceil)$-code, that is, C_n has length n and rate at least R.
2) The maximum probability of error of f_n approaches 0 as $n \to \infty$, that is,

$$p_e^{max}(n) \to 0$$

Proof. We prove this theorem for a binary symmetric channel ($s = 2$), with crossover probability $p < \frac{1}{2}$. The plan of the proof was outlined in the previous section, so let us go directly to the details.

Let ϵ be any positive number for which $p + \epsilon < \frac{1}{2}$, and let $\rho = n(p+\epsilon)$. Let $C_n = \{\mathbf{c}_1, \ldots, \mathbf{c}_m\}$ be any code of length n and size m. In order to perform random coding, we must allow the possibility that some of the codewords $\mathbf{c}_i \in C_n$ are identical. (Thus, the code C_n is a *multiset*, rather than a set.) We leave it as an exercise to show that this presents no problems.

Now let us define a decision scheme f_n as follows. Suppose that the string $\mathbf{d}$ is received. If there is a *unique* codeword $\mathbf{c}_i$ in the closed ball $\overline{B}(\mathbf{d}, \rho)$, we decide on that codeword. Otherwise, we simply admit a decision error. Thus, if $\mathbf{c}_i, \mathbf{c}_k \in \overline{B}(\mathbf{d}, \rho)$ for $k \neq i$, we admit an error, even if $\mathbf{c}_i = \mathbf{c}_k$.

According to this decision scheme, if $\mathbf{c}_i$ is sent and $\mathbf{d}$ is received, a decision error will occur if $\mathbf{c}_i$ is not in $\overline{B}(\mathbf{d}, \rho)$, or if there is

a codeword $\mathbf{c}_k$, with $k \neq i$, for which $\mathbf{c}_k \in \overline{B}(\mathbf{d},\rho)$. Thus

$$P(\text{error} \mid \mathbf{c}_i) \leq \sum_{\substack{\mathbf{d} \in 2^n: \\ \mathbf{c}_i \notin \overline{B}(\mathbf{d},\rho)}} p(\mathbf{d} \mid \mathbf{c}_i) + \sum_{\substack{\mathbf{d} \in 2^n: \\ \exists \mathbf{c}_j \in \overline{B}(\mathbf{d},\rho) \\ (j \neq i)}} p(\mathbf{d} \mid \mathbf{c}_i)$$

$$\leq \sum_{\substack{\mathbf{d} \in 2^n: \\ \mathbf{c}_i \notin \overline{B}(\mathbf{d},\rho)}} p(\mathbf{d} \mid \mathbf{c}_i) + \sum_{\substack{j=1 \\ j \neq i}}^{m} \sum_{\substack{\mathbf{d} \in 2^n: \\ \mathbf{c}_j \in \overline{B}(\mathbf{d},\rho)}} p(\mathbf{d} \mid \mathbf{c}_i)$$

The first sum on the right is the probability that the received word $\mathbf{d}$ is not in the closed ball $\overline{B}(\mathbf{d},\rho)$, which is the probability that at least ρ bit errors have been made by the channel. But, according to the law of large numbers, since $\rho > np$, this probability tends to 0 as $n \to \infty$. It follows that

$$\sum_{\substack{\mathbf{d} \in 2^n: \\ \mathbf{c}_i \notin \overline{B}(\mathbf{d},\rho)}} p(\mathbf{d} \mid \mathbf{c}_i) \leq \delta_n$$

where $\delta_n \to 0$ as $n \to \infty$. Note that, since the set of all binary strings of length n is finite, there are only a finite number of possibilities for the $\mathbf{c}_i$. Hence, we can assume that δ_n is *independent* of the code C. We now have

$$P(\text{error} \mid \mathbf{c}_i) \leq \delta_n + \sum_{\substack{j=1 \\ j \neq i}}^{m} \sum_{\substack{\mathbf{d} \in 2^n: \\ \mathbf{c}_j \in \overline{B}(\mathbf{d},\rho)}} p(\mathbf{d} \mid \mathbf{c}_i)$$

In view of Theorem 3.4.2, we can assume that each codeword is equally likely to be input into the channel and find a bound on the *average* probability of error. Since, in this case, $p(\mathbf{c}_i) = \frac{1}{m}$ for all i, we have

$$p_e^{av}(\mathbf{c}_1,\ldots,\mathbf{c}_m) = \frac{1}{m}\sum_{i=1}^{m} P(\text{error} \mid \mathbf{c}_i)$$

$$\leq \delta_n + \frac{1}{m}\sum_{i=1}^{m} \sum_{\substack{j=1 \\ j \neq i}}^{m} \sum_{\substack{\mathbf{d} \in 2^n: \\ \mathbf{c}_j \in \overline{B}(\mathbf{d},\rho)}} p(\mathbf{d} \mid \mathbf{c}_i)$$

Letting

$$\delta_\rho(\mathbf{x},\mathbf{y}) = \begin{cases} 1 & \text{if } d(\mathbf{x},\mathbf{y}) \leq \rho \\ 0 & \text{otherwise} \end{cases}$$

this can be written

$$p_e^{av}(\mathbf{c}_1,\ldots,\mathbf{c}_m) \le \delta_n + \frac{1}{m}\sum_{i=1}^{m}\sum_{\substack{j=1\\ j\ne i}}^{m}\sum_{\mathbf{d}\in 2^n} p(\mathbf{d}\mid\mathbf{c}_i)\delta_\rho(\mathbf{d},\mathbf{c}_j)$$

or, after rearranging the summations,

$$(3.4.3)\qquad p_e^{av}(\mathbf{c}_1,\ldots,\mathbf{c}_m) \le \delta_n + \frac{1}{m}\sum_{\mathbf{d}\in 2^n}\sum_{i=1}^{m}\sum_{\substack{j=1\\ j\ne i}}^{m} p(\mathbf{d}\mid\mathbf{c}_i)\delta_\rho(\mathbf{d},\mathbf{c}_j)$$

Now, suppose that the codewords $\mathbf{c}_i$ are chosen independently and at random, with each binary string equally likely to occur. If ω_i represents the i-th codeword chosen, then $\omega_1,\ldots,\omega_m$ are independent random variables, uniformly distributed over the set of all 2^n binary words of length n. Further, from (3.4.3), we deduce that

$$p_e^{av}(\omega_1,\ldots,\omega_m) \le \delta_n + \frac{1}{m}\sum_{\mathbf{d}\in 2^n}\sum_{i=1}^{m}\sum_{\substack{j=1\\ j\ne i}}^{m} p(\mathbf{d}\mid\omega_i)\delta_\rho(\mathbf{d},\omega_j)$$

Taking expected values, and noting that for $i\ne j$, the random variables $p(\mathbf{d}\mid\omega_i)$ and $\delta(\mathbf{d},\omega_j)$ are independent, we get

(3.4.4)

$$\mathcal{E}(p_e^{av}(\omega_1,\ldots,\omega_m)) \le \delta_n + \frac{1}{m}\sum_{\mathbf{d}\in 2^n}\sum_{i=1}^{m}\sum_{\substack{j=1\\ j\ne i}}^{m} \mathcal{E}(p(\mathbf{d}\mid\omega_i))\mathcal{E}(\delta_\rho(\mathbf{d},\omega_j))$$

But, since each ω_j is uniformly distributed,

$$\mathcal{E}(\delta_\rho(\mathbf{d}\mid\omega_j)) = \sum_{\mathbf{x}\in 2^n}\delta_\rho(\mathbf{d},\mathbf{x})P(\omega_j=\mathbf{x}) = \frac{1}{2^n}\sum_{\mathbf{x}\in 2^n}\delta_\rho(\mathbf{d},\mathbf{x}) = \frac{|\overline{B}(\mathbf{d},\rho)|}{2^n}$$

and

$$\mathcal{E}(p(\mathbf{d}\mid\omega_i)) = \sum_{\mathbf{x}\in 2^n} p(\mathbf{d}\mid\mathbf{x})P(\omega_i=\mathbf{x}) = \frac{1}{2^n}\sum_{\mathbf{x}\in 2^n} p(\mathbf{d}\mid\mathbf{x})$$

Substituting these values into (3.4.4), we get

$$\mathcal{E}(p_e^{av}(\omega_1,\ldots,\omega_m)) \le \delta_n + \frac{1}{m}\sum_{\mathbf{d}\in 2^n}\sum_{i=1}^{m}\sum_{\substack{j=1\\ j\ne i}}^{m}\frac{1}{2^n}\sum_{\mathbf{x}\in 2^n} p(\mathbf{d}\mid\mathbf{x})\,\frac{|\overline{B}(\mathbf{d},\rho)|}{2^n}$$

Rearranging terms, and noting that for any $\mathbf{d}$, $|\overline{B}(\mathbf{d},\rho)| = |\overline{B}(\mathbf{0},\rho)|$, where $\mathbf{0}$ is the binary word consisting of all 0's, we get

(3.4.5)

$$\mathcal{E}(p_e^{av}(\omega_1,\ldots,\omega_m)) \le \delta_n + \frac{|\overline{B}(\mathbf{0},\rho)|}{m2^n2^n} \sum_{i=1}^{m} \sum_{\substack{j=1 \\ j\neq i}}^{m} \sum_{\mathbf{x}\in 2^n} \sum_{\mathbf{d}\in 2^n} p(\mathbf{d}\,|\,\mathbf{x})$$

Now,

$$\sum_{\mathbf{x}\in 2^n} \sum_{\mathbf{d}\in 2^n} p(\mathbf{d}\,|\,\mathbf{x}) = \sum_{\mathbf{x}\in 2^n} 1 = 2^n$$

and so

$$\sum_{i=1}^{m} \sum_{\substack{j=1 \\ j\neq i}}^{m} \sum_{\mathbf{x}\in 2^n} \sum_{\mathbf{d}\in 2^n} p(\mathbf{d}\,|\,\mathbf{x}) = \sum_{i=1}^{m} \sum_{\substack{j=1 \\ j\neq i}}^{m} 2^n = m(m-1)2^n$$

Substituting this into (3.4.5) and simplifying gives

$$\mathcal{E}(p_e^{av}(\omega_1,\ldots,\omega_m)) \le \delta_n + \frac{m-1}{2^n}|\overline{B}(\mathbf{0},\rho)|$$

Since $\rho = n(p+\epsilon)$, Lemma 3.4.3, with $\lambda = p+\epsilon$, implies that

$$|\overline{B}(\mathbf{0},\rho)| = \sum_{k=0}^{n(p+\epsilon)} \binom{n}{k} \le 2^{nH(p+\epsilon)}$$

and so

$$\mathcal{E}(p_e^{av}(\omega_1,\ldots,\omega_m)) \le \delta_n + \frac{m}{2^n}2^{nH(p+\epsilon)}$$

Since this holds for all sufficiently small $\epsilon > 0$ (so that $p+\epsilon < \frac{1}{2}$), and since the entropy function H is continuous, we have

$$\mathcal{E}(p_e^{av}(\omega_1,\ldots,\omega_m)) \le \delta_n + \frac{m}{2^n}2^{nH(p)}$$

Writing

$$\frac{m}{2^n}2^{nH(p)} = 2^{\log_2 m - n(1-H(p))} = 2^{\log_2 m - n\mathcal{C}}$$

this becomes

$$\mathcal{E}(p_e^{av}(\omega_1,\ldots,\omega_m)) \le \delta_n + 2^{\log_2 m - n\mathcal{C}}$$

Now, since $R < \mathcal{C}$, we can find an R' for which

$$R + \frac{1}{n} < R' < \mathcal{C}$$

for n sufficiently large. Therefore, we have for such n,

$$\lceil 2^{nR} \rceil \le 2^{nR} + 1 \le 2^{nR+1} = 2^{n(R+\frac{1}{n})} < 2^{nR'}$$

Therefore, taking the code size to be

$$m = \lceil 2^{nR} \rceil$$

we have

$$\log_2 m - n\mathcal{C} < \log_2 (2^{nR'}) - n\mathcal{C} \le n(R' - \mathcal{C})$$

and

$$\mathcal{E}(p_e^{av}(\omega_1,\ldots,\omega_m)) \le \delta_n + 2^{-n(\mathcal{C}-R')}$$

Finally, we observe that, for each n, there must exist at least one value

$$(\omega_1,\ldots,\omega_m) = (c_1,\ldots,c_m)$$

of the random vector $(\omega_1,\ldots,\omega_m)$, that is, one code $C_n = (c_1,\ldots,c_m)$, for which

$$p_e^{av}(c_1,\ldots,c_m) \le \delta_n + 2^{-n(\mathcal{C}-R')}$$

Since both terms on the right side of this inequality tend to 0 as $n \to \infty$, the proof is complete. ∎

PROOF OF THE STRONG CONVERSE

In order to prove the strong converse of the Noisy Coding Theorem, we need a preliminary lemma. Stirling's formula is

$$n! = \sqrt{2\pi n}\, n^n e^{-n} \exp\left(\frac{1}{12n} - \frac{1}{360n^3} + \cdots\right)$$

This formula underestimates n! when an even number (including zero) of terms of the series in parentheses are taken, and overestimates n! when an odd number of terms are taken. We will indicate in the exercises how this formula can be used to prove the following lemma.

Lemma 3.4.5 If $0 < a < \frac{1}{2}$, then

$$\binom{n}{an} \ge 2^{n[H(a) - \delta_n]}$$

where $H(a) = a \log \frac{1}{a} + (1-a) \log \frac{1}{1-a}$ is the entropy function, and $\delta_n \to 0$ as $n \to \infty$. □

Now we are ready for the proof of the strong converse. We will prove the theorem with the words *average probability* replaced by *maximum probability*, and leave it as an exercise to show that the two versions are equivalent. Thus, we prove the following result.

Theorem 3.4.6 Consider a discrete memoryless channel, with capacity $\mathfrak{C}$. Suppose that C_n is a sequence of $(n,\lceil s^{nR}\rceil)$-codes, with corresponding decision schemes f_n, and that the maximum probability of error of f_n is $p_e^{max}(n)$. Then if $R > \mathfrak{C}$, we must have

$$p_e^{max}(n) \to 1$$

as $n \to \infty$.

Proof. (For the binary symmetric channel, with crossover probability $p < \frac{1}{2}$.) We will prove the theorem by assuming that $p_e^{max}(n)$ does not converge to 1, and showing that this implies $R \le \mathfrak{C}$. In particular, we assume that there exists a number $\lambda < 1$ for which $p_e^{max}(n) \le \lambda$, for all sufficiently large n, and show that $R \le \mathfrak{C}$.

Intuitively speaking, the inequalities

$$p(\text{error} \mid c_i) \le p_e^{max}(n) \le \lambda < 1 \tag{3.4.6}$$

which tell us that the probability of error is not too large, also tell us that the sets $B_i = f_n^{-1}(c_i)$ cannot be too small. For if B_i is small, then not enough errors will be corrected, and the probability of an error will therefore be large. On the other hand, the sets B_i cannot be too large, for they are disjoint subsets of the 2^n possible binary words of length n.

Our plan then is to first get a lower bound on the size of the sets B_i, say

$$|B_i| \ge L$$

and then use the fact that the $\lceil 2^{nR}\rceil$ sets B_i are disjoint to deduce that

$$\lceil 2^{nR}\rceil \cdot L \le 2^n$$

This tells us that R cannot be too large and will lead us to the desired conclusion.

To make all of this more precise, consider the set B_i, as shown in Figure 3.4.1.

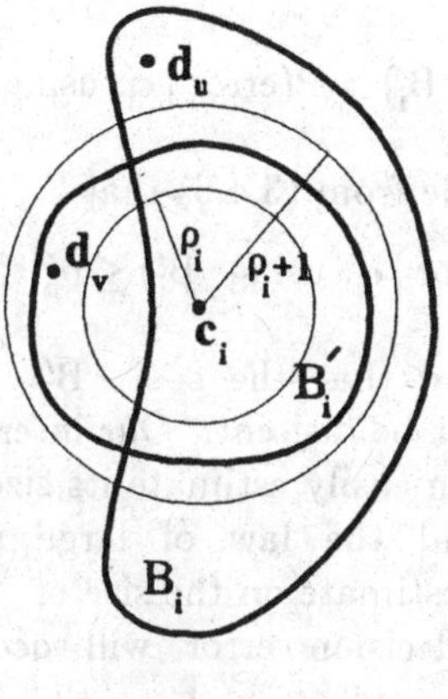

Figure 3.4.1

Because of the nature of the channel, and because $p < \frac{1}{2}$, a given string with fewer errors is more likely to be received than a given string with more errors. In symbols,

$$d(\mathbf{d}_v,\mathbf{c}_i) < d(\mathbf{d}_u,\mathbf{c}_i) \ \Rightarrow\ p(\mathbf{d}_v \mid \mathbf{c}_i) > p(\mathbf{d}_u \mid \mathbf{c}_i)$$

Hence, we will not increase the probability $P(\text{error} \mid \mathbf{c}_i)$ if we replace B_i by a set B_i', of the same size as B_i, but satisfying

$$B(\mathbf{c}_i,\rho_i) \subset B_i' \subsetneq B(\mathbf{c}_i,\rho_i+1)$$

for some $\rho_i \geq 0$. (The set B_i' is more tightly packed around $\mathbf{c}_i$ than is B_i.) To verify this, suppose that $\mathbf{d}_u \in B_i$, $\mathbf{d}_v \notin B_i$, but that $\mathbf{d}_v$ is closer to $\mathbf{c}_i$ than is $\mathbf{d}_u$. Then,

$$\begin{aligned}
P(\text{no error} \mid \mathbf{c}_i, \text{ using } B_i) &= \sum_{\mathbf{d}_j \in B_i} p(\mathbf{d}_j \mid \mathbf{c}_i) \\
&= \sum_{\mathbf{d}_j \in B_i - \{\mathbf{d}_u\}} p(\mathbf{d}_j \mid \mathbf{c}_i) + p(\mathbf{d}_u \mid \mathbf{c}_i) \\
&< \sum_{\mathbf{d}_j \in B_i - \{\mathbf{d}_u\}} p(\mathbf{d}_j \mid \mathbf{c}_i) + p(\mathbf{d}_v \mid \mathbf{c}_i) \\
&= \sum_{\mathbf{d}_j \in B_i - \{\mathbf{d}_u\} \cup \{\mathbf{d}_v\}} p(\mathbf{d}_j \mid \mathbf{c}_i) \\
&= P(\text{no error} \mid \mathbf{c}_i, \text{ using } B_i - \{\mathbf{d}_u\} \cup \{\mathbf{d}_v\})
\end{aligned}$$

Hence,

$$P(\text{error} \mid \mathbf{c}_i, \text{ using } B_i) > P(\text{error} \mid \mathbf{c}_i, \text{ using } B_i - \{\mathbf{d}_u\} \cup \{\mathbf{d}_v\})$$

We can therefore conclude from (3.4.6) that

$$P(\text{error} \mid \mathbf{c}_i, \text{ using } B_i') \le p_e^{max}(n) \tag{3.4.7}$$

We do not contend that the sets $B_1', \ldots, B_n'$ form a decision scheme, for they may not be disjoint. Our interest in the set B_i' comes from the fact that we can easily estimate its size, by estimating the size of ρ_i, using (3.4.7) and the law of large numbers. Then, since $|B_i'| = |B_i|$, we get an estimate on the size of B_i.

In particular, a decision error will occur using the set B_i' whenever the channel makes at least ρ_i+1 bit errors, since in that case, the received word will lie outside the ball $B(\mathbf{c}_i, \rho_i+1)$, and so also outside the set B_i'. Hence, (3.4.7) implies that

$$P(\text{at least } \rho_i+1 \text{ bit errors}) \le p_e^{max}(n) \le \lambda < 1$$

Now, if $\rho_i+1 \le np'$ for any $p' < p$, then

$$P(\text{at least } np' \text{ bit errors}) \le P(\text{at least } \rho_i+1 \text{ bit errors}) \le \lambda < 1$$

which cannot happen for n large, since it would contradict the law of large numbers, which says that the probability on the far left must approach 1 as $n \to \infty$. Thus, given any $p' < p$, we must have

$$\rho_i + 1 > np' \tag{3.4.8}$$

for n sufficiently large (and depending on p'). Writing (3.4.8) in the form

$$\rho_i > n(p' - \tfrac{1}{n}) = np''$$

we see that, for any $p'' < p$,

$$\rho_i > np''$$

for n sufficiently large.

Now we can estimate the size of B_i', and so also of B_i. For any $p'' < p$, if n is sufficiently large, we have $\rho_i > np''$, and so

$$|B_i| = |B_i'| \ge |B(\mathbf{c}_i, \rho_i)| = \sum_{j=0}^{\rho_i} \binom{n}{j} \ge \sum_{j=0}^{np''} \binom{n}{j} \ge \binom{n}{np''}$$

But then, since $p'' < p < \frac{1}{2}$, Lemma 3.4.5 implies that

$$|B_i| \ge 2^{n[H(p'') - \delta_n]} \tag{3.4.9}$$

where $\delta_n \to 0$ as $n \to \infty$. This is the lower bound that we have been seeking.

Since the $\lceil 2^{nR} \rceil$ decision sets B_i are disjoint, and since there are 2^n binary strings of length n, we have

$$\sum_{i=1}^{\lceil 2^{nR} \rceil} |B_i| \le 2^n$$

Then (3.4.9) implies

$$\lceil 2^{nR} \rceil \cdot 2^{n[H(p'')-\delta_n]} \le 2^n$$

and so

$$2^{nR} \le 2^{n[1-H(p'')+\delta_n]}$$

which implies that

$$R \le 1 - H(p'') + \delta_n$$

where $\delta_n \to 0$ as $n \to \infty$, and where p'' is any number satisfying $p'' < p$. But since the entropy function H is continuous, and since $\delta_n \to 0$, we deduce that R cannot be strictly greater than $1 - H(p) = \mathfrak{C}$. In other words, $R \le \mathfrak{C}$. This concludes the proof. ∎

EXERCISES

1. Show that, if the average of a set of nonnegative numbers is equal to a, then at least half of the numbers must be less than or equal to 2a.
2. If X is a discrete random variable, prove that at least one value of X must be less than or equal to $\mathcal{E}(X)$.
3. Show that the use of multisets in the proof of the Noisy Coding Theorem does not cause any problems.
4. Show that, for $0 < a < 1$, and $b = 1 - a$,

$$\binom{n}{an} \ge \frac{1}{\sqrt{8nab}} 2^{nH(a)}$$

as follows.

(a) Use Stirling's formula to get a lower bound on

$$\binom{n}{an} = \frac{n!}{(an)!(bn)!}$$

by underestimating n! and overestimating (an)! and (bn)!, as indicated in the text.

(b) Use in the result of part (a), the fact that, for $an \geq 1$ and $bn \geq 3$,

$$\frac{1}{12an} + \frac{1}{12bn} \leq \frac{1}{9}$$

and so

$$\exp\left(-\frac{1}{12an} - \frac{1}{12bn}\right) \geq \frac{\sqrt{\pi}}{2}$$

(c) Do the cases not covered in part (b) separately.

5. Use the results of the previous exercise to prove Lemma 3.4.5.
6. Show that Theorem 3.4.6 is equivalent to the strong converse of the Noisy Coding Theorem (Theorem 3.3.7).

Part 2

Coding Theory

CHAPTER 4

General Remarks on Codes

4.1 Error Detection and Correction

The jumping off point for coding theory is Shannon's Noisy Coding Theorem. Let us undertake a review of some of the topics that we covered in the previous chapter, so that we can restate this important theorem. This will also help to set our terminology, for those readers who have studied the information theory portion of this book, as well as for those who have not.

BLOCK CODES

We begin with the definition of a block code.

Definition Let $\mathcal{A} = \{a_1, \ldots, a_q\}$ be a finite set, called a **code alphabet,** and let $\mathcal{A}^n$ be the set of all strings of length n over $\mathcal{A}$. Any nonempty subset C of $\mathcal{A}^n$ is called a **q-ary block code.** Each string in C is called a **codeword.** If $C \subset \mathcal{A}^n$ contains M codewords, then it is customary to say that C has **length** n and **size** M, or is an **(n,M)-code.** The **rate** of a q-ary (n,M)-code is

$$R = \frac{\log_q M}{n}$$ □

THE CHANNEL

Now let us define a communications channel. Since our intention is to think of the channel as accepting codewords $\mathbf{c} = c_1 \cdots c_n$ from a code C of length n over a code alphabet $\mathcal{A}$ and outputting strings

$\mathbf{d} = d_1 \cdots d_n$ of the same length over an alphabet that contains $\mathcal{A}$, we will assume that the input alphabet is a subset of the output alphabet.

Definition A **discrete memoryless channel** consists of an **input alphabet** $\mathcal{A} = \{a_1, \ldots, a_q\}$, an **output alphabet** $\mathcal{O} = \{b_1, \ldots, b_t\}$ containing $\mathcal{A}$, and a set of **channel probabilities**, or **transition probabilities**, $p(b_j \mid a_i)$, satisfying

$$\sum_{j=1}^{t} p(b_j \mid a_i) = 1$$

for all i. Intuitively, we think of $p(b_j \mid a_i)$ as the probability that b_j is received, given that a_i is sent through the channel.

Furthermore, if $\mathbf{c} = c_1 \cdots c_n$ and $\mathbf{d} = d_1 \cdots d_n$ are words of length n over $\mathcal{A}$ and $\mathcal{O}$, respectively, the probability $p(\mathbf{d} \mid \mathbf{c})$ that $\mathbf{d}$ is received, given that $\mathbf{c}$ is sent, is

$$p(\mathbf{d} \mid \mathbf{c}) = \prod_{i=1}^{n} p(d_i \mid c_i) \qquad \square$$

Example 4.1.1 One of the most important discrete memoryless channels is the **binary symmetric channel**, which has input and output alphabets $\{0,1\}$ and channel probabilities

$$p(1 \mid 0) = p(0 \mid 1) = p \quad \text{and} \quad p(0 \mid 0) = p(1 \mid 1) = 1 - p$$

Thus, the probability of a bit error, or **crossover probability**, is p. This channel is pictured in Figure 4.1.1. $\square$

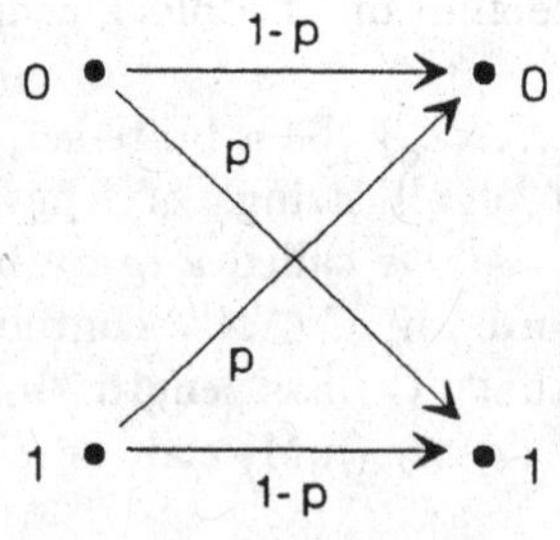

Figure 4.1.1 **A binary symmetric channel**

The **channel matrix** of a channel is defined to be the matrix of channel probabilities

$$\begin{bmatrix} p(b_1 \mid a_1) & p(b_2 \mid a_1) & \cdots & p(b_t \mid a_1) \\ p(b_1 \mid a_2) & p(b_2 \mid a_2) & \cdots & p(b_t \mid a_2) \\ & & \vdots & \\ p(b_1 \mid a_q) & p(b_2 \mid a_q) & \cdots & p(b_t \mid a_q) \end{bmatrix}$$

Since input to a channel is generally probabilistic in nature, we think of the input as the values of a random variable X, with *input distribution* defined by $P(X{=}a_i) = p(a_i)$. As we did in Section 3.1, we can now define the output and joint distributions, as well as the backward channel probabilities. However, since our primary interest is in codeword input and not just symbol input, we think of the input as a random vector **X**, with **input distribution** defined by

$$P(\mathbf{X}{=}\mathbf{c}) = p(\mathbf{c})$$

Each input **X** induces an output **Y** (which need not be a codeword) with **output distribution** defined by

$$P(\mathbf{Y}{=}\mathbf{d}) = \sum_{\mathbf{c}} p(\mathbf{d} \mid \mathbf{c})P(\mathbf{X}{=}\mathbf{c})$$

The **joint distribution** of **X** and **Y** is given by

$$P(\mathbf{X}{=}\mathbf{c},\mathbf{Y}{=}\mathbf{d}) = p(\mathbf{d} \mid \mathbf{c})P(\mathbf{X}{=}\mathbf{c})$$

and the **backward channel probabilities** are

$$P(\mathbf{X}{=}\mathbf{c} \mid \mathbf{Y}{=}\mathbf{d}) = \frac{P(\mathbf{X}{=}\mathbf{c},\mathbf{Y}{=}\mathbf{d})}{P(\mathbf{Y}{=}\mathbf{d})}$$

When no confusion can arise, we will adopt the following notation

$$\begin{aligned} p(\mathbf{c}) &= P(\mathbf{X}{=}\mathbf{c}) \\ p(\mathbf{d}) &= P(\mathbf{Y}{=}\mathbf{d}) \\ p(\mathbf{c},\mathbf{d}) &= P(\mathbf{X}{=}\mathbf{c},\mathbf{Y}{=}\mathbf{d}) \\ p(\mathbf{c} \mid \mathbf{d}) &= P(\mathbf{X}{=}\mathbf{c} \mid \mathbf{Y}{=}\mathbf{d}) \\ p(\mathbf{d} \mid \mathbf{c}) &= P(\mathbf{Y}{=}\mathbf{d} \mid \mathbf{X}{=}\mathbf{c}) \end{aligned}$$

When a received *symbol* is different from the symbol that was input to the channel, we say that a **symbol error**, or **channel error**, has occurred. When a *codeword* **c** is input into the channel, but the output **d** is not equal to **c**, we say that a **word error** has occurred.

BURST ERRORS

Much of coding theory assumes that errors in transmission are independent. However, this can be an unrealistic assumption. For instance, electrical interferences often last longer than the time it takes to transmit a single code symbol, and defects on magnetic tape or disk are usually larger than the space required to store a single code symbol. Hence, errors often occur in *bursts.*

Accordingly, we will also discuss the question of designing codes that are effective in correcting *burst errors.* As we will see, the quality of a code may be quite different when it is used for burst error correction, rather than independent error correction.

THE DECISION SCHEME

A **decision scheme** is a partial function f from the set of output strings to the set of codewords. The word *partial* refers to the fact that f may not be defined for *all* output strings. The intention is that, if an output string **d** is received, and if f(**d**) is defined, then the decision scheme decides that f(**d**) is the codeword that was sent. If f(**d**) is not the codeword that was sent, we say that a **decision error**, or a **decoding error**, has been made.

By letting

$$B_i = f^{-1}(\mathbf{c}) = \{\mathbf{d} \mid f(\mathbf{d})=\mathbf{c}\}$$

be the set of all outputs for which we decide that the correct input was **c**, we can also think of a decision scheme as a collection $\{B_1,\ldots,B_m\}$ of disjoint subsets of the set of output strings. This is pictured in Figure 4.1.2.

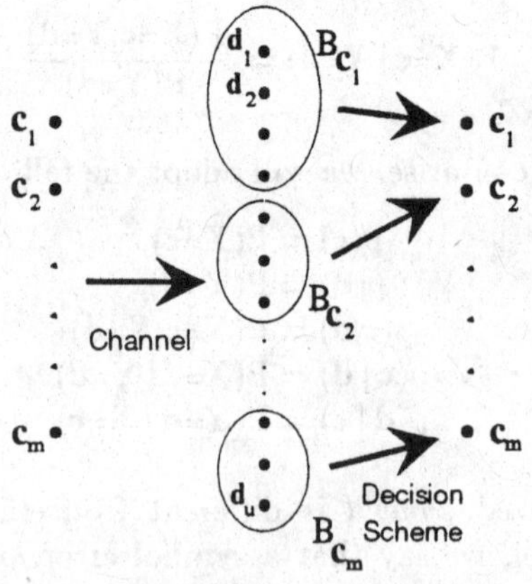

Figure 4.1.2

Let us clear up an issue related to terminology. With regard to a communications channel, such as the one shown in Figure 2 of the Introduction, the term *decoding* refers to the processes of attempting to correct any errors in transmission, as then reclaiming the original source message from the corrected codeword. However, for coding theory purposes, the term decoding refers simply to the process of error correction, which is implemented by means of a decision scheme.

PROBABILITIES ASSOCIATED WITH ERROR DETECTION

If errors occur in transmission, they will be *detected* if and only if the received word is *not* another codeword. Thus, if a codeword $\mathbf{c} \in C$ is sent, we have

$$P(\text{undetected error} \mid \mathbf{c} \text{ sent}) = \sum_{\substack{\mathbf{d} \in C \\ \mathbf{d} \neq \mathbf{c}}} p(\mathbf{d} \mid \mathbf{c})$$

Hence, the **probability of undetected error** is given by

$$p_{undet\ err} = \sum_{\mathbf{c} \in C} \sum_{\substack{\mathbf{d} \in C \\ \mathbf{d} \neq \mathbf{c}}} p(\mathbf{d} \mid \mathbf{c}) p(\mathbf{c})$$

PROBABILITIES ASSOCIATED WITH ERROR CORRECTION

For any decision scheme, if a codeword $\mathbf{c}$ is sent through the channel, then the probability of a decision error is given by

$$P(\text{error} \mid \mathbf{c}) = \sum_{\mathbf{d} \notin f^{-1}(\mathbf{c})} p(\mathbf{d} \mid \mathbf{c})$$

Hence, the (unconditional) probability p_e of a decision error is

(4.1.1) $$p_e = \sum_{\mathbf{c}} P(\text{error} \mid \mathbf{c}) p(\mathbf{c}) = \sum_{\mathbf{c}} \sum_{\mathbf{d} \notin f^{-1}(\mathbf{c})} p(\mathbf{d} \mid \mathbf{c}) p(\mathbf{c})$$

Notice that this probability depends on the input distribution $p(\mathbf{c})$, as well as on the decision scheme.

In order to see how we can determine a decision scheme that minimizes the probability of decision error, let us compute this error by conditioning on the output, rather than the input. If the output of the channel is $\mathbf{d}$, then the correct decision will be made if and only if $f(\mathbf{d})$ was the actual input. Hence,

$$P(\text{error} \mid \mathbf{d}) = 1 - p(f(\mathbf{d}) \mid \mathbf{d})$$

Averaging over all possible outputs, we have

$$p_e = \sum_{\mathbf{d}} P(\text{error} \mid \mathbf{d}) p(\mathbf{d}) = 1 - \sum_{\mathbf{d}} p(f(\mathbf{d}) \mid \mathbf{d}) p(\mathbf{d})$$

Now, this probability can be minimized by choosing a decision scheme that maximizes the sum on the far right. But since each term in this sum is nonnegative, and since the factors $p(\mathbf{d})$ do not depend on the decision scheme, we maximize this sum by choosing $f(\mathbf{d})$ so that $p(f(\mathbf{d}) \mid \mathbf{d})$ is as large as possible, for all $\mathbf{d}$. Let us summarize.

Definition For a given input distribution, any decision scheme f for which $f(\mathbf{d})$ has the property that

$$p(f(\mathbf{d}) \mid \mathbf{d}) = \max_{\mathbf{c}} p(\mathbf{c} \mid \mathbf{d})$$

for all output strings $\mathbf{d}$, is called an **ideal observer.** In words, an ideal observer is one for which $f(\mathbf{d})$ is a codeword most likely to have been sent, given that $\mathbf{d}$ was received. □

Notice that the probabilities $p(\mathbf{c} \mid \mathbf{d})$ are *backward* channel probabilities, and so they depend on the input distribution.

Theorem 4.1.1 For any given input distribution, an ideal observer decision scheme will minimize the probability p_e of a decision error, among all decision schemes. □

The ideal observer decision scheme has advantages as well as disadvantages. Perhaps its main disadvantage is that it depends on the input distribution. Thus, if the input distribution is changed, the ideal observer may also change. We can eliminate this dependency by considering the *maximum* probability of a decision error, defined by

$$p_e^{max} = \max_{\mathbf{c}} P(\text{error} \mid \mathbf{c})$$

This probability depends only on the decision scheme (and the channel, of course), and has the virtue that if $p_e^{max} < \epsilon$, then we have a *uniform* bound on the error probability, in the sense that the probability of error is small no matter what the input is, and so for *all* input distributions, we have

$$p_e = \sum_{\mathbf{c}} P(\text{error} \mid \mathbf{c}) p(\mathbf{c}) \le p_e^{max} < \epsilon$$

The difficulty here is that, unfortunately, we do not have a general method for finding decision schemes that make p_e^{max} small.

Another way to remove the dependency on the input distribution is to consider a uniform input distribution

$$p(\mathbf{c}) = \frac{1}{M}$$

where M is the size of the code. Then, according to (4.1.1), the probability of a decision error is given by

$$p_e^{av} = \frac{1}{M}\sum_{\mathbf{c}} P(\text{error} \mid \mathbf{c})$$

This is referred to as the **average probability of error.** (This terminology is a bit misleading, since (4.1.1) is also an average probability of error, the average being taken over the input distribution $p(\mathbf{c})$. Perhaps a better term would have been *uniform probability of error.*)

Since, in the uniform case

$$p(\mathbf{c} \mid \mathbf{d}) = \frac{p(\mathbf{d} \mid \mathbf{c})\, p(\mathbf{c})}{p(\mathbf{d})} = \frac{1}{mp(\mathbf{d})}\, p(\mathbf{d} \mid \mathbf{c})$$

we have

$$\max_{\mathbf{c}} p(\mathbf{c} \mid \mathbf{d}) = \max_{\mathbf{c}} \frac{1}{mp(\mathbf{d})}\, p(\mathbf{d} \mid \mathbf{c}) = \frac{1}{mp(\mathbf{d})} \max_{\mathbf{c}} p(\mathbf{d} \mid \mathbf{c})$$

In other words, for the uniform input distribution, maximizing the backward channel probabilities $p(\mathbf{c} \mid \mathbf{d})$ is equivalent to maximizing the (forward) channel probabilities $p(\mathbf{d} \mid \mathbf{c})$.

Definition Any decision scheme f for which $f(\mathbf{d})$ has the property that

$$p(\mathbf{d} \mid f(\mathbf{d})) = \max_{\mathbf{c}} p(\mathbf{d} \mid \mathbf{c})$$

for all output strings **d**, is called a **maximum likelihood decision scheme.** In words, $f(\mathbf{d})$ is an input string with the property that for no other input string would it be more likely that the output **d** was received. ▯

Theorem 4.1.2 For the uniform input distribution, an ideal observer (which minimizes the *average* probability of error) is the same as a maximum likelihood decision scheme. ▯

We will assume from now on that the input to a channel is uniform. Thus, maximum likelihood decision making is an ideal observer, and we have

(4.1.2) $$p_{undet\ err} = \sum_{c \in C} \sum_{\substack{d \in C \\ d \neq c}} \frac{1}{M} p(d \mid c)$$

and

(4.1.3) $$p_{decode\ err} = \sum_{c \in C} \sum_{d \notin f^{-1}(c)} \frac{1}{M} p(d \mid c)$$

THE NOISY CODING THEOREM

The *capacity* $\mathfrak{C}$ of a channel is the maximum amount of information that can be sent through the channel, the maximum being taken over all input distributions. More details on this can be found in Section 3.2. The Noisy Coding Theorem tells us that, as long as we are willing to settle for a rate that is *below* the capacity of the channel, we can compensate for the loss of information by the channel to any desired degree of accuracy.

Theorem 4.1.3 (The Noisy Coding Theorem) Consider a discrete memoryless channel with capacity $\mathfrak{C}$. For any positive number $R < \mathfrak{C}$, there exists a sequence C_n of r-ary codes, and corresponding decision schemes f_n, with the following properties.

1) C_n is an $(n, \lceil r^{nR} \rceil)$-code, that is, C_n has length n and rate at least R.
2) The maximum probability of error of f_n approaches 0 as $n \to \infty$, that is,
$$p_e^{max}(n) \to 0$$
□

The proof of the Noisy Coding Theorem given in Chapter 3, and indeed all known proofs, are non-constructive, that is, they do not tell us how to construct the codes promised in the theorem. Furthermore, despite the fact that the proof given in Chapter 3 says that there are a lot of codes that perform as promised in the theorem, no one has yet actually been able to construct such codes.

There are some additional practical problems to be considered in searching for desirable codes. In particular, a sequence of codes that fulfills the promise of the Noisy Coding Theorem would not be of much use unless the corresponding decision schemes were relatively easy to implement. Also, in order to bring the probability of error down to a desired level, the Noisy Coding Theorem implies that we may have to increase the length of the code to a perhaps unworkable size. Thus, simply finding codes that satisfy the Noisy Coding Theorem does not solve the problem of reliable communication.

As we are about to see, in an effort to find decision schemes that are relatively easy to implement, coding theorists have been led to search for codes that have considerable algebraic or geometric structure.

We have now set the stage for a discussion of coding theory. The first order of business will be to describe minimum distance decoding in a more intuitive fashion, by introducing a metric on the set C of codewords in a code.

EXERCISES

(*These exercises are taken from Section 3.3.*)

1. Consider the channel with channel matrix

$$\begin{bmatrix} \frac{1}{6} & \frac{1}{3} & \frac{1}{2} \\ \frac{1}{3} & \frac{1}{2} & \frac{1}{6} \\ \frac{1}{2} & \frac{1}{6} & \frac{1}{3} \end{bmatrix}$$

For the input distribution $p(x_1) = \frac{1}{2}$, $p(x_2) = p(x_3) = \frac{1}{4}$, find the best decision procedure and the associated average and maximum probabilities of error.

2. Let BSC be a binary symmetric channel, with crossover probability p. Consider the following three channels, each with input and output alphabets consisting of the set 2^3 of all binary words of length 3. Let $i_1i_2i_3$ be the input and $o_1o_2o_3$ be the output.
   (a) Output o_1 is determined by sending i_1 through BSC. Outputs o_2 and o_3 are chosen randomly, with each possibility being equally likely.
   (b) Output o_1 is determined by sending the *majority* bit from among i_1, i_2, and i_3 through BSC. Then $o_2 = o_3 = o_1$.
   (c) Output o_k is determined by sending i_k through BSC.
   For each channel, assume a uniform input distribution and a decision scheme that simply declares that the input is the same as the output. Find the probability of a decision error in each case.

3. The *Hamming codes*, which we will study in detail later in the book, have the following properties.
   (i) For $r = 1,2,\ldots$ the Hamming code $\mathcal{H}_2(r)$ has length $n = 2^r - 1$ and size

$$|\mathcal{H}_2(r)| = 2^{2^r - r - 1}$$

   (The codeword alphabet is binary.)
   (ii) The Hamming distance between any pair of codewords is at

least 3.

(iii) Any binary word of length n is either a codeword, or has distance 1 from exactly one codeword.

(a) Calculate the rate of the Hamming code $\mathcal{H}_2(r)$. What happens to the rate as $n \to \infty$?

(b) What is the ideal observer decision scheme for this code, assuming a uniform input distribution into a binary symmetric channel with crossover probability $p < \frac{1}{2}$?

(c) Compute $P(\text{error} \mid \mathbf{d})$, under the ideal observer decision scheme.

(d) Compute p_e, under the ideal observer decision scheme.

4.2 Minimum Distance Decoding

MINIMUM DISTANCE DECODING

In general, the problem of finding good codes is a very difficult one. However, by making certain assumptions about the channel, we can at least give the problem a highly intuitive flavor. Let us begin with a definition.

Definition Let **x** and **y** be strings of the same length, over the same alphabet. The **Hamming distance** $d(\mathbf{x},\mathbf{y})$ between **x** and **y** is the number of positions in which **x** and **y** differ. ▯

For instance, if $\mathbf{x} = 10112$ and $\mathbf{y} = 20110$ then $d(\mathbf{x},\mathbf{y}) = 2$, since these words differ in two positions (the first and fifth). The following result says that Hamming distance is a metric. We leave the proof as an exercise.

Theorem 4.2.1 Let $\mathcal{A}^n$ be the set of all words of length n over the alphabet $\mathcal{A}$. Then the Hamming distance function $d{:}\mathcal{A}^n \times \mathcal{A}^n \to \mathbb{N}$ satisfies the following properties. For all **x**, **y**, and **z** in $\mathcal{A}^n$,

1) **(positive definiteness)**

$$d(\mathbf{x},\mathbf{y}) \geq 0, \text{ and } d(\mathbf{x},\mathbf{y}) = 0 \text{ if and only if } \mathbf{x} = \mathbf{y}$$

2) **(symmetry)**

$$d(\mathbf{x},\mathbf{y}) = d(\mathbf{y},\mathbf{x})$$

3) **(triangle inequality)**

$$d(\mathbf{x},\mathbf{y}) \leq d(\mathbf{x},\mathbf{z}) + d(\mathbf{z},\mathbf{y})$$

Hence, the pair $(\mathcal{A}^n, d)$ is a metric space. ▯

Now let us consider the binary symmetric channel, with crossover probability $p < 1/2$, and channel matrix

$$\begin{bmatrix} 1-p & p \\ p & 1-p \end{bmatrix}$$

If the codeword **c** is sent through this channel and the word **d** is received, then the number of symbol errors made by the channel is equal to the Hamming distance $d(\mathbf{c},\mathbf{d})$. Hence, we have

$$p(\mathbf{d} \mid \mathbf{c}) = p^{d(\mathbf{c},\mathbf{d})}(1-p)^{n-d(\mathbf{c},\mathbf{d})}$$

Since $p < 1/2$, this probability is greatest when $d(\mathbf{c},\mathbf{d})$ is smallest. Thus, maximum likelihood decoding is equivalent to choosing a codeword **c** that is *closest* to the received word **d**. We refer to this as **minimum distance decoding.**

Example 4.2.1 The binary **repetition code** of length 3 is the code

$$C = \{000, 111\}$$

For this code, using minimum distance decoding, a decoding error will occur if and only if at least two channel errors are made, and so according to (4.1.3)

$$p_{decode\ err} = 3p^2(1-p) + p^3 = 3p^2 - 2p^3 \qquad \square$$

When more than one codeword has minimum distance from the received word, we will refer to this situation as a **tie.** In practice, the procedure for handling ties usually depends on the seriousness of making a decoding error. In some cases, we may wish to choose randomly from among the nearest codewords. In other cases, it might be more desirable simply to admit a decoding error, thereby reducing the chance of letting an undetected error slip by. The term **complete decoding** refers to the case where all received words are decoded, and the term **incomplete decoding** refers to the case where we prefer occasionally to admit an error, rather than always decode. (Recall that we used incomplete decoding in the proof of the Noisy Coding Theorem.)

There are many other channels for which maximum likelihood decoding takes the intuitive form of minimum distance decoding. For instance, the binary erasure channel, with channel matrix

$$\begin{bmatrix} 1-p & p & 0 \\ 0 & p & 1-p \end{bmatrix}$$

has this property, for $p < 1/2$.

Also, maximum likelihood decoding is equivalent to minimum distance decoding for any channel in which, if a symbol error does occur, then the received symbol is equally likely to be any of the other possible symbols. When the input alphabet agrees with the output alphabet, we have for such a channel

$$p(a_i \mid a_i) = 1 - p$$

and

$$p(a_i \mid a_j) = \frac{p}{q-1}, \quad i \neq j$$

where C is a q-ary code. (Again, we require that $p < 1/2$.) Hence, the channel matrix has size $q \times q$, and has the form

$$(4.2.1)\qquad \begin{bmatrix} 1-p & \frac{p}{q-1} & \cdots & \frac{p}{q-1} \\ \frac{p}{q-1} & 1-p & \cdots & \frac{p}{q-1} \\ \vdots & \vdots & \ddots & \vdots \\ \frac{p}{q-1} & \frac{p}{q-1} & \cdots & 1-p \end{bmatrix}$$

with $1-p$ on the main diagonal and $\frac{p}{q-1}$ everywhere else.

t-ERROR-CORRECTING AND t-ERROR-DETECTING CODES

In implementing minimum distance decoding, the following concepts are useful.

Definition The **minimum distance** of a code C is defined to be

$$d(C) = \min_{c,d \in C} d(c,d)$$

An **(n,M,d)-code** is a code with length n, size M, and minimum distance d. ▯

Definition A code C is **t-error-detecting** if whenever at most t, but at least one, error is made in a codeword, the resulting word is *not* a codeword. A code C is **exactly t-error-detecting** if it is t-error-detecting, but not (t+1)-error-detecting. ▯

We leave the proof of the next result as an exercise.

Theorem 4.2.2 A code C is exactly t-error-detecting if and only if $d(C) = t+1$. ▯

If t errors occur in the transmission of a codeword, we will say that an error of **size** t has occurred.

Definition A code C is **t-error-correcting** if minimum distance decoding is able to correct all errors of size t or less in any codeword, assuming that all ties are reported as errors. A code C is **exactly t-error-correcting** if it is t-error-correcting, but not (t+1)-error-correcting. (In other words, all errors of size t are corrected, but at least one error of size t+1 is decoded incorrectly.) ▯

It is not hard to make the connection between t-error-correction and the minimum distance of a code.

Theorem 4.2.3 A code C is exactly t-error-correcting if and only if $d(C) = 2t+1$ or $2t+2$.

Proof. Suppose first that $d(C) = 2t+1$ or $2t+2$. Suppose also that the received word **d** differs from the original codeword **c** in at most t positions, that is, $d(\mathbf{d},\mathbf{c}) \leq t$. Then **d** is closer to **c** than to any other codeword, for if $d(\mathbf{d},\mathbf{c}') \leq t$, we would have, by the triangle inequality

$$d(\mathbf{c},\mathbf{c}') \leq d(\mathbf{c},\mathbf{d}) + d(\mathbf{d},\mathbf{c}') \leq t + t = 2t < d(C)$$

which is a contradiction. Therefore, minimum distance decoding will correct t or fewer errors.

Furthermore, if $d(C) = 2t+1$, then there are codewords **c** and **c'** for which $d(\mathbf{c},\mathbf{c}') = 2t+1$. In other words, **c** and **c'** differ in exactly $2t+1$ positions. Suppose that the codeword **c** is sent and that the received word **d** has exactly $t+1$ errors, all of which are located in the aforementioned $2t+1$ positions, and that **d** now agrees with **c'** in those $t+1$ error positions. Then $d(\mathbf{d},\mathbf{c}) = t+1$ but

$$d(\mathbf{d},\mathbf{c}') = 2t+1 - (t+1) = t$$

and so maximum likelihood decoding would decode **d** as **c'**, which is incorrect. Hence, C is not $(t+1)$-error-correcting. We leave the case $d(C) = 2t+2$ as an exercise.

For the converse, if C is exactly t-error-correcting, we could not have $d(\mathbf{c},\mathbf{c}') \leq 2t$, for then it would be possible for the received word **d** to have precisely t errors, placing it at least as close to **c'** as to the codeword **c** that was originally sent. Hence, $d(C) \geq 2t+1$. On the other hand, if $d(C) \geq 2t+3 = 2(t+1)+1$, then by our earlier argument, the code C would be $(t+1)$-error-correcting. Hence, $d(C) = 2t+1$ or $2t+2$. ∎

Corollary 4.2.4 $d(C) = d$ if and only if C is exactly $\lfloor (d-1)/2 \rfloor$-error-correcting. □

USING A CODE FOR SIMULTANEOUS ERROR CORRECTION AND DETECTION

In a given situation, the error *detecting* quality of a code with *even* minimum distance depends on whether or not the code is used for error detection only or for both error detection and error correction.

To see this, suppose that C is an (n,M,d)-code C, where $d = 2t+2$. As a strictly error-detecting code, C can detect up to $d-1 = 2t+1$ errors.

On the other hand, suppose we wish to use C for simultaneous error correction and error detection. Since C is exactly t-error-correcting, if we wish to maximize the error-correcting capabilities of

C, we must "allow" it to correct any t errors. That is, if a word **x** is received and if there is a codeword **c** for which $d(\mathbf{c},\mathbf{x}) \leq t$, then we *assume* that at most t errors have occurred, and correct **x** to **c**. If no such codeword **c** exists, then we may declare that some errors have occurred.

Now, if exactly $t+1$ errors have occurred in transmitting **c**, then the received word **x** cannot be within t of any codeword **d**, for if it were, then

$$d(\mathbf{c},\mathbf{d}) \leq d(\mathbf{c},\mathbf{x}) + d(\mathbf{x},\mathbf{d}) \leq t+1+t = 2t+1 < d$$

Hence, our strategy requires us to *detect* that errors have occurred, and so C can detect $t+1$ errors.

However, if $t+2$ errors have occurred, then since $d = 2t+2$, in at least one case the received word **x** will have distance t from the *wrong* codeword, and we will "correct" the received word incorrectly, rather than detect the errors. Hence, C does not always detect $t+2$ errors.

Theorem 4.2.5 If an (n,M,d)-code with even minimum distance $d = 2t+2$ is used for error detection *only*, then it is exactly $(2t+1)$-error-detecting. On the other hand, if C is used for maximum error correction, as well as error detection, then C is exactly t-error-correcting and can simultaneously detect $t+1$ errors, but it cannot always detect more than $t+1$ errors. ▯

We leave it to the reader to show that, in case the minimum distance is odd, when the quality of error correction is maximized, error detection does not take place.

Example 4.2.2 In January 1979, the Mariner 9 spacecraft took black-and-white photographs of Mars. A grid of size 600×600 was placed over each photograph, and each of the resulting 360,000 grid components was assigned one of 64 shades of gray. Thus, the source information consisted of 64 different source symbols. Each source symbol was then encoded using a binary (32,64,16)-code, known as a *Reed-Muller code.* (We will study Reed-Muller codes in detail later in the book.) Since the minimum distance of this code is 16, it is exactly 7-error-correcting. Furthermore, in addition to correcting up to 7 errors, this code can simultaneously detect 8 errors. The rate of the Reed-Muller code is

$$R = \frac{\log_2 64}{32} = \frac{6}{32} = \frac{3}{16}$$ ▯

Example 4.2.3 In the period from 1979 through 1981, the Voyager spacecrafts took *color* photographs of Jupiter and Saturn. This required a source alphabet of size 4096 to represent various shades of color. The source information was then encoded using a binary (24,4096,8)-code, known as a *Golay code.* (We will also study Golay codes in detail.) This code is exactly 3-error-correcting and can simultaneously detect 4 errors. It has rate

$$R = \frac{\log_2 4096}{24} = \frac{12}{24} = \frac{1}{2}$$ □

Example 4.2.4 The q-ary **repetition code** of length n is the code

$$C = \{00\cdots0,\ 11\cdots1, \ldots, (q-1)(q-1)\cdots(q-1)\}$$

This code has minimum distance n, and so it is exactly $\lfloor (n-1)/2 \rfloor$-error-correcting. On the other hand, for strict error detection, it is (n−1)-error-detecting. □

THE RELATIONSHIP BETWEEN MINIMUM DISTANCE AND THE PROBABILITY OF ERROR

The code $C = \{000,111,333\}$, over the alphabet $\mathcal{A} = \{0,1,2,3\}$, can easily be made more efficient by adding the codeword 222, which would increase the rate of the code, but not change its minimum distance. This leads us to make the following definition.

Definition An (n,M,d)-code is said to be **maximal** if it is not contained in any (n,M+1,d)-code. □

An (n,M,d)-code C code is maximal if and only if, for all strings **x**, there is a codeword **c** with the property that $d(\mathbf{x},\mathbf{c}) < d$. In this case, if a codeword **c** is transmitted, and if d or more errors are made, so that the received word **x** has the property that $d(\mathbf{x},\mathbf{c}) \geq d$, then **x** will be closer to a *different* codeword, and so a decoding error will be made. This establishes the following lower bound on the probability of a decoding error, for the binary symmetric channel

$$p_{decode\ err} \geq \sum_{k=d}^{n} \binom{n}{k} p^k (1-p)^{n-k}$$

On the other hand, an (n,M,d)-code C is exactly t-error-correcting, where $t = \lfloor (d-1)/2 \rfloor$. Hence, the probability of correct decoding satisfies

$$p_{correct} \geq \sum_{k=0}^{t} \binom{n}{k} p^k (1-p)^{n-k}$$

and so

$$p_{decode\ err} = 1 - p_{correct} \le 1 - \sum_{k=0}^{t} \binom{n}{k} p^k (1-p)^{n-k}$$

For convenience, let us set

$$B_p(n,m) = \sum_{k=0}^{m} \binom{n}{k} p^k (1-p)^{n-k}$$

Then we have the following.

Theorem 4.2.6 For the binary symmetric channel, the probability of a decoding error for a maximal (n,M,d)-code satisfies

$$1 - B_p(n,d-1) \le p_{decode\ err} \le 1 - B_p(n,\lfloor \tfrac{d-1}{2} \rfloor) \qquad \square$$

Let us consider the situation as n gets large, which may be necessary to obtain a good code. If the probability of a channel error is p, then the expected number of errors in a codeword of length n is np. Therefore, if the minimum distance d does not keep pace with np, we cannot hope to correct many errors. We can get more specific information from Theorem 4.2.6. For an (n,M,d)-code C, let

$$\delta = \delta(C) = \frac{d}{n}$$

Corollary 4.2.7 Assume a binary symmetric channel, with crossover probability $p < 1$. Suppose that C_n is a sequence of (n,M_n,d_n)-codes, with $\delta_n = d_n/n$. If for some constant r, we have $\delta_n \le r < p$ for all sufficiently large n, then the probability of a decoding error for C_n tends to 1 as $n \to \infty$.

Proof. By adding codewords to each C_n if possible without changing d_n, we may assume that each C_n is maximal. Let us set $\lambda = (d_n-1)/n$. The condition $\delta_n \le r < p$ for n sufficiently large is equivalent to $\lambda \le s < p$ for n sufficiently large. According to Theorem A.3.8 (in the appendix), if $0 \le \lambda \le s < p < 1$, then

$$\begin{aligned} B_p(n,d_n-1) = B_p(n,\lambda n) &\le p^{\lambda n}(1-p)^{(1-\lambda)n}\, 2^{nH(\lambda)} \\ &\le 2^{[\lambda n \log p + (1-\lambda)n \log(1-p)] + nH(\lambda)} \\ &= 2^{n[h(\lambda)+H(\lambda)]} \end{aligned}$$

where $h(\lambda) = \lambda \log p + (1-\lambda)\log(1-p)$. Now, $f(\lambda) = h(\lambda)+H(\lambda)$ is a continuous function on $[0,s]$, and according to Lemma 1.2.2, if $\lambda \ne p$, then $f(\lambda) < 0$. Hence, on $[0,s]$, $f(\lambda)$ is bounded away from 0, that is, there exists a constant k such that $f(\lambda) \le k < 0$ for all $0 \le \lambda \le s$.

Therefore,

$$B_p(n,d_n-1) \le 2^{nk} \to 0 \quad \text{as } n \to \infty$$

An application of Theorem 4.2.6 completes the proof. ▮

Corollary 4.2.7 tells us, in particular, that we cannot expect good error correction if the ratio d/n of minimum distance to code length approaches 0 as n gets large, as it does (unfortunately) for many of the known families of codes.

THE PACKING AND COVERING RADII OF A CODE

We can get further geometric insight into the error correcting capabilities of codes by introducing some additional concepts.

Definition Let $\mathbf{x}$ be a word in $\mathcal{A}^n$, where $|\mathcal{A}| = q$, and let r be any nonnegative real number. The **sphere** of radius r about $\mathbf{x}$ is the set

$$S_q(\mathbf{x},r) = \{\mathbf{y} \in \mathcal{A}^n \mid d(\mathbf{x},\mathbf{y}) \le r\}$$

The **volume** $V_q(n,r)$ of the sphere $S_q(\mathbf{x},r)$ is the number of elements in $S_q(\mathbf{x},r)$. This volume is independent of the center and is given by

$$V_q(n,r) = \sum_{k=0}^{r} \binom{n}{k} (q-1)^k \qquad \square$$

For the binary alphabet $\mathcal{A} = \{0,1\}$, the set $\mathcal{A}^3$ is shown in Figure 4.2.1. The words that lie in the sphere of radius 1 about the word 111 are shown as solid dots in this figure. (Thus, spheres in $\mathcal{A}^n$ do not look very round.)

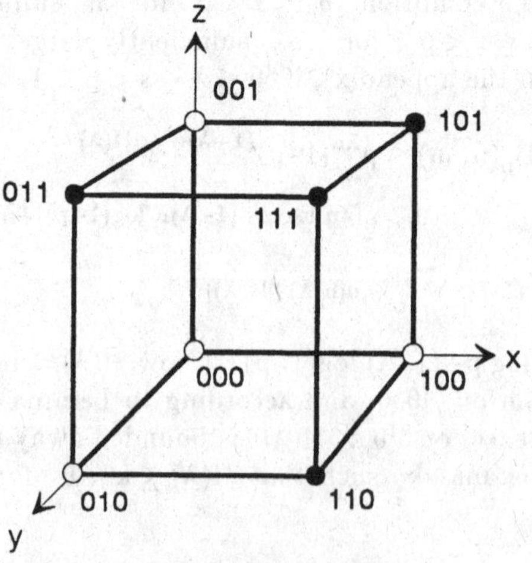

Figure 4.2.1

Unfortunately, for $n > 3$, it is not possible to draw realistic pictures like Figure 4.2.1. However, we can get some representation of the situation for larger values of n by looking at Figure 4.2.2. (Of course, this figure is not metrically accurate.)

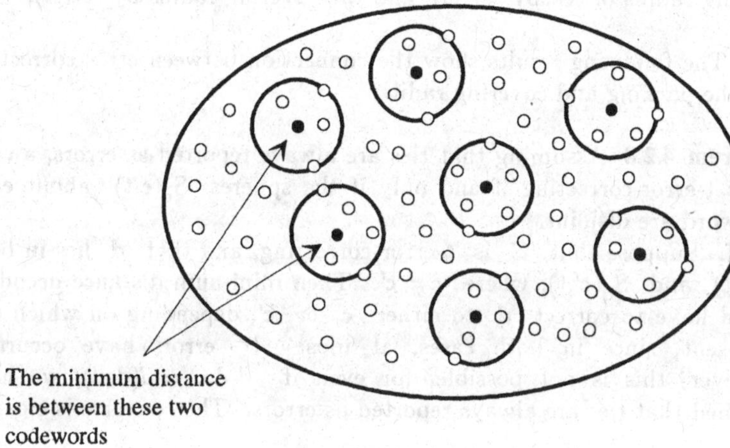
The minimum distance
is between these two
codewords

Figure 4.2.2

The solid dots in this figure represent codewords, and the open dots represent all other words of length n. The radius of the spheres shown in Figure 4.2.2 is determined by the minimum distance of the code, by simply increasing the radius of the spheres until just *before* two spheres become "tangent," which will happen when $d(C) = 2t+2$ (see left side of Figure 4.2.3), or just before two spheres "overlap," which will happen when $d(C) = 2t+1$ (see right side of Figure 4.2.3). In either case, the radius of each sphere is t, and C is *exactly* t-error-correcting.

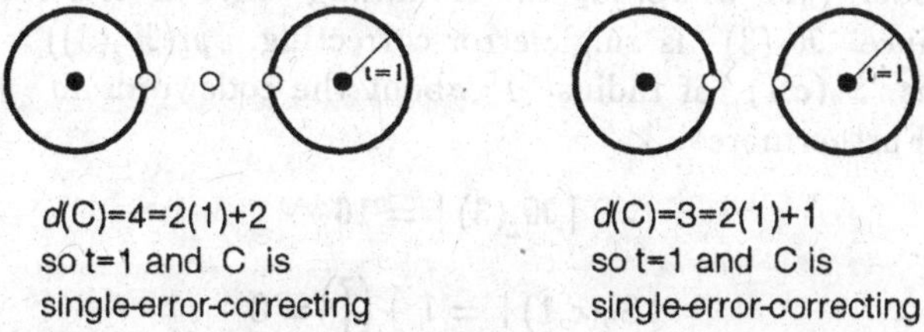
t=1
t=1
d(C)=4=2(1)+2
so t=1 and C is
single-error-correcting
d(C)=3=2(1)+1
so t=1 and C is
single-error-correcting

Figure 4.2.3

Definition Let $C \subset \mathcal{A}^n$ be a code. The **packing radius** of C is the *largest* integer r for which the spheres $S_q(\mathbf{c},r)$ about each codeword **c** are disjoint. The **covering radius** of C is the *smallest* integer s for which the spheres $S_q(\mathbf{c},s)$ about each codeword **c** *cover* $\mathcal{A}^n$, that is, for which the union of the spheres $S_q(\mathbf{c},s)$ is $\mathcal{A}^n$. We will denote the packing radius of C by *pr*(C) and the covering radius by *cr*(C). □

The following results show the connection between error correction and the packing and covering radii.

Theorem 4.2.8 Assuming that ties are always reported as errors, a code C is t-error-correcting if and only if the spheres $S_q(\mathbf{c},t)$ about each codeword are disjoint.
Proof. Suppose that C is t-error-correcting, and that **d** lies in both $S_q(\mathbf{c},t)$ and $S_q(\mathbf{c}',t)$, where $\mathbf{c} \neq \mathbf{c}'$. Then minimum distance decoding would have to correct **d** to either **c** or **c**′, depending on which one was sent, since in both cases, at most t errors have occurred. However, this is not possible, for even if $d(\mathbf{d},\mathbf{c}) = d(\mathbf{d},\mathbf{c}')$, we have assumed that ties are always reported as errors. The converse is clear. ∎

Corollary 4.2.9 Assuming that ties are always reported as errors, a code C is exactly t-error-correcting if and only if $pr(C) = t$. □

PERFECT AND QUASI-PERFECT CODES

The following concept plays a major role in coding theory.

Definition A code C is said to be **perfect** if $pr(C) = cr(C)$. In words, a code $C \subset \mathcal{A}^n$ is perfect if their exists a number r for which the spheres $S_q(\mathbf{c},r)$ about each codeword **c** are disjoint and cover $\mathcal{A}^n$. (See Figure 4.2.4.) □

Example 4.2.5 One of the so-called *Hamming codes* $\mathcal{H}_2(3)$ is a binary (7,16,3)-code. (We will study the Hamming codes in detail later in the book.) Since $\mathcal{H}_2(3)$ is single-error-correcting, $pr(\mathcal{H}_2(3)) = 1$, and so the spheres $S_2(\mathbf{c},1)$ of radius 1 about the codewords in $\mathcal{H}_2(3)$ are disjoint. Furthermore

$$|\mathcal{H}_2(3)| = 16$$

$$|S_2(\mathbf{c},1)| = 1 + \binom{7}{1} = 8$$

and

$$|\mathcal{A}^7| = 2^7 = 128$$

and since $16 \cdot 8 = 128$, we deduce that the spheres $S_2(\mathbf{c},1)$ cover $\mathcal{A}^7$. Hence, $\mathcal{H}_2(3)$ is perfect. □

The size of a perfect code is uniquely determined by the length and the minimum distance. The following result is known as the **sphere-packing condition.**

Theorem 4.2.10 (The Sphere-Packing Condition) Let C be a q-ary (n,M,d)-code. Then C is perfect if and only if $d = 2t+1$ is odd and

$$M \cdot V_q(n,t) = q^n$$

that is,

$$M = q^n \Big/ \sum_{k=0}^{t} \binom{n}{k}(q-1)^k$$

Proof. We leave it as an exercise to show that a perfect code must have odd minimum distance. If C is perfect, then the packing and covering radii of C are both equal to t, and so the set of M spheres of radius t about the codewords in C form a partition of the q^n strings of length n. Hence, the sphere-packing condition holds. On the other hand, if the sphere-packing condition holds for some (n,M,2t+1)-code C, then since the spheres of radius t about each codeword are disjoint, and the sphere-packing condition implies that they cover $\mathcal{A}^n$, we have $pr(C) = cr(C) = t$. ∎

It is important to emphasize that the existence of numbers n, M, and t for which the sphere-packing condition holds does *not* imply that a code with these parameters must exist, and we shall see an example of this in the sequel.

The problem of determining all perfect codes has not yet been solved. However, as we will see in the next section, a great deal is known about perfect codes over alphabets whose size is a power of a prime.

We leave it to the reader to show that the sphere-packing condition is satisfied by the following families of parameters

1) $(n,M,d) = (n,q^n,1)$,

2) $(n,M,d) = (n,1,2n+1)$,

3) $(n,M,d) = (2m+1,2,2m+1)$,

4) $(n,M,d) = (\frac{q^r-1}{q-1},q^{n-r},3)$, $r \geq 2$.

The first solution corresponds to the entire space V(n,q). The

second solution corresponds to codes with a single codeword (whose minimum distance is undefined), the third solution corresponds to the binary repetition codes of odd length, and the fourth solution corresponds to the famous q-ary Hamming codes, which we will study later in the book.

A computer search was conducted by van Lint in 1967, which showed that the only solutions to the sphere-packing condition for

$$n \le 1000,\ d \le 2001\ (t \le 1000),\ q \le 100$$

besides those given by 1)-3) above are

5) $(n,M,d) = (23,2^{11},7)$,

6) $(n,M,d) = (90,2^{78},5)$,

7) $(n,M,d) = (11,3^{6},5)$.

As we will see when we discuss designs and codes, there is no code with parameters corresponding to solution 6. Solutions 5 and 7 correspond to the famous Golay codes, which we will also study later in the book.

Thus, there are not very many perfect codes. The following definition describes the "next best thing."

Definition A code C is said to be **quasi-perfect** if $cr(C) = pr(C)+1$. In words, a code $C \subset \mathcal{A}^n$ is quasi-perfect if their exists a number r for which the spheres $S_q(\mathbf{c},r)$ of radius r are disjoint, and the spheres $S_q(\mathbf{c},r+1)$ of radius $r+1$ cover $\mathcal{A}^n$. (See Figure 4.2.4.) □

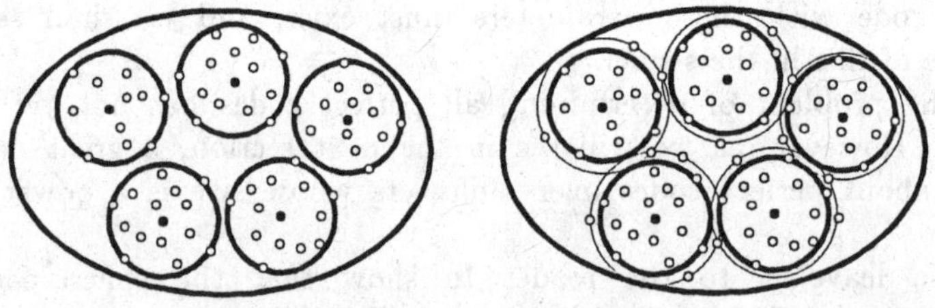

This code is perfect   This code is quasi-perfect

Figure 4.2.4

EXERCISES

1. Calculate the following Hamming distances.
   (a) $d(01001,10110)$ (b) $d(12345,54321)$
2. Calculate the probability of a decoding error for the binary repetition code of length 4. How many errors will this code

detect? How many will it correct?

3. Calculate the probability of a decoding error for the binary repetition code of length 5. How many errors will this code detect? How many will it correct?
4. Consider the binary code $C = \{11100, 01001, 10010, 00111\}$.
   (a) Compute the minimum distance of C.
   (b) Decode the words 10000, 01100, and 00100.
   (c) Compute the rate of this code.
5. Construct a binary (8,4,5)-code.
6. Does a binary (7,3,5)-code exist? Justify your answer.
7. Prove Theorem 4.2.1.
8. Prove Theorem 4.2.2.
9. Finish the proof of Theorem 4.2.3. (Do the case $d(C) = 2t+2$.)
10. Prove Corollary 4.2.4.
11. Prove Corollary 4.2.9.
12. With reference to the discussion of simultaneous error correction and detection, show that, in case the minimum distance is odd, when the quality of error correction is maximized, error detection does not take place.
13. Consider the binary code
   $$C = \{00000000, 00001111, 00110011, 00111100\}$$
   (a) Compute the distance between all codewords in C.
   (b) We define the *complement* of a binary word to be the word obtained by changing all 0's to 1's and all 1's to 0's. Describe the distance features of the code obtained by taking all codewords in C and the complements of these codewords.
   (c) Generalize the results of part (b).
14. Define the *maximum distance* of a code. Can you state and prove any results about maximum distance that are analogous to any results in this section?
15. Prove that, for the binary erasure channel, maximum likelihood decoding is equivalent to minimum distance decoding.
16. Prove that, for the channel whose matrix is given by (4.2.1), maximum likelihood decoding is equivalent to minimum distance decoding.
17. Find a *symmetric* channel for which maximum likelihood decoding is *not* equivalent to minimum distance decoding.
18. Prove that
   $$V_q(n,r) = \sum_{k=0}^{r} \binom{n}{k} (q-1)^k$$
19. Show that a perfect code must have odd minimum distance.
20. Show that the sphere-packing condition is satisfied for the following families of parameters

a) $(n,M,d) = (n,1,2n+1)$ b) $(n,M,d) = (2m+1,2,2m+1)$

c) $(n,M,d) = (\frac{q^r-1}{q-1},q^{n-r},3)$ $r \geq 2$

21. Prove that a binary $(n,M,2t+1)$-code exists if and only if a binary $(n+1,M,2t+2)$-code exists. *Hint.* Think about parity check bits.

22. Let $\mathbf{c},\mathbf{d} \in \mathcal{A}^n$, and consider the sets

$$S = \{\mathbf{x} \in \mathcal{A}^n \mid d(\mathbf{x},\mathbf{c}) < d(\mathbf{x},\mathbf{d})\}, \quad T = \{\mathbf{x} \in \mathcal{A}^n \mid d(\mathbf{x},\mathbf{d}) < d(\mathbf{x},\mathbf{c})\}$$

Prove that $|S| = |T|$.

23. Let C be a perfect, binary $(n,M,7)$-code. Use the sphere-packing condition to show that $n = 7$ or $n = 23$. *Hint.* Show that the sphere packing condition is

$$(n+1)[(n+1)^2 - 3(n+1)+8] = 3 \cdot 2^k$$

for some $k \geq 0$. If $2^i \mid (n+1)$ show that $i \leq 3$.

4.3 Families of Codes

SYSTEMATIC CODES

Since a q-ary (n,M)-code has M codewords, it can be used to encode up to M source messages. However, certain types of codes are more easily used for encoding than others.

Definition A q-ary (n,q^k)-code is called **systematic** if there are k positions $i_1,i_2,\ldots,i_k$ with the property that, by restricting the codewords to these positions, we get all of the q^k possible q-ary words of length k. The set $\{i_1,i_2,\ldots,i_k\}$ is called an **information set**, and the codeword symbols in these positions are called **information symbols.** ▯

If a source can be represented as the set of all q-ary words of length k, then a q-ary systematic code of size q^k can be used to encode each source word, by embedding it without change, into a codeword, as the following example shows.

Example 4.3.1 The binary code

$$C = \{0000,\ 0110, 1001, 1010\}$$

is systematic on the first and third positions. Hence, if the source is $\{00,01,10,11\}$, we can encode as follows

$$00 \to \underline{0}0\underline{0}0, \quad 01 \to \underline{0}1\underline{1}0, \quad 10 \to \underline{1}0\underline{0}1, \quad 11 \to \underline{1}0\underline{1}0 \qquad ▯$$

The type of encoding shown in the previous example is called **systematic encoding.** Clearly, systematic encoding makes the reverse process extremely easy, since we can simply read the source word directly from the codeword.

Example 4.3.2 The binary code $C = \{000,100,010,001\}$ is not systematic. ▯

FINITE FIELDS

Up to now, we have not said much about the code alphabet $\mathcal{A}$. However, if we expect to endow our codes with any meaningful structure, we must generally assume that $\mathcal{A}$ has some structure, which is usually taken to be that of a *finite field.*

We will have a great deal to say about finite fields in a later chapter. For now, let us simply note the following key fact, which will be proved at a later time.

Theorem 4.3.1 If p is a prime number and n is a positive integer, then there is (up to isomorphism) exactly one field of size $q = p^n$, which is denoted by F_q or $GF(q)$. Furthermore, all finite fields have size p^n, for some prime number p and positive integer n. ▯

The set $(F_q)^n$ of all n-tuples whose components belong to F_q is a vector space over F_q of dimension n. We will denote $(F_q)^n$ by $V(n,q)$, and usually write the vector $(x_1,x_2,\ldots,x_n)$ in the form $x_1x_2\cdots x_n$.

EQUIVALENCE OF CODES

As far as error correcting properties are concerned, there is no difference between the codes $C_1 = \{000,011\}$ and $C_2 = \{000,101\}$, and so, in this sense, they should be considered as equivalent.

There are various definitions of equivalence of codes in the literature. We will adopt the following definitions.

Definition Two q-ary (n,M)-codes C_1 and C_2 are **equivalent** if there exists a permutation σ of the n coordinate positions and permutations $\pi_1,\pi_2,\ldots,\pi_n$ of the code alphabet for which

$$c_1c_2\cdots c_n \in C_1 \text{ if and only if } \pi_1(c_{\sigma(1)})\pi_2(c_{\sigma(2)})\cdots\pi_n(c_{\sigma(n)}) \in C_2$$

In words, two codes are equivalent if one can be turned into the other by permuting the coordinate positions of each codeword (via σ) and by permuting the code symbols in each position of each codeword (via $\pi_1,\ldots,\pi_n$). Of course, σ or any π_i may be the identity permutation. ▯

The following definition of equivalence is useful for special types of codes. We begin with a preliminary definition.

Definition Let σ be a permutation of size n. For $i = 1,\ldots,n$, let $\pi_i{:}F_q \to F_q$ be multiplication by a nonzero scalar α_i in F_q, that is,

$$\pi_i s = \alpha_i s$$

Then the map $\mu{:}V(n,q) \to V(n,q)$ defined by

$$\mu(c_1\cdots c_n) = \pi_1(c_{\sigma(1)})\pi_2(c_{\sigma(2)})\cdots\pi_n(c_{\sigma(n)})$$

is called a **monomial transformation** of degree n. In words, a monomial transformation acting on n coordinates is a permutation of those coordinates, followed by multiplication of each coordinate by a nonzero scalar. ▯

Definition Two (n,M)-codes C_1 and C_2 over F_q are **scalar multiple equivalent** if there is a monomial transformation μ of degree n for which $\mu(C_1) = C_2$, where $\mu(C_1) = \{\mu c \mid c \in C_1\}$. Thus, C_1 and C_2 are scalar multiple equivalent if they are equivalent in the sense of the previous definition, where each permutation π_i is multiplication by a nonzero scalar. □

Of course, if two codes are scalar multiple equivalent, they are also equivalent. We leave proof of the next two results as exercises.

Lemma 4.3.2 If $0 \in \mathcal{A}$, then any code over $\mathcal{A}$ is equivalent (but not necessarily scalar multiple equivalent) to a code that contains the zero codeword $\mathbf{0} = 0 \cdots 0$. □

Theorem 4.3.3 If C_1 and C_2 are equivalent, then $d(C_1) = d(C_2)$. Also, under minimum distance decoding, and assuming a channel of the form (4.2.1) [which includes the binary symmetric channel], the probability of a decoding error is the same for equivalent codes. □

TYPES OF CODES

We wish now to discuss some of the more important types of error-correcting codes. Each of these codes will be studied in detail in later chapters.

Linear Codes

One of the great advantages of using a finite field F_q as code alphabet is that we can perform vector space operations on the codewords. However, unless the code is a *subspace* of V(n,q), we cannot be certain that the sum of two codewords (or a scalar multiple of a codeword) is another codeword. This prompts us to define perhaps the most important type of codes.

Definition A code $L \subset V(n,q)$ is a **linear code** if it is a subspace of V(n,q). If L has dimension k over V(n,q), we say that L is an **[n,k]-code** (note the square brackets), and if L has minimum distance d, we say that L is an **[n,k,d]-code.** □

(In the literature, one often finds (n,k) in place of [n,k], which can be a bit confusing.)

Note that all linear codes contain the **zero codeword**, denoted by $\mathbf{0} = 00 \cdots 0$. Note also that the size and rate of a q-ary [n,k]-code are

$$M = q^k, \quad R = \frac{k}{n}$$

To explore some of the advantages of linear codes, we need a simple definition and a few basic facts.

Definition The **weight** $w(\mathbf{x})$ of a word $\mathbf{x} \in V(n,q)$ is the number of *nonzero* positions in $\mathbf{x}$. The **(minimum) weight** $w(C)$ of a code C is the minimum weight of all *nonzero* codewords in C. ▯

Definition If $\mathbf{x} = x_1x_2\cdots x_n$ and $\mathbf{y} = y_1y_2\cdots y_n$ are *binary* words, we define the **intersection** of $\mathbf{x}$ and $\mathbf{y}$ by

$$\mathbf{x} \cap \mathbf{y} = (x_1y_1, x_2y_2, \ldots, x_ny_n)$$

Thus, $\mathbf{x} \cap \mathbf{y}$ has a 1 in the i-th position if and only if both $\mathbf{x}$ and $\mathbf{y}$ have a 1 in the i-th position. ▯

We leave proof of the following as an exercise.

Lemma 4.3.4

1) For all $\mathbf{x},\mathbf{y} \in V(n,q)$,

$$d(\mathbf{x},\mathbf{y}) = w(\mathbf{x} - \mathbf{y})$$

2) For all $\mathbf{x},\mathbf{y} \in V(n,2)$ [note $q = 2$],

$$d(\mathbf{x},\mathbf{y}) = w(\mathbf{x}) + w(\mathbf{y}) - 2w(\mathbf{x} \cap \mathbf{y})$$ ▯

In order to find the minimum distance of an arbitrary (n,M)-code, it is, in general, necessary to compute all $\binom{M}{2}$ Hamming distances. However, as the next result shows, for a *linear* code, we need only compute $M - 1$ weights!

Theorem 4.3.5 For a linear code L, we have $d(L) = w(L)$.
Proof. Since $\mathbf{c} - \mathbf{d}$ runs through all codewords in L, as $\mathbf{c}$ and $\mathbf{d}$ run through all codewords, we have

$$d(L) = \min_{\mathbf{c} \neq \mathbf{d} \in L} \{d(\mathbf{c},\mathbf{d})\} = \min_{\mathbf{c} \neq \mathbf{d} \in L} \{w(\mathbf{c} - \mathbf{d})\} = \min_{\mathbf{0} \neq \mathbf{c} \in L} \{w(\mathbf{c})\} = w(L)$$ ∎

Cyclic Codes

Many important codes have considerably more structure than that of a vector space. Consider a linear code $L \subset V(n,q)$. We may associate to each codeword $\mathbf{c} = c_0c_1\cdots c_{n-1}$ a polynomial in $F_q[x]$ as follows

$$\phi: c_0c_1\cdots c_{n-1} \to c_0 + c_1x + \cdots + c_{n-1}x^{n-1} \tag{4.3.1}$$

In fact, the map ϕ is a vector space isomorphism from L onto the subspace $\phi(L)$ of $F_q[x]$. As is customary, we can ignore this map and simply think of the codewords in L as polynomials, or vice-versa.

The advantage of this approach is that $F_q[x]$ is more than just a vector space – it is also an algebra. In other words, by thinking of codewords as polynomials, we can multiply two codewords. However, the product of two codewords may not be another codeword, and so we must make some refinements in our thinking.

Let $p(x)$ be a polynomial of degree n, and let $\langle p(x)\rangle$ denote the ideal generated by $p(x)$. The quotient algebra is the set of all polynomials $f(x)$ of degree less than n

$$R = \frac{F_q[x]}{\langle p(x)\rangle} = \{f(x) \in F_q(x) \mid \deg(f(x)) < n\}$$

with the operations of ordinary addition and scalar multiplication, and multiplication modulo $p(x)$. Thus, in R, the product of two polynomials of degree less than n is another polynomial of degree less than n.

The map ϕ defined in (4.3.1) can be thought of as a vector space isomorphism from L onto the subspace $\phi(L)$ of R. In other words, we can think of the codewords in L as polynomials in the algebra R, where multiplication is more suited to our present needs.

In fact, we now have two new sets of structures that can be used as codes, namely *subalgebras* and *ideals*. If the code L is an *ideal* of R, then not only is it an algebra, but it has the property that the product of any codeword and *any* other polynomial is a codeword.

Let us specialize one step further, by taking $p(x) = x^n - 1$, and letting R_n be the quotient algebra

$$R_n = \frac{F_q[x]}{\langle x^n - 1\rangle}$$

If L is an ideal in R_n, then L is closed under multiplication by *any* polynomial in R_n. But this is equivalent (for a vector subspace) to being closed under multiplication by x, and since

$$\begin{aligned} x \cdot (c_0 + c_1 x + \cdots + c_{n-1}x^{n-1}) &= (c_0 x + c_1 x^2 + \cdots + c_{n-1}x^n) \bmod (x^n - 1) \\ &= c_{n-1} + c_0 x + c_1 x^2 + \cdots + c_{n-2}x^{n-1} \end{aligned}$$

we see that if

$$c_0 + c_1 x + \cdots + c_{n-1}x^{n-1} \in L$$

then

$$c_{n-1} + c_0 x + c_1 x^2 + \cdots + c_{n-2}x^{n-1} \in L$$

In the symbolism of strings,

$$c_0c_1 \cdots c_{n-1} \in L \quad \text{implies} \quad c_{n-1}c_0c_1 \cdots c_{n-2} \in L$$

We can now define the most important subclass of linear codes.

Definition A linear code $L \subset V(n,q)$ is **cyclic** if

$$c_0c_1 \cdots c_{n-1} \in L \quad \text{implies} \quad c_{n-1}c_0c_1 \cdots c_{n-2} \in L$$

Put another way, when codewords are thought of as polynomials using (4.3.1), a code L is **cyclic** if it is an ideal of $R_n = F_q[x]/\langle x^n - 1\rangle$. $\square$

We will devote two entire chapters to cyclic codes. For the moment, let us make a few brief remarks, which will be discussed thoroughly at the appropriate time.

As it happens, the algebra R_n is a *principal ideal domain*, which means that each ideal (cyclic code) C is generated by a single polynomial $g(x)$, called the *generator polynomial* of C, that is the unique monic polynomial of smallest degree in C. Thus,

$$C = \{f(x)g(x) \mid f(x) \in R\}$$

The dimension of the code C is equal to $n - \deg(g(x))$. Furthermore, $g(x)$ must divide $x^n - 1$, and so the roots of $g(x)$ are all roots of $x^n - 1$, that is, *n-th roots of unity*.

Therefore, we can define a cyclic code by specifying its generator polynomial $g(x)$ or, equivalently, by specifying which n-th roots of unity are the roots of $g(x)$. This leads to several families of codes, depending on how the choice of roots is made.

For instance, we shall see that the n-th roots of unity form a cyclic group under multiplication, and so they have the form

$$1, \omega, \omega^2, \ldots, \omega^{n-1}$$

where ω is known as a *primitive n-th root of unity*. Thus, for example, we could define a cyclic code by requiring that the generator polynomial be the polynomial of smallest degree that has roots that include the "consecutive" numbers

$$\omega, \omega^2, \ldots, \omega^{e-1}$$

whose exponents are the $e-1$ *consecutive* integers $1,2,\ldots,e-1$, for some $e \geq 1$. Of course, this may mean that $g(x)$ has additional roots as well. Such codes are called *BCH codes* and form one of the most important families of codes.

The point we wish to emphasize here is that the rich algebraic structure of the algebra R_n makes it possible to define cyclic codes

that have a rich structure as well. This allows us to learn a great deal about these codes and also allows for the application of sophisticated decoding techniques.

Nonlinear Codes

Nonlinear codes can be constructed by a variety of means. For instance, some nonlinear codes come from certain types of combinatorial objects, such as block designs, or Latin Squares. Here is an example.

Example 4.3.3 The sets of points and lines shown in Figure 4.3.1 is referred to as the *projective plane of order 2* or the *Fano plane.* It is an example of a *combinatorial design.*

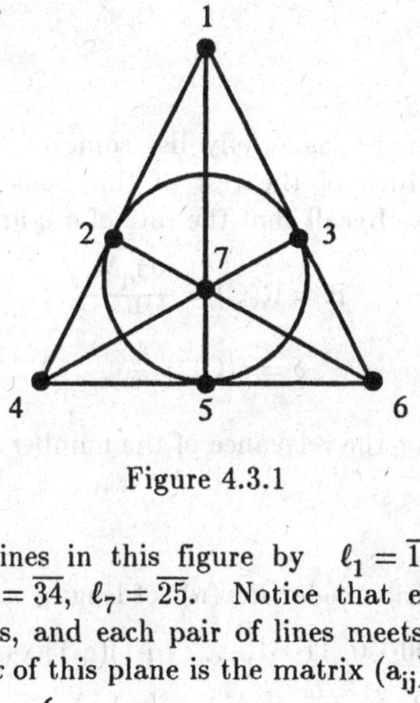

Figure 4.3.1

Let us denote the lines in this figure by $\ell_1 = \overline{14}$, $\ell_2 = \overline{16}$, $\ell_3 = \overline{46}$, $\ell_4 = \overline{15}$, $\ell_5 = \overline{26}$, $\ell_6 = \overline{34}$, $\ell_7 = \overline{25}$. Notice that each line is incident with exactly 3 points, and each pair of lines meets at exactly 1 point. The *incidence matrix* of this plane is the matrix (a_{ij}) for which

$$a_{ij} = \begin{cases} 1 & \text{if line } \ell_i \text{ contains point j} \\ 0 & \text{otherwise} \end{cases}$$

This matrix is the 7×7 matrix

$$\mathfrak{I} = \begin{bmatrix} 1 & 1 & 0 & 1 & 0 & 0 & 0 \\ 1 & 0 & 1 & 0 & 0 & 1 & 0 \\ 0 & 0 & 0 & 1 & 1 & 1 & 0 \\ 1 & 0 & 0 & 0 & 1 & 0 & 1 \\ 0 & 1 & 0 & 0 & 0 & 1 & 1 \\ 0 & 0 & 1 & 1 & 0 & 0 & 1 \\ 0 & 1 & 1 & 0 & 1 & 0 & 0 \end{bmatrix}$$

Notice that any two rows of $\mathcal{I}$, thought of as binary words, have distance 4 from each other, since they each have 3 ones, whose positions agree exactly once. If we denote these rows by $\mathbf{r}_1,\ldots,\mathbf{r}_7$, then

$$d(\mathbf{r}_i,\mathbf{r}_j) = 4$$

Now, consider the *complements* $\mathbf{s}_1,\ldots,\mathbf{s}_7$ of the rows, where $\mathbf{s}_i$ is obtained from $\mathbf{r}_i$ by interchanging 0's and 1's. Clearly,

$$d(\mathbf{s}_i,\mathbf{s}_j) = 4$$

We leave it as an exercise to show that

$$d(\mathbf{r}_i,\mathbf{s}_j) = 3$$

and so the code $C = \{\mathbf{0},\mathbf{1},\mathbf{r}_1,\ldots,\mathbf{r}_7,\mathbf{s}_1,\ldots,\mathbf{s}_7\}$ is a (7,16,3)-code. We will also leave it as an exercise to show that this code is perfect, but not linear. □

FAMILIES OF CODES

For concreteness, let us briefly list some of the more important families of codes. Much of the rest of this book will be devoted to discussing these codes. Recall that the rate of a q-ary (n,M,d)-code is

$$R = R(C) = \frac{\log_q M}{n}$$

and that

$$\delta = \delta(C) = \frac{d}{n}$$

(See Corollary 4.2.7 for the relevance of the number δ.)

Repetition Codes

The q-ary repetition code *Rep*(n) of length n is

$$C = \{00\cdots0,\ 11\cdots1,\ \ldots,\ (q\text{-}1)(q\text{-}1)\cdots(q\text{-}1)\}$$

These very simple codes are q-ary linear [n,1,n]-codes, with

$$R = \frac{1}{n} \quad \text{and} \quad \delta = 1$$

Note that while δ is maximum possible, we have $R \to 0$ as $n \to \infty$.

Hamming Codes

The family of *Hamming codes* $\mathcal{H}_q(r)$ is probably the most famous of all error-correcting codes. These codes were discovered independently by Marcel Golay in 1949 and Richard Hamming in 1950. They are perfect, linear, and very easy to decode. All *binary* Hamming

codes are equivalent to cyclic codes, and some, but not all, non-binary Hamming codes are equivalent to cyclic codes.

Specifically, for each $r > 0$, $\mathcal{H}_q(r)$ is a q-ary [n,k,d]-code, where

$$n = \frac{q^r-1}{q-1}, \quad k = n-r, \quad d = 3$$

The most common case by far is the binary case, where

$$n = 2^r - 1, \quad k = n-r, \quad d = 3$$

The Hamming codes are exactly single-error-correcting, with

$$R = 1 - \frac{r}{n} \quad \text{and} \quad \delta = \frac{3}{n}$$

Note that $R \to 1$, but $\delta \to 0$ as $n \to \infty$.

Golay Codes

In 1948, Marcel Golay introduced some linear codes, denoted by $\mathcal{G}_{23}$, $\mathcal{G}_{24}$, $\mathcal{G}_{11}$ and $\mathcal{G}_{12}$ that are now called *Golay codes.* The code $\mathcal{G}_{24}$ is a binary linear (24,4096,8)-code which, as we mentioned before, was used by the Voyager spacecraft to transmit color photographs of Jupiter and Saturn. The related code $\mathcal{G}_{23}$ is a binary perfect cyclic (23,4096,7)-code.

The code $\mathcal{G}_{11}$ is a ternary perfect cyclic (11,729,5)-code, and $\mathcal{G}_{12}$ is a ternary linear (12,729,6)-code. MacWilliams and Sloane (1977), in their monumental treatise on coding theory, refer to the binary Golay codes as "probably the most important of all codes, for both practical and theoretical reasons."

Reed-Muller Codes

The *Reed-Muller codes* are a family of binary linear codes that have good practical value and nice decoding properties. For each positive integer m, and each integer r for which $0 \leq r \leq m$, the r-th order Reed-Muller code $\mathcal{R}(r,m)$ has parameters

$$n = 2^m, \quad k = 1+\binom{m}{1}+\cdots+\binom{m}{r}, \quad d = 2^{m-r}$$

The first order Reed-Muller codes $\mathcal{R}(1,m)$ are $(2^m, 2^{m+1}, 2^{m-1})$-codes. The code $\mathcal{R}(1,5)$ was used by Mariner 9 to transmit black and white photographs of Mars in 1972.

For the Reed-Muller codes, we have

$$R = \frac{1 + \binom{m}{1} + \cdots + \binom{m}{r}}{2^m} \quad \text{and} \quad \delta = \frac{1}{2^r}$$

Hence, if r is fixed we have $R \to 0$ as $n \to \infty$, and if $r \to \infty$ then $\delta \to 0$ as $n \to \infty$.

BCH Codes and Reed-Solomon Codes

The *BCH codes*, named after Bose and Ray-Chaudhuri, who discovered them in 1960, and Hocquenghem, who discovered them in 1959, are generalizations of the Hamming codes. They are cyclic q-ary codes, which have great practical importance for error-correction.

We saw earlier that the BCH codes are defined by specifying that the generator polynomial have roots that include a list of $e-1$ "consecutive" roots

$$\omega, \omega^2, \ldots, \omega^{e-1}$$

The number e is referred to as the **designed distance** of the code. A BCH code C with designed distance e has minimum distance $d \geq e$. Hence, if the generator polynomial of C is $g(x)$, then C has parameters

$$n, \quad k = n - \deg(g(x)), \quad d \geq e$$

Many BCH codes seem to have $d = e$ and so, in some sense, we can "design" our own minimum distance.

For any prime power q, the *q-ary Reed-Solomon code* is the BCH code with length $q-1$. Thus, reasonably long Reed-Solomon codes have large radix, which may seem impractical. However, as we will see, there are ways to "map" a q-ary code onto a binary code. We will also see that the Reed-Solomon codes are very important for burst-error-correction. In fact, NASA uses Reed-Solomon codes extensively in their deep space programs, (for instance, on their *Galileo*, *Magellan*, and *Ulysses* missions).

Quadratic Residue Codes

The quadratic residue codes are another class of cyclic codes that have prime length p. An integer a is *quadratic residue mod p* if the equation

$$x^2 \equiv a \bmod p$$

has a solution, that is, if a has a square root modulo p. It is known from number theory that 2 is a quadratic residue mod p if and only if p has the form $8m \pm 1$, and so the binary quadratic residue codes must have prime length of the form $p = 8m \pm 1$.

As we will see, when p has this form, we can define a generator polynomial by specifying that its roots be either

$$\{\omega^r \mid r \text{ is a quadratic residue mod } p\}$$

or

$$\{\omega^u \mid u \text{ is } \textit{not} \text{ a quadratic residue mod } p\}$$

In other words, the generator polynomial may be either

$$g_1(x) = \prod_{r \in QR} (x - \omega^r) \quad \text{or} \quad g_2(x) = \prod_{u \in NQR} (x - \omega^u)$$

where $QR \subset \{1,\ldots,p-1\}$ is the set of quadratic residues mod p and $NQR \subset \{1,\ldots,p-1\}$ is the set of quadratic *non*residues mod p.

Since there are precisely $(p-1)/2$ elements in each of the sets QR and NQR, the quadratic residue codes have dimension

$$k = n - \deg(g(x)) = p - \frac{p-1}{2} = \frac{p+1}{2}$$

Unfortunately, it is difficult to determine the minimum distance of a quadratic residue code. We do know that the minimum distance does satisfy the *square root bound*

$$d \geq \sqrt{p}$$

but this bound is often not very good. At any rate, the parameters of the binary quadratic residue codes are

$$n = p, \quad k = \frac{p+1}{2}, \quad d \geq \sqrt{p}$$

Notice that the rate of the quadratic residue codes satisfies $R \geq \frac{1}{2}$.

The overall quality of quadratic residue codes is still uncertain. There are some rather good quadratic residue codes of small length (including the perfect Golay codes $\mathcal{G}_{23}$ and $\mathcal{G}_{11}$), which is promising, but it is not known whether there are good quadratic residue codes of large length.

Goppa Codes

Just as cyclic codes are specified by their generator polynomial, *Goppa codes* are specified by a special *Goppa polynomial* $G(x)$, defined over a field F_{q^m}, along with a set $L \subset F_{q^m}$ of *nonzeros* of $G(x)$. The Goppa codes are linear but not, in general, cyclic.

One of the nicest features of Goppa codes is that the minimum distance is bounded below by $1 + \deg(G(x))$. In fact, if $g = \deg(G(x))$, then the parameters of the Goppa code satisfy

$$n = |L|, \quad k \geq n - mg, \quad d \geq 1 + g$$

(In the binary case, if $G(x)$ has no multiple roots, we can improve the bound on minimum distance to $d \geq 1+2g$.)

Justesen Codes

Notice that in each of the previous cases where we give specific values for the rate R and for the number δ, we have either $R \to 0$ or $\delta \to 0$ as $n \to \infty$. In other words, either the rate tends to 0, or the probability of a decoding error tends to 1 as $n \to \infty$. (See Corollary 4.2.7.) We will discuss a family of linear codes, known as *Justesen codes*, that have constant rate $R = \frac{1}{2}$, and for which

$$\delta \to H^{-1}(\tfrac{1}{2}) \approx 0.110$$

where $H(\lambda) = -\lambda \log \lambda - (1-\lambda)\log(1-\lambda)$ is the entropy function. Thus, the Justesen codes are "asymptotically good" codes.

PERFECT CODES

For quite some time, it was thought that the Hamming codes $\mathcal{H}_q(r)$ and the Golay codes $\mathcal{G}_{23}$ and $\mathcal{G}_{11}$ were the only nontrivial perfect codes. The *trivial* perfect codes are binary repetition codes of odd length, codes consisting only of the zero codeword and all of $V(n,q)$. (Compare with the discussion of perfect codes in Section 4.2.) However, as we shall see later in the book, Vasil'ev found a family of *nonlinear* perfect codes with the same parameters as the Hamming codes.

The following theorem, due to Tietäväinen (1973), with a large contribution by van Lint, describes the situation with regard to perfect codes over alphabets of prime power. We will not prove this theorem, however.

Theorem 4.3.6 A nontrivial perfect q-ary code C, where q is a prime power, must have the same parameters (length, size and minimum distance) as either a Hamming code $\mathcal{H}_q(r)$ or one of the Golay codes $\mathcal{G}_{23}$ or $\mathcal{G}_{11}$. Furthermore,

1) if C has the parameters of one of the Golay codes, then it is equivalent to that Golay code,
2) if C is *linear* and has the parameters of one of the Hamming codes, then it is equivalent to that Hamming code. ☐

OBTAINING NEW CODES FROM OLD CODES

There are several useful techniques that can be used to obtain new codes from old ones.

Extending a Code

The process of adding one or more additional coordinate positions to the codewords in a code is referred to as *extending the code.* The most common way to extend a code is by *adding an overall parity check,* which is done as follows. If C is a q-ary (n,M,d)-code, we define the **extended code** $\widehat{C}$ by

$$\widehat{C} = \left\{c_1c_2\cdots c_nc_{n+1} \mid c_1c_2\cdots c_n \in C \text{ and } \sum_{k=1}^{n+1} c_k = 0\right\}$$

If $\widehat{C}$ is an $(\widehat{n},\widehat{M},\widehat{d})$-code, then

$$\widehat{n} = n+1, \quad \widehat{M} = M, \quad \widehat{d} = d \text{ or } d+1$$

Example 4.3.4 Adding an overall parity check to the code $C = \{00,01,10,11\}$ gives the extended code $\widehat{C} = \{000,011,101,110\}$. Notice that C has minimum distance 1, but $\widehat{C}$ has minimum distance 2. ▯

Example 4.3.5 Consider the binary Hamming code $\mathcal{H}_2(r)$, which has parameters

$$n = 2^r - 1, \quad M = 2^{(2^r-1-r)}, \quad d = 3$$

The **extended Hamming code** $\widehat{\mathcal{H}}_2(r)$, obtained by adding an overall parity check to $\mathcal{H}_2(r)$, has parameters

$$\widehat{n} = 2^r, \quad \widehat{M} = 2^{(2^r-1-r)}, \quad \widehat{d} = 4$$

Thus, while the extended code has no better error *correcting* capabilities than the original code, it does have better error *detecting* capabilities, since it can detect two errors (while also correcting a single error). ▯

Puncturing a Code

The opposite process to extending a code is *puncturing a code,* in which one or more coordinate positions are removed from the codewords. If C is a q-ary (n,M,d)-code, and if $d \geq 2$, then the code C^* obtained by puncturing C once has parameters

$$n^* = n-1, \quad M^* = M, \quad d^* = d \text{ or } d-1$$

Example 4.3.6 The $(23,4096,7)$-code $\mathcal{G}_{23}$ is obtained from the $(24,4096,8)$-code $\mathcal{G}_{24}$ by puncturing the latter in the last position. (In fact, puncturing $\mathcal{G}_{24}$ in any position gives a code equivalent to $\mathcal{G}_{23}$.) Notice that the code $\mathcal{G}_{24}$ does not satisfy the sphere-packing condition, and so it is not perfect. However, the code $\mathcal{G}_{23}$ is perfect. Hence, puncturing a non-perfect code can result in a perfect code. ▯

For binary codes, the process of extending and puncturing can be used to prove the following useful result.

Theorem 4.3.7 A binary (n,M,2t+1)-code exists if and only if a binary (n+1,M,2t+2)-code exists.

Proof. Suppose that C is a binary (n,M,2t+1)-code and that $\widehat{C}$ is the code obtained from C by adding an overall parity check. Then since each codeword in $\widehat{C}$ has even weight, Lemma 4.3.4 tells us that the distance between any two codewords in $\widehat{C}$ is even. Hence, $d(\widehat{C})$ is even, and so it must be 2t+2.

Conversely, suppose that C is a binary (n+1,M,2t+2)-code and that $d(\mathbf{c},\mathbf{d}) = 2t+2$. If we puncture C in a position at which **c** and **d** *disagree*, the resulting code C^* has minimum distance 2t+1. ∎

Expunging a Code

Expunging a code refers to the process of simply deleting some of the codewords in the code. (Some authors use the term *expurgate.*) As an example, suppose that L is a binary linear (n,M,d)-code. If L contains at least one codeword of odd weight, then exactly one-half of the codewords in L must have odd weight. (We leave proof of this as an exercise.) By throwing away the codewords of odd weight, we get an (n,M/2,d')-code, where $d' \geq d$. Furthermore, since each remaining codeword has even weight, the minimum distance of the expunged code must be even, and so if d is odd then $d' > d$.

Augmenting a Code

The opposite process of expunging a code is *augmenting a code*, which simply means adding additional words to the code. A common way to augment a *binary* code C is to include the *complements* of each codeword in C, where the **complement** $\mathbf{c}^c$ of a binary word **c** is the word obtained from **c** by interchanging all 0's and 1's. Let us denote the set of all complements of the words in C by C^c.

Lemma 4.3.8 If $\mathbf{x},\mathbf{y} \in V(n,2)$, then $d(\mathbf{x},\mathbf{y}^c) = n - d(\mathbf{x},\mathbf{y})$.

Proof. $d(\mathbf{x},\mathbf{y}^c)$ is the number of positions in which **x** and $\mathbf{y}^c$ disagree, which is precisely the same as the number of positions in which **x** and **y** agree (since q = 2), which in turn is equal to $n - d(\mathbf{x},\mathbf{y})$. ∎

Theorem 4.3.9 Let C be a binary (n,M,d)-code. Then

$$d(C \cup C^c) = \min\{d, n - d_{max}\}$$

where d_{max} is the *maximum* distance between codewords in C.

Proof. We have

$$d(C \cup C^c) = \min\left\{d(C),\ d(C^c),\ \min_{\substack{\mathbf{c} \in C \\ \mathbf{d} \in C^c}} d(\mathbf{c},\mathbf{d})\right\}$$

But $d(C) = d(C^c) = d$, and according to the previous lemma,

$$\min_{\substack{\mathbf{c} \in C \\ \mathbf{d} \in C^c}} d(\mathbf{c},\mathbf{d}) = \min_{\substack{\mathbf{c} \in C \\ \mathbf{d} \in C}} d(\mathbf{c},\mathbf{d}^c) = \min_{\substack{\mathbf{c} \in C \\ \mathbf{d} \in C}} \{n - d(\mathbf{c},\mathbf{d})\} = n - \max_{\substack{\mathbf{c} \in C \\ \mathbf{d} \in C}} d(\mathbf{c},\mathbf{d})$$

from which the result follows. ∎

If L is a binary *linear* code, and if $\mathbf{1} = 11\cdots1 \in L$, then $L = L^c$. However, if $\mathbf{1} \notin L$, then $L \cap L^c = \emptyset$. In fact, we have the following.

Theorem 4.3.10 If L is a binary linear (n,M,d)-code that does not contain the codeword $\mathbf{1} = 11\cdots1$, then $L \cup L^c$ is a binary linear (n,2M,d')-code, where

$$d' = \min\{d, n - w_{max}\}$$

and where w_{max} is the *maximum* weight of any codeword in L. □

Shortening a Code

Shortening a code refers to the process of keeping only those codewords in a code that have a given symbol in a given position (for instance, a 0 in the first position), and then deleting that position. If C is an (n,M,d)-code then a shortened code has length n–1 and minimum distance d. The shortened code formed by taking codewords with an s in the i-th position is referred to as the **cross-section** $x_i = s$.

The proof of the following result is left to the reader.

Theorem 4.3.11 If C is a binary *linear* (n,M,d)-code, then the cross-section $x_i = 0$ is a binary linear $(n-1,\frac{1}{2}M,d)$-code. □

The Direct Sum Construction

If C_1 is a q-ary (n_1,M_1,d_1)-code and C_2 is a q-ary (n_2,M_2,d_2)-code, the **direct sum** C_3 is the code

$$C_3 = \{\mathbf{cd} \mid \mathbf{c} \in C_1,\ \mathbf{d} \in C_2\}$$

Clearly, C_3 has parameters

$$n = n_1+n_2, \quad M = M_1M_2, \quad d = \min\{d_1,d_2\}$$

The (u,u+v)-Construction

A much more useful construction than the direct sum is the following. If C_1 is an (n,M_1,d_1)-code and C_2 is an (n,M_2,d_2)-code, both of which are over the alphabet F_q (note that C_1 and C_2 have the same length), then we can define a code $C_1 \oplus C_2$ by

$$C_1 \oplus C_2 = \{\mathbf{c}(\mathbf{c}+\mathbf{d}) \mid \mathbf{c} \in C_1,\ \mathbf{d} \in C_2\}$$

Certainly, the length of $C_1 \oplus C_2$ is $2n$, and the size is M_1M_2. As for the minimum distance, consider two distinct codewords $\mathbf{u}_1 = \mathbf{c}_1(\mathbf{c}_1+\mathbf{d}_1)$ and $\mathbf{u}_2 = \mathbf{c}_2(\mathbf{c}_2+\mathbf{d}_2)$. If $\mathbf{d}_1 = \mathbf{d}_2$, then

$$d(\mathbf{u}_1,\mathbf{u}_2) = 2\,d(\mathbf{c}_1,\mathbf{c}_2) \geq 2d_1$$

On the other hand, if $\mathbf{d}_1 \neq \mathbf{d}_2$, then

$$\begin{aligned} d(\mathbf{u}_1,\mathbf{u}_2) &= w(\mathbf{u}_1 - \mathbf{u}_2) \\ &= w(\mathbf{c}_1 - \mathbf{c}_2) + w(\mathbf{c}_1 - \mathbf{c}_2 + \mathbf{d}_1 - \mathbf{d}_2) \\ &\geq w(\mathbf{d}_1 - \mathbf{d}_2) \\ &= d(\mathbf{d}_1,\mathbf{d}_2) \geq d_2 \end{aligned}$$

Hence, $d(C_1 \oplus C_2) \geq \min\{2d_1,d_2\}$. Since equality can hold in all cases, we get the following result.

Theorem 4.3.12 If C_1 is an (n,M_1,d_1)-code and C_2 is an (n,M_2,d_2)-code, both of which are over the same alphabet F_q, then $C_1 \oplus C_2$ is a $(2n,M_1M_2,d')$-code, where $d' = \min\{2d_1,d_2\}$. ▯

Example 4.3.7 Recall that the first order Reed-Muller codes $\mathcal{R}(1,m)$ are binary $(2^m,2^{m+1},2^{m-1})$-codes, and that $Rep(2^m)$ denotes the binary repetition code of length 2^m, which is a $(2^m,2,2^m)$-code. According to Theorem 4.3.11, the code

$$\mathcal{R}(1,m) \oplus Rep(2^m)$$

is a binary $(2^{m+1},2^{m+2},2^m)$-code. As we will see when we study Reed-Muller codes, $\mathcal{R}(1,m) \oplus Rep(2^m)$ is the Reed-Muller code $\mathcal{R}(1,m+1)$. ▯

THE AUTOMORPHISM GROUP OF A CODE

Associated with every code over F_q is a certain group, called the *automorphism group* of the code. This group can be useful in studying the structure of the code, as well as in decoding.

Definition Let C be an (n,M)-code over F_q. The **automorphism group** $Aut(C)$ of C is the set of all monomial transformations μ of degree n for which $\mu(C) \subset C$. ▯

When $q = 2$, a monomial transformation is nothing more than a permutation π of the coordinate positions, that is,

$$\pi \mathbf{c} = c_{\pi(1)} c_{\pi(2)} \cdots c_{\pi(n)}$$

Thus, for binary codes, we have the following.

Definition Let C be a binary (n,k)-code. The **automorphism group** of C is

$$Aut(C) = \{\pi \in S_n \mid \pi \mathbf{c} \in C \text{ for all } \mathbf{c} \in C\} \qquad ▯$$

Recall that if μ is a monomial transformation, then the code $\mu(C) = \{\pi \mathbf{c} \mid \mathbf{c} \in C\}$ is, by definition, scalar multiple equivalent to C. The monomial transformations for which $\mu C \subset C$ (and hence $\mu C = C$) are precisely the ones in $Aut(C)$.

We leave proof of the following as an exercise.

Theorem 4.3.13 The set $Aut(C)$ is a group. ▯

For a binary code C, the group $Aut(C)$ is a subgroup of the symmetric group S_n.

Example 4.3.8 The automorphism group of the code

$$C = \{0000,1100,0011,1111\}$$

is the following subgroup of S_4 of size 8

$$Aut(C) = \{id,(12),(34),(12)(34),(13)(24),(14)(23),(1324),(1423)\} \qquad ▯$$

****Transitive Permutation Groups***

A group G of permutations $\pi \in S_n$ is said to be **transitive** if, given $i,j \in \{1,\ldots,n\}$, there is a $\pi \in G$ for which $\pi(i) = j$. A group G of permutations is said to **fix** a code C, or C is said to be **invariant** under G, if $G \subset Aut(C)$. It turns out that many important codes are fixed by a transitive permutation group.

If C is invariant under a transitive permutation group, then the *multiset* $\mathcal{M}_i$ of all symbols appearing in the i-th position is the same for all i. (In other words, the same symbols appear in each position, with the same multiplicity.)

Here is a small sampling of the consequences of being invariant under a transitive permutation group.

Theorem 4.3.14 If an (n,M)-code C is invariant under a transitive permutation group G, and if A_i denotes the number of codewords of weight i, then $n \mid iA_i$.

Proof. Let B be the matrix whose rows consist of the codewords in C of weight i. Hence, B has size $A_i \times n$. Fix $j \in \{1,\ldots,n\}$, and let $\pi \in G$ be such that $\pi(1) = j$. Applying π to the codewords in B results in moving the first column of B to the j-th column. However, since $\pi \in Aut(C)$, the resulting matrix B′ has the same rows as B, but perhaps in a different order. In other words, the entries in the first column of B are simply a rearrangement of the entries in the j-th column. This shows that all columns of B have the same number of nonzero entries. If we denote this number by r_i, then the total number of nonzero entries in B is equal to both $r_i n$ and iA_i. Hence, $r_i n = iA_i$, which completes the proof. ∎

Corollary 4.3.15 Let C be an (n+1,M)-code in which all codewords have even weight. Let C^* be the (n,M)-code obtained from C by puncturing C in any position. Let A_i^* be the number of codewords of weight i in C^*. If C is invariant under a transitive permutation group G, then

$$2kA_{2k}^* = (n+1-2k)A_{2k-1}^*$$

In particular, C^* has odd minimum weight.

Proof. Since C is invariant under a transitive permutation group G, Theorem 4.3.14 applies, and we have

$$2kA_{2k} = r_{2k}(n+1)$$

where r_{2k} is the number of nonzero entries in any column of the matrix B formed from the codewords in C of weight 2k, as in the proof of Theorem 4.3.14.

Now, A_{2k-1}^* is the number of codewords in C^* of weight 2k−1. But these codewords came from the codewords in C of weight 2k that have a nonzero entry in the column that was punctured, and since there are r_{2k} such codewords, we deduce that $r_{2k} = A_{2k-1}^*$. Hence,

$$2kA_{2k} = A_{2k-1}^*(n+1) \tag{4.3.2}$$

Since C has only even weight codewords, we also deduce that

$$A_{2k}^* + A_{2k-1}^* = A_{2k}$$

Substituting this into (4.3.2) gives the desired result. The final statement follows from the fact that if $A_{2k}^* \neq 0$, then $A_{2k-1}^* \neq 0$. ∎

Corollary 4.3.16 Let C be a code, and suppose that the extended code $\widehat{C}$ is invariant under a transitive permutation group. Then C has odd weight. ▯

EXERCISES

1. Is the code $C = \{000,110,011,111\}$ systematic?
2. Is $\mathbb{Z}_{p^n}$ a field for all positive integers n?
3. Prove Lemma 4.3.2.
4. Prove Theorem 4.3.3.
5. Prove Lemma 4.3.4.
6. Show that if L_1 and L_2 are linear, then so is $L_1 \oplus L_2$.
7. Compute the rate of the Golay codes.
8. Referring to Example 4.3.3, show that $d(\mathbf{r}_i,\mathbf{s}_j) = 3$, and so the code $C = \{\mathbf{0},\mathbf{1},\mathbf{r}_1,\ldots,\mathbf{r}_7,\mathbf{s}_1,\ldots,\mathbf{s}_7\}$ is a (7,16,3)-code.
9. Show that the code in Example 4.3.3 is perfect, but not linear.
10. Let E_n denote the set of all even weight words in $V(n,2)$. Show that E_n is linear. What are its parameters n, k, and d?
11. Describe scalar multiple equivalence for binary codes.
12. Prove Theorem 4.3.11. Does a corresponding theorem hold for cross-sections of the form $x_i = 1$? Explain.
13. If C is an (n,M,d)-code, show that the extended code $\widehat{C}$ is an $(\widehat{n},\widehat{M},\widehat{d})$-code, with $\widehat{n} = n+1$, $\widehat{M} = M$, and $\widehat{d} = d$ or $d+1$.
14. Show that the condition that the codes be binary is necessary in Theorem 4.3.7.
15. Show that Golay $\mathcal{G}_{24}$ is not perfect, but that $\mathcal{G}_{23}$ is perfect. Show that $\mathcal{G}_{11}$ is perfect.
16. Prove that if L is a binary linear code, and if $\mathbf{1} = 11\cdots1 \in L$, then $L = L^c$. However, if $\mathbf{1} \notin L$, then $L \cap L^c = \emptyset$. Prove Theorem 4.3.10.
17. Show that a linear code can be equivalent to a nonlinear code.
18. Show that, up to equivalence, there is exactly one binary (8,4,5)-code.
19. Show that any q-ary (n,q,n)-code is equivalent to a repetition code.
20. Is any code equivalent to a code containing the string $\mathbf{1} = 11\cdots1$?
21. Prove that any binary $(5,M,3)$-code must satisfy $M \le 4$. Furthermore, there is, up to equivalence, exactly one (5,4,3)-code.
22. How many inequivalent binary $(n,2)$-codes are there?
23. Consider the binary words $c_1 = 11010000$, $c_2 = 11100100$, and $c_3 = 10101010$. Let C be the code formed by including c_1, c_2, c_3, all cyclic shifts of these words, and the words $\mathbf{0}$ and $\mathbf{1}$. Show that C is a (8,20,3)-code.

24. Use the (u,u+v)-construction and the code of the previous exercise to construct a (16,2560,3)-code.
25. Suppose that L is a *linear* (n,q^k,d)-code that is systematic on *all* choices of k positions. Show that $d = n - k + 1$.
26. Show that a q-ary $(q+1,q^2,q)$-code, where q is odd, is perfect if and only if $q = 3$.
27. Prove Theorem 4.3.13.
28. What is $Aut(E_n)$, where E_n is the set of all even weight words in V(n,2)?
29. What is the automorphism group of the extended code $\widehat{C}$ in Example 4.3.4?
30. Show that $C_1 \subset C_2$ does not necessarily imply that $Aut(C_1) \supset Aut(C_2)$.
31. Suppose that L′ is the binary linear code obtained from a binary linear code L by adjoining the vector **1**. Show that $Aut(L') \supset Aut(L)$. If, in addition, L has odd length and all weights in L are even, then $Aut(L') = Aut(L)$.

4.4 Codes and Designs

We will see in this brief section that there is a connection between codes and certain other combinatorial structures known as *combinatorial designs.* We begin this section with some general remarks on combinatorial designs. Readers interested in learning more about designs may wish to consult the book by Anderson (1990).

A set of size k will be referred to as a **k-set.**

t-DESIGNS

Definition Let S be a v-set, whose elements will be called **points.** A **t-(v,k,λ) design** is a collection $\mathfrak{D}$ of distinct k-subsets of S, called **blocks**, with the property that every t-subset of S is contained in exactly λ blocks. ▯

When the parameters v, k, and λ do not require emphasis, it is customary to refer to $\mathfrak{D}$ simply as a **t-design.** Some special cases of t-designs are worth singling out.

Definition

1) A 2-(v,k,λ) design with $k < v$ is referred to as a **balanced incomplete block design** or **BIBD.**
2) A t-(v,k,1) design is referred to as a **Steiner system** and denoted by S(t,k,v). Thus, a Steiner system is a collection of k-subsets (blocks) of a v-set with the property that every t-subset lies in exactly one block. A **Steiner triple system** is a Steiner system S(2,3,v), that is, a collection of triplets with the property that each pair of elements lies in exactly one triplet.
3) A 2-(v,k,1) design is called a **projective plane.** The k-subsets (blocks) are the *lines*, and every pair of *points* lies on exactly one line. It is possible to show that a 2-(v,k,1) design is also an $S(2,n+1,n^2+n+1)$. Such a design is a projective plane of **order** n. ▯

Example 4.4.1 An example of a Steiner triple system is given by Figure 4.4.1, which we have met before in connection with nonlinear codes. In this case, $S = \{1,2,3,4,5,6,7\}$, and the blocks are represented by the line segments and the circle. Each block is a 3-subset of S, and each pair of points in S is in exactly one block. Hence, this is a Steiner triple system S(2,3,7), better known as the projective plane of order 2. ▯

Relatively little is known about the existence of Steiner systems in general. In fact, no Steiner systems with $t > 5$ have ever been discovered, and only a handful have been discovered for $t = 4$ or 5. However, we do have the following results, whose proofs can be found in Anderson (1990).

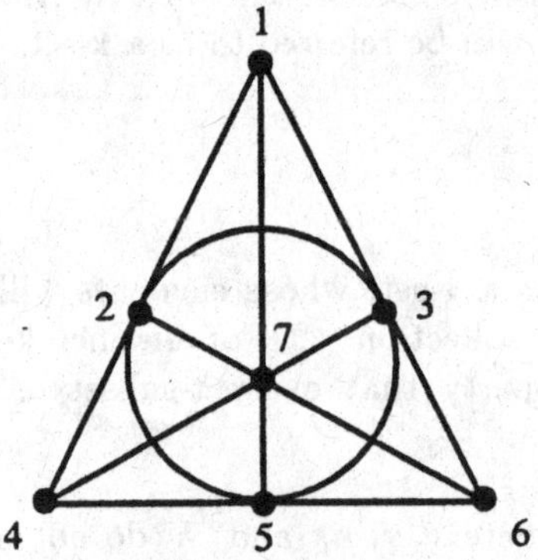

Figure 4.4.1

Theorem 4.4.1
1) **A Steiner triple system** $S(2,3,v)$ exists if and only if

$$v \equiv 1 \text{ or } 3 \pmod 6$$

2) **A Steiner quadruple system** $S(3,4,v)$ exists if and only if

$$v \equiv 2 \text{ or } 4 \pmod 6$$ □

It is customary to let b denote the number of blocks in a t-(v,k,λ) design $\mathfrak{D}$, and let r denote the number of blocks containing a given point in $\mathfrak{D}$. We will see in a moment that r is well defined.

Theorem 4.4.2 A t-design is also an s-design, for any $s \leq t$.
Proof. Let A be an s-subset of the t-(v,k,λ) design $\mathfrak{D}$, and suppose that λ_s is the number of blocks containing A. We must show that λ_s is independent of A. To this end, let us count the number of pairs (D,B), where B is a block, D is a t-subset, and $A \subset D \subset B$. On the one hand, for each of the $\binom{v-s}{t-s}$ choices for D, there are λ blocks B containing D. On the other hand, for each of the λ_s blocks B containing A, there are $\binom{k-s}{t-s}$ choices for D. Hence, we have

$$\lambda\binom{v-s}{t-s} = \lambda_s\binom{k-s}{t-s}$$

This implies that λ_s is independent of the set A, and so $\mathfrak{D}$ is an s-design. In fact, if $\mathfrak{D}$ is a t-(v,k,λ) design, then it is also an s-(v,k,λ_s) design, where

$$\lambda_s = \lambda\binom{v-s}{t-s}\Big/\binom{k-s}{t-s}$$ ∎

We leave proof of the following result as an exercise.

Theorem 4.4.3 Let $\mathfrak{D}$ be a t-(v,k,λ) design.

1) Each point of $\mathfrak{D}$ is contained in r blocks, where

$$r = \lambda\binom{v-1}{t-1}\Big/\binom{k-1}{t-1}$$

2) $\mathfrak{D}$ has b blocks, where

$$b = \lambda\binom{v}{t}\Big/\binom{k}{t}$$

3) The parameters v, k, b, and r are related by

$$rv = bk$$ □

The Intersection Numbers of a t-Design

We have seen that, for a t-design $\mathfrak{D}$, the number of blocks containing a given point p_1 is independent of the point p_1. Let us generalize this fact.

Let $M = \{p_1,\ldots,p_m\}$ and $N = \{q_1,\ldots,q_n\}$ be disjoint sets of points, with $0 \le m, n \le t$, and let $\mu(M,N)$ be the number of blocks that contain M but are disjoint from N. Note that $\mu(M,\emptyset)$ is the number of blocks that contain M and $\mu(\emptyset,N)$ is the number of blocks that are disjoint from N. Also, $\mu(\emptyset,\emptyset) = b$. We want to show that $\mu(M,N)$ depends only on the sizes of M and N.

According to Theorem 4.4.2, $\mu(M,\emptyset)$ depends only on the size of M. Let us assume that $\mu(M,N)$ depends only on the sizes of M and N, whenever $|N| \le n-1$. Let $N = \{p_1,\ldots,p_n\}$. Then the $\mu(M,N-\{p_n\})$ blocks that contain M but are disjoint from $N-\{p_n\}$ fall into two groups – those that contain p_n and those that do not. Hence, we have

$$\mu(M,N-\{p_n\}) = \mu(M\cup\{p_n\},N-\{p_n\}) + \mu(M,N)$$

or

$$\mu(M,N) = \mu(M,N-\{p_n\}) - \mu(M\cup\{p_n\},N-\{p_n\})$$

But by assumption, the terms on the right depend only on the sizes of M and N, and therefore so does $\mu(M,N)$.

Thus, we may let $\mu_{m,n} = \mu(M,N)$, for any sets M and N with $|M| = m$ and $|N| = n$. It is customary in the literature to treat the numbers

$$\lambda_{i,j} = \mu_{i-j,j}$$

where $0 \le j \le i \le t$. These are defined as follows. If $\{p_1,\ldots,p_i\}$ is a set of points, then $\lambda_{i,j}$ is the number of blocks that contain the points $p_1,\ldots,p_j$, but *not* the points $p_{j+1},\ldots,p_i$.

Let us summarize.

Theorem 4.4.4 Let $\mathfrak{D}$ be a t-(v,k,λ) design, and let $\{p_1,\ldots,p_i\}$ be a set of points in $\mathfrak{D}$, and suppose that $0 \le j \le i \le t$. The number $\lambda_{i,j}$ of blocks that contain $p_1,\ldots,p_j$ but *not* $p_{j+1},\ldots,p_i$ depends only on the numbers i and j, and is called a **block intersection number** of $\mathfrak{D}$. Furthermore, we have the *Pascal property*

(4.4.1) $$\lambda_{i,j-1} = \lambda_{i-1,j-1} - \lambda_{i,j}$$

as well as

$$\lambda_{0,0} = b, \quad \lambda_{i,i} = \lambda \binom{v-i}{t-i} \Big/ \binom{k-i}{t-i}$$ □

Figure 4.4.2 shows a portion of the block intersection numbers, along with the recurrence (4.4.1).

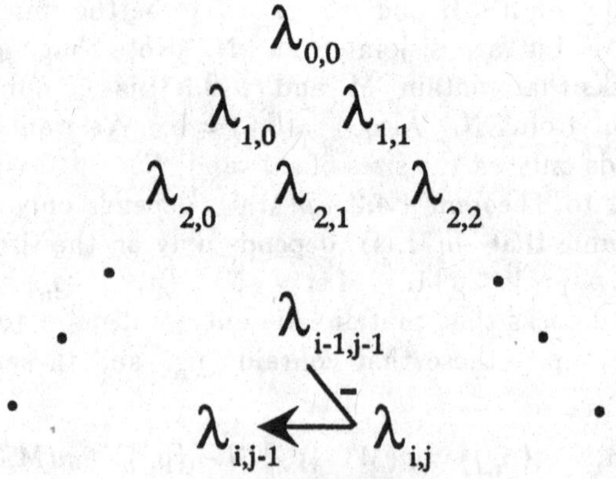

Figure 4.4.2

DESIGNS AND CODES

To explain the connection between designs and codes, we need a few simple definitions. We will restrict our attention to the binary case.

Definition Let $\mathbf{x}$ and $\mathbf{y}$ be binary words of length n. The set of positions in which $\mathbf{x}$ has nonzero entries is called the **support** of $\mathbf{x}$. We say that $\mathbf{x}$ **covers** $\mathbf{y}$ if the support of $\mathbf{y}$ is a subset of the support of $\mathbf{x}$. □

For example, the support of $\mathbf{x} = 10110$ is $\{1,3,4\}$, and $\mathbf{x}$ covers $\mathbf{y} = 00100$.

Definition Let C be a binary code of length n. Let S_w be the set of codewords in C of weight w. We say that S_w **holds a t-(n,w,λ) design** if the supports of codewords in S_w form the blocks of a t-(n,w,λ) design, that is, if for any t-set $T \subset \{1,2,\ldots,n\}$ there are exactly λ codewords of weight w in C with 1's in the positions given by T. □

Our first result on codes and designs shows that perfect codes always hold designs.

Theorem 4.4.5 Let C be a perfect binary (n,M,d)-code. Then the set S_d of all codewords of minimum weight d holds a Steiner system S(t+1,d,n), where $t = (d-1)/2$.

Proof. Since C is perfect, the spheres of radius t are disjoint and cover V(n,2). Hence, any binary word $\mathbf{x}$ of weight $t+1$ is contained in exactly one sphere, say $\mathbf{x} \in S(\mathbf{c},t)$. But $d(\mathbf{c},\mathbf{x}) \leq t$ implies that $w(\mathbf{c}) \leq d(\mathbf{c},\mathbf{x}) + w(\mathbf{x}) \leq t + t + 1 = d$, and so $\mathbf{c}$ must be in S_d.

Now, we have

$$w(\mathbf{x}) = t+1, \quad w(\mathbf{c}) = d = 2t+1, \quad \text{and} \quad d(\mathbf{x},\mathbf{c}) \leq t$$

Then according to Lemma 4.3.4,

$$2w(\mathbf{x} \cap \mathbf{c}) = w(\mathbf{x}) + w(\mathbf{c}) - d(\mathbf{x},\mathbf{c}) \geq 2t+2$$

and so $w(\mathbf{x} \cap \mathbf{c}) \geq t+1 = w(\mathbf{x})$, which implies that $\mathbf{c}$ covers $\mathbf{x}$. ∎

Example 4.4.2 Since the perfect Hamming code $\mathcal{H}_2(r)$ has parameters

$$n = 2^r - 1, \quad M = 2^{(2^r-1-r)}, \quad d = 3$$

the codewords S_3 of weight 3 in $\mathcal{H}_2(r)$ hold a Steiner triple system $S(2,3,2^r-1)$. □

Corollary 4.4.6 Let C be a perfect binary (n,M,d)-code. Then the number A_d of codewords of minimum weight $d = 2t + 1$ is

$$A_d = \binom{n}{t+1} \Big/ \binom{d}{t+1}$$

Example 4.4.3 For the Hamming codes $\mathcal{H}_2(r)$, we have

$$A_3 = \binom{2^r-1}{2} \Big/ \binom{3}{2} = \frac{(2^r-1)(2^{r-1}-1)}{3}$$ □

Theorem 4.4.5 can be used to give necessary conditions on the existence of perfect (n,M,d)-codes. (Recall that the sphere-packing condition also gives necessary conditions for the existence of perfect codes.)

Corollary 4.4.7 If C is a perfect binary (n,M,d)-code, where $d = 2t + 1$, then the numbers

$$\lambda_s = \binom{n-s}{t+1-s} \Big/ \binom{2t+1-s}{t+1-s}$$

must be integers for all $1 \le s \le t$.
Proof. This follows from Theorem 4.4.5 and the value of λ_s given in the proof of Theorem 4.4.2. ▮

Example 4.4.4 Recall from Section 4.2 that one of the solutions to the sphere-packing condition is $(n,M,d) = (90,2^{78},5)$. If a code with these parameters were to exist, then the number

$$\lambda_2 = \binom{88}{1} \Big/ \binom{3}{1} = \frac{88}{3}$$

would have to be an integer, which it is not. Hence, no such code can exist. □

EXERCISES

1. Prove Theorem 4.4.3. *Hint.* For part 2), count the number of pairs (A,B), where A is a t-subset, B is a block of the design, and $A \subset B$.
2. Let $S(t,k,\nu)$ be a Steiner system, with underlying set S, and let $s \in S$. Suppose we *puncture* $S(t,k,\nu)$ by taking only those blocks containing s, and then removing s from these blocks. Show that the resulting collection of blocks forms an $S(t-1,k-1,\nu-1)$ Steiner system.

3. Generalize the previous exercise to t-designs. The resulting design is called a *derived design.*
4. Using (4.4.1), compute the block intersection numbers for the Steiner triple system S(2,3,7) in Figure 4.4.1.
5. Prove Theorem 4.4.4 in detail.
6. Show that $\mathbf{x}$ covers $\mathbf{y}$ if and only if $w(\mathbf{x}+\mathbf{y}) = w(\mathbf{x}) - w(\mathbf{y})$.
7. Show that $\mathbf{x}$ covers $\mathbf{y}$ if and only if $\mathbf{x} \cap \mathbf{y} = \mathbf{y}$.
8. Show that $\mathbf{x}$ covers $\mathbf{y}$ if and only if $w(\mathbf{x} \cap \mathbf{y}) \geq w(\mathbf{y})$.
9. Prove Corollary 4.4.6.
10. Give the details of the proof of Corollary 4.4.7.

4.5 The Main Coding Theory Problem

A good (n,M,d)-code should have a relatively large size so that it can be used to encode a large number of source messages and a relatively large minimum distance, so that it can be used to correct a large number of errors. Not surprisingly, these goals are conflicting.

It is customary to let $A_q(n,d)$ denote the largest possible size M for which there exists a q-ary (n,M,d)-code. The numbers $A_q(n,d)$ play a central role in coding theory, and much effort has been expended in attempting to determine their values. In fact, determining the values of $A_q(n,d)$ has come to be known as *the main coding theory problem.*

Most of the results obtained thus far center on determining $A_q(n,d)$ for small values of q, n, and d, or finding upper bounds on $A_q(n,d)$. There has also been considerable work done on determining the asymptotic behavior of $A_q(n,d)$ as a function of $\delta = d/n$ as $n \to \infty$. An (n,M,d)-code for which $M = A_q(n,d)$ is said to be **optimal**.

We begin with a summary of some of the main results of this section. This summary will be sufficient for those readers who would like to move on as quickly as possible. The remainder of the section contains the details, along with complete proofs. Since some of these details are a bit technical, the reader may wish to skip the starred sections upon first reading.

OVERVIEW

Elementary Results

Theorem 4.5.1 For any $n \geq 1$,

1) $A_q(n,1) = q^n$

2) $A_q(n,n) = q$ □

Theorem 4.5.2 For any $n \geq 2$,

$$A_q(n,d) \leq q\,A_q(n-1,d)$$ □

Theorem 4.5.3 For binary codes,

$$A_2(n,2t+1) = A_2(n+1,2t+2)$$

Put another way, if d is even, then $A_2(n,d) = A_2(n-1,d-1)$. Thus, for *binary* codes, it is enough to determine $A_2(n,d)$ for all *odd* values of d (or for all even values). □

A Lower Bound on $A_q(n,d)$

Theorem 4.5.4 (The Gilbert-Varshamov Bound)

$$A_q(n,d) \geq \frac{q^n}{\sum_{k=0}^{d-1}\binom{n}{k}(q-1)^k}$$

Theorem 4.5.5 (The Gilbert-Varshamov Bound) For q a prime power, there exists a q-ary linear [n,k]-code with minimum distance at least d provided that

$$q^k < \frac{q^n}{\sum_{i=0}^{d-2}\binom{n-1}{i}(q-1)^i}$$

Hence, if k is the largest integer for which this inequality holds, then $A_q(n,d) \geq k$. □

Upper Bounds on $A_q(n,d)$

Theorem 4.5.6 (The Singleton Bound) $A_q(n,d) \leq q^{n-d+1}$ □

The Singleton bound is often not very good. However, there are cases where the Singleton bound gives equality. See Example 4.5.1.

Theorem 4.5.7 (The Sphere-Packing Bound or The Hamming Bound)

$$A_q(n,d) \leq \frac{q^n}{\sum_{k=0}^{t}\binom{n}{k}(q-1)^k}, \qquad t = \left\lfloor \frac{d-1}{2} \right\rfloor$$ □

Recall that a perfect code is a code for which equality holds in the sphere-packing bound.

Theorem 4.5.16 (The Plotkin Bound) Let $\theta = \frac{q-1}{q}$. If $d > \theta n$, then

$$A_q(n,d) \leq \frac{d}{d - \theta n}$$ □

Notice that the Plotkin bound applies only when the minimum distance d is rather large. This bound can easily be refined a bit when $q = 2$.

Theorem 4.5.18 (The Plotkin Bound)
1) If d is even and $2d > n \geq d$, then

$$A_2(n,d) \leq 2\left\lfloor \frac{d}{2d-n} \right\rfloor$$

Also, if d is even,

$$A_2(2d,d) \leq 4d$$

2) If d is odd and $2d+1 > n \geq d$, then

$$A_2(n,d) \leq 2\left\lfloor \frac{d+1}{2d+1-n} \right\rfloor$$

Also, if d is odd,

$$A_2(2d+1,d) \leq 4d+4$$

Theorem 4.5.24 (The Elias Bound) Let $\theta = \frac{q-1}{q}$. If r is a positive integer satisfying $r < \theta n$ and $r^2 - 2\theta nr + \theta nd > 0$, then

$$A_q(n,d) \leq \frac{\theta nd}{r^2 - 2\theta nr + \theta nd} \cdot \frac{q^n}{V_q(n,r)}$$ □

This completes the overview. Now we present the details.

ELEMENTARY RESULTS

The following result is easily established.

Theorem 4.5.1 For any $n \geq 1$,
1) $A_q(n,1) = q^n$
2) $A_q(n,n) = q$ □

Theorem 4.5.2 For any $n \geq 2$,

$$A_q(n,d) \leq q\,A_q(n-1,d)$$

Proof. Let C be an optimal q-ary (n,M,d)-code. Thus, $M = A_q(n,d)$. Clearly, one of the q cross-sections $x_1 = i$ of C must contain at least M/q codewords, and so

$$A_q(n-1,d) \geq \frac{M}{q} = \frac{A_q(n,d)}{q}$$ ∎

According to Theorem 4.3.7, a binary $(n,M,2t+1)$-code exists if and only if a binary $(n+1,M,2t+2)$-code exists. Hence, we immediately have the following.

Theorem 4.5.3 For binary codes,

$$A_2(n,2t+1) = A_2(n+1,2t+2)$$

Put another way, if d is even, then $A_2(n,d) = A_2(n-1,d-1)$. ☐

Thus, for *binary* codes, it is enough to determine $A_2(n,d)$ for all *odd* values of d (or for all even values).

SMALL VALUES OF $A_q(n,d)$

The following table of values of $A_2(n,d)$ is taken from Hill (1986), which in turn comes from Sloane (1982).

TABLE 4.5.1 – $A_2(n,d)$

n	*d=3*	*d=5*	*d=7*
5	4	2	–
6	8	2	–
7	16	2	2
8	20	4	2
9	40	6	2
10	72-79	12	2
11	144-158	24	4
12	256	32	4
13	512	64	8
14	1024	128	16
15	2048	256	32
16	2560-3276	256-340	36-37

As the exercises show, many of the lower bounds on $A_2(n,d)$ in Table 4.5.1 come from codes that we discussed in the previous section, such as the Hamming codes and the repetition codes, along with shortened versions of these codes.

A LOWER BOUND ON $A_q(n,d)$

If C is a *maximal* (n,M,d)-code, then no word in V(n,q) has distance d or more from all codewords in C. In other words, the spheres $S_q(\mathbf{c}_i,d-1)$ about the codewords in C must cover V(n,q), and so

$$M \cdot V_q(n,d-1) \geq q^n$$

This gives the following important lower bound on $A_q(n,d)$.

Theorem 4.5.4 (The Gilbert-Varshamov Bound)

$$A_q(n,d) \geq \frac{q^n}{\sum_{i=0}^{d-1}\binom{n}{i}(q-1)^i}$$ □

This bound can actually be improved by considering linear codes. However, unlike the previous result, the bound applies only when q is a prime power. We postpone the proof of the following result until the next chapter, when we consider linear codes in more detail.

Theorem 4.5.5 (The Gilbert-Varshamov Bound) There exists a q-ary linear [n,k]-code with minimum distance at least d provided that

$$q^k < \frac{q^n}{\sum_{i=0}^{d-2}\binom{n-1}{i}(q-1)^i}$$

Hence, if k is the largest integer for which this inequality holds, then $A_q(n,d) \geq k$. □

UPPER BOUNDS ON $A_q(n,d)$

Now let us consider various upper bounds on $A_q(n,d)$.

The Singleton Bound

The Singleton bound is one of the simplest of all upper bounds on $A_q(n,d)$.

Theorem 4.5.6 (The Singleton Bound)

$$A_q(n,d) \leq q^{n-d+1}$$

Proof. Let C be a q-ary (n,M,d)-code. If we remove the last $d-1$ coordinate positions from each codeword in C, the resulting M words must be distinct. Since these words have length $n-d+1$, we get $M \leq q^{n-d+1}$. ∎

The Singleton bound is often not very good. However, there are cases where the Singleton bound gives equality. Here is an example.

Example 4.5.1 The following code is a (3,16,2)-code over $F_4 = \{0,1,a,b\}$

$$C = \{000 \quad 1a0 \quad b0a \quad ba1$$
$$01a \quad ab0 \quad 10b \quad 111$$
$$0ab \quad b10 \quad 1ba \quad aaa$$
$$0b1 \quad a01 \quad a1b \quad bbb\}$$

Hence, $A_4(3,2) \geq 16$. But the Singleton bound gives $A_4(3,2) \leq 4^{3-2+1} = 4^2 = 16$, and so equality holds. ☐

The Singleton bound implies that any [n,k,d]-code must satisfy

$$q^k \leq q^{n-d+1}$$

or

$$k \leq n - d + 1$$

or

$$d \leq n - k + 1$$

A linear code for which equality holds in this inequality is called a **maximum distance separable code**, or **MDS code**, since it has the largest possible minimum distance for any code with given length and dimension. Thus, an MDS code has parameters [n,n–d+1,d], or equivalently, [n,k,n–k+1]. MDS codes have some very nice properties, as we will see in Chapter 5.

For now, let us observe that if we remove *any* set of $n - k = d - 1$ coordinate positions from an MDS code C, then we must get q^k distinct strings of length k, that is, we get all of V(k,q). Thus, any set of k coordinate positions in C contains all possible k-tuples, and C is systematic on *any* set of k positions.

The Sphere-Packing Bound

Since the spheres of radius *pr*(C) about each codeword of C are disjoint, we immediately obtain the following *sphere-packing bound*, also known as the *Hamming bound.*

Theorem 4.5.7 (The Sphere-Packing Bound)

$$A_q(n,d) \leq \frac{q^n}{\sum_{k=0}^{t}\binom{n}{k}(q-1)^k}, \qquad t = \left\lfloor \frac{d-1}{2} \right\rfloor \qquad \square$$

Recall that a perfect code is a code for which equality holds in the sphere-packing bound.

∗ The Numbers A(n,d,w)

The sphere-packing bound can be improved upon by taking a more careful look at just what the spheres of radius $pr(C)$ actually cover. To this end, we restrict attention to the binary case $q = 2$, and introduce the notation $A(n,d,w)$ to denote the maximum size of a binary code of length n and minimum distance *at least* d, for which *all* codewords have weight w. Let us consider some of the more basic properties of the numbers $A(n,d,w)$. We leave proof of the following lemma as an exercise.

Lemma 4.5.8
1) $A(n,2k,k) = \lfloor \frac{n}{k} \rfloor$
2) $A(n,2k,w) = A(n,2k,n-w)$
3) $A(n,2k-1,w) = A(n,2k,w)$ □

The next result gives an upper bound on $A(n,2k,w)$.

Theorem 4.5.9 $A(n,2k,w) \le \left\lfloor \frac{kn}{w^2 - wn + kn} \right\rfloor$

Proof. Let C be an $(n,M,2k)$-code, each of whose codewords has weight w, where $M = A(n,2k,w)$. Arrange the codewords in C as rows of an $M \times n$ matrix B, and let k_i be the number of 1's in the i-th column of B. We compute the sum S of the scalar products $\mathbf{c}_i \cdot \mathbf{c}_j$ of all pairs of distinct rows of B.

On the one hand, if $i \neq j$, then

$$\mathbf{c}_i \cdot \mathbf{c}_j = w(\mathbf{c}_i \cap \mathbf{c}_j) = \tfrac{1}{2}[w(\mathbf{c}_i) + w(\mathbf{c}_j) - d(\mathbf{c}_i,\mathbf{c}_j)] \le \tfrac{1}{2}[2w - 2k] = w - k$$

and so

$$S = \sum_{i=1}^{M} \sum_{\substack{j=1 \\ j \neq i}}^{M} \mathbf{c}_i \cdot \mathbf{c}_j \le (w-k)M(M-1)$$

On the other hand, the contribution to S from the i-th column is $k_i(k_i - 1)$, and so we have

$$S = \sum_{i=1}^{n} k_i(k_i - 1) = \sum_{i=1}^{n} k_i^2 - \sum_{i=1}^{n} k_i = \sum_{i=1}^{n} k_i^2 - wM$$

But the last sum above is minimized when $k_i = wM/n$, and so we get

$$\frac{w^2M^2}{n} - wM \le S \le (w-k)M(M-1)$$

which gives the desired result upon solving for M. ∎

Example 4.5.2 According to Theorem 4.5.9,

$$A(9,6,4) \le \left\lfloor \frac{27}{16-36+27} \right\rfloor = 3$$

Furthermore, the code

$$C = \{1111100000, 100011100, 010010011\}$$

shows that $A(9,6,4) = 3$. □

Next, we have a recursive inequality for $A(n,2k,w)$.

Theorem 4.5.10 $A(n,2k,w) \le \left\lfloor \frac{n}{w} A(n-1,2k,w-1) \right\rfloor$.

Proof. Let C be an $(n,M,2k)$-code, each of whose codewords has weight w, where $M = A(n,2k,w)$. The cross-section $x_1 = 1$ of C has length $n-1$, minimum distance at least $2k$, constant codeword weight $w-1$, and hence size at most $A(n-1,2k,w-1)$. As a result, there are at most $A(n-1,2k,w-1)$ 1's in the first position of all codewords in C. Hence, the total number of 1's in all codewords in C is at most $nA(n-1,2k,w)$. But this number is equal to wM, and so we get $wM \le nA(n-1,2k,w)$, from which the result follows. ∎

Corollary 4.5.11

$$A(n,2k-1,w) = A(n,2k,w) \le \left\lfloor \frac{n}{k} \left\lfloor \frac{n-1}{k-1} \right\lfloor \cdots \left\lfloor \frac{n-w+k}{k} \right\rfloor \cdots \right\rfloor \right\rfloor \right\rfloor \qquad \square$$

Example 4.5.3 According to Corollary 4.5.11,

$$A(13,5,5) \le \left\lfloor \frac{13}{5} \left\lfloor \frac{12}{4} \left\lfloor \frac{11}{3} \right\rfloor \right\rfloor \right\rfloor = 23 \qquad \square$$

Here are some additional values of $A(n,2k,d)$, which we quote without proof.

Theorem 4.5.12

1) Schönheim (1966)

$$A(n,4,3) = \left\lfloor \frac{n}{3} \left\lfloor \frac{n-1}{2} \right\rfloor \right\rfloor - \epsilon$$

where $\epsilon = 1$ if $n \equiv 5 \bmod 6$ and 0 otherwise.

2) Kalbfleisch and Stanton (1968)

$$A(n,4,4) = \begin{cases} \dfrac{n(n-1)(n-2)}{24} & \text{for } n \equiv 2 \text{ or } 4 \bmod 6 \\ \dfrac{n(n-1)(n-3)}{24} & \text{for } n \equiv 1 \text{ or } 3 \bmod 6 \\ \dfrac{n(n-1)(n-6)}{24} & \text{for } n \equiv 0 \bmod 6 \end{cases}$$ □

∗ The Johnson Bound

Now we are ready to improve upon the sphere-packing bound, for binary codes. We begin with a lemma, whose proof is left as an exercise.

Lemma 4.5.13 If **c** and **d** are binary words, with $d(\mathbf{c},\mathbf{d}) = 2t+1$, then there exist exactly $\binom{2t+1}{t}$ binary words **x** with the property that $d(\mathbf{c},\mathbf{x}) = t+1$ and $d(\mathbf{d},\mathbf{x}) = t$. □

Now, if C is a binary (n,M,2t+1)-code, the sphere-packing bound says that the spheres of radius $t = pr(C)$ about the codewords in C are disjoint, and so

$$M \cdot V_2(n,t) \le 2^n$$

However, we may improve upon this inequality as follows.

Let T_{t+1} be the set of all binary words that have minimum distance exactly $t+1$ from the codewords in C. In other words, $\mathbf{x} \in T_{t+1}$ if and only if $d(\mathbf{x},\mathbf{c}) = t+1$ for some codeword **c** and $d(\mathbf{x},\mathbf{d}) \ge t+1$ for all codewords **d**. Since none of the words in T_{t+1} are in any of the spheres of radius t about the codewords of C, we have

$$M \cdot V_2(n,t) + |T_{t+1}| \le 2^n \tag{4.5.1}$$

To find a lower bound on $|T_{t+1}|$, consider the set

$$\begin{aligned} U &= \{(\mathbf{c},\mathbf{x}) \mid \mathbf{c} \in C,\ \mathbf{x} \in T_{t+1},\ d(\mathbf{c},\mathbf{x}) = t+1\} \\ &= \{(\mathbf{c},\mathbf{x}) \mid \mathbf{c} \in C,\ d(\mathbf{c},\mathbf{x}) = t+1,\ d(\mathbf{d},\mathbf{x}) \ge t+1 \text{ for all } \mathbf{d} \in C\} \end{aligned}$$

First, fix a codeword **c**, and consider the **c**-section

$$\begin{aligned} U_{\mathbf{c}} &= \{\mathbf{x} \mid (\mathbf{c},\mathbf{x}) \in U\} \\ &= \{\mathbf{x} \mid d(\mathbf{c},\mathbf{x}) = t+1,\ d(\mathbf{d},\mathbf{x}) \ge t+1 \text{ for all } \mathbf{d} \in C\} \end{aligned}$$

Notice that if $d(\mathbf{c},\mathbf{x}) = t+1$, then $d(\mathbf{d},\mathbf{x}) \geq t$ for all $\mathbf{d} \in C$, since otherwise $d(\mathbf{c},\mathbf{d}) \leq d(\mathbf{c},\mathbf{x})+d(\mathbf{x},\mathbf{d}) < 2t+1$. Hence,

$$U_{\mathbf{c}} = \{\mathbf{x} \mid d(\mathbf{c},\mathbf{x}) = t+1,\ d(\mathbf{d},\mathbf{x}) \neq t \text{ for all } \mathbf{d} \in C\}$$

Thus, in order to determine the size of $U_{\mathbf{c}}$, we must subtract from the $\binom{n}{t+1}$ binary words whose distance from $\mathbf{c}$ is $t+1$, those words whose distance from some other codeword $\mathbf{d}$ is equal to t.

Observe, however, that if $d(\mathbf{x},\mathbf{c}) = t+1$ and $d(\mathbf{x},\mathbf{d}) = t$, then $d(\mathbf{c},\mathbf{d}) = 2t+1$. Furthermore, no two distinct codewords can have distance t from a single word $\mathbf{x}$. Hence, according to Lemma 4.5.13,

$$|U_{\mathbf{c}}| = \binom{n}{t+1} - [\#\text{ codewords at distance } 2t+1 \text{ from } \mathbf{c}] \cdot \binom{2t+1}{t}$$

If we subtract $\mathbf{c}$ from each codeword in C that has distance exactly $2t+1$ from $\mathbf{c}$, the result is a code with length n, each of whose codewords has weight exactly $2t+1$. This implies that there are at most $A(n,2t+1,2t+1)$ codewords in C with distance exactly $2t+1$ from $\mathbf{c}$. Hence,

$$|U_{\mathbf{c}}| \geq \binom{n}{t+1} - A(n,2t+1,2t+1)\binom{2t+1}{t}$$

and so

$$|U| \geq M \cdot \left[\binom{n}{t+1} - A(n,2t+1,2t+1)\binom{2t+1}{t}\right] \tag{4.5.2}$$

Now fix a word $\mathbf{x} \in T_{t+1}$, and consider the $\mathbf{x}$-section

$$V_{\mathbf{x}} = \{\mathbf{c} \mid (\mathbf{c},\mathbf{x}) \in U\} = \{\mathbf{c} \mid \mathbf{c} \in C,\ d(\mathbf{c},\mathbf{x}) = t+1\}$$

If $\mathbf{c}, \mathbf{d} \in V_{\mathbf{x}}$, then $d(\mathbf{c},\mathbf{x}) = d(\mathbf{d},\mathbf{x}) = t+1$. Hence,

$$w(\mathbf{c}-\mathbf{x}) = w(\mathbf{d}-\mathbf{x}) = t+1$$

and so

$$\begin{aligned} 2t+1 \leq d(\mathbf{c},\mathbf{d}) &= d(\mathbf{c}-\mathbf{x},\mathbf{d}-\mathbf{x}) \\ &= w(\mathbf{c}-\mathbf{x})+w(\mathbf{d}-\mathbf{x}) - 2w([\mathbf{c}-\mathbf{x}] \cap [\mathbf{d}-\mathbf{x}]) \\ &= 2t+2 - 2w([\mathbf{c}-\mathbf{x}] \cap [\mathbf{d}-\mathbf{x}]) \end{aligned}$$

which implies that

$$w([\mathbf{c}-\mathbf{x}] \cap [\mathbf{d}-\mathbf{x}]) = 0$$

and so $\mathbf{c}-\mathbf{x}$ and $\mathbf{d}-\mathbf{x}$ have no 1's in common. Hence, there can be at most $\lfloor \frac{n}{t+1} \rfloor$ words of the form $\mathbf{c}-\mathbf{x}$, where $\mathbf{c} \in V_{\mathbf{x}}$, that is, at most $\lfloor \frac{n}{t+1} \rfloor$ words in $V_{\mathbf{x}}$. Therefore,

(4.5.3) $$|U| \le \left\lfloor \frac{n}{t+1} \right\rfloor |T_{t+1}|$$

Putting (4.5.2) and (4.5.3) together, we get

$$|T_{t+1}| \ge \frac{M}{\left\lfloor \frac{n}{t+1} \right\rfloor}\left[\binom{n}{t+1} - A(n,2t+1,2t+1)\binom{2t+1}{t}\right]$$

This bound, together with (4.5.1), gives the *Johnson bound.*

Theorem 4.5.14 (The Johnson bound)

$$A_2(n,2t+1) \le \frac{2^n}{\sum_{k=0}^{t}\binom{n}{k} + \frac{1}{\left\lfloor \frac{n}{t+1} \right\rfloor}\left[\binom{n}{t+1} - A(n,2t+1,2t+1)\binom{2t+1}{t}\right]}$$ □

Example 4.5.4 According to the results of Example 4.5.3, $A(13,5,5) \le 23$, and so the Johnson bound gives

$$A_2(13,5) \le \frac{2^{13}}{\sum_{k=0}^{2}\binom{13}{k} + \frac{1}{\left\lfloor \frac{13}{3} \right\rfloor}\left[\binom{13}{3} - 23\binom{5}{2}\right]} = \frac{2^{13}}{106} = 77.28$$

Thus, $A_2(13,5) \le 77$. (Note that this bound is not good enough to give the true value of $A_2(13,5)$, which is 64.) □

Using Corollary 4.5.11, we get the following corollary to Theorem 4.5.14.

Corollary 4.5.15

$$A_2(n,2t+1) \le \frac{2^n}{\sum_{k=0}^{t}\binom{n}{k} + \frac{1}{\left\lfloor \frac{n}{t+1} \right\rfloor}\binom{n}{t}\left(\frac{n-t}{t+1} - \left\lfloor \frac{n-t}{t+1} \right\rfloor\right)}$$ □

Example 4.5.5 A code is **nearly perfect** if it gives equality in Corollary 4.5.15. The shortened Hamming $[2^r-2,2^r-2-r,3]$-code is nearly perfect, and so

$$A_2(2^r-2,3) = 2^{2^r-2-r}$$

Hence, the shortened Hamming codes are optimal.

The so-called *punctured Preparata code* is a nearly perfect nonlinear $(2^m-1,2^{2^m-2m},5)$-code, and so

$$A_2(2^m-1,5) = 2^{2^m-2m}$$

Lindström (1975) has shown that there are no other nearly perfect binary codes. □

The Plotkin Bound

The Plotkin bound applies only when the minimum distance d is rather large. However, as we will see, it may be a superb bound. We begin with a preliminary version.

Theorem 4.5.16 **(Plotkin)** Let $\theta = \frac{q-1}{q}$. If $d > \theta n$, then

$$A_q(n,d) \leq \frac{d}{d - \theta n}$$

Proof. Let C be a q-ary (n,M,d)-code, and consider the sum of the distances between codewords, which is given by

$$S = \sum_{c \in C} \sum_{d \in C} d(c,d)$$

Since the minimum distance of C is d, we have

$$S \geq M(M-1)d$$

On the other hand, suppose that the number of j's in the i-th position of all codewords in C is k_{ij}, where $j = 0,\ldots,q-1$. Then the i-th position contributes a total of

$$\sum_{j=0}^{q-1} k_{ij}(M - k_{ij}) = M^2 - \sum_{j=0}^{q-1} k_{ij}^2 \leq M^2 - \frac{M^2}{q} = \theta M^2$$

to S, since the last sum above is smallest when $k_{ij} = M/q$. Since there are n positions, we have

$$M(M-1)d \leq S \leq \theta M^2 n$$

Solving for M gives the desired result. ▮

The Plotkin bound can easily be refined a bit when $q = 2$.

Theorem 4.5.17 If $d > \frac{1}{2}n$, then

$$A_2(n,d) \leq 2\left\lfloor \frac{d}{2d-n} \right\rfloor$$

Proof. As in the previous proof, let C be an (n,M,d)-code, and let S be defined as before. Hence,

$$S \geq M(M-1)d \tag{4.5.4}$$

Suppose that there are k_i 1's in the i-th position of all codewords. Then this position contributes $k_i(M-k_i)$ to S, and so

$$S = \sum_{i=1}^{n} k_i(M-k_i) \tag{4.5.5}$$

If M is even, this sum is maximized for $k_i = M/2$, in which case

$$S \leq \tfrac{1}{2}M^2 n$$

This, together with (4.5.4), gives

$$\binom{M}{2} d \leq \tfrac{1}{2}M^2 n$$

which is equivalent to

$$M \leq \frac{2d}{2d-n}$$

from which the result follows. If M is odd, then the sum in (4.5.5) is maximized for $k_i = \frac{1}{2}(M-1)$, in which case

$$S \leq \tfrac{1}{2}(M-1)^2 n$$

This, together with (4.5.4), gives

$$\binom{M}{2} d \leq \tfrac{1}{2}(M-1)^2 n$$

which is equivalent to

$$M \leq \frac{2d}{2d-n} - 1$$

which again implies the desired result. ∎

We can improve upon this result by separating the cases where d is even and where d is odd.

Theorem 4.5.18 (The Plotkin Bound)
1) If d is even and $d > \frac{1}{2}n$, then

$$A_2(n,d) \leq 2\left\lfloor \frac{d}{2d-n} \right\rfloor$$

Also, if d is even,

$$A_2(2d,d) \leq 4d$$

2) If d is odd and $d > \frac{1}{2}(n-1)$, then

$$A_2(n,d) \le 2\left\lfloor \frac{d+1}{2d+1-n} \right\rfloor$$

Also, if d is odd,

$$A_2(2d+1,d) \le 4d + 4$$

Proof. The first inequality in part 1) is Theorem 4.5.17. For the second inequality, let $d = 2k$. Then, from Theorem 4.5.2 and the first inequality, we have

$$A_2(4k,2k) \le 2A_2(4k-1,2k) \le 4\left\lfloor \frac{2k}{4k-(4k-1)} \right\rfloor = 8k$$

For the first inequality in part 2), we have

$$A_2(n,d) = A_2(n+1,d+1) \le 2\left\lfloor \frac{d+1}{2d+1-n} \right\rfloor$$

Finally, $A_2(2d+1,d) = A_2(2d+2,d+1) \le 4(d+1)$. ∎

Example 4.5.6 The Plotkin bound can also be used, in conjunction with Theorem 4.5.2, to give an upper bound when $d \le \frac{q-1}{q}n$. For example, we have

$$A_2(13,5) = 2^3 A_2(10,5) \le 8 \cdot 2 \cdot \left\lfloor \frac{6}{11-10} \right\rfloor = 96$$ □

Equality in the Plotkin Bound – Hadamard codes

The question of equality in the Plotkin bound is an interesting (and unfinished) one.

Definition A **Hadamard matrix** H_n of order n is an $n \times n$ matrix whose entries are all equal to 1 or -1, and for which

$$H_n H_n^T = nI_n$$

A Hadamard matrix whose first row and first column consist entirely of 1's is said to be **normalized.** □

For example, the following are normalized Hadamard matrices.

$$H_4 = \begin{bmatrix} 1 & 1 & 1 & 1 \\ 1 & -1 & 1 & -1 \\ 1 & 1 & -1 & -1 \\ 1 & -1 & -1 & 1 \end{bmatrix}$$

$$H_{12} = \begin{bmatrix} 1 & 1 & 1 & 1 & 1 & 1 & 1 & 1 & 1 & 1 & 1 & 1 \\ 1 & -1 & 1 & -1 & 1 & 1 & 1 & -1 & -1 & -1 & 1 & -1 \\ 1 & -1 & -1 & 1 & -1 & 1 & 1 & 1 & -1 & -1 & -1 & 1 \\ 1 & 1 & -1 & -1 & 1 & -1 & 1 & 1 & 1 & -1 & -1 & -1 \\ 1 & -1 & 1 & -1 & -1 & 1 & -1 & 1 & 1 & 1 & -1 & -1 \\ 1 & -1 & -1 & 1 & -1 & -1 & 1 & -1 & 1 & 1 & 1 & -1 \\ 1 & -1 & -1 & -1 & 1 & -1 & -1 & 1 & -1 & 1 & 1 & 1 \\ 1 & 1 & -1 & -1 & -1 & 1 & -1 & -1 & 1 & -1 & 1 & 1 \\ 1 & 1 & 1 & -1 & -1 & -1 & 1 & -1 & -1 & 1 & -1 & 1 \\ 1 & 1 & 1 & 1 & -1 & -1 & -1 & 1 & -1 & -1 & 1 & -1 \\ 1 & -1 & 1 & 1 & 1 & -1 & -1 & -1 & 1 & -1 & -1 & 1 \\ 1 & 1 & -1 & 1 & 1 & 1 & -1 & -1 & -1 & 1 & -1 & -1 \end{bmatrix}$$

It is known that a necessary condition for the existence of a Hadamard matrix H_n is that $n = 1$ or 2, or $4 \mid n$, but it is not known whether this condition is also sufficient. (It is a simple exercise to show that if a Hadamard matrix H_n exists, then so does a *normalized* Hadamard matrix of the same size.)

Given a normalized Hadamard matrix H_n, we can form several different codes as follows. Recall that the complement $\mathbf{x}^c$ of a binary word is the word obtained from $\mathbf{x}$ by interchanging 0's and 1's. Also, $C^c = \{\mathbf{c}^c \mid \mathbf{c} \in C\}$.

Theorem 4.5.19 The matrix A_n obtained from a normalized Hadamard matrix H_n, by replacing all 1's by 0's and all -1's by 1's, is called a **binary Hadamard matrix.** Furthermore, from A_n, we can construct the following **Hadamard codes.**

1) The rows of A_n, with the first coordinate removed, form an $(n-1, n, \frac{1}{2}n)$-code, which we denote by $\mathit{Had1}_n$.

2) The shortened Hadamard code $\mathit{SHad1}_n$, obtained by taking the cross-section $x_1 = 0$ of $\mathit{Had1}_n$, is an $(n-2, \frac{1}{2}n, \frac{1}{2}n)$-code.

3) The set $\mathit{Had1}_n \cup (\mathit{Had1}_n)^c$ is an $(n-1, 2n, \frac{1}{2}n-1)$-code, denoted by $\mathit{Had2}_n$.

4) The rows of A_n, together with the complements of these rows, form an $(n, 2n, \frac{1}{2}n)$-code, denoted by $\mathit{Had3}_n$. □

Proof.

1) Since the rows of H_n are orthogonal, the number of positions in which any two rows agree must equal the number of positions in which they disagree. Hence, the distance between any two distinct rows of A_n is $\frac{1}{2}n$.

2) This follows from the fact that, except for the first column, each column of H_n has the same number of 1's and -1's.

3) This follows from Theorem 4.3.9, since

$$d = \min\{\tfrac{1}{2}n,(n-1)-\tfrac{1}{2}n\} = \tfrac{1}{2}n - 1$$

4) This also follows from Theorem 4.3.9. ∎

Our plan now is to discuss a theorem of Levenshtein that says if Hadamard matrices H_m exist for certain values of m (an open question at present), then equality must hold in Plotkin's bound.

We begin by discussing some new constructions. Let C be an (n,M,d)-code. If u is a nonnegative integer and $\mathbf{c}$ is a codeword, then $u\mathbf{c}$ denotes the word $\mathbf{c}\mathbf{c}\cdots\mathbf{c}$, formed by juxtaposing u copies of $\mathbf{c}$. Also, uC denotes the code

$$uC = \{u\mathbf{c} \mid \mathbf{c} \in C\}$$

which is a (un,M,ud)-code. If $u = 0$, then uC is taken to be the empty set.

Now suppose that C_1 is an (n_1,M_1,d_1)-code and C_2 is an (n_2,M_2,d_2)-code, both over the same alphabet. Suppose further that we fix the order of the codewords in each code, writing

$$C_1 = \langle \mathbf{c}_{11},\mathbf{c}_{12},\ldots,\mathbf{c}_{1M_1}\rangle \quad \text{and} \quad C_2 = \langle \mathbf{c}_{21},\mathbf{c}_{22},\ldots,\mathbf{c}_{2M_2}\rangle$$

where the angle brackets indicate an *ordered* set. Then we can *interleave* codewords of C_1 and C_2, to form a new code $C_1 \odot C_2$, by juxtaposing corresponding codewords in the *ordered* codes C_1 and C_2, until we run out of codewords in either code. In symbols,

$$C_1 \odot C_2 = \begin{cases} \{\mathbf{c}_{11}\mathbf{c}_{21},\mathbf{c}_{12}\mathbf{c}_{22},\ldots,\mathbf{c}_{1M_1}\mathbf{c}_{2M_1}\} & \text{if } M_1 \le M_2 \\ \{\mathbf{c}_{11}\mathbf{c}_{21},\mathbf{c}_{12}\mathbf{c}_{22},\ldots,\mathbf{c}_{1M_2}\mathbf{c}_{2M_2}\} & \text{if } M_1 > M_2 \end{cases}$$

We will leave it as an exercise to show that $C_1 \odot C_2$ has parameters

$$(n_1+n_2,\ \min\{M_1,M_2\},\ d)$$

where $d \ge d_1+d_2$. If either of the codes C_i is the empty set, then by definition, $C_1 \odot C_2$ is equal to the other code.

We can combine the two previous constructions to get the following result.

Theorem 4.5.20 Let C_1 be an ordered (n_1,M_1,d_1)-code, and let C_2 be an ordered (n_2,M_2,d_2)-code, both over the same alphabet. If u,v are positive integers, then $uC_1 \odot vC_2$ has parameters

$$(un_1+vn_2,\ \min\{M_1,M_2\},\ d)$$

where $d \ge ud_1+vd_2$. □

Now we are ready to discuss Levenshtein's theorem. If n is any multiple of 4, we know that the Hadamard codes of Theorem 4.5.19 exist. Let us consider a code of the form

$$C = u\mathit{SHad1}_{4k} \odot v\mathit{SHad1}_{4k+4} \tag{4.5.6}$$

where the Hadamard codes are ordered by the rows of the binary Hadamard matrices from whence they come. According to the previous theorem, if $u,v > 0$, then C has parameters

$$N = (4k-2)u+(4k+2)v, \quad M = 2k, \quad D \geq 2ku+(2k+2)v$$

Also, if $u > 0$ and $v = 0$, then $C = u\mathit{SHad1}_{4k}$ has parameters

$$N = (4k-2)u, \quad M = 2k, \quad D = 2ku$$

Since

$$2\left\lfloor \frac{D}{2D-N} \right\rfloor = 2\left\lfloor \frac{2ku+(2k+2)v}{2[2ku+(2k+2)v] - [(4k-2)u+(4k+2)v]} \right\rfloor$$

$$= 2\left\lfloor k + \frac{v}{u+v} \right\rfloor = 2k = M$$

we see that equality holds in the Plotkin bound for the code (4.5.6).

Now, if n and d are both even, with $2d > n \geq d$, then setting

$$n = (4k-2)u+(4k+2)v$$

and

$$d = 2ku+(2k+2)v$$

and solving for u and v gives

$$u = \tfrac{1}{2}[d(2k+1) - n(k+1)], \quad v = \tfrac{1}{2}[nk - d(2k-1)] \tag{4.5.7}$$

which implies, using $2d > n \geq d$, that $u > 0$ and $v \geq 0$.

We have therefore shown that, if n and d are both even and $2d > n \geq d$, and *if* Hadamard matrices H_{4k} and H_{4k+4} exist for

$$k = \left\lfloor \frac{d}{2d-n} \right\rfloor$$

then equality holds in the Plotkin bound, that is,

$$A_2(n,d) = 2\left\lfloor \frac{d}{2d-n} \right\rfloor$$

and the code (4.5.6), where u and v are given by (4.5.7), is optimal.

The case where n is odd and k is even is handled in an entirely

analogous manner by considering codes of the form

$$C = u\,\mathit{Had1}_{2k} + v\,\mathit{SHad1}_{4k+4}$$

(Notice that the first code is not shortened and has length 2k.) The case where both n and k are odd is handled with codes of the form

$$C = u\,\mathit{SHad1}_{4k} + v\,\mathit{Had1}_{4k+2}$$

Let us now state Levenshtein's Theorem.

Theorem 4.5.21 (Levenshtein)

1) Let d be even, $2d > n \geq d$, and $k = \lfloor \frac{d}{2d-n} \rfloor$.

(i) Then equality holds in Plotkin's bound, that is,

$$A_2(n,d) = 2\left\lfloor \frac{d}{2d-n} \right\rfloor$$

whenever the Hadamard matrices H_{4k} and H_{4k+4} exist, and in addition, H_{2k} exists if n is odd and k is even, and H_{2k+2} exists if both n and k are odd. Furthermore, if

$$u = d(2k+1) - n(k+1), \quad v = nk - d(2k-1)$$

then the following codes are optimal:

n even: $\frac{u}{2}\mathit{SHad1}_{4k} \odot \frac{v}{2}\mathit{SHad1}_{4k+4}$

n odd, k even: $u\,\mathit{Had1}_{2k} \odot \frac{v}{2}\mathit{SHad1}_{4k+4}$

n odd, k odd: $\frac{u}{2}\mathit{SHad1}_{4k} \odot v\,\mathit{Had1}_{2k+2}$

(ii) We also have

$$A_2(2d,d) = 4d$$

whenever a Hadamard matrix H_{2d} exists, where the code $\mathit{Had3}_{2d}$ is optimal.

2) Let d be odd, $2d+1 > n \geq d$, and $k = \lfloor \frac{d+1}{2d+1-n} \rfloor$.

(iii) Then equality holds in Plotkin's bound, that is,

$$A_2(n,d) = 2\left\lfloor \frac{d+1}{2d+1-n} \right\rfloor$$

whenever the Hadamard matrices H_{4k} and H_{4k+4} exist, and in addition, H_{2k} exists if n odd and k even, and H_{2k+2} exists if both n and k are odd.

(iv) We also have

$$A_2(2d+1,d) = 4(d+1)$$

whenever a Hadamard matrix H_{2d} exists.

Optimal codes in case d is odd are obtained by shortening optimal codes for d+1.

Proof. Part 2) follows from part 1) using the fact that, if d is odd, then $A_2(n,d) = A_2(n+1,d+1)$. We have already established part (i) for n even. We leave the rest of the proof as an exercise. ∎

Example 4.5.7 Let us apply Levenshtein's Theorem to the case $n = 19$ and $d = 12$. In this case, $2d > n \geq d$, and

$$k = \left\lfloor \frac{d}{2d-n} \right\rfloor = 2$$

Furthermore, $u = 12(5) - 19(3) = 3$ and $v = 19(2) - 12(3) = 2$. Hence, the code

$$C = 3Had1_4 \odot SHad1_{12}$$

is optimal. Now, from H_4, we get the ordered code

$$Had1_4 = \begin{bmatrix} 0 & 0 & 0 \\ 1 & 0 & 1 \\ 0 & 1 & 1 \\ 1 & 1 & 0 \end{bmatrix}$$

(the codewords are the rows), and from H_{12}, we get the ordered code

$$SHad1_{12} = \begin{bmatrix} 0&0&0&0&0&0&0&0&0 \\ 1&1&0&1&0&0&0&1&1&1 \\ 1&1&1&0&1&1&0&1&0&0 \\ 0&1&1&1&0&1&1&0&1&0 \end{bmatrix}$$

Pasting together 3 copies of $Had1_4$ and 1 copy of $SHad_{12}$ gives

$$C = \begin{bmatrix} 0\,0\,0 & 0\,0\,0 & 0\,0\,0 & 0\,0\,0\,0\,0\,0\,0\,0\,0\,0 \\ 1\,0\,1 & 1\,0\,1 & 1\,0\,1 & 1\,1\,0\,1\,0\,0\,0\,1\,1\,1 \\ 0\,1\,1 & 0\,1\,1 & 0\,1\,1 & 1\,1\,1\,0\,1\,1\,0\,1\,0\,0 \\ 1\,1\,0 & 1\,1\,0 & 1\,1\,0 & 0\,1\,1\,1\,0\,1\,1\,0\,1\,0 \end{bmatrix}$$ □

***The Elias Bound**

The Elias bound is one of the best known bounds (although the linear programming bound discussed in the next chapter can be better). It is proved in a manner similar to the Plotkin bound. Let C be a q-ary (n,M)-code, and let r be a positive integer. Consider the spheres

$S_q(\mathbf{x},r)$ of radius r about all words in $V(n,q)$, and suppose that the sphere $S_q(\mathbf{x},r)$ contains $K_\mathbf{x}$ codewords. Let us count the number of pairs $(\mathbf{c},S_q(\mathbf{x},r))$, where the codeword $\mathbf{c}$ lies in the sphere $S_q(\mathbf{x},r)$. On the one hand, each codeword is in exactly $V_q(n,r)$ of these spheres, and on the other hand, each sphere $S_q(\mathbf{x},r)$ contains $K_\mathbf{x}$ codewords. Hence,

$$\sum_{\mathbf{x}\in V(n,q)} K_\mathbf{x} = M\cdot V_q(n,r)$$

Dividing both sides by q^n gives the average number of codewords in the spheres, and since there must be at least one sphere that meets or exceeds the average, we have established the following theorem.

Theorem 4.5.22 Let C be a q-ary (n,M)-code. For any integer $r>0$, there is a sphere $S_q(\mathbf{x},r)$ about some word $\mathbf{x}$ in $V(n,q)$ with the property that the number $K_\mathbf{x}$ of codewords in $S_q(\mathbf{x},r)$ satisfies

$$K_\mathbf{x} \geq \frac{M\cdot V_q(n,r)}{q^n}$$

Such a sphere $S_q(\mathbf{x},r)$ is called a **critical sphere** for C. ▯

Now suppose that C is a q-ary (n,M,d)-code, and let $S_q(\mathbf{x},r)$ be a critical sphere. The code C' formed by subtracting $\mathbf{x}$ from all codewords in C is also a q-ary (n,M,d)-code, and so we may as well assume that $\mathbf{x}=\mathbf{0}$, and so the critical sphere is $S_q(\mathbf{0},r)$.

Let us consider the code $C_1 = C\cap S_q(\mathbf{0},r)$, which consists of the codewords in C that have weight at most r. The code C_1 is an (n,K,e)-code, where by Theorem 4.5.22,

$$K \geq \frac{M\cdot V_q(n,r)}{q^n}$$

and $e\geq d$.

Our desire is to use a refinement of the method of proof for the Plotkin bound on the code C_1. Since we know something about the maximum weight of each codeword in C_1, we have some additional advantages in this case. Thus, we consider the sum of the distances between codewords in C_1

$$S=\sum_{\mathbf{c}\in C_1}\sum_{\mathbf{d}\in C_1} d(\mathbf{c},\mathbf{d})$$

As before, we have

$$S\geq K(K-1)e \tag{4.5.8}$$

Again, we let $k_{i,j}$ be the number of j's in the i-th position in all codewords in C_1. Hence,

$$S = \sum_{i=1}^{n} \sum_{j=0}^{q-1} k_{ij}(K - k_{ij})$$

Our task is to estimate this sum from above. Let us collect some facts about the numbers k_{ij}. Since there are K codewords in C_1,

$$\sum_{j=0}^{q-1} k_{ij} = K$$

Also, since $w(\mathbf{c}) \le r$, the number of 0's in $\mathbf{c}$ must be at least $n - r$. Hence, the total number of 0's in all codewords is at least $K(n-r)$, that is,

$$T = \sum_{i=1}^{n} k_{i0} \ge K(n-r) \tag{4.5.9}$$

Now we are ready to estimate S. Since we have some special knowledge about k_{i0}, we separate these terms from the others,

$$S = \sum_{i=1}^{n} \sum_{j=0}^{q-1} k_{ij}(K - k_{ij}) = nK^2 - \sum_{i=1}^{n}\left(k_{i0}^2 + \sum_{j=1}^{q-1} k_{ij}^2 \right)$$

Using the Cauchy-Schwarz inequality, we have

$$\sum_{j=1}^{q-1} k_{ij}^2 \ge \frac{1}{q-1}\left(\sum_{j=1}^{q-1} k_{ij} \right)^2 = \frac{1}{q-1}(K - k_{i0})^2$$

and so

$$\begin{aligned}
S &\le nK^2 - \sum_{i=1}^{n}\left(k_{i0}^2 + \frac{1}{q-1}(K - k_{i0})^2 \right) \\
&= nK^2 - \frac{1}{q-1}\sum_{i=1}^{n}\left((q-1)k_{i0}^2 + K^2 - 2Kk_{i0} + k_{i0}^2 \right) \\
&= nK^2 - \frac{1}{q-1}\sum_{i=1}^{n}\left(qk_{i0}^2 + K^2 - 2Kk_{i0} \right) \\
&= nK^2 - \frac{n}{q-1}K^2 - \frac{q}{q-1}\sum_{i=1}^{n} k_{i0}^2 + \frac{2}{q-1}K\sum_{i=1}^{n} k_{i0} \\
&= \frac{q-2}{q-1}nK^2 - \frac{q}{q-1}\sum_{i=1}^{n} k_{i0}^2 + \frac{2}{q-1}KT
\end{aligned}$$

Another application of the Cauchy-Schwarz inequality gives

$$\sum_{i=1}^{n} k_{i0}^2 \geq \frac{1}{n}\left(\sum_{i=1}^{n} k_{i0}\right)^2 = \frac{1}{n}T^2$$

and so

$$S \leq \frac{q-2}{q-1}nK^2 - \frac{q}{q-1}\frac{1}{n}T^2 + \frac{2}{q-1}KT$$
$$= \frac{1}{q-1}\left((q-2)nK^2 - \frac{q}{n}T^2 + 2KT\right)$$

The expression in parentheses is a quadratic in T, whose maximum value occurs at

$$T = \frac{nK}{q}$$

Thus, if we wish to bound this quadratic from above by using (4.5.9), we must choose r so that

$$K(n-r) > \frac{nK}{q}$$

that is,

$$r < \frac{q-1}{q}n$$

Assuming this to be the case, we get using (4.5.9)

$$S \leq \frac{1}{q-1}\left((q-2)nK^2 - \frac{q}{n}K^2(n-r)^2 + 2K^2(n-r)\right)$$
$$= \frac{1}{q-1}K^2 r\left(2(q-1) - \frac{q}{n}r\right)$$

Therefore, together with (4.5.8), we have

$$K(K-1)e \leq \frac{1}{q-1}K^2 r\left(2(q-1) - \frac{q}{n}r\right)$$

Solving this for K gives the following theorem, which is the main portion of the Elias bound.

Theorem 4.5.23 Let C_1 be a q-ary (n,K,e)-code, and let $\theta = \frac{q-1}{q}$. If r is a positive integer satisfying $r < n\theta$, then

(4.5.10) $$K \leq \frac{\theta ne}{r^2 - 2\theta nr + \theta ne}$$

provided that the denominator $r^2 - 2\theta nr + \theta ne$ is positive. ☐

Recall that C_1 came from an (n,M,d)-code C, and that

$$K \geq \frac{M \cdot V_q(n,r)}{q^n}$$

and $e \geq d$. Hence, replacing e by d in (4.5.10), we have

$$\frac{M \cdot V_q(n,r)}{q^n} \leq K \leq \frac{\theta nd}{r^2 - 2\theta nr + \theta nd}$$

Solving for M, we get the Elias bound.

Theorem 4.5.24 **(The Elias Bound)** Let $\theta = \frac{q-1}{q}$. If r is a positive integer satisfying $r < \theta n$ and $r^2 - 2\theta nr + \theta nd > 0$, then

$$A_q(n,d) \leq \frac{\theta nd}{r^2 - 2\theta nr + \theta nd} \cdot \frac{q^n}{V_q(n,r)} \qquad \square$$

ASYMPTOTIC BOUNDS

We conclude this section with a brief discussion of asymptotic bounds. Recalling our discussion relating to Corollary 4.2.7, we set $\delta = d/n$ and consider the quantity

$$\alpha_q(\delta) = \limsup_{n \to \infty} \left[\frac{1}{n} \log_q A_q(n, \delta n)\right]$$

which is the limit superior of the maximum rate of a code of length n and minimum distance d, when $n \to \infty$ in such as way that $\delta = d/n$ remains constant.

Each of the bounds we have discussed in this section leads to an asymptotic bound, that is, a bound on $\alpha(\delta)$. We give only a sampling here. For more on this subject, we suggest the books of van Lint (1982) and MacWilliams and Sloane (1977).

For $0 < \lambda < 1$ and $\mu = 1 - \lambda$, let

$$H_q(\lambda) = \lambda \log_q \frac{1}{\lambda} + \mu \log_q \frac{1}{\mu}$$

Note that $H_2(\lambda) = H(\lambda)$ is the entropy function of Chapter 1. The following result is proved in the appendix. (See Corollary A.3.11.)

Lemma 4.5.25 Let $\theta = \frac{q-1}{q}$. For $0 \leq \delta \leq \theta$,

$$\lim_{n \to \infty} \left[\frac{1}{n} \log_q V_q(n, \lfloor \delta n \rfloor)\right] = H_q(\delta) + \delta \log_q(q-1) \qquad \square$$

Our first result is an asymptotic lower bound, which comes from the Gilbert-Varshamov bound.

Theorem 4.5.26 Let $\theta = \frac{q-1}{q}$. If $0 \le \delta \le \theta$, then

$$\alpha_q(\delta) \ge 1 - H_q(\delta) - \delta \log_q(q-1)$$

Proof. The Gilbert-Varshamov bound (Theorem 4.5.4) is

$$A_q(n,d) \ge \frac{q^n}{V_q(n,d-1)}$$

Hence,

$$\frac{1}{n}\log_q A_q(n,d) = 1 - \frac{1}{n}\log_q V_q(n,d-1) \ge 1 - \frac{1}{n}\log_q V_q(n,d)$$

and so

$$\alpha_q(\delta) = \limsup_{n\to\infty}\left[\frac{1}{n}\log_q A_q(n,\delta n)\right] \ge 1 - \lim_{n\to\infty}\frac{1}{n}\log_q V_q(n,\delta n)$$

Applying Lemma 4.5.25 completes the proof. ∎

Corollary 4.5.27 Let $\theta = \frac{q-1}{q}$ and $0 \le \delta \le \theta$. There is a sequence C_m of (n_m, M_m, d_m)-codes with the following property. Given any $\epsilon > 0$, there is an m_ϵ such that $m > m_\epsilon$ implies that

$$\delta(C_m) = \frac{d_m}{n_m} \ge \delta$$

and

$$R(C_m) \ge 1 - H_q(\delta) - \delta \log_q(q-1) - \epsilon$$ □

Any sequence of codes that meets the conditions of the previous corollary is said to *meet the Gilbert-Varshamov bound.*

Now let us consider an asymptotic upper bound.

Theorem 4.5.28 Let $\theta = \frac{q-1}{q}$. Then

$$\alpha_q(\delta) \le 1 - H_q(\theta - \sqrt{\theta(\theta-\delta)}) \qquad \text{if } 0 \le \delta \le \theta$$
$$\alpha_q(\delta) = 0 \qquad \text{if } \theta < \delta < 1$$

Proof. The second part follows from Plotkin's bound (Theorem 4.5.16), which tells us that if $d > \theta n$ then

$$A_q(n,d) \le \frac{d}{d - \theta n} = \frac{\delta}{\delta - \theta}$$

where the expression on the right is constant (since δ is constant). For the first half, we use the Elias bound (Theorem 4.5.24), which tells us that if $0 < r < \theta n$ and $f(r) = r^2 - 2\theta nr + \theta nd > 0$, then

$$A_q(n,d) \le \frac{\theta nd}{r^2 - 2\theta nr + \theta nd} \cdot \frac{q^n}{V_q(n,r)}$$

Now, the first factor on the right side can be rewritten, to give

$$A_q(n,\delta n) \le \frac{\theta\delta}{(\frac{r}{n})^2 - 2\theta(\frac{r}{n}) + \theta\delta} \cdot \frac{q^n}{V_q(n,r)} \tag{4.5.11}$$

We wish to choose a positive integer $r < \theta n$ so that $f(r) > 0$. But the zeros of $f(r)$ are $n(\theta \pm \sqrt{\theta(\theta-\delta)})$, and so if we choose r so that

$$\tfrac{r}{n} < \theta - \sqrt{\theta(\theta-\delta)} \le \theta$$

we will have $r < \theta n$ and $f(r) > 0$.

In particular, we may take $\lambda < \theta - \sqrt{\theta(\theta-\delta)}$, and choose $r = \lfloor \lambda n \rfloor$. Then (4.5.11) holds, and so by Lemma 4.5.24,

$$\tfrac{1}{n}\log_q A_q(n,\delta n) \le \tfrac{1}{n}\log_q\left(\frac{\theta\delta}{(\frac{r}{n})^2 - 2\theta(\frac{r}{n}) + \theta\delta}\right) + 1 - \tfrac{1}{n}\log_q V_q(n,\lfloor\lambda n\rfloor)$$

$$\to 1 - H_q(\lambda) - \lambda\log_q(q-1) \quad \text{as } n\to\infty$$

Since this holds for any $\lambda < \theta - \sqrt{\theta(\theta-\delta)}$, the result follows. ∎

EXERCISES

1. Prove Theorem 4.5.1.
2. Use Theorem 4.5.2 to establish the Singleton bound.
3. Prove Lemma 4.5.8.
4. Prove Corollary 4.5.11.
5. Show that $A(8,6,4) = 2$.
6. Prove that $\binom{n}{w}A_2(n,2k) \le 2^n A(n,2k,w)$.
7. Prove that
$$A(n,2k,w) \le \left\lfloor \tfrac{n}{n-w} A(n-1,2k,w) \right\rfloor$$
8. Prove Lemma 4.5.13.
9. Prove Corollary 4.5.15.
10. Verify that the codes in Example 4.5.5 are nearly perfect.
11. Is a perfect code nearly perfect?

12. Prove that if we multiply any row or column of a Hadamard matrix by -1, the result is another Hadamard matrix. What does this have to do with normalized Hadamard matrices?
13. Show that the columns of a Hadamard matrix are orthogonal.
14. If H_n is a Hadamard matrix, show that

$$\begin{bmatrix} H_n & H_n \\ H_n & -H_n \end{bmatrix}$$

is also a Hadamard matrix.
15. Use the results of the previous exercise to construct H_8.
16. Show that $C_1 \oplus C_2$ is an $(n_1+n_2,\min\{M_1,M_2\},d)$-code, where $d \geq d_1+d_2$. Prove Theorem 4.5.20.
17. Finish the proof of Theorem 4.5.21.
18. Using Levenshtein's Theorem, find an optimal binary code of length 6 and minimum distance 4.
19. Using Levenshtein's Theorem, find an optimal binary code of length 9 and minimum distance 6.
20. Using Levenshtein's Theorem, find an optimal binary code of length 17 and minimum distance 10.
21. Using Levenshtein's Theorem, find an optimal binary code of length 27 and minimum distance 16.
22. Use Hamming codes and their shortened versions to verify the lower bounds on $A_2(5,3)$, $A_2(6,3)$ and $A_2(7,3)$ in Table 4.5.1.
23. Use Hamming codes and their shortened versions to verify the lower bounds on $A_2(12,3)$, $A_2(13,3)$, $A_2(14,3)$ and $A_2(15,3)$ in Table 4.5.1.
24. Use repetition codes and codes obtained from the repetition codes to verify the lower bounds on $A_2(3,3)$, $A_2(4,3)$, $A_2(5,5)$, $A_2(6,5)$, and $A_2(7,5)$ in Table 4.5.1.
25. Let L be a binary linear $[n,k,d]$-code with generator matrix G. We may assume that the first row of G has the form $1\cdots10\cdots0$, where there are d 1's. Why? Hence, G has the form

$$G = \begin{bmatrix} 1\cdots1 & 0\cdots0 \\ G_1 & G_2 \end{bmatrix}$$

Let d_2 be the minimum distance of the $[n-d,k-1]$-code with generator matrix G_2. Show that $d_2 \geq d/2$.

CHAPTER 5

Linear Codes

5.1 Linear Codes and Their Duals

In this chapter, we look more closely at the most important class of codes – the linear codes. For completeness, we restate the definition.

Definition A code $L \subset V(n,q)$ is a **linear code** if it is a subspace of $V(n,q)$. If L has dimension k over $V(n,q)$, we say that L is an **[n,k]-code**, and if L has minimum distance d, we say that L is an **[n,k,d]-code.** ☐

Recall also that, according to Theorem 4.3.5, the minimum distance of a linear code is equal to the minimum weight of the code, in symbols, $d(L) = w(L)$.

THE GENERATOR MATRIX OF A LINEAR CODE

Since a linear code is a vector space, we can describe it by giving a basis. It is customary to arrange the basis vectors as rows of a matrix.

Definition Let L be an $[n,k]$-code. A $k \times n$ matrix G whose rows form a basis for L is called a **generator matrix** for L. ☐

If L is an $[n,k]$-code, with generator matrix G, then the codewords in L are precisely the linear combinations of the rows of G. Put another way,

$$L = \{xG \mid x \in V(k,q)\}$$

This provides a very simple method for encoding source data. For if the source can be represented as the set of all q-ary words of length k, then we may encode the source word $\mathbf{x} \in V(k,q)$ as the codeword $\mathbf{x}G$.

Example 5.1.1 Consider the binary code with generator matrix

$$G = \begin{bmatrix} 1 & 1 & 0 & 0 \\ 0 & 1 & 1 & 1 \\ 1 & 0 & 1 & 0 \end{bmatrix}$$

This code can encode source symbols from V(3,2). In particular, for each $\mathbf{x} = (x_1,x_2,x_3) \in V(3,2)$, we associate the codeword

$$\begin{bmatrix} x_1 & x_2 & x_3 \end{bmatrix} \begin{bmatrix} 1 & 1 & 0 & 0 \\ 0 & 1 & 1 & 1 \\ 1 & 0 & 1 & 0 \end{bmatrix} = (x_1+x_3, x_1+x_2, x_2+x_3, x_2) \qquad \square$$

Since performing elementary row operations (interchanging rows, multiplying a row by a nonzero scalar, and adding a multiple of one row to another) does not change the row space of a matrix, any matrix that is row equivalent to a generator matrix for a code L is also a generator matrix for L. This implies the following theorem.

Theorem 5.1.1 Let L be a linear [n,k]-code. Given any k coordinate positions, there is a code equivalent to L that is systematic on those positions. $\square$

A generator matrix of the form $G = (I_k \mid A)$, where I_k is the identity matrix of size k, is said to be in **standard form**. The generator matrix in Example 5.1.2 below has this form. In view of the previous remarks, every linear code has a generator matrix in standard form.

When a $k \times n$ generator matrix is in standard form, it is systematic on its first k coordinate positions. This makes both encoding and the reverse process very simple.

Example 5.1.2 As we will see in the next chapter, the matrix

$$G = \begin{bmatrix} 1 & 0 & 0 & 0 & 0 & 1 & 1 \\ 0 & 1 & 0 & 0 & 1 & 0 & 1 \\ 0 & 0 & 1 & 0 & 1 & 1 & 0 \\ 0 & 0 & 0 & 1 & 1 & 1 & 1 \end{bmatrix}$$

is a generator matrix for the Hamming code $\mathcal{H}_2(3)$. Notice that G is in standard form. The Hamming code $\mathcal{H}_2(3)$ can encode source words

from V(4,2) as follows

$$\mathbf{x}G = \begin{bmatrix} x_1 \, x_2 \, x_3 \, x_4 \end{bmatrix} \begin{bmatrix} 1 & 0 & 0 & 0 & 0 & 1 & 1 \\ 0 & 1 & 0 & 0 & 1 & 0 & 1 \\ 0 & 0 & 1 & 0 & 1 & 1 & 0 \\ 0 & 0 & 0 & 1 & 1 & 1 & 1 \end{bmatrix}$$

$$= [x_1 \quad x_2 \quad x_3 \quad x_4 \quad x_2{+}x_3{+}x_4 \quad x_1{+}x_3{+}x_4 \quad x_1{+}x_2{+}x_4]$$

Since G is in standard form, the original source message appears as the first k symbols of its codeword. ▯

Generator matrices can also be used to describe scalar multiple equivalence.

Theorem 5.1.2 Two linear codes L_1 and L_2, with generating matrices G_1 and G_2, respectively, are scalar multiple equivalent if and only if G_1 can be transformed into G_2 by elementary row operations, by permuting the columns of G_1, and by multiplying the columns of G_1 by nonzero scalars. ▯

THE DUAL OF A LINEAR CODE

The vector space V(n,q) has a natural inner product defined on it. In particular, if $\mathbf{x} = x_1 \cdots x_n$ and $\mathbf{y} = y_1 \cdots y_n$ are in V(n,q), we define the **dot product** or **scalar product** of **x** and **y** by

$$\mathbf{x} \cdot \mathbf{y} = x_1 y_1 + \cdots + x_n y_n$$

(We will also use the notation $\langle \mathbf{x}, \mathbf{y} \rangle$ for $\mathbf{x} \cdot \mathbf{y}$.) The following concept plays a key role in the theory of linear codes.

Definition Let L be a linear [n,k]-code. The set

$$L^{\perp} = \{\mathbf{x} \in V(n,q) \mid \mathbf{x} \cdot \mathbf{c} = 0 \text{ for all } \mathbf{c} \in L\}$$

is called the **dual code** of L. ▯

Theorem 5.1.3

1) If G is a generator matrix for L, then

$$L^{\perp} = \{\mathbf{x} \in V(n,q) \mid \mathbf{x}G^{\mathsf{T}} = \mathbf{0}\}$$

2) The dual $L^{\perp}$ of a linear [n,k]-code is a linear [n,n−k]-code.
3) For any linear code L, we have $L^{\perp\perp} = L$.

Proof. Part 1) follows from the fact that **x** is orthogonal to every codeword in L if and only if it is orthogonal to every codeword in a basis for L. Part 2) follows from part 1), which says that $L^{\perp}$ is the solution space of a system of k equations in n unknowns, with

rank k. We leave the details of this as an exercise. For part 3), we first observe that $L \subset L^{\perp\perp}$. But L and $L^{\perp\perp}$ have the same dimension, and so they must be equal. ∎

We should remark that the properties of the dual of a linear code over a *finite* field can be quite different from those of the dual space of a vector space over the real numbers. For instance, if W is a subspace of a finite dimensional *real* vector space V, then $W \cap W^{\perp} = \{\mathbf{0}\}$, since no vector is orthogonal to itself. This is not always the case for linear codes, however. In fact, as the next example illustrates, we can even have $L^{\perp} = L$!

Example 5.1.3 For the binary [4,2]-code

$$L = \{0000, 1100, 0011, 1111\}$$

we have $L \subset L^{\perp}$, and since $L^{\perp}$ is also a [4,2]-code, we get $L = L^{\perp}$. A linear code L for which $L = L^{\perp}$ is said to be ***self-dual.*** We will discuss self-dual codes in more detail later in this section. □

Despite the fact that $L \cap L^{\perp}$ need not be the zero code, part 2 of Theorem 5.1.3 does tells us that

$$dim(L) + dim(L^{\perp}) = n$$

Let L be a linear code with $k \times n$ generator matrix $G = (I_k \mid A)$ in standard form, and let H be the matrix

$$H = (-A^T \mid I_{n-k})$$

where A^T is the transpose of A. Then

$$GH^T = (I_k \mid A)\begin{pmatrix} -A \\ I_{n-k} \end{pmatrix} = -A+A = 0$$

Hence, the rows of H are orthogonal to the rows of G, and since $\mathrm{rank}(H) = n-k = dim(L^{\perp})$, we deduce that H is a generator matrix for the dual code $L^{\perp}$.

The matrix H is also called a **parity check matrix** for L. The reason for this terminology is that

$$\mathbf{x} \in L \quad \text{if and only if} \quad \mathbf{x}H^T = \mathbf{0}$$

and if $H = (h_{ij})$ and $\mathbf{x} = (x_i)$, then $\mathbf{x}H^T = \mathbf{0}$ has the form

$$\begin{aligned} h_{11}x_1 + h_{12}x_2 + \cdots + h_{1n}x_n &= 0 \\ h_{21}x_1 + h_{22}x_2 + \cdots + h_{2n}x_n &= 0 \\ &\vdots \\ h_{n-k,1}x_1 + h_{n-k,2}x_2 + \cdots + h_{n-k,n}x_n &= 0 \end{aligned}$$

Thus, the rows of H are the coefficients of a system of linear equations whose solutions are precisely the codewords in L. These linear equations are called **parity check equations.** (Appending an even *parity check bit* x_{n+1} to a binary codeword $x_1 \cdots x_n$ is done in such a way that $\sum x_i = 0$. This is the origin of the term parity check equation.)

The parity check matrix H is *not* in standard form *as a generator matrix* for L. However, *as a parity check matrix*, a matrix of the form $(B \mid I_m)$ is said to be in **standard form.**

Example 5.1.4 Since the generator matrix G for the code $\mathcal{H}_2(3)$ of Example 5.1.2 is in standard form, $\mathcal{H}_2(3)$ has parity check matrix (in standard form)

$$H = \begin{bmatrix} 0 & 1 & 1 & 1 & 1 & 0 & 0 \\ 1 & 0 & 1 & 1 & 0 & 1 & 0 \\ 1 & 1 & 0 & 1 & 0 & 0 & 1 \end{bmatrix}$$

In this case, the parity check equations are

$$\begin{aligned} x_2 + x_3 + x_4 + x_5 &= 0 \\ x_1 + x_3 + x_4 + x_6 &= 0 \\ x_1 + x_2 + x_4 + x_7 &= 0 \end{aligned}$$

□

There does not seem to be an efficient way to determine the minimum distance (weight) of a linear code *directly* from a generator matrix. However, we can do so from a parity check matrix H for L. In particular, let the columns of H be $k_1, \ldots, k_n$, and suppose that a particular choice of w of the columns are linearly dependent. Then there exist coefficients $c_1, \ldots, c_n$, with exactly w being nonzero, for which

$$c_1 k_1 + \cdots + c_n k_n = 0$$

This is equivalent to $\mathbf{c}H^{\mathsf{T}} = \mathbf{0}$, where $\mathbf{c} = c_1 \cdots c_n$, and so $\mathbf{c} \in L$. Furthermore, since $\mathbf{c}$ has weight w, we have $d(L) = w(L) \leq w$.

On the other hand, if $\mathbf{c}$ is any codeword of weight w, then $\mathbf{c}H^{\mathsf{T}} = \mathbf{0}$, and therefore some w columns of H are linearly dependent. We have established the following very useful result.

Theorem 5.1.4 Let L be a linear [n,k,d]-code, with parity check matrix H. Then d is the smallest integer r for which there are r linearly dependent columns in H. (Thus, H has d linearly dependent columns, but any $d-1$ columns are linearly independent.) □

This theorem can be used to give the promised proof of the Gilbert-Varshamov bound (Theorem 4.5.5).

Theorem 5.1.5 (The Gilbert-Varshamov Bound) There exists a q-ary linear [n,k]-code with minimum distance at least d provided that

$$q^k < \frac{q^n}{\sum_{i=0}^{d-2} \binom{n-1}{i} (q-1)^i}$$

Hence, if k is the largest integer for which this inequality holds, then $A_q(n,d) \geq k$.

Proof. According to Theorem 5.1.4, we can establish this theorem by constructing an $(n-k) \times n$ parity check matrix for which any set of $d-1$ columns is linearly independent. To this end, we may choose any nonzero (n–k)-tuple for the first column of H. Then we may choose the second column to be any nonzero (n–k)-tuple that is not a scalar multiple of the first column. In general, we want to choose the i-th column to be any nonzero (n–k)-tuple that is not in the linear span of any set of $d-2$ previously chosen columns. Now, the number of linear combinations of $d-2$ or fewer columns, from among the $i-1$ existing columns, is

$$N_i = \binom{i-1}{1}(q-1) + \binom{i-1}{2}(q-1)^2 + \cdots + \binom{i-1}{d-2}(q-1)^{d-2}$$

Since we also cannot choose the zero column, there are

$$q^{n-k} - N_i - 1$$

available choices. As long as $q^{n-k} - N_i - 1 > 0$ for $i = n$ (and hence for all $i \leq n$), we may complete the matrix H. This completes the proof as well. ∎

SYNDROME DECODING

An efficient decoding process for linear codes can be obtained through the use of parity check matrices.

Definition Let L be an [n,k]-code, with parity check matrix H. For any $\mathbf{x} \in V(n,q)$, the word $\mathbf{x}H^{\top}$ is called the **syndrome** of **x**. □

Thus, $\mathbf{x} \in L$ if and only if the syndrome of **x** is **0**.

Let us recall a few simple facts about quotient spaces. If $L \subset V(n,q)$ is a linear code (i.e., subspace), the *quotient space* of V(n,q) *modulo* L is defined by

$$\frac{V(n,q)}{L} = \{\mathbf{x} + L \mid \mathbf{x} \in V(n,q)\}$$

The set $\mathbf{x} + L = \{\mathbf{x} + \mathbf{c} \mid \mathbf{c} \in L\}$ is called a *coset* of L. The quotient

space is also a vector space over F_q, where

$$a(\mathbf{x}+L) = a\mathbf{x}+L \quad \text{and} \quad (\mathbf{x}+L)+(\mathbf{y}+L) = (\mathbf{x}+\mathbf{y})+L$$

Recall also that $\mathbf{x}+L = \mathbf{y}+L$ if and only if $\mathbf{x}-\mathbf{y} \in L$.

Now we can state the following result.

Theorem 5.1.6 Let L be an $[n,k]$-code, with parity check matrix H. Then $\mathbf{x}$ and $\mathbf{y}$ in $V(n,q)$ have the same syndrome if and only if they lie in the same coset of the quotient space $V(n,q)/L$.

Proof. We have

$$\mathbf{x}+L = \mathbf{y}+L \text{ iff } \mathbf{x}-\mathbf{y} \in L \text{ iff } (\mathbf{x}-\mathbf{y})H^T = \mathbf{0} \text{ iff } \mathbf{x}H^T = \mathbf{y}H^T \quad \square$$

Now let us suppose that a word $\mathbf{x}$ is received. Minimum distance decoding requires that we decode $\mathbf{x}$ as a codeword $\mathbf{c}$ for which $\mathbf{a} = \mathbf{x}-\mathbf{c}$ has smallest weight. But as $\mathbf{c}$ ranges over L, $\mathbf{a}$ ranges over the coset $\mathbf{x}+L$. Hence, minimum distance decoding requires that we decode $\mathbf{x}$ as the codeword

$$\mathbf{c} = \mathbf{x}-\mathbf{a}$$

where $\mathbf{a}$ is a word in $\mathbf{x}+L$ of smallest weight, that is, a word of smallest weight among those words with the same syndrome as $\mathbf{x}$.

Theorem 5.1.7 Let L be a linear code with parity check matrix H. Then minimum distance decoding is equivalent to decoding a received word $\mathbf{x}$ as a word $\mathbf{c} = \mathbf{x}-\mathbf{a}$, where $\mathbf{a}$ is a word of smallest weight in the coset $\mathbf{x}+L$, or equivalently, where $\mathbf{a}$ is a word of smallest weight with the same syndrome as $\mathbf{x}$. $\square$

The decoding process in Theorem 5.1.7 can be described in terms of a so-called **standard array** for L,

$$\begin{array}{ccccc}
\mathbf{0} & \mathbf{c}_1 & \mathbf{c}_2 & \cdots & \mathbf{c}_m \\
\mathbf{a}_1 & \mathbf{c}_1+\mathbf{a}_1 & \mathbf{c}_2+\mathbf{a}_1 & \cdots & \mathbf{c}_m+\mathbf{a}_1 \\
\mathbf{a}_2 & \mathbf{c}_1+\mathbf{a}_2 & \mathbf{c}_2+\mathbf{a}_2 & \cdots & \mathbf{c}_m+\mathbf{a}_2 \\
 & & \vdots & & \\
\mathbf{a}_s & \mathbf{c}_1+\mathbf{a}_s & \mathbf{c}_2+\mathbf{a}_s & \cdots & \mathbf{c}_m+\mathbf{a}_s
\end{array}$$

The first row of the array consists of the codewords in L. To form the second row, we choose a word $\mathbf{a}_1$ of smallest weight that is not in the first row, and add it to each word of the first row. This forms the coset

$\mathbf{a}_1 + L$. In general, the i-th row of the array is formed by choosing a word $\mathbf{a}_i$ of smallest weight that is not yet in the array, and adding it to each word of the first row, to form the coset $\mathbf{a}_i + L$. This process continues until the array contains all words in $V(n,q)$. The elements $\mathbf{a}_i$ are called the **coset leaders** of the array.

Since each row of the standard array is a coset of L, two words in $V(n,q)$ have the same syndrome if and only if they lie in the same row of the array.

In view of the way that the coset leaders were selected, if a received word $\mathbf{x}$ is in the j-th column of the array, then $\mathbf{x} = \mathbf{c}_j + \mathbf{a}_i$ for some i, where $\mathbf{a}_i$ is a word of smallest weight in the coset $\mathbf{x} + L$. Hence, $\mathbf{x}$ is decoded as the codeword $\mathbf{c}_j$. Put another way, we decode $\mathbf{x}$ as the codeword at the top of the column containing $\mathbf{x}$.

Fortunately, we can avoid having to maintain the entire standard array by determining the coset leader $\mathbf{a}_i$, rather than the codeword $\mathbf{c}_j$. Of course, determining the coset leader is equivalent to determining the codeword, since $\mathbf{c}_j = \mathbf{x} - \mathbf{a}_i$. The point is that, since each row is uniquely determined by the syndrome of its members, we need only maintain a table of coset leaders and their syndromes. Then if $\mathbf{x}$ is received, we compute its syndrome, find the coset leader $\mathbf{a}_i$ with the same syndrome, and decode $\mathbf{x}$ as $\mathbf{c} = \mathbf{x} - \mathbf{a}_i$. This process is referred to as **syndrome decoding**.

Example 5.1.5 Let L be the binary [4,2]-code with generator matrix

$$G = \begin{bmatrix} 1 & 1 & 0 & 1 \\ 0 & 1 & 0 & 0 \end{bmatrix}$$

The cosets of L are

$$\begin{aligned} \mathbf{0} + C &= \{0000, 0100, 1101, 1001\} \\ 1000 + C &= \{1000, 1100, 0101, 0001\} \\ 0010 + C &= \{0010, 0110, 1111, 1011\} \\ 1010 + C &= \{1010, 1110, 0111, 0011\} \end{aligned}$$

Since the coset leaders were chosen with minimum weight, the standard array is

0000	0100	1101	1001
1000	1100	0101	0001
0010	0110	1111	1011
1010	1110	0111	0011

Now, by adding the second row of G to the first row, we obtain a generating matrix in standard form

$$G' = \begin{bmatrix} 1 & 0 & 0 & 1 \\ 0 & 1 & 0 & 0 \end{bmatrix}$$

which gives the parity check matrix

$$H = \begin{bmatrix} 0 & 0 & 1 & 0 \\ 1 & 0 & 0 & 1 \end{bmatrix}$$

From this we can create a table of coset leaders and their syndromes.

Coset leader	*Syndrome*
0000	00
1000	01
0010	10
1010	11

To decode the received word $\mathbf{x} = 1110$, for instance, we compute its syndrome

$$1110 \cdot H^T = 11$$

Hence, according to the syndrome table, the coset leader is 1010, and so we decode $\mathbf{x}$ as

$$1110 + 1010 = 0100$$

□

We should observe that errors in transmission will be corrected if and only if those errors correspond to one of the coset leaders. To see this, suppose that $\mathbf{c}$ is the codeword that was actually sent, and $\mathbf{x} = \mathbf{c} + \mathbf{e}$ is received, where $\mathbf{e}$ is the **error vector**. If $\mathbf{e}$ is a coset leader, then $\mathbf{x}$ lies in the row headed by $\mathbf{e}$, and so decoding $\mathbf{x}$ will give the correct codeword $\mathbf{x} - \mathbf{e} = \mathbf{c}$. On the other hand, if $\mathbf{e}$ is not a coset leader but lies in the j-th row of the standard array, then $\mathbf{x}$ is in the j-th row as well and will be decoded *incorrectly* as $\mathbf{x} - \mathbf{a}_j \neq \mathbf{x} - \mathbf{e} = \mathbf{c}$.

In particular, if L has minimum distance d, then all of the words in V(n,q) of weight at most $t = \lfloor (d-1)/2 \rfloor$ must be coset leaders. This can also be seen by noting that if two words $\mathbf{x}$ and $\mathbf{y}$ of weight at most t were in the same coset, then their difference $\mathbf{x} - \mathbf{y}$ would be a codeword of weight at most $2t \leq d-1$, which is not possible.

THE PROBABILITY OF CORRECT DECODING

Let L be a linear code. We have seen that syndrome decoding will result in the correct codeword if and only if the error made in transmission is one of the coset leaders. Assuming a binary symmetric channel, with crossover probability p, if we let α_i be the number of

coset leaders that have weight i, for $i = 0,\ldots,n$, then the probability of correct decoding is the probability that the error is one of these coset leaders, which is

$$p_{corr\ decode} = \sum_{i=0}^{n} \alpha_i p^i (1-p)^{n-i}$$

THE PROBABILITY OF ERROR DETECTION

An error in the transmission of a codeword **c** will go *undetected* if and only if the received word **d** is a *codeword* that is different from **c**. Thus, for a linear code L, an undetected error occurs if and only if the error vector $\mathbf{d}-\mathbf{c}$ is a nonzero codeword. Hence, if A_k denotes the number of codewords in L of weight k, the probability of an undetected error, for the binary symmetric channel with crossover probability p, is

$$p_{undet\ err} = \sum_{k=1}^{n} A_k p^k (1-p)^{n-k}$$

We will have much more to say about the important numbers A_k in the next section.

MAJORITY LOGIC DECODING

A procedure referred to as *majority logic decoding* often provides a simple method for decoding a linear code. Let us describe this procedure using an example.

Recall that the matrix

$$G = \begin{bmatrix} 1 & 0 & 0 & 0 & 0 & 1 & 1 \\ 0 & 1 & 0 & 0 & 1 & 0 & 1 \\ 0 & 0 & 1 & 0 & 1 & 1 & 0 \\ 0 & 0 & 0 & 1 & 1 & 1 & 1 \end{bmatrix}$$

of Example 5.1.2 is the generating matrix of the Hamming code $\mathcal{H}_2(3)$. Therefore, it is also the parity check matrix of the dual code, which we simply denote by C.

Now, we may perform any elementary row operations on the rows of G and still have a parity check matrix for C. Adding row 1 to the other rows gives

$$G_1 = \begin{bmatrix} 1 & 0 & 0 & 0 & 0 & 1 & 1 \\ 1 & 1 & 0 & 0 & 1 & 1 & 0 \\ 1 & 0 & 1 & 0 & 1 & 0 & 1 \\ 1 & 0 & 0 & 1 & 1 & 0 & 0 \end{bmatrix}$$

Adding rows 1 and 2 to row 3 gives

$$G_2 = \begin{bmatrix} 1 & 0 & 0 & 0 & 0 & 1 & 1 \\ 1 & 1 & 0 & 0 & 1 & 1 & 0 \\ 1 & 1 & 1 & 0 & 0 & 0 & 0 \\ 1 & 0 & 0 & 1 & 1 & 0 & 0 \end{bmatrix}$$

The parity check equations that correspond to rows 3, 4, and 1, respectively, are

$$(5.1.1) \qquad \begin{array}{l} x_1 + x_2 + x_3 = 0 \\ x_1 + x_4 + x_5 = 0 \\ x_1 + x_6 + x_7 = 0 \end{array}$$

Notice that the coefficient of x_1 in each equation is 1 and that each of the other variables appears with a nonzero coefficient in one and only one equation. Let us pause for a definition.

Definition A system of parity check equations for a binary linear code is said to be **orthogonal** with respect to the variable x_i provided x_i appears in *every* equation of the system with coefficient 1, but all other variables appear in *exactly one* equation with coefficient 1. □

Thus (5.1.1) is orthogonal with respect to x_1.

Now suppose that a single error occurs in transmission. If the error is in the first position, then x_1 is incorrect, but all other x_j are correct. Hence, each of the equations (5.1.1) will be unsatisfied (that is, we will get $1=0$ upon substitution of the variables). On the other hand, if the error is in any position other than the first, then exactly one of the equations (5.1.1) will be unsatisfied. Thus, the number of unsatisfied equations will tell us whether or not the first position in the received word is correct (assuming a single error). (If exactly two equations are unsatisfied, we deduce that at least two errors have occurred.)

Thus, we have a simple method for correcting the first position in the received word. Assuming a single error, if the *majority* of equations (5.1.1) are satisfied, then position 1 is correct – otherwise it is not. This is majority logic decoding.

We leave it as an exercise to show that, by judicious choice of row operations, we can perform majority logic decoding on all 7 positions in the code C.

More generally, suppose we have r parity check equations for a binary [n,k]-code L. Suppose that these equations are orthogonal with respect to the variable x_i. Suppose further that $t \le r/2$ errors have

occurred in transmission. If one of the errors is in the i-th position, then at most $t-1$ of the equations can be "corrected" by the remaining errors, and so at least $r-(t-1) \geq r/2+1$ equations will be unsatisfied. On the other hand, if the i-th position does not suffer an error, then at most $t \leq r/2$ equations will be unsatisfied. Therefore, the i-th position in the received word is in error *if and only if* the majority of equations is unsatisfied.

SELF-DUAL CODES

A linear code L is said to be **self-orthogonal** if $L \subset L^{\perp}$. We leave proof of the following simple result as an exercise.

Theorem 5.1.8 Let G be a generator matrix for a q-ary linear code L. Then L is self-orthogonal if and only if distinct rows of G are orthogonal and have weight divisible by q. ▯

Theorem 5.1.9 If distinct rows of a generator matrix for a *binary* linear code L are orthogonal and have weight divisible by 4, then L is self-orthogonal and all weights in L are divisible by 4.
Proof. Self-orthogonality follows from the previous theorem. The second statement follows by induction, using the fact that

$$w(\mathbf{u}+\mathbf{v}) = w(\mathbf{u}) + w(\mathbf{v}) - 2w(\mathbf{u} \cap \mathbf{v})$$

and that $w(\mathbf{u} \cap \mathbf{v})$ is even, since **u** and **v** are orthogonal. ∎

Example 5.1.6 According to the previous theorem, the binary [7,3]-code L with generator matrix

$$G = \begin{bmatrix} 1 & 1 & 1 & 1 & 0 & 0 & 0 \\ 0 & 0 & 1 & 1 & 0 & 1 & 1 \\ 0 & 1 & 0 & 1 & 1 & 0 & 1 \end{bmatrix}$$

is self-orthogonal and all codeword weights are divisible by 4. Hence, all seven nonzero codewords in L have weight 4. Furthermore, since $L^{\perp}$ is a [7,4]-code that contains L, as well as the string **1**, we deduce that $L^{\perp}$ is generated by the rows of G, along with the string **1**. ▯

A linear code L is said to be **self-dual** if $L^{\perp} = L$. In this case, L must be an [n,n/2]-code, where n is even. In fact, a linear [n,k]-code L is self-dual if and only if it is self-orthogonal and $k = n/2$. We will encounter some important self-dual codes in the sequel. In the meanwhile, here is a simple example of a self-dual code.

Example 5.1.7 The code L in Example 5.1.6 is self-orthogonal, and its dual code $L^{\perp}$ has generator matrix

$$G^{\perp} = \begin{bmatrix} 1 & 1 & 1 & 1 & 0 & 0 & 0 \\ 0 & 0 & 1 & 1 & 0 & 1 & 1 \\ 0 & 1 & 0 & 1 & 1 & 0 & 1 \\ 1 & 1 & 1 & 1 & 1 & 1 & 1 \end{bmatrix}$$

By adding an overall parity check to this code, we obtain a code L′ with generator matrix

$$G' = \begin{bmatrix} 1 & 1 & 1 & 1 & 0 & 0 & 0 & 0 \\ 0 & 0 & 1 & 1 & 0 & 0 & 1 & 1 \\ 0 & 1 & 0 & 1 & 1 & 0 & 1 & 0 \\ 1 & 1 & 1 & 1 & 1 & 1 & 1 & 1 \end{bmatrix}$$

This code is a self-orthogonal [8,4]-code, and so it is self-dual. □

One of the reasons that self-dual codes are important is that there are arbitrarily long self-dual codes that meet the Gilbert-Varshamov lower bound (see Section 4.5 on asymptotic results). Hence, there are "reasonably good" long self-dual codes. (See MacWilliams, Sloane, and Thompson (1972) and Pless and Pierce (1973).)

If a code L is self-dual, then any parity check matrix for L is also a generating matrix, and any generating matrix is also a parity check matrix. Thus, if $G = (I_{n/2} \mid A)$ is a generating matrix for L, so is $H = (-A^{T} \mid I_{n/2})$.

The following theorem, whose proof we shall omit (see MacWilliams and Sloane (1977), p. 633 and Pless (1968)) describes the conditions under which a self-dual code exists.

Theorem 5.1.10 A q-ary self-dual [n,n/2]-code exists if and only if one of the following holds

1) q and n are both even
2) $q \equiv 1 \bmod 4$ and n is even
3) $q \equiv 3 \bmod 4$ and n is divisible by 4. □

In particular, we note that a binary self-dual [n,n/2]-code exists for all positive even integers n, and a ternary self-dual [n,n/2]-code exists if and only if n is divisible by 4.

A binary self-dual code L has the property that all codeword weights are even. If, in addition, all codeword weights in L are divisible by 4, then L is said to be an **even code.** (Some authors use

the term **doubly even.**) For a proof of the following result, see Gleason (1971).

Theorem 5.1.11 An even [n,n/2]-code exists if and only if n is divisible by 8. □

*THE NUMBER OF BINARY SELF-DUAL CODES

A relatively simple counting argument can be used to count the number of binary self-dual [n,n/2]-codes. These results have been generalized to q-ary codes by Pless (1968). We begin with a definition and a preliminary result, which is of independent interest.

Definition A linear code L is **weakly self-dual** if $\mathbf{1} \in L \subset L^{\perp}$. □

Theorem 5.1.12 If q is a prime power, then there are

$$\binom{n}{k}_q = \frac{(q^n-1)\cdots(q-1)}{(q^k-1)\cdots(q-1)(q^{n-k}-1)\cdots(q-1)}$$

subspaces of V(n,q) of dimension k. The expressions $\binom{n}{k}_q$ are called **Gaussian coefficients** and have properties similar to those of the binomial coefficients.

Proof. Let S(n,k) be the number of k-dimensional subspaces of V(n,q). Let N(n,k) be the number of k-tuples of linearly independent vectors $(v_1,\ldots,v_k)$ in V(n,q). To determine N(n,k), we observe that v_1 can be chosen in q^n-1 ways, v_2 can be chosen in q^n-q ways, and so on. Hence

$$N(n,k) = (q^n-1)(q^n-q)\cdots(q^n-q^{n-k+1})$$

Now, each of these k-tuples can be obtained by first choosing a subspace of V(n,q) of dimension k, and then selecting the vectors from this subspace. By a reasoning similar to that given above, for any k-dimensional subspace of V(n,q), the number of k-tuples of independent vectors in this subspace is

$$(q^k-1)(q^k-q)\cdots(q^k-q^{k-1})$$

Hence, we have

$$N(n,k) = S(n,k)(q^k-1)(q^k-q)\cdots(q^k-q^{k-1})$$

from which the result follows. ▮

Theorem 5.1.13 Let L be a binary weakly self-dual [n,k]-code. The total number of binary weakly self-dual [n,n/2]-codes containing L is

$$\prod_{i=1}^{n/2-k} (2^i + 1)$$

Proof. Let $\sigma_{n,m}$ be the number of weakly self-dual [n,m]-codes C containing L, for $k \le m \le n/2$. Let us count the number of ordered pairs (D,E), where

$$L \subset D \subset E$$

and where D is a weakly self-dual [n,m]-code and E is a weakly self-dual [n,m+1]-code.

First, we fix D. Since E has dimension one greater than D, it must be generated by D and a single nonzero vector $\mathbf{x}$ not in D. In fact, if $\mathbf{x} \in E - D$, then

$$E = D \cup (\mathbf{x}+D)$$

is the code generated by D and $\mathbf{x}$. Furthermore, since E is weakly self-dual, we have $\mathbf{x} \in E \subseteq E^{\perp} \subset D^{\perp}$. Thus, E is obtained from D by adjoining a string $\mathbf{x} \in D^{\perp} - D$. But $D \cup (\mathbf{x}+D) = D \cup (\mathbf{y}+D)$ if and only if $\mathbf{x}$ and $\mathbf{y}$ lie in the same coset of D in $D^{\perp}$. Hence, the number of distinct codes of the form $E = D \cup (\mathbf{x}+D)$ is the number of distinct nontrivial cosets of D in $D^{\perp}$, which is

$$\frac{2^{n-m}}{2^m} - 1 = 2^{n-2m} - 1$$

Thus, the number of pairs (D,E) is $\sigma_{n,m}(2^{n-2m} - 1)$.

On the other hand, let us fix E. Since every subcode (subspace) of a weakly self-dual code that contains 1 is also weakly self-dual, any [n,m]-code D for which $L \subset D \subset E$ will form a pair (D,E). Now, the number of m-dimensional subspaces D of E that contain the k-dimensional subspace L is the same as the number of (m−k)-dimensional subspaces of an (m+1−k)-dimensional space, which is equal to the Gaussian coefficient

$$\binom{m+1-k}{m-k}_2 = 2^{m+1-k} - 1$$

Hence, the number of pairs (D,E) is also equal to $\sigma_{n,m+1}(2^{m+1-k} - 1)$. Equating the two expressions for this number gives

$$\sigma_{n,m}(2^{n-2m} - 1) = \sigma_{n,m+1}(2^{m+1-k} - 1)$$

or

$$\sigma_{n,m+1} = \sigma_{n,m} \frac{2^{n-2m} - 1}{2^{m+1-k} - 1} \quad \text{for } k \le m < n/2$$

Since $\sigma_{n,k} = 1$, the result follows. ∎

By taking $L = \{0,1\}$ and $k = 1$ in Theorem 5.1.13, we arrive at our goal.

Corollary 5.1.14 The total number of binary self-dual [n,n/2]-codes is

$$\prod_{i=1}^{n/2-1} (2^i + 1) \qquad \square$$

BURST ERROR DETECTION AND CORRECTION

We mentioned in Section 4.1 that errors in transmission often occur in clusters or *bursts*. For instance, electrical interferences often last longer than the time it takes to transmit a single code symbol, and defects on magnetic tape or disk are usually larger than the space required to store a single code symbol. We will discuss the issue of burst error correction in some detail in Chapter 7 and again in Chapter 8. For now, let us present a few simple facts on the subject.

Definition A **burst** in $V(n,q)$ of length b is a string in $V(n,q)$ whose nonzero coordinates are confined to b consecutive positions, the first and last of which must be nonzero. $\square$

The following lemma will be useful.

Lemma 5.1.15 Let L be a linear [n,k]-code. If L contains no bursts of length b or less, then we must have $k \leq n - b$.
Proof. Consider the set S of all strings in $V(n,q)$ with 0's in the last $n-b$ positions. (The first b positions may contain any values, including 0.) If any two distinct strings in S lie in the same coset of L, then their difference would be a nonzero burst of length at most b, which is not possible. Hence, the number of cosets of L, which is q^{n-k}, must be greater than or equal to the size of S, which is q^b. In other words, $q^{n-k} \geq q^b$, from which the result follows. ∎

We have seen that the more errors we expect a code to detect (or correct), the smaller must be the code. The situation for burst error *detection* is settled quite easily by the following result.

Theorem 5.1.16 If a linear [n,k]-code L can detect all burst errors of length b or less, then we must have $k \leq n - b$. Furthermore, there is a linear [n,n−b]-code that will detect all burst errors of length b or less.
Proof. In order for L to detect all burst errors of length b or less, no burst of length b or less can be in L. Hence, by Lemma 5.1.15, we

have $k \le n - b$. To prove the second statement, consider the parity check matrix H of size $b \times n$ whose first row is

$$\underbrace{100\cdots0}_{b\ symbols}\ \underbrace{100\cdots0}_{b\ symbols}\ \cdots\ \cdots\ \underbrace{100\cdots0}_{b\ symbols}\ \underbrace{100\cdots0}_{\le b\ symbols}$$

and each of whose remaining rows is a right shift of the previous row (adding a 0 to the beginning of the row). For instance, when $b = 3$ and $n = 11$, we have

$$H = \begin{bmatrix} 1 & 0 & 0 & 1 & 0 & 0 & 1 & 0 & 0 & 1 & 0 \\ 0 & 1 & 0 & 0 & 1 & 0 & 0 & 1 & 0 & 0 & 1 \\ 0 & 0 & 1 & 0 & 0 & 1 & 0 & 0 & 1 & 0 & 0 \end{bmatrix}$$

Let e be a burst of length b or less in $V(n,q)$. Since any b consecutive columns of H form a $b \times b$ permutation matrix, the syndrome eH^T is just a reordering of the vector e, and so is nonzero. Hence, any burst error of length b or less will be detected. ∎

Now let us consider burst error correction.

Theorem 5.1.17 If a linear [n,k]-code L can correct all burst errors of length b or less (using minimum distance decoding) then we must have $k \le n - 2b$.

Proof. If $2 \le \ell \le 2b$, then any burst e of length ℓ can be written as the difference $e_1 - e_2$ of two distinct bursts of length at most b. Since L can correct e_1 and e_2, they cannot lie in the same coset of L, and so their difference e cannot be a codeword. Since a burst of length 1 cannot be a codeword, we may apply Lemma 5.1.15 (with b replaced by 2b), to get $k \le n - 2b$. ∎

We observed in the previous proof that if a code L can correct any burst error of length b or less, then no two such bursts can lie in the same coset of L. By counting the number of bursts of length b or less, we get a lower bound on the number of cosets of L, and hence an upper bound on the dimension of L. We leave proof of the following result as an exercise.

Theorem 5.1.18 If a linear [n,k]-code L can correct all burst errors of length b or less, then we must have

$$k \le n - b + 1 - \log_q[(q-1)(n-b+1) + 1]$$ □

EXERCISES

1. Prove Theorem 5.1.1.
2. Prove Theorem 5.1.2.
3. Finish the proof of Theorem 5.1.3.
4. Find the generator matrix for a q-ary repetition code.
5. Can a binary (11,24,5)-code be linear? Explain.
6. If L is a linear code, is the extended code $\widehat{L}$, defined by adding an overall parity check to L, also linear?
7. Find the minimum distance of a binary linear code with generator matrix

$$G = \begin{bmatrix} 1 & 0 & 0 & 1 & 1 \\ 0 & 0 & 1 & 0 & 1 \\ 0 & 1 & 1 & 1 & 1 \end{bmatrix}$$

8. Find the minimum distance of a ternary linear code with generator matrix

$$G = \begin{bmatrix} 0 & 1 & 2 & 1 \\ 1 & 0 & 2 & 2 \end{bmatrix}$$

9. Construct a standard array for the binary linear code with generator matrix

$$G = \begin{bmatrix} 1 & 1 & 0 & 1 & 0 \\ 0 & 1 & 0 & 1 & 0 \end{bmatrix}$$

Then decode the words 11111 and 10000.

10. Show that for a linear code L with generator matrix G, the coordinate positions $i_1, i_2, \ldots, i_k$ form an information set if and only if the corresponding columns of G are linearly independent.
11. Show that, for a binary linear code, Hx^T is the sum of the columns of H where an error has occurred.
12. Repeat Example 5.1.5 for the binary code with generating matrix

$$G = \begin{bmatrix} 1 & 0 & 1 & 1 \\ 1 & 1 & 0 & 1 \end{bmatrix}$$

13. Construct a syndrome table, containing coset leaders and their syndromes, for the Hamming code $\mathcal{H}_2(3)$. Then decode the words 0000010, 1111111, 1100110, and 1010101.
14. If a linear code L has parity check matrix H, what is the parity check matrix for the extended code $\widehat{H}$ obtained by adding an overall parity check to H?
15. Show that if a binary linear code L contains at least one codeword of odd weight, then exactly one-half of the codewords in L must have odd weight. If L does contain a word of odd weight, describe the set of words of even weight.
16. Show that, by judicious choice of row operations, majority logic

decoding can be performed on all 7 positions in the code C discussed in the subsection on majority logic decoding.

17. Prove Theorem 5.1.8.
18. Show that if L is a binary self-orthogonal code, then all codeword weights are even and $\mathbf{1} \in L^{\perp}$.
19. Prove that if L is a binary [n,(n−1)/2]-code (hence n is odd), then $L^{\perp}$ is generated by any basis for L, together with the string $\mathbf{1}$.
20. Show that a linear [n,k]-code L is self-dual if and only if it is self-orthogonal and $k = n/2$.
21. Can a subspace of a vector space over a field of characteristic zero be self-dual? Explain.
22. Prove that a *binary* self-dual [n,n/2]-code exists for all positive even integers n. *Hint.* Construct a generating matrix.
23. If L is a linear code, prove that $Aut(L) = Aut(L^{\perp})$.
24. If a monomial transformation μ sends a basis for a linear code L to another basis for L, does this imply that μ is in $Aut(L)$?
25. Prove Theorem 5.1.18.
26. Let A and B be mutually orthogonal *subsets* of V(n,q), that is, $a \cdot b = 0$ for all $a \in A$, $b \in B$. Suppose further that $|A| = q^k$ and $|B| \geq q^{n-k-1} + 1$. Show that A is a linear code. *Hint.* Consider the subspaces $\langle A \rangle$ and $\langle B \rangle$ generated by A and B. What can you say about the sum of their dimensions?

5.2 Weight Distributions

In this section, we delve more deeply into the structure of codes, paying particular attention to the relationship between a linear code and its dual. We will also discover an upper bound on the numbers $A_q(n,d)$ that is often superior to the bounds of Section 4.5.

For any (n,M)-code C, we let A_k denote the number of codewords in C of weight k. The numbers $A_0,\ldots,A_n$ are referred to as the **weight distribution** of C, and the formal sum

$$W_C(s) = \sum_{k=0}^{n} A_k s^k$$

is called the **weight enumerator** of C.

Weight distributions play an important role in various ways. As a simple example, it can be shown that any code with the same length, size, and weight distribution as a Golay code must be equivalent to that Golay code.

In general, it is difficult to determine the weight distribution of a given code. One of the most important tools for this purpose is the *MacWilliams identity*, which relates the weight enumerator of a linear code L to the weight enumerator of the dual code $L^{\perp}$. Our approach to deriving this identity will be rather algebraic, since this will enable us to generalize to nonlinear codes as well. However, we should mention that we will not have further use of the relatively sophisticated algebraic techniques set forth in this section, and so it is not essential that the reader master these techniques before going on to subsequent material.

CHARACTERS

Let $(G,+)$ be a group, and let $\mathbb{C}_1$ be the multiplicative group of complex numbers of absolute value 1. A homomorphism $\chi:G\to\mathbb{C}_1$ is called a **character** of G. Note that, since χ is a homomorphism, we have

$$\chi(g+h) = \chi(g)\chi(h)$$

and

$$\chi(0) = 1$$

The **principal character** of G is the character

$$\Xi(g) = 1 \quad \text{for all } g \in G$$

Theorem 5.2.1 Let G be a group, and let χ be a character of G. Then

$$\sum_{g \in G} \chi(g) = \begin{cases} |G| & \text{if } \chi \text{ is principal} \\ 0 & \text{otherwise} \end{cases}$$

Proof. If χ is the principal character, the result follows easily. If not, then there exists an $h \in G$ for which $\chi(h) \neq 1$. In this case, we have

$$\chi(h) \sum_{g \in G} \chi(g) = \sum_{g \in G} \chi(g+h) = \sum_{g \in G} \chi(g)$$

and so $\sum_{g \in G} \chi(g) = 0.$ ∎

Now, let $\chi: F_q \to \mathbb{C}_1$ be a non-principal character on $(F_q, +)$, and let $\mathbf{u} \in V(n,q)$. (We will indicate in the exercises how to show that such a character always exists.) Then for any linear code $L \subset V(n,q)$, we can define a mapping $\chi_{\mathbf{u}}: L \to \mathbb{C}_1$ by

$$\chi_{\mathbf{u}}(\mathbf{c}) = \chi(\langle \mathbf{c},\mathbf{u} \rangle) = \chi(x_1 c_1 + \cdots + x_n c_n)$$

where $\mathbf{c} = c_1 \cdots c_n$ and $\mathbf{u} = u_1 \cdots u_n$. We will abuse the notation somewhat and write $\chi\langle \mathbf{x},\mathbf{y} \rangle$ for $\chi(\langle \mathbf{x},\mathbf{y} \rangle)$. Then

$$\begin{aligned} \chi_{\mathbf{u}}(\mathbf{c}+\mathbf{d}) &= \chi\langle \mathbf{c}+\mathbf{d},\mathbf{u} \rangle = \chi(\langle \mathbf{c},\mathbf{u} \rangle + \langle \mathbf{d},\mathbf{u} \rangle) \\ &= \chi\langle \mathbf{c},\mathbf{u} \rangle \chi\langle \mathbf{d},\mathbf{u} \rangle) = \chi_{\mathbf{u}}(\mathbf{c}) \chi_{\mathbf{u}}(\mathbf{d}) \end{aligned}$$

and so $\chi_{\mathbf{u}}$ is a character on L. Note also that

$$\chi_{\mathbf{u}}(\mathbf{x}) = \chi_{\mathbf{x}}(\mathbf{u})$$

The following theorem tells us when this character is principal.

Theorem 5.2.2 The character $\chi_{\mathbf{u}}: L \to \mathbb{C}_1$ is principal if and only if $\mathbf{u} \in L^{\perp}$. In particular, $\chi_{\mathbf{u}}: V(n,q) \to \mathbb{C}_1$ is principal if and only if $\mathbf{u} = \mathbf{0}$.

Proof. Suppose that $\mathbf{u} \in L^{\perp}$. Then $\langle \mathbf{c},\mathbf{u} \rangle = 0$ for all $\mathbf{c} \in L$, and so

$$\chi_{\mathbf{u}}(\mathbf{c}) = \chi\langle \mathbf{c},\mathbf{u} \rangle = 1$$

for all $\mathbf{c} \in L$, showing that $\chi_{\mathbf{u}}$ is principal. Conversely, suppose that $\chi_{\mathbf{u}}$ is principal, that is,

$$\chi_{\mathbf{u}}(\mathbf{c}) = \chi\langle \mathbf{c},\mathbf{u} \rangle = \chi(0) = 1 \quad \text{for all } \mathbf{c} \in L$$

Now, if $\mathbf{u} \notin L^{\perp}$, then as $\mathbf{c}$ ranges over L, the values of $\langle \mathbf{c},\mathbf{u} \rangle$ range over all of F_q, and so $\chi(\alpha) = 1$ for all $\alpha \in F_q$. But this contradicts the assumption that $\chi: F_q \to \mathbb{C}_1$ is not principal. Hence, $\mathbf{u} \in L^{\perp}$. ∎

From now on, we will assume that χ *is a non-principal character on* F_q. We will find it convenient to use the notation δ_P to equal 1 if P is a true statement, and 0 if P is not true. For example, $\delta_{\mathbf{x}\in L}$ equals 1 if $\mathbf{x}$ is in L and 0 otherwise.

Combining Theorems 5.2.1 and 5.2.2, we get the following.

Corollary 5.2.3 Let $L \subset V(n,q)$ be a linear code. Then for $\mathbf{u} \in V(n,q)$

$$\sum_{\mathbf{c}\in L} \chi_{\mathbf{u}}(\mathbf{c}) = |L|\,\delta_{\mathbf{u}\in L^{\perp}} \qquad \square$$

THE GROUP ALGEBRA

As we know, the set V(n,q) is an additive group. We can obtain an algebra based on this group by using the elements of V(n,q) as exponents of formal sums. More specifically, let t be an independent variable, and let $\mathbb{C}V(n,q)$ be the set of all formal sums of the form

$$g = g(t) = \sum_{\mathbf{x}\in V(n,q)} \alpha_{\mathbf{x}} t^{\mathbf{x}}$$

where $\alpha_{\mathbf{x}}$ are complex numbers. When n and q are understood, it will be convenient to simply use V for V(n,q) and $\mathbb{C}V$ for $\mathbb{C}V(n,q)$.

The set $\mathbb{C}V$ can be made into an algebra over the complex numbers by defining addition, scalar multiplication, and multiplication as follows

$$\sum_{\mathbf{x}\in V} \alpha_{\mathbf{x}} t^{\mathbf{x}} + \sum_{\mathbf{x}\in V} \beta_{\mathbf{x}} t^{\mathbf{x}} = \sum_{\mathbf{x}\in V} (\alpha_{\mathbf{x}} + \beta_{\mathbf{x}}) t^{\mathbf{x}}$$

and

$$\beta \sum_{\mathbf{x}\in V} \alpha_{\mathbf{x}} t^{\mathbf{x}} = \sum_{\mathbf{x}\in V} (\beta \cdot \alpha_{\mathbf{x}}) t^{\mathbf{x}}$$

$$\left(\sum_{\mathbf{x}\in V} \alpha_{\mathbf{x}} t^{\mathbf{x}}\right)\left(\sum_{\mathbf{y}\in V} \beta_{\mathbf{y}} t^{\mathbf{y}}\right) = \sum_{\mathbf{x},\mathbf{y}\in V} (\alpha_{\mathbf{x}}\beta_{\mathbf{y}}) t^{\mathbf{x}+\mathbf{y}} = \sum_{\mathbf{z}\in V}\left(\sum_{\mathbf{x}+\mathbf{y}=\mathbf{z}} \alpha_{\mathbf{x}}\beta_{\mathbf{y}}\right) t^{\mathbf{z}}$$

With these operations, $\mathbb{C}V$ is known as the **group algebra** of V over the complex field $\mathbb{C}$.

While it makes no sense to evaluate an element of the group algebra by substituting a complex number for the variable t, we can apply characters to elements of the group algebra. In particular, we set

$$\chi_{\mathbf{u}}(g) = \chi_{\mathbf{u}}\left(\sum_{\mathbf{x}\in V} \alpha_{\mathbf{x}} t^{\mathbf{x}}\right) = \sum_{\mathbf{x}\in V} \alpha_{\mathbf{x}} \chi_{\mathbf{u}}(\mathbf{x}) = \sum_{\mathbf{x}\in V} \alpha_{\mathbf{x}} \chi\langle \mathbf{x},\mathbf{u}\rangle$$

It is easy to verify that $\chi_{\mathbf{u}}(gh) = \chi_{\mathbf{u}}(g)\chi_{\mathbf{u}}(h)$.

THE TRANSFORM OF AN ELEMENT OF THE GROUP ALGEBRA

Given an element $g(t) = \sum \alpha_{\mathbf{x}} t^{\mathbf{x}}$ of the group algebra CV, we define the **transform** of g by

$$\widehat{g}(t) = \sum_{\mathbf{x} \in V} \chi_{\mathbf{x}}(g) t^{\mathbf{x}}$$

The following inversion theorem shows us how we can recover g from its transform $\widehat{g}$.

Theorem 5.2.4 If $g(t) = \sum_{\mathbf{x} \in V} \alpha_{\mathbf{x}} t^{\mathbf{x}}$ then

1) $\alpha_{\mathbf{x}} = q^{-n} \chi_{-\mathbf{x}}(\widehat{g})$

2) $\widehat{\widehat{g}}(t) = q^n g(t^{-1})$

Proof. First, we have

$$\widehat{\widehat{g}}(t) = \sum_{\mathbf{x} \in V} \chi_{\mathbf{x}}(\widehat{g}) t^{\mathbf{x}} \tag{5.2.1}$$

and so we are led to compute, using Corollary 5.2.3,

$$\begin{aligned}
\chi_{\mathbf{x}}(\widehat{g}) &= \chi_{\mathbf{x}}\Big(\sum_{\mathbf{y} \in V} \chi_{\mathbf{y}}(g) t^{\mathbf{y}} \Big) \\
&= \sum_{\mathbf{y} \in V} \chi_{\mathbf{y}}(g) \chi_{\mathbf{x}}(\mathbf{y}) \\
&= \sum_{\mathbf{y} \in V} \sum_{\mathbf{z} \in V} \alpha_{\mathbf{z}} \chi_{\mathbf{y}}(\mathbf{z}) \chi_{\mathbf{x}}(\mathbf{y}) \\
&= \sum_{\mathbf{y} \in V} \sum_{\mathbf{z} \in V} \alpha_{\mathbf{z}} \chi_{\mathbf{y}}(\mathbf{z}) \chi_{\mathbf{y}}(\mathbf{x}) \\
&= \sum_{\mathbf{z} \in V} \alpha_{\mathbf{z}} \sum_{\mathbf{y} \in V} \chi_{\mathbf{y}}(\mathbf{z} + \mathbf{x}) \\
&= \sum_{\mathbf{z} \in V} \alpha_{\mathbf{z}} \sum_{\mathbf{y} \in V} \chi_{\mathbf{z}+\mathbf{x}}(\mathbf{y}) \\
&= \sum_{\mathbf{z} \in V} \alpha_{\mathbf{z}}\, q^n \delta_{\mathbf{x}+\mathbf{z}=0} \\
&= q^n \alpha_{-\mathbf{x}}
\end{aligned}$$

from which part 1) follows. Using this in (5.2.1), we have

$$\widehat{\widehat{g}}(t) = q^n \sum_{\mathbf{x} \in V} \alpha_{-\mathbf{x}} t^{\mathbf{x}} = q^n \sum_{\mathbf{x} \in V} \alpha_{\mathbf{x}} t^{-\mathbf{x}} = q^n g(t^{-1})$$

which proves part 2). ∎

WEIGHT ENUMERATORS AND WEIGHT DISTRIBUTIONS

For each element

$$g(t) = \sum_{\mathbf{x} \in V} \alpha_{\mathbf{x}} t^{\mathbf{x}}$$

of the group algebra $\mathbb{C}V(n,q)$, we can collect terms $\alpha_{\mathbf{x}}$ of the same weight $w(\mathbf{x})$ by defining the **weight enumerator** of g as the formal sum

$$W_g(s) = \sum_{\mathbf{x} \in V} \alpha_{\mathbf{x}} s^{w(\mathbf{x})} = \sum_{k=0}^{n} \left[\sum_{w(\mathbf{x})=k} \alpha_{\mathbf{x}} \right] s^k = \sum_{k=0}^{n} A_k s^k$$

where the coefficients

$$A_k = \sum_{w(\mathbf{x})=k} \alpha_{\mathbf{x}}$$

form the **weight distribution** of g.

Notice that, if $C \subset V(n,q)$ is a code, and if

$$g_C(t) = \sum_{\mathbf{c} \in C} t^{\mathbf{c}}$$

is the **generating function** of C, then the numbers A_k form the weight distribution of C. Hence the terminology.

The weight enumerator of the transform $\widehat{g}$ of g is

$$W_{\widehat{g}}(s) = \sum_{\mathbf{x} \in V} \chi_{\mathbf{x}}(g) s^{w(\mathbf{x})} = \sum_{k=0}^{n} \left[\sum_{w(\mathbf{x})=k} \chi_{\mathbf{x}}(g) \right] s^k = \sum_{k=0}^{n} \widehat{A}_k s^k$$

where the coefficients

(5.2.2) $$\widehat{A}_k = \sum_{w(\mathbf{x})=k} \chi_{\mathbf{x}}(g)$$

form the weight distribution of $\widehat{g}$.

Now we are ready for the MacWilliams identity, which describes the relationship between $W_g(s)$ and $W_{\widehat{g}}(s)$.

Theorem 5.2.5 (The MacWilliams Identity) Let g be an element of the group algebra $\mathbb{C}V(n,q)$. Then

$$W_{\widehat{g}}(s) = [1 + (q-1)s]^n W_g\left(\frac{1-s}{1+(q-1)s}\right)$$

Proof. Since

$$\widehat{g}(t) = \sum_{\mathbf{x} \in V} \chi_{\mathbf{x}}(g) t^{\mathbf{x}}$$

we have

$$W_{\hat{g}}(s) = \sum_{\mathbf{x} \in V} \chi_{\mathbf{x}}(g) s^{w(\mathbf{x})} = \sum_{\mathbf{x} \in V} \sum_{\mathbf{y} \in V} \alpha_{\mathbf{y}} \chi\langle \mathbf{x},\mathbf{y}\rangle s^{w(\mathbf{x})}$$

$$= \sum_{\mathbf{y} \in V} \alpha_{\mathbf{y}} \sum_{\mathbf{x} \in V} \chi\langle \mathbf{x},\mathbf{y}\rangle s^{w(\mathbf{x})}$$

Now let us write $\mathbf{x} = x_1 \cdots x_n$ and $\mathbf{y} = y_1 \cdots y_n$. If we let $w(x_i) = 0$ if $x_i = 0$ and $w(x_i) = 1$ if $x_i \neq 0$, then

$$\sum_{\mathbf{x} \in V} \chi\langle \mathbf{x},\mathbf{y}\rangle s^{w(\mathbf{x})} = \sum_{\mathbf{x} \in V} \chi(x_1 y_1 + \cdots + x_n y_n)\, s^{w(x_1) + \cdots + w(x_n)} =$$

$$\sum_{x_1 \in F_q} \sum_{x_2 \in F_q} \cdots \sum_{x_n \in F_q} \left(\chi(x_1 y_1) s^{w(x_1)}\right)\left(\chi(x_2 y_2) s^{w(x_2)}\right) \cdots \left(\chi(x_n y_n) s^{w(x_n)}\right)$$

which is the expansion of the product

$$\sum_{x \in F_q} \chi(x y_1)\, s^{w(x)} \times \sum_{x \in F_q} \chi(x y_2)\, s^{w(x)} \times \cdots \times \sum_{x \in F_q} \chi(x y_n)\, s^{w(x)}$$

$$= \prod_{i=1}^{n} \sum_{x \in F_q} \chi(x y_i)\, s^{w(x)}$$

Now, if $y_i = 0$, then

$$\sum_{x \in F_q} \chi(x y_i) s^{w(x)} = \sum_{x \in F_q} s^{w(x)} = 1 + (q-1)s$$

and if $y_i \neq 0$, then by Theorem 5.2.1

$$\sum_{x \in F_q} \chi(x y_i) s^{w(x)} = 1 + \sum_{x \neq 0} \chi(x y_i) s = 1 + s(-1) = 1 - s$$

Hence

$$\sum_{\mathbf{x} \in V} \chi\langle \mathbf{x},\mathbf{y}\rangle s^{w(\mathbf{x})} = [1 - s]^{w(\mathbf{y})}[1 + (q-1)s]^{n - w(\mathbf{y})}$$

and so

$$W_{\hat{g}}(s) = \sum_{\mathbf{y} \in V} \alpha_{\mathbf{y}}\, [1 - s]^{w(\mathbf{y})}[1 + (q-1)s]^{n - w(\mathbf{y})}$$

$$= [1 + (q-1)s]^n \sum_{\mathbf{y} \in V} \alpha_{\mathbf{y}} \left(\frac{1-s}{1+(q-1)s}\right)^{w(\mathbf{y})}$$

$$= [1 + (q-1)s]^n\, W_g\left(\frac{1-s}{1+(q-1)s}\right) \qquad \blacksquare$$

THE KRAWTCHOUK POLYNOMIALS

To obtain an explicit relationship between the weight distribution A_k of g and the weight distribution $\widehat{A}_k$ of the transform $\widehat{g}$, we write the MacWilliams identity in the form

$$\sum_{k=0}^{n} \widehat{A}_k s^k = \sum_{k=0}^{n} A_k (1-s)^k [1+(q-1)s]^{n-k} \tag{5.2.3}$$

Thus, we are interested in expanding the expression

$$(1-s)^k[1+(q-1)s]^{n-k}$$

as a power series in s. This is done by considering the polynomials $K_i(x) = K_i(x;n,q)$ that are uniquely defined by the generating function

$$(1-s)^x[1+(q-1)s]^{n-x} = \sum_{i=0}^{\infty} K_i(x)s^i \tag{5.2.4}$$

These polynomials are known as the **Krawtchouk polynomials** and have been studied a good deal. (The Krawtchouk polynomials, with slight modification, are polynomials of Sheffer type. See Roman (1984) for more details.)

By expanding the left side of (5.2.4) and equating coefficients of powers of s, we get the following explicit expression for the Krawtchouk polynomials

$$K_k(x) = \sum_{i=0}^{k} \binom{x}{i}\binom{n-x}{k-i}(-1)^i(q-1)^{k-i} \tag{5.2.5}$$

Setting $x = k$ in (5.2.4), and substituting into the right side of (5.2.3), we get

$$\sum_{k=0}^{n} \widehat{A}_k s^k = \sum_{k=0}^{n} A_k \sum_{i=0}^{\infty} K_i(k)s^i = \sum_{i=0}^{\infty}\sum_{k=0}^{n} A_k K_i(k)s^i = \sum_{k=0}^{\infty}\sum_{i=0}^{n} A_i K_k(i)s^k$$

Equating coefficients of s^k on both sides gives

$$\widehat{A}_k = \sum_{i=0}^{n} A_i K_k(i)$$

Let us summarize.

Theorem 5.2.6 (The MacWilliams Identity) Let

$$g = \sum_{\mathbf{x}\in V} \alpha_{\mathbf{x}} t^{\mathbf{x}} \quad \text{and} \quad \widehat{g}(t) = \sum_{\mathbf{x}\in V} \chi_{\mathbf{x}}(g)t^{\mathbf{x}}$$

be an element of $\mathbb{C}V$ and its transform, respectively. Let

$$W_g(s) = \sum_{k=0}^{n} A_k s^k \quad \text{and} \quad W_{\widehat{g}}(s) = \sum_{k=0}^{n} \widehat{A}_k s^k$$

be the corresponding weight enumerators, where

$$A_k = \sum_{w(\mathbf{x})=k} \alpha_{\mathbf{x}} \quad \text{and} \quad \widehat{A}_k = \sum_{w(\mathbf{x})=k} \chi_{\mathbf{x}}(g)$$

Then

$$W_{\widehat{g}}(s) = [1+(q-1)s]^n W_g\left(\frac{1-s}{1+(q-1)s}\right)$$

or, equivalently,

$$\widehat{A}_k = \sum_{i=0}^{n} A_i K_k(i)$$

where $K_k(x)$ are the Krawtchouk polynomials. □

LINEAR CODES

Now let us apply the previous results to linear codes. If $C \subset V(n,q)$ is any code, the sum

$$g_C = g_C(t) = \sum_{\mathbf{c} \in C} t^{\mathbf{c}}$$

is an element of the group algebra $\mathbb{C}V$, which has already been defined as the generating function of C. Since

$$W_{g_C}(s) = \sum_{k=0}^{n} \left[\sum_{w(\mathbf{c})=k} 1 \right] s^k = \sum_{k=0}^{n} A_k s^k$$

the weight enumerator of the generating function g_C is the same as the weight enumerator of C.

The following result shows what happens when we apply the character $\chi_{\mathbf{u}}$ to the generating function g_L of a linear code.

Theorem 5.2.7 Let $L \subset V(n,q)$ be a linear code. Then

$$\chi_{\mathbf{u}}(g_L) = |L| \delta_{\mathbf{u} \in L^{\perp}}$$

Proof. We have

$$\chi_{\mathbf{u}}(g_L) = \chi_{\mathbf{u}}\left(\sum_{\mathbf{c} \in L} t^{\mathbf{c}} \right) = \sum_{\mathbf{c} \in L} \chi_{\mathbf{u}}(\mathbf{c}) = |L| \delta_{\mathbf{u} \in L^{\perp}}$$

by Corollary 5.2.3. ∎

There is a simple relationship between the transfer of the generating function of a linear code and the generating function of the dual code. If $L \subset V(n,q)$ is a linear code, then

$$\widehat{g}_L = \sum_{\mathbf{x}\in V} \chi_{\mathbf{x}}(g_L)t^{\mathbf{x}} = \sum_{\mathbf{x}\in V} |L|(\delta_{\mathbf{x}\in L^{\perp}})\, t^{\mathbf{x}} = |L| \sum_{\mathbf{x}\in L^{\perp}} t^{\mathbf{x}} = |L|\, g_{L^{\perp}}$$

Thus, we have the following.

Theorem 5.2.8 If $L \subset V(n,q)$ is a linear code, then

$$\widehat{g}_L = |L|\, g_{L^{\perp}}$$

□

Example 5.2.1 Consider the binary linear code

$$L = \{000,111\}$$

which has generating function

$$g_L(t) = t^{000} + t^{111}$$

Since $\langle \mathbf{x},\mathbf{y}\rangle \in F_2$ for $\mathbf{x},\mathbf{y} \in V(n,2)$, we may take χ to be the nonprincipal character defined by

$$\chi(\alpha) = (-1)^{\alpha}$$

for $\alpha \in F_2$, and so

$$\chi_{\mathbf{u}}(\mathbf{x}) = (-1)^{\langle \mathbf{x},\mathbf{u}\rangle}$$

Then

$$\chi_{\mathbf{x}}(g_L) = (-1)^{\langle 000,\mathbf{x}\rangle} + (-1)^{\langle 111,\mathbf{x}\rangle}$$

and so

$$\chi_{000}(g_L) = (-1)^{\langle 000,000\rangle} + (-1)^{\langle 111,000\rangle} = 1+1 = 2$$

$$\chi_{001}(g_L) = (-1)^{\langle 000,001\rangle} + (-1)^{\langle 111,001\rangle} = 1+(-1) = 0$$

$$\chi_{010}(g_L) = (-1)^{\langle 000,010\rangle} + (-1)^{\langle 111,010\rangle} = 1+(-1) = 0$$

$$\chi_{011}(g_L) = (-1)^{\langle 000,011\rangle} + (-1)^{\langle 111,011\rangle} = 1+1 = 2$$

Continuing in this way, we get

$$\chi_{100}(g_L) = 0, \quad \chi_{101}(g_L) = 2, \quad \chi_{110}(g_L) = 2, \quad \chi_{111}(g_L) = 0$$

Hence,

$$\widehat{g}_L(t) = \sum_{\mathbf{x}\in V} \chi_{\mathbf{x}}(g_L)t^{\mathbf{x}} = 2t^{000} + 2t^{011} + 2t^{101} + 2t^{110}$$

Theorem 5.2.8 then gives

$$g_{L^{\perp}}(t) = t^{000} + t^{011} + t^{101} + t^{110}$$

and so $L^{\perp} = \{000,011,101,110\}$. □

According to Theorem 5.2.8, if L is a linear code, then

$$\hat{g}_L = |L|\, g_{L^\perp}$$

and so

$$W_{\hat{g}_L}(s) = |L|\, W_{g_{L^\perp}}(s) = |L|\, W_{L^\perp}(s)$$

Hence, we obtain the following MacWilliams identity for linear codes.

Corollary 5.2.9 (The MacWilliams Identity for Linear Codes) Let $L \subset V(n,q)$ be a linear code, let $L^\perp$ be its dual, and let

$$W_L(s) = \sum_{k=0}^{n} A_k s^k, \quad \text{and} \quad W_{L^\perp}(s) = \sum_{k=0}^{n} A_k^\perp s^k$$

be the weight enumerators of L and $L^\perp$, respectively. Then

$$W_{L^\perp}(s) = \frac{1}{|L|}[1+(q-1)s]^n W_L\left(\frac{1-s}{1+(q-1)s}\right)$$

or equivalently,

$$A_k^\perp = \frac{1}{|L|}\sum_{i=0}^{n} A_i K_k(i)$$

where $K_k(x)$ are the Krawtchouk polynomials. ☐

Of course, since $L^{\perp\perp} = L$, we may use the MacWilliams identity to express the weight enumerator of L in terms of the weight enumerator of $L^\perp$.

Example 5.2.2 Recall the binary code $L = \{000,111\}$ of Example 5.2.1. Here $A_0 = A_3 = 1$, $A_1 = A_2 = 0$, and so

$$W_L(s) = 1 + s^3$$

The MacWilliams identity then gives

$$W_{L^\perp}(s) = \frac{1}{2}(1+s)^3\left[1+\left(\frac{1-s}{1+s}\right)^3\right] = \frac{1}{2}[(1+s)^3-(1-s)^3] = 1+3s^2$$

and so $L^\perp$ has 1 word of weight 0 (the zero word) and 3 words of weight 2. This implies that $L^\perp = \{000,011,101,110\}$, as seen in Example 5.2.1. ☐

MOMENTS OF THE WEIGHT DISTRIBUTION

The MacWilliams identity (Corollary 5.2.9) can be used to derive various types of moments of the weight distribution A_i of a linear code. Let us sketch an example, leaving the details of the computations

to the reader.

Let L be a linear [n,k]-code. Reversing the roles of L and $L^{\perp}$, the MacWilliams identity can be written in the form

$$(5.2.6)\qquad \sum_{i=0}^{n} A_i s^i = \frac{1}{q^{n-k}} \sum_{i=0}^{n} A_i^{\perp}(1-s)^i(1+(q-1)s)^{n-i}$$

Differentiating this r times and setting $s=1$ gives (Leibniz's rule is helpful here)

$$(5.2.7)\qquad \frac{1}{q^k} \sum_{i=r}^{n} \binom{i}{r} A_i = \frac{1}{q^r} \sum_{i=0}^{r} (-1)^i (q-1)^{r-i} \binom{n-i}{n-r} A_i^{\perp}$$

The sum on the left-hand side of this equation is the *r-th binomial moment* of the weight distribution $A_0,\ldots,A_n$.

Replacing s by s^{-1} and then multiplying by s^n in (5.2.6) gives

$$(5.2.8)\qquad \sum_{i=0}^{n} A_i s^{n-i} = \frac{1}{q^{n-k}} \sum_{i=0}^{n} A_i^{\perp}(s-1)^i(s+q-1)^{n-i}$$

Differentiating this r times and setting $s=1$ gives the following formula, which we will use in Chapter 6

$$(5.2.9)\qquad \frac{1}{q^k} \sum_{i=0}^{n-r} \binom{n-i}{r} A_i = \frac{1}{q^r} \sum_{i=0}^{r} \binom{n-i}{n-r} A_i^{\perp}$$

*DISTANCE DISTRIBUTIONS

We may generalize the previous discussion as follows. For any (n,M)-code C, let

$$B_k = \frac{1}{M}\{(\mathbf{c},\mathbf{d}) \mid \mathbf{c},\mathbf{d} \in C,\ d(\mathbf{c},\mathbf{d}) = k\}$$

The numbers $B_0,\ldots,B_n$ are referred to as the **distance distribution** of C, and the formal sum

$$B_C(s) = \sum_{k=0}^{n} B_k s^k$$

is the **distance enumerator** of C. We leave it as an exercise to show that, for a linear code L, the weight distribution and the distance distribution coincide.

As we have seen, the MacWilliams identity applies to weight enumerators of elements of the group algebra. Fortunately, it happens that the distance enumerator of a code is the weight enumerator of a group algebra element. To see this, consider the product

$$h_C(t) = \frac{1}{M} g_C(t) g_C(t^{-1})$$

where $g_C(t)$ is the generating function for an (n,M)-code $C \subset V(n,q)$. We have

$$h_C(t) = \frac{1}{M} g_C(t) g_C(t^{-1})$$

$$= \frac{1}{M} \sum_{c \in C} t^{c} \cdot \sum_{d \in C} t^{-d} = \frac{1}{M} \sum_{c,d \in C} t^{c-d}$$

Hence, the weight enumerator of $h_C(t)$ is

$$W_{h_C}(s) = \frac{1}{M} \sum_{k=0}^{n} \left| \{(c,d) \mid w(c-d) = k\} \right| s^k = \sum_{k=0}^{n} B_k s^k$$

which is the distance enumerator of C. This proves the following.

Theorem 5.2.10 If $C \subset V(n,q)$ is an (n,M)-code, then the distance enumerator of C is the weight enumerator of

$$h_C(t) = \frac{1}{M} g_C(t) g_C(t^{-1}) \qquad \square$$

Thus, we may apply the MacWilliams identity (Theorem 5.2.6) to $h_C(t)$ to obtain the following result.

Theorem 5.2.11 Let $C \subset V(n,q)$ be a code, and let

$$B_C(s) = \sum_{k=0}^{n} B_k s^k = W_{h_C}(s)$$

be its distance enumerator (see Theorem 5.2.10), where

$$h_C(t) = \frac{1}{M} g_C(t) g_C(t^{-1})$$

Let

$$W_{\widehat{h}_C}(s) = \sum_{k=0}^{n} \widehat{B}_k s^k = \sum_{k=0}^{n} \left[\sum_{w(\mathbf{x})=k} \chi_{\mathbf{x}}(h_C) \right] s^k$$

be the weight enumerator of the transform $\widehat{h}_C(t)$. Then

$$\widehat{B}_k = \sum_{i=0}^{n} B_i K_k(i) \qquad \square$$

To see the significance of this result, we need a lemma, whose proof is left as an exercise.

Lemma 5.2.12 If $C \subset V(n,q)$ is a code, then

$$\chi_{\mathbf{c}}(g_C(t^{-1})) = \overline{\chi_{\mathbf{c}}(g_C(t))}$$

where the bar denotes complex conjugate. ▯

Now, since

$$\widehat{B}_k = \sum_{w(\mathbf{x})=k} \chi_{\mathbf{x}}(h_C)$$

and since

$$\begin{aligned}\chi_{\mathbf{x}}(h_C(t)) &= \frac{1}{M}\chi_{\mathbf{x}}(g_C(t)g_C(t^{-1})) = \frac{1}{M}\chi_{\mathbf{x}}(g_C(t))\chi_{\mathbf{x}}(g_C(t^{-1}))\\ &= \frac{1}{M}\chi_{\mathbf{x}}(g_C(t))\overline{\chi_{\mathbf{x}}(g_C(t))} = \frac{1}{M}|\chi_{\mathbf{x}}(g_C(t))|^2\end{aligned}$$

we have

$$\widehat{B}_k = \frac{1}{M}\sum_{w(\mathbf{x})=k} |\chi_{\mathbf{x}}(g_C)|^2 \ge 0 \tag{5.2.10}$$

This, together with Theorem 5.2.11, gives the following result, which will lead us later in this section to an important bound on the numbers $A_q(n,d)$.

Theorem 5.2.13 For a code $C \subset V(n,q)$, we have for $k = 0,\ldots,n$

$$\sum_{i=0}^{n} B_i K_k(i) \ge 0$$ ▯

**THE FOUR FUNDAMENTAL PARAMETERS OF A CODE*

The distributions

$$B_0,\ldots,B_n \quad\text{and}\quad \widehat{B}_0,\ldots,\widehat{B}_n$$

can be used to define what are referred to as the four fundamental parameters of a code, denoted by d, s, $\widehat{d}$, and $\widehat{s}$.

The parameter d is the minimum distance of the code. In other words, it is the smallest *positive* index k for which B_k is nonzero. The parameter s is the *number* of distinct nonzero distances between codewords, in other words, the number of *positive* indices k for which B_k is nonzero. Similarly, $\widehat{d}$ is the smallest positive index k for which $\widehat{B}_k$ is nonzero, and $\widehat{s}$ is the number of positive indices k for which $\widehat{B}$ is nonzero.

For linear codes, we have remarked that $B_k = A_k$, and so $\widehat{d}$ is the minimum distance of the dual code, and $\widehat{s}$ is the number of distinct nonzero distances in the dual code.

For any code C, the parameter $\widehat{d}$ is referred to as the **dual distance** of C, and $\widehat{s}$ is called the **external distance** of C, because

any string is at distance less than $\hat{s}$ from some codeword in C. (We will not prove this result here, but refer the interested reader to MacWilliams and Sloane (1977), p. 172.)

Let us give an example of the type of information one might gain from knowledge of the fundamental parameters. We begin with a simple lemma, whose proof is left as an exercise.

Lemma 5.2.14 Let C be a code.

1) If $\mathbf{0} \in C$, then $B_k = 0 \Rightarrow A_k = 0$.

2) $\widehat{B}_k = 0 \Leftrightarrow \chi_{\mathbf{x}}(g_C) = 0$ for all $\mathbf{x}$ with $w(\mathbf{x}) = k$.

3) $\widehat{B}_k = 0 \Rightarrow \widehat{A}_k = 0$. □

Now let us consider a *binary* code C. We know that $\widehat{B}_k = 0$ for $k < \hat{d}$, and that $\widehat{B}_{\hat{d}} \neq 0$. Hence, according to the previous lemma, if $w(\mathbf{x}) < \hat{d}$, then

$$\chi_{\mathbf{x}}(g_C) = \sum_{c \in C} \chi\langle \mathbf{c}, \mathbf{x} \rangle = 0$$

Since $\langle \mathbf{x}, \mathbf{y} \rangle \in F_2$ for $\mathbf{x}, \mathbf{y} \in V(n,2)$, we may take χ to be the non-principal character defined by

$$\chi(\alpha) = (-1)^{\alpha}$$

for $\alpha \in F_2$.

Assuming $\hat{d} \geq 2$, if $\mathbf{x} = e_i = 0 \cdots 010 \cdots 0$, with the 1 in the i-th position, we deduce that

$$\sum_{c \in C} (-1)^{c_i} = 0$$

where $\mathbf{c} = c_1 c_2 \cdots c_n$. Hence, one-half of the codewords in C have a 0 in the i-th position, and the other half have a 1.

If $\hat{d} \geq 3$, then we may let $\mathbf{x}$ range over all strings of weight 2. For instance, if $\mathbf{x} = 110 \cdots 0$, then

$$\sum_{c \in C} (-1)^{c_1 + c_2} = 0$$

and so one-half of the codewords in C begin with 01 or 10 and the other half begin with either 00 or 11. This, together with the fact that exactly one-half of the codewords begin with a 0, and exactly one-half have a 0 in the second position, imply that exactly one-fourth of the codewords begin with 00, 01, 10, or 11. Continuing in this way, we get the following result.

Theorem 5.2.15 Let C be a *binary* (n,M)-code. By restricting the codewords in C to any set of $k < \widehat{d}$ coordinate positions, we get each possible binary string of length k exactly $M/2^k$ times. Furthermore, $\widehat{d}$ is the largest number with this property. □

*THE LINEAR PROGRAMMING BOUND

Let us conclude this section with the promised upper bound on $A_q(n,d)$ that comes from Theorem 5.2.13. If C is an (n,M)-code over F_q, then by counting the number of pairs (**c**,**d**) in $C \times C$, we deduce that the distance distribution $\{B_k\}$ of C satisfies

$$\sum_{i=0}^{n} B_i = M^2$$

If we let $E_i = \frac{1}{M} B_i$, then

$$M = \sum_{i=0}^{n} E_i$$

(Some authors refer to the numbers E_i as the distance distribution of C.) Furthermore, by Theorem 5.2.12, we have

$$\sum_{i=0}^{n} E_i K_k(i) \geq 0$$

and

$$E_i \geq 0$$

and

$$E_0 = 1,\ E_i = 0 \ \text{ for } 1 < i < d$$

This leads to the so-called *Linear Programming Bound*, discovered by P. Delsarte (1972, 1973).

Theorem 5.2.16 (The Linear Programming Bound)

$$A_q(n,d) \leq \max\left\{ \sum_{i=0}^{n} E_i \right\}$$

where the maximum is taken over all E_i for which

$$E_i \geq 0$$

$$E_0 = 1,\ E_i = 0 \ \text{ for } 1 < i < d$$

and

(5.2.11) $$\sum_{i=0}^{n} E_i K_k(i) \geq 0, \quad \text{for } k = 0, \ldots, n$$ □

In the binary case q = 2, we can make a few improvements on this

result. First, we observe that if a binary (n,M,d)-code exists for which d is even, then a binary (n,M,d)-code exists for which each codeword has even weight, and hence all distances are even. We leave proof of this fact as an exercise. As a result, if $q = 2$ and d is even, we may assume that $E_i = 0$ for all *odd* values of i.

Second, we observe that, for $q = 2$,

$$K_k(2i) = K_{n-k}(2i)$$

(Proof of this is also left as an exercise.) Hence, for d even, since we assume that $E_{2i+1} = 0$, the constraints (5.2.11) need only be considered for $k = 0,\ldots,\lfloor n/2 \rfloor$. Finally, we observe that $K_0(i) = 1$ for all i, and so the first constraint in (5.2.11) follows from $E_i \geq 0$.

Let us summarize.

Theorem 5.2.17 **(The Linear Programming Bound for q = 2)** If d is even, then

$$A_2(n,d) \leq \max\left\{\sum_{i=0}^{n} E_i\right\}$$

where the maximum is taken over all E_i for which

$$E_i \geq 0$$

$$E_0 = 1,\ E_i = 0 \text{ for } 1 < i < d$$

$$E_i = 0 \text{ for } i \text{ odd}$$

and

$$\binom{n}{k} + \sum_{\substack{i=d \\ i \text{ even}}}^{n} E_i K_k(i) \geq 0, \quad \text{for } k = 1,\ldots,\lfloor \tfrac{n}{2} \rfloor \tag{5.2.12}$$ □

Since $A_2(n,2t+1) = A_2(n+1,2t+2)$, we may apply Theorem 5.2.17 even when the minimum distance is odd.

Example 5.2.3 Let us use the linear programming bound to obtain an upper bound on $A_2(13,5) = A_2(14,6)$. In this case, $n = 14$ and $d = 6$. Hence, we have

$$E_0 = 1$$

$$E_1 = E_2 = E_3 = E_4 = E_5 = E_7 = E_9 = E_{11} = E_{13} = 0$$

Also, from the values of the Krawtchouk polynomials given in the appendix, we find that (5.2.12) is

$$14 + 2E_6 - 2E_8 - 6E_{10} - 10E_{12} - 14E_{14} \geq 0$$

$$91 - 5E_6 - 5E_8 + 11E_{10} + 43E_{12} + 91E_{14} \geq 0$$

$$364 - 12E_6 + 12E_8 + 4E_{10} - 100E_{12} - 364E_{14} \geq 0$$

$$1001 + 9E_6 + 9E_8 - 39E_{10} + 121E_{12} + 1001E_{14} \geq 0$$

$$2002 + 30E_6 - 30E_8 + 38E_{10} - 22E_{12} - 2002E_{14} \geq 0$$

$$3003 + 5E_6 - 5E_8 + 27E_{10} - 165E_{12} + 3003E_{14} \geq 0$$

$$3432 + 40E_6 + 40E_8 - 72E_{10} + 264E_{12} - 3432E_{14} \geq 0$$

Using a computer to maximize the objective function $S = E_6 + E_8 + E_{10} + E_{12} + E_{14}$, subject to these constraints, gives the optimal solution

$$E_6 = 42, \ E_8 = 7, \ E_{10} = 14, \ E_{12} = E_{14} = 0$$

and so S has maximum value 63. Adding $E_0 = 1$ gives

$$A_2(13,5) = A_2(14,6) \leq 64$$

Now, we will see in Section 6.1 that the so-called *Nordstrom-Robinson code* is a binary (16,256,6)-code. If we shorten this code twice, we obtain a (14,M,6)-code, where $M \geq 64$. Puncturing this code once gives a (13,M,5)-code, and so $A_2(13,5) \geq 64$. Thus, we see that the linear programming bound is sharp in this case. (None of the bounds mentioned in Section 4.5 are sharp in this case.) □

EXERCISES

1. Show that, for a linear code L, the weight distribution and the distance distribution coincide.
2. Prove that distinct characters are linearly independent.
3. Show that $\chi_{\mathbf{u}}$ is a character of V(n,q). Furthermore, if L is a linear code in V(n,q), then show that $\chi_{\mathbf{u}}$ is also a character of L.
4. Verify that $\chi_{\mathbf{x}}(\mathbf{y}) = \chi_{\mathbf{y}}(\mathbf{x})$.
5. Verify that $\chi_{\mathbf{u}}(gh) = \chi_{\mathbf{u}}(g)\chi_{\mathbf{u}}(h)$.
6. Show that $\widehat{g}_L(t) = q^n g_L(t)$.
7. Find the generating function $g_L(t)$ of the binary code $L = \{0000,1100,0011,1111\}$. Compute $\widehat{g}_L(t)$.
8. Find the generating function $g_L(t)$ of the binary code $L = \{0000,1100,0011,1111\}$. Compute $\widehat{g}_L(t)$.
9. Compute the weight enumerator $W_{L^\perp}(s)$ of the dual $L^\perp$ of the Hamming code $L = \mathcal{H}_2(3)$, by first listing each codeword in $L^\perp$. Then use the MacWilliams identity to find the weight generating function of $\mathcal{H}_2(3)$.
10. Show that the probability of an *undetected* error for a binary linear [n,k]-code L is given by

$$p_{undet\ err}(L) = (1-p)^n\left[W_L\left(\frac{p}{1-p}\right) - 1\right]$$

Hint. First express the probability in terms of the weight distribution $\{A_k\}$.

11. Let $L \subset V(n,q)$ be a linear code. Let
$$g_k = \sum_{\substack{c \in L \\ w(c)=k}} t^c$$
be the characteristic function of the set of codewords of weight k. Express the characteristic function of the sphere $S_q(\mathbf{x},r)$ in terms of the g_k. Describe the fact that L is perfect in terms of characteristic functions.

12. Describe the direct sum and (u,u+v)-construction in terms of the group algebra.

13. Let L be a linear code, and let L_0 be the code consisting of all codewords of L that have even weight. Show that
$$W_{L_0}(s) = \tfrac{1}{2}[W_L(s) + W_L(-s)]$$

14. Let L be a binary linear code, and let $\hat{L}$ be obtained from L by adding an overall parity check. Show that
$$W_{\hat{L}}(s) = \tfrac{1}{2}[(1+s)W_L(s) + (1-s)W_L(-s)]$$

15. Verify the expression (5.2.5) for the Krawtchouk polynomials.

16. Use Theorem 5.2.6 to show that
$$\sum_{w(\mathbf{x})=k} \chi_{\mathbf{u}}(\mathbf{x}) = K_k(w(\mathbf{u}))$$

17. Prove that if a binary (n,M,d)-code exists for which d is even, then a binary (n,M,d)-code exists for which each codeword has even weight. *Hint.* Do some puncturing and extending.

18. a) Show that for $q = 2$,
$$K_k(x) = (-1)^k K_k(n-k)$$
b) Show that
$$(q-1)^i\binom{n}{i}K_k(i) = (q-1)^k\binom{n}{k}K_i(k)$$
c) Use (a) and (b) to show that, for $q = 2$
$$K_k(2i) = K_{n-k}(2i)$$

19. Verify equations (5.2.6) through (5.2.9).

20. What is the mean weight of a linear code L?

21. With reference to equation (5.2.7), suppose that the dual code $L^\perp$ has minimum distance d'. Show that, for $r < d'$, the r-th

binomial moment of L depends only on the length n.

22. Let D_s denote differentiation with respect to s, let sD_s denote differentiation with respect to s, followed by multiplication by s, and let $(sD_s)^r$ be sD_s applied r times. It is well known that

$$(sD_s)^r = \sum_{j=0}^{r} S(r,j)x^j D_s^j$$

where S(r,j) are the *Stirling numbers of the second kind.* Apply the operator $(sD_s)^r$ to equation (5.2.6), and then set $s = 1$, to obtain the so-called **Pless power moments**

$$\sum_{i=0}^{n} i^r A_i = \sum_{i=0}^{n} (-i)^i A_i^{\perp} \left\{ \sum_{u=0}^{r} u!S(r,u)q^{k-u}(q-1)^{u-i}\binom{n-i}{n-u} \right\}$$

23. Use the linear programming bound to obtain an upper bound on $A_2(5,3)$. How good is this bound?

24. Set up the constraints for the linear programming bound for $A_2(7,3)$. If you have access to a computer program for solving LP problems, find the upper bound in this case. How good is this bound? *Hint.* Think Hamming.

25. Prove Lemma 5.2.11.

26. Prove Lemma 5.2.13. *Hint.* Use (5.2.2) and (5.2.10).

27. Show that for any prime power $q = p^m$, there exists a non-principal character $\chi: F_q \to C_1$. *Hint.* We will show in Chapter 7 that F_q is a vector space over F_p. Let $\beta_1, \beta_2, \ldots, \beta_m$ be a basis for F_q over F_p. Thus, $\alpha \in F_q$ implies $\alpha = a_1\beta_1 + \cdots + a_m\beta_m$, where $a_i \in F_p$. Let $\chi(\alpha) = \omega^{a_1}$, where ω is a complex primitive p-th root of unity. (That is, $\omega^n = 1$ if and only if $n \equiv 0 \bmod p$.)

5.3 Maximum Distance Separable Codes

The main coding theory problem for linear codes can be stated as follows. Given the length n and minimum distance d, find the largest dimension k for which there exists an [n,k,d]-code. As we have seen, this problem is quite difficult in general.

On the other hand, we may instead fix n and k and ask for the largest minimum distance d among all codes of length n and dimension k. This problem has a very simple answer, and leads to some fascinating theory. The Singleton bound (Theorem 4.5.6)

$$A_q(n,d) \leq q^{n-d+1}$$

implies the following theorem for linear codes.

Theorem 5.3.1 For a linear [n,k]-code, we must have $d \leq n-k+1$. ▯

Definition A linear [n,k,n−k+1]-code, that is, an [n,k]-code with largest possible minimum distance, is called a **maximum distance separable** or **MDS** code. ▯

THE TRIVIAL MDS CODES

It is not hard to see that q-ary MDS codes exist with parameters [n,n,1], [n,1,n], and [n,n−1,2]. These codes are referred to as the **trivial MDS codes.** Thus, any *nontrivial* MDS [n,k] code must satisfy $2 \leq k \leq n-2$.

CHARACTERIZATIONS OF MDS CODES

MDS codes can be characterized in several very elegant ways. We begin with parity check matrices.

Recall that a linear code has minimum distance *d* if and only if any $d-1$ columns of a parity check matrix H are linearly independent, but some *d* columns are linearly dependent. Thus, we have the following.

Theorem 5.3.2 Let L be an [n,k]-code, with parity check matrix H. Then L is MDS if and only if any $n-k$ columns of H are linearly independent. ▯

Theorem 5.3.3 If a code L is MDS, then so is its dual code $L^{\perp}$.

Proof. Suppose that L is an [n,k]-code. Let H be a parity check matrix for L and, therefore, a generator matrix for $L^{\perp}$. Since any $(n-k)\times(n-k)$ submatrix of H is invertible, all nontrivial linear

combinations of the rows of H have at most $n-k-1$ zero coordinates. Therefore, all non trivial linear combinations of the rows of H have weight at least $n-(n-k-1)=k+1$. In other words, the minimum weight of $L^{\perp}$ is $k+1=n-(n-k)+1$, which says that $L^{\perp}$ is an MDS code. ∎

We can use Theorems 5.3.2 and 5.3.3 to characterize MDS codes in terms of their generator matrices. For a code L is an MDS [n,k]-code if and only if $L^{\perp}$ is an MDS [n,n–k]-code, which holds if and only if any k columns of a parity check matrix for $L^{\perp}$ (i.e., generator matrix for L) are linearly independent.

Theorem 5.3.4 Let L be an [n,k]-code, with generator matrix G. Then L is MDS if and only if any k columns of G are linearly independent. □

Theorem 5.3.4 tells us that any k positions of an MDS code form an information set, and so an MDS code is systematic on any k positions.

Here is another beautiful characterization of MDS codes.

Theorem 5.3.5 Let L be an [n,k]-code with generator matrix $G=(I_k \mid A)$ in standard form. Then L is an MDS code if and only if every square submatrix of A is nonsingular.

Proof. Let L be an MDS code, and let B_u be a $u \times u$ submatrix of A. By interchanging rows and columns of G, we may assume that B_u occupies the upper left corner of A. Now let us consider the matrix M consisting of the last $k-u$ columns of I_k and the columns of G containing B. This is a $k \times k$ matrix of the form

$$M=\begin{bmatrix} \mathbf{0}_{u,k-u} & \mathbf{B}_u \\ \mathbf{I}_{k-u} & * \end{bmatrix}$$

Since L is MDS, Theorem 5.3.4 implies that $\det(M) \neq 0$, and since $\det(B_u)=\pm\det(M)$, the result follows. We leave proof of the converse as an exercise. ∎

The **support** of a vector $\mathbf{x} \in V(n,q)$ is the set of all coordinate positions where $\mathbf{x}$ is nonzero. Our next result characterizes MDS codes in yet another way.

Theorem 5.3.6 A linear [n,k,d]-code L is an MDS code if and only if given any d coordinate positions, there is a (minimum weight)

codeword whose support is precisely these positions.

Proof. Suppose that L is an MDS [n,k]-code. Choose $d = n - k + 1$ coordinate positions, say $i_1, \ldots, i_d$. Consider the position i_1, together with the $k - 1$ non-chosen positions, say $j_1, \ldots, j_{k-1}$. Since the k positions $i_1, j_1, \ldots, j_{k-1}$ are information symbols, there is a codeword **c** that has a 1 in position i_1 and 0's in the positions $j_1, \ldots, j_{k-1}$. Hence, **c** has support $i_1, \ldots, i_d$.

For the converse, consider the $n - d + 1$ rows of the matrix

$$M = \left[I_{n-d+1} \middle| \; 1_{d-1} \right]$$

where I_{n-d+1} is an identity matrix and 1_{d-1} is the $(d-1) \times (d-1)$ matrix all of whose entries are 1's. These rows are linearly independent and have weight d. Since there is a codeword $\mathbf{c}_i$ in L that has the same support as the i-th row of M, we deduce that $k \geq n - d + 1$, which shows that L is MDS. ∎

Let us summarize our results on the characterization of MDS codes.

Theorem 5.3.7 Let L be an [n,k,d]-code over F_q. The following statements are equivalent.

1) L is an MDS code.
2) Any k columns of a generator matrix for L are linearly independent.
3) Any $n - k$ columns of a parity check matrix for L are linearly independent.
4) $L^{\perp}$ is an MDS code.
5) If $G = (I \mid A)$ is a standard form generator matrix for L, then every square submatrix of A is nonsingular.
6) Given any d coordinate positions, there is a (minimum weight) codeword whose support is precisely these positions. □

EXISTENCE OF NONTRIVIAL MDS CODES

Since MDS codes are very special, it is not surprising that the existence of such a code puts strong constraints on the possible values of the parameters of the code.

Theorem 5.3.8 There are no nontrivial MDS [n,k]-codes for which $1 \leq k \leq n - q$.

Proof. Certainly, we cannot have $k = 1$, for then the MDS code would be trivial. Assume that $1 < k \leq n - q$ and that L is a q-ary [n,k] MDS code, with generator matrix G in standard form. Thus, $G =$

$(I_k \mid A)$, where A has $n-k \geq q$ columns.

If any column of A contains a 0 entry, then that column is a linear combination of $k-1$ columns from I_k, which contradicts Theorem 5.3.4. Hence, A has no 0 entries. In particular, we may multiply each column by a scalar if necessary to obtain an equivalent code for which the first row of A consists of all 1's, as shown below. (The property of being MDS is invariant under equivalence of codes.)

Now, let us consider the second row of A. Since it contains no 0 entries and has $n-k \geq q$ columns, there must be two columns with the same nonzero entry a. Thus, G has the form

$$G = \left[\begin{array}{c|c} \mathbf{I}_k & \begin{matrix} 1 & \cdots & 1 & \cdots & 1 & \cdots & 1 \\ * & \cdots & a & \cdots & a & \cdots & * \\ * & \cdots & * & \cdots & * & \cdots & * \\ \vdots & & \vdots & & \vdots & & \vdots \\ * & \cdots & * & \cdots & * & \cdots & * \end{matrix} \end{array} \right]$$

But then the codeword formed by subtracting a^{-1} times the second row from the first row of G has weight at most $n-k$, which implies that L is not MDS. This contradiction shows that no q-ary MDS [n,k]-codes exist with $2 \leq k \leq n-q$. ∎

By applying the previous corollary to the dual code $L^{\perp}$, we get the dual result.

Corollary 5.3.9 There are no nontrivial MDS [n,k]-codes for which $q \leq k \leq n$. □

Corollaries 5.3.8 and 5.3.9 can be restated as follows.

Theorem 5.3.10 If a nontrivial MDS [n,k]-code exists, then

$$n-q+1 \leq k \leq q-1$$

Put another way, if a nontrivial MDS [n,k]-code exists, then

$$2 \leq k \leq q-1 \quad \text{and} \quad 2 \leq n-k \leq q-1$$ □

This theorem spells apathy for binary MDS code.

Corollary 5.3.11 The only *binary* MDS [n,k]-codes are the trivial codes. □

One of the most important problems related to MDS codes is the following. Given k and q, find the largest value of n for which there

exists a q-ary MDS [n,k]-code. Let us denote this value of n by $m(k,q)$. According to Theorem 5.3.10,

$$m(k,q) \leq k+q-1$$

however, the evidence seems to suggest that $m(k,q)$ is actually $q+1$.

This problem can be rephrased in a variety of ways, corresponding to the variety of characterizations of MDS codes. For example, it is equivalent to each of the following.

1) Given k and q, find the largest value of n for which there is a $k \times n$ matrix G over F_q with the property that any k columns of G are linearly independent.
2) Given a k-dimensional vector space over F_q, find the largest number of vectors with the property that any k of these vectors forms a basis for $V(n,q)$.
3) Given k and q, find the largest n such that there exists an $k \times (n-k)$ matrix A over F_q with the property that every square submatrix of A is nonsingular.

THE WEIGHT DISTRIBUTION OF AN MDS CODE

It happens that there is an explicit formula for the weight distribution of an MDS code. The next result follows from Theorem 5.3.6.

Corollary 5.3.12 Let L be an MDS [n,k]-code. The number of codewords in L of minimum weight $n-k+1$ is

$$(q-1)\binom{n}{n-k+1}$$ □

We can determine the complete weight distribution of an MDS code using the MacWilliams identity, in the form of equation (5.2.9), that is,

$$\frac{1}{q^k}\sum_{i=0}^{n-r}\binom{n-i}{r}A_i = \frac{1}{q^r}\sum_{i=0}^{r}\binom{n-i}{n-r}A_i^\perp$$

First, we use the fact that $A_0 = 1$, $A_i = 0$ for $i = 1,\ldots,n-k$ and, in view of Theorem 5.3.3, $A_0^\perp = 1$, $A_i^\perp = 0$ for $i = 1,\ldots,k$. Thus, for $r \leq k$,

$$\text{(5.3.1)} \qquad \sum_{i=n-k+1}^{n-r}\binom{n-i}{r}A_i = \binom{n}{r}(q^{k-r}-1) \qquad r = 0,\ldots,k-1$$

We leave it as an exercise to solve this triangular system for the A_j, to get the following.

Theorem 5.3.13 Let L be a q-ary MDS [n,k]-code. If A_w denotes the number of codewords in L of weight w, then

$$A_w = \binom{n}{w} \sum_{j=0}^{w-d} (-1)^j \binom{w}{j} (q^{w-d+1-j} - 1)$$

$$= \binom{n}{w} (q-1) \sum_{j=0}^{w-d} (-1)^j \binom{w-1}{j} q^{w-d-j} \qquad \square$$

MDS CODES FROM VANDERMONDE MATRICES

It is not difficult to construct a family of MDS codes. Let $\alpha_1, \ldots, \alpha_u$ be nonzero elements from a field. The **Vandermonde matrix** based on these elements is

$$V(\alpha_1, \ldots, \alpha_u) = \begin{bmatrix} 1 & 1 & \cdots & 1 \\ \alpha_1 & \alpha_2 & \cdots & \alpha_u \\ \alpha_1^2 & \alpha_2^2 & \cdots & \alpha_u^2 \\ \vdots & \vdots & \cdots & \vdots \\ \alpha_1^{u-1} & \alpha_2^{u-1} & \cdots & \alpha_u^{u-1} \end{bmatrix}$$

(The transpose of this matrix is also called a Vandermonde matrix.)

Lemma 5.3.14 The determinant of the Vandermonde matrix is

$$\det[V(\alpha_1, \ldots, \alpha_u)] = \prod_{1 \le i < j \le u} (\alpha_j - \alpha_i)$$

In particular, if the α_i are distinct, then $V(\alpha_1, \ldots, \alpha_u)$ is nonsingular.

Proof. Let $p(x) = \det[V(\alpha_1, \ldots, \alpha_{u-1}, x)]$. Then $p(x)$ is a polynomial in x of degree $u-1$. Since each of $\alpha_1, \ldots, \alpha_{u-1}$ is a root of $p(x)$, each of $(x - \alpha_1), \ldots, (x - \alpha_{u-1})$ is a factor. Hence,

$$p(x) = \beta \prod_{i=1}^{u-1} (x - \alpha_i)$$

where β is a constant (with respect to x). But β is the coefficient of x^{u-1} in $p(x)$, and this is clearly $\det[V(\alpha_1, \ldots, \alpha_{u-1})]$. Hence, letting $x = \alpha_u$, we have

$$\det[V(\alpha_1,\ldots,\alpha_u)] = \det[V(\alpha_1,\ldots,\alpha_{u-1})]\prod_{i=1}^{u-1}(\alpha_u - \alpha_i)$$

from which the result follows by induction. ∎

Now let $F_q = \{0,\alpha_1,\ldots,\alpha_{q-1}\}$ and consider the following $(q-k+1)\times(q+1)$ matrix, obtained from a Vandermonde matrix by adding two additional columns

$$H_1 = \begin{bmatrix} 1 & \cdots & 1 & 1 & 0 \\ \alpha_1 & \cdots & \alpha_{q-1} & 0 & 0 \\ \alpha_1^2 & \cdots & \alpha_{q-1}^2 & 0 & 0 \\ \vdots & \cdots & \vdots & \vdots & \vdots \\ \alpha_1^{q-k} & \cdots & \alpha_{q-1}^{q-k} & 0 & 1 \end{bmatrix}$$

where $1 \le k \le q$. (For $k = q$, the matrix H_1 is the row matrix, all of whose entries are 1's.) We leave it to the reader to show, using Lemma 5.3.14, that any $q-k+1$ columns of H_1 form a nonsingular matrix. Therefore, we have the following.

Theorem 5.3.15 For $1 \le k \le q$, the matrix H_1 is the parity check matrix of a q-ary MDS [q+1,k]-code. □

Notice that we cannot, in general, add additional columns to the matrix H_1 and expect a parity check matrix for an MDS code. For instance, consider the matrix

$$H_2 = \begin{bmatrix} 1 & \cdots & 1 & 1 & 0 & 0 \\ \alpha_1 & \cdots & \alpha_{q-1} & 0 & 1 & 0 \\ \alpha_1^2 & \cdots & \alpha_{q-1}^2 & 0 & 0 & 1 \end{bmatrix}$$

Let us choose 2 columns from among the first $q-1$, along with the $(q+1)$-st column,

$$\begin{bmatrix} 1 & 1 & 0 \\ \alpha_i & \alpha_j & 1 \\ \alpha_i^2 & \alpha_j^2 & 0 \end{bmatrix}$$

This matrix has determinant $\alpha_i^2 - \alpha_j^2$. In order for this to be nonzero

for any choice of distinct α_i and α_j, the field F_q must satisfy

$$x \neq y \Rightarrow x^2 \neq y^2$$

We leave it to the reader to show that this is the case if and only if the characteristic of F_q is 2, that is, if and only if q is a power of 2.

Thus, only in the case where $q = 2^m$ is the matrix H_2 a parity check matrix for an MDS code, and we have the following result.

Theorem 5.3.16 For $q = 2^m$, the matrix H_2 is the parity check matrix of a q-ary MDS [q+2,q−1]-code. ▯

Taking into account the dual codes, Theorems 5.3.15 and 5.3.16 give the following.

Corollary 5.3.17 For $1 \leq k \leq q$, there exist q-ary MDS [q+1,k] and [q+1,q−k+1] codes. For $q = 2^m$, there exist q-ary MDS [q+2,q−1] and [q+2,3] codes. ▯

We have already seen (Theorem 5.3.10) that if $k \geq 2$, then $n \leq q + k - 1$. For $k \geq 3$ and q odd, we can improve upon this slightly.

Theorem 5.3.18 If L is a nontrivial q-ary MDS [n,k]-code, where $k \geq 3$ and q is odd, then $n \leq q + k - 2$.

Proof. We know that $n \leq q + k - 1$. Suppose that $n = q + k - 1$. Then, according to Theorem 5.3.13,

$$A_{n-k+2} = \binom{n}{k-2}(q-1)\sum_{j=0}^{1}(-1)^j\binom{n-k+1}{j}q^{1-j} = 0$$

and so the weight of every codeword in L is one of n−k+1, n−k+3, n−k+4,.... In other words, the number of zeros in any codeword in L is one of k−1,k−3,k−4,....

Let $\mathbf{c}$ be a codeword in L of weight $n - k + 3$. (Such a codeword exists since $A_{n-k+3} > 0$.) We may assume for convenience that $\mathbf{c}$ has its 0's in the first $k - 3$ positions. Since any k positions form an information set, there exists a nonzero codeword $\mathbf{d}$ in L, with 0's also in its first $k - 3$ positions, that is not a scalar multiple of $\mathbf{c}$. Thus, $\mathbf{c}$ and $\mathbf{d}$ have the form

$$\mathbf{c} = \underbrace{00\cdots 0}_{k-3}\, c_1 c_2 \cdots c_{n-k+3}$$

and

$$\mathbf{d} = \underbrace{00\cdots 0}_{k-3}\, d_1 d_2 \cdots d_{n-k+3}$$

where the c_i are nonzero. Let Σ be the subspace of L spanned by $\{\mathbf{c},\mathbf{d}\}$. Note that any codeword $\mathbf{e} \in \Sigma$ has 0's in its first k−3 positions. Hence, if $\mathbf{e}$ has any more 0's, it must have *exactly* two more.

Choose any position $i \in \{1,\ldots,n-k+3\}$, and let Σ_i be the subspace of Σ consisting of all linear combinations of $\mathbf{c}$ and $\mathbf{d}$ that have a 0 in position i. Since

$$\mathbf{c}_i = \frac{d_i}{c_i}\mathbf{c} - \mathbf{d} \in \Sigma_i$$

but $\mathbf{c} \notin \Sigma_i$, we see that $dim(\Sigma_i) = 1$. Hence,

$$\Sigma_i = \{\lambda \mathbf{c}_i \mid \lambda \in F_q\}$$

In particular, all nonzero codewords in Σ_i have zeros in precisely the same positions as $\mathbf{c}_i$, which are the first $k-3$ positions, and position i, and hence exactly one more position, say j.

By the same reasoning, Σ_j is a one-dimensional subspace of Σ, and we have just seen that it contains Σ_i, and so $\Sigma_j = \Sigma_i$. In other words, $\mathbf{e} \in \Sigma_i$ if and only if $\mathbf{e} \in \Sigma$ and $\mathbf{e}$ has a 0 in either position i or j, in which case it has a 0 in both positions and nowhere else (except for the first $k-3$ positions, of course).

Now we choose a position $i' \in \{1,\ldots,n-k+3\} - \{i,j\}$ and form the subspace $\Sigma_{i'}$, whose codewords have 0's in positions i' and j'. From the remarks in the previous paragraph, we see that $\{i,j\}$ and $\{i',j'\}$ are disjoint.

Continuing to choose new positions in this way, we get a complete list of the positions $\{1,\ldots,n-k+3\}$ in the form of disjoint pairs. Hence, $n-k+3 = q+2$ must be even, which implies that q is even. Since this is contrary to assumption, we must have $n < q+k-1$. ∎

We have gathered enough information at this point to determine the value of m(3,q), which is the largest value of n for which there exists a q-ary MDS [n,3]-code. If q is odd, Theorem 5.3.18 implies that $n \leq q+1$, and the codes MDS [q+1,3]-codes of Theorem 5.3.15 show that $m(3,q) = q+1$. If q is even, Theorem 5.3.10 implies that $n \leq q+2$, and the MDS [q+2,3]-codes of Corollary 5.3.17 show that $m(3,q) = q+2$. In short,

$$m(3,q) = \begin{cases} q+1 & \text{if } q \text{ is odd} \\ q+2 & \text{if } q \text{ is even} \end{cases}$$

It has been conjectured that, except for the case $k = 3$, q even, if there exists a nontrivial MDS [n,k]-code, then $m(k,q) = q+1$. This

conjecture is known to be true for $k \le 5$ or $q \le 11$ or $q > (4k-9)^2$.

EXERCISES

1. Prove Theorem 5.3.1.
2. Is the even weight code E_n MDS?
3. Show that q-ary codes exist with parameters [n,n,1], [n,1,n], and [n,n−1,2].
4. Verify the statements in Example 5.3.1.
5. Let $F_4 = \{0,1,\alpha,\beta\}$, where $\beta = \alpha^2$. Find a generator matrix for a [4,2,3] MDS code over F_4. Is the dual code also MDS?
6. In the proof of Theorem 5.3.5, exactly why are we allowed to assume that B_u occupies the upper left corner of A?
7. Prove the converse of Theorem 5.3.5.
8. Prove, using Lemma 5.3.14, that any $q-k+1$ columns of the matrix H_1 forms a nonsingular matrix.
9. Let F_q be a finite field. Prove that in order for $x \ne y \Rightarrow x^2 \ne y^2$ to hold for all $x,y \in F_q$, we must have $q = 2^m$ for some m.
10. Prove Corollary 5.3.12 using Theorem 5.3.6.
11. Prove Theorem 5.3.13. *Hint.* Set $r = k-1$, $k-2$, and $k-3$ in (5.3.1). Solve the resulting system and guess the general solution.
12. Show that any square submatrix of a Vandermonde matrix with real positive entries is nonsingular. Show that this is false in general for entries over a finite field.

5.4 Invariant Theory and Self-Dual Codes

INTRODUCTION

We can get additional insight into the structure of self-dual codes through the use of the nineteenth century technique of invariant theory, which has, of late, enjoyed a strong comeback. Our plan here is to touch very briefly on these matters – enough only to give the flavor of the technique. For more on this subject, we refer the reader to MacWilliams and Sloane (1977), and to Sloane (1977), on which our discussion relies heavily.

We begin by altering the definition of the weight enumerator of a linear code. Let L be a linear [n,k]-code, and let A_i denote, as usual, the number of codewords of weight i in L. Then we take the weight enumerator of L to be

$$W_L(x,y) = \sum_{i=0}^{n} A_i x^{n-i} y^i$$

Notice that W(x,y) is a homogeneous polynomial of degree n. (*Homogeneous* means that all terms have the same degree.)

It is easy to see that, for a q-ary linear code, the MacWilliams identity is equivalent to the identity

$$(5.4.1) \qquad W_{L^\perp}(x,y) = \frac{1}{|L|} W_L(x+(q-1)y, x-y)$$

and for a self-dual q-ary [n,k]-code, this can be written in the form (recall that $k = n/2$)

$$(5.4.2) \qquad W_L(x,y) = W_L\left(\frac{x+(q-1)y}{\sqrt{q}}, \frac{x-y}{\sqrt{q}}\right)$$

Moreover, since L is self-dual, it must have even length, and so

$$(5.4.3) \qquad W_L(x,y) = W_L(-x,-y)$$

Equations (5.4.2) and (5.4.3) tell us a great deal about the possibilities for the weight enumerator $W_L(x,y)$ of a self-dual code. To explore this further, we introduce some notation.

For any polynomial p(x,y), we will think of the pair (x,y) as a column vector. Thus, if A is a 2×2 complex matrix, we can define the product Ap(x,y) by

$$Ap(x,y) = p[A \cdot (x,y)^T]$$

For instance, if $p(x,y) = x^2 + y^2$ and

$$A = \begin{bmatrix} 1 & 1 \\ 0 & i \end{bmatrix}$$

then

$$A(x,y)^T = \begin{bmatrix} 1 & 1 \\ 0 & i \end{bmatrix} \begin{bmatrix} x \\ y \end{bmatrix} = \begin{bmatrix} x+y \\ iy \end{bmatrix}$$

and so

$$Ap(x,y) = p(x+y,iy) = (x+y)^2 + (iy)^2 = x^2 + 2xy$$

With this in mind, equations (5.4.2) and (5.4.3) become

$$\begin{bmatrix} \frac{1}{\sqrt{q}} & \frac{q-1}{\sqrt{q}} \\ \frac{1}{\sqrt{q}} & \frac{-1}{\sqrt{q}} \end{bmatrix} W_L(x,y) = W_L(x,y) \tag{5.4.4}$$

and

$$\begin{bmatrix} -1 & 0 \\ 0 & -1 \end{bmatrix} W_L(x,y) = W_L(x,y) \tag{5.4.5}$$

Thus, $W_L(x,y)$ is *invariant* under the complex matrices

$$A_1 = \begin{bmatrix} \frac{1}{\sqrt{q}} & \frac{q-1}{\sqrt{q}} \\ \frac{1}{\sqrt{q}} & \frac{-1}{\sqrt{q}} \end{bmatrix} \quad \text{and} \quad -I_2 = \begin{bmatrix} -1 & 0 \\ 0 & -1 \end{bmatrix}$$

The question we must ask ourselves now is, "What homogeneous polynomials are invariant under these matrices?" Let us take a more general viewpoint while answering this question.

INVARIANT THEORY

The following definition is basic.

Definition A polynomial $p(x_1,\ldots,x_m)$ is said to be **invariant** under an $m \times m$ complex matrix A if

$$Ap(x_1,\ldots,x_m) = p(x_1,\ldots,x_m)$$ □

We leave proof of the following elementary result as an exercise.

Theorem 5.4.1

1) If $p(x_1,\ldots,x_m)$ is invariant under the matrices S and T, then it is invariant under any product of these matrices (such as ST, S^2T, $TSTS$, etc.).

2) If $p(x_1,\ldots,x_m)$ is invariant under each of the *invertible* matrices $M_1,\ldots,M_s$, then it is also invariant under any matrix in the (multiplicative) group generated by these matrices. □

Example 5.4.1 Since $A_1^2 = A_1$, the group G generated by the matrices

$$A_1 = \begin{bmatrix} \frac{1}{\sqrt{q}} & \frac{q-1}{\sqrt{q}} \\ \frac{1}{\sqrt{q}} & \frac{-1}{\sqrt{q}} \end{bmatrix} \quad \text{and} \quad -I_2 = \begin{bmatrix} -1 & 0 \\ 0 & -1 \end{bmatrix}$$

is $G = \{I_2, -I_2, A_1, -A_1\}$. □

Definition Let G be a finite group of $m \times m$ complex matrices. Any polynomial $p(x_1,\ldots,x_m)$ that is invariant under every matrix in G is called an **invariant** of the group G. □

The following result is easily proved.

Theorem 5.4.2 Let G be a finite group of $m \times m$ complex matrices. The set of all invariant polynomials is an algebra over the complex numbers. We denote this algebra by $\mathfrak{I}(G)$. □

The next result can be used to find invariants of a group.

Theorem 5.4.3 Let $G = \{A_1,\ldots,A_s\}$ be a finite group of $m \times m$ matrices. Let $p(x_1,\ldots,x_m)$ be any polynomial. Then the *average* polynomial

$$\overline{p}(x_1,\ldots,x_m) = \frac{1}{s}\sum_{i=1}^{s} A_i p(x_1,\ldots,x_m)$$

is an invariant of G.

Proof. Applying A_k to the polynomial $\overline{p}(x_1,\ldots,x_m)$ gives

$$A_k\overline{p}(x_1,\ldots,x_m) = \frac{1}{s}\sum_{i=1}^{s} A_kA_i p(x_1,\ldots,x_m)$$

But as A_i runs through all elements of the group G, so does A_kA_i, and so this is equal to $\overline{p}(x_1,\ldots,x_m)$. ∎

Example 5.4.2 Let us try averaging a few monomials, using the group G from Example 5.4.1. First, we average the monomial x^2, to get

$$\frac{1}{4}\left[x^2+x^2+[\frac{1}{\sqrt{q}}(x+(q-1)y)]^2+[\frac{1}{\sqrt{q}}(x+(q-1)y)]^2\right] = \frac{1}{2}\left[x^2+\frac{1}{q}(x+(q-1)y)^2\right]$$

Averaging y^2 gives

$$\frac{1}{4}\left[y^2+y^2+[\frac{1}{\sqrt{q}}(x-y)]^2+[\frac{1}{\sqrt{q}}(x-y)]^2\right]=\frac{1}{2}\left[y^2+\frac{1}{q}(x-y)^2\right]$$

Multiplying both of these polynomials by $2q$ and expanding gives the invariants

$$(q+1)x^2+2(q-1)xy+(q-1)^2y^2 \quad \text{and} \quad x^2-2xy+(q-1)y^2$$

Now we can eliminate x^2 by multiplying the second polynomial by $-(q+1)$ and adding the result to the first polynomial to get (after dividing by $4q$) the homogeneous invariant

$$p(x,y)=y(x-y)$$

Similarly, by eliminating the xy terms, we get the homogeneous invariant

$$q(x,y)=x^2+(q-1)y^2$$

□

One of the main problems in invariant theory is to describe the invariants $\mathfrak{I}(G)$ of a given group G. Since applying a matrix in G to a polynomial does not change its degree, it suffices to find the homogeneous invariants of G.

Definition Let G be a finite group of $m\times m$ complex matrices. A set $\{p_1,\ldots,p_s\}$ of invariants of G is said to form a **polynomial basis** for $\mathfrak{I}(G)$ if any invariant of G is a polynomial in $p_1,\ldots,p_s$, that is, if the set

$$S=\{p_1^{i_1}\cdots p_s^{i_s} \mid i_1,\ldots,i_s\geq 0\}$$

spans $\mathfrak{I}(G)$. □

Notice that in the previous definition, we do not require that S be linearly independent. Of course, S is linearly independent if and only if no nontrivial linear combination of monomials of the form $p_1^{i_1}\cdots p_s^{i_s}$ is the zero polynomial, that is, if and only if there is no nonzero polynomial $f(x_1,\ldots,x_s)$ for which $f(p_1,\ldots,p_s)=0$. This is the same as saying that $p_1,\ldots,p_s$ are *algebraically independent.*

In 1916, Emmy Noether proved that for any finite group G of $m\times m$ matrices, the algebra $\mathfrak{I}(G)$ of invariants has a polynomial basis of size at most $\binom{m+s}{m}$, where s is the order of G. Noether also showed that such a basis can be obtained by averaging monomials. However, we will take a different approach to finding a polynomial basis, using a theorem of Molien (1897).

Theorem 5.4.4 (Molien's Theorem) Let G be a finite group of $m \times m$ complex matrices. Let a_k denote the number of linearly independent homogeneous invariants of degree k in $\mathfrak{I}(G)$, and let

$$\Phi_G(t) = \sum_{i=0}^{\infty} a_i t^i$$

Then

$$\Phi_G(t) = \frac{1}{|G|} \sum_{A \in G} \frac{1}{\det(I - tA)}$$ □

To see how we might use Molien's Theorem, we need the following combinatorial result.

Theorem 5.4.5 Let $p_1, \ldots, p_s$ be polynomials in m variables, with $\deg(p_i) = \alpha_i$. If b_k is the number of elements of

$$S = \{p_1^{i_1} \cdots p_s^{i_s} \mid i_1, \ldots, i_s \geq 0\}$$

of degree k, then b_k is the coefficient of t^k in the expansion of

$$\frac{1}{(1-t^{\alpha_1}) \cdots (1-t^{\alpha_s})}$$

Proof. We prove the result for $s = 2$ and leave generalization to the reader. Suppose that $\deg(p) = \alpha$ and $\deg(q) = \beta$. Let b_k be the number of elements of

$$S = \{p^i q^j \mid i,j \geq 0\}$$

of degree k. Since $\deg(p^i q^j) = \alpha i + \beta j$, we see that b_k is also the number of nonnegative integral solutions (i,j) to the equation

$$\alpha i + \beta j = k$$

which, in turn, is the coefficient of t^k in the expansion of

$$(1 + t^\alpha + t^{2\alpha} + \cdots)(1 + t^\beta + t^{2\beta} + \cdots) = \frac{1}{(1-t^\alpha)(1-t^\beta)}$$ ∎

Thus, if we find algebraically independent, homogeneous invariants $p_1, \ldots, p_s$, and if we verify through Molien's Theorem that

$$\Phi_G(t) = \frac{1}{(1-t^{\alpha_1}) \cdots (1-t^{\alpha_s})}$$

then we will know that $b_k = a_k$. In other words, the number of (linearly independent) homogeneous invariants of the form $p_1^{i_1} \cdots p_s^{i_s}$, with degree k is the same as the total number of linearly independent homogeneous invariants of G of degree k. Hence, $\{p_1, \ldots, p_s\}$ is a polynomial basis for $\mathfrak{I}(G)$.

THE WEIGHT ENUMERATOR OF A SELF-DUAL CODE

We are now in a position to answer the question posed at the end of the introduction to this section. Recall that the weight enumerator $W_L(x,y)$ of a q-ary self-dual code L is invariant under the matrices

$$A_1 = \frac{1}{\sqrt{q}}\begin{bmatrix} 1 & q-1 \\ 1 & -1 \end{bmatrix} \quad \text{and} \quad -I_2 = \begin{bmatrix} -1 & 0 \\ 0 & -1 \end{bmatrix}$$

that generate the group $G = \{I_2, -I_2, A_1, -A_1\}$. We leave it for the reader to show, using Molien's Theorem, that

$$\Phi_G(t) = \frac{1}{|G|}\sum_{A \in G} \frac{1}{\det(\mathbf{I} - tA)} = \frac{1}{(1-t^2)^2} = \frac{1}{(1-t^2)(1-t^2)}$$

Thus, we seek two algebraically independent, homogeneous invariants of G, each of degree 2. But we obtained the homogeneous invariants

$$p(x,y) = y(x-y) \quad \text{and} \quad q(x,y) = x^2 + (q-1)y^2$$

in Example 5.4.2, by the method of averaging.

These polynomials are seen to be algebraically independent as follows. Suppose they are algebraically dependent and that

$$\sum_{i,j} c_{ij} p^i q^j = 0 \tag{5.4.6}$$

where the c_{ij} are all nonzero. By the *leading term* in a polynomial, we mean the term with the largest power of x, after collecting all terms with like powers of x. Let us examine the leading term in the sum on the left side of (5.4.6), which must be 0 by (5.4.6). Thus, if this term is $f(y)x^m$, then $f(y)$ must be the zero polynomial.

Clearly, the leading term in (5.4.6) must come from leading terms in the expansions of the individual terms

$$c_{ij}p^iq^j = c_{ij}(yx - y^2)^i(x^2 + (q-1)y^2)^j \tag{5.4.7}$$

Since the leading term of (5.4.7) is

$$c_{ij}x^{i+2j}y^i$$

we see that the leading term of (5.4.6) is the same as the leading term of

$$\sum_{i,j} c_{ij}x^{i+2j}y^i$$

But the leading term of this sum is

$$\sum_j c_{m-2j,j}x^m y^{m-2j}$$

Hence,

$$f(y) = \sum_j c_{m-2j,j} y^{m-2j}$$

If this is to be the zero polynomial, then the coefficients $c_{m-2j,j}$ must equal 0. This contradiction implies that the original polynomials are algebraically independent.

Thus, we have proved the following result.

Theorem 5.4.6 The weight enumerator of any q-ary self-dual code is a polynomial in $p(x,y) = y(x-y)$ and $q(x,y) = x^2 + (q-1)y^2$. □

THE WEIGHT ENUMERATOR OF AN EVEN SELF-DUAL CODE

Let us briefly describe another example that shows that we can sometimes determine the weight enumerator of a code completely, using invariant theory methods. Recall that a *binary* self-dual code is even if all codeword weights are divisible by 4. In this case, the MacWilliams identity

$$\frac{1}{\sqrt{q}}\begin{bmatrix} 1 & q-1 \\ 1 & -1 \end{bmatrix} W_L(x,y) = W_L(x,y)$$

and the fact that

$$\begin{bmatrix} 1 & 0 \\ 0 & i \end{bmatrix} W_L(x,y) = W_L(x,y)$$

(see Exercise 10) lead to a group G that contains 192 elements. The Molien series in this case turns out to be

$$\Phi_L(t) = \frac{1}{(1-t^8)(1-t^{24})}$$

Some of the details of this can be found in MacWilliams and Sloane (1977). This leads to the following theorem of Gleason (1971).

Theorem 5.4.7 The weight enumerator of an even self-dual code is a polynomial in $r(x,y) = x^8 + 14x^4y^4 + y^8$ and $s(x,y) = x^4y^4(x^4-y^4)^4$. □

As we will learn later in the book, there exists an even self-dual [48,24,12]-code. (This is a quadratic residue code.) Let us denote this code by $\mathcal{Q}$. Since $\mathcal{Q}$ has length 48 and minimum distance 12, the weight enumerator $W_{\mathcal{Q}}(x,y)$ is a homogeneous polynomial of degree 48 of the form

$$W_{\mathcal{Q}}(x,y) = x^{48} + A_{12}x^{36}y^{12} + \cdots + y^{48} \tag{5.4.8}$$

Now, it happens that (5.4.8), together with Theorem 5.4.7, are enough to determine completely the weight enumerator $W_{\mathbb{Q}}(x,y)$.

We know that $W_{\mathbb{Q}}(x,y)$ is a polynomial in $r(x,y)$ and $s(x,y)$, that is,

$$W_{\mathbb{Q}}(x,y) = \sum_{i,j} c_{ij} r^i s^j$$

But $r^i s^j$ is a homogeneous polynomial of degree $8i + 24j$, and $8i + 24j = 48$ if and only if $(i,j) = (6,0)$, $(3,1)$ or $(0,2)$. Hence, we have

$$\begin{aligned} W_{\mathbb{Q}}(x,y) &= a[r(x,y)]^6 + b[r(x,y)]^3 s(x,y) + c[s(x,y)]^2 \\ &= a(x^{48} + 84x^{44}y^4 + 2946x^{40}y^8 + \cdots) \\ &\quad + b(x^{44}y^4 + 38x^{40}y^8 + \cdots) \\ &\quad + c(x^{40}y^8 - \cdots) \end{aligned}$$

Equating coefficients with (5.4.8), we get $a = 1$, $b = -84$, $c = 246$. A little algebra (hand or computer) then gives

$$\begin{aligned} W_{\mathbb{Q}}(x,y) = x^{48} &+ 17296x^{36}y^{12} + 535095x^{32}y^{16} \\ &+ 3995376x^{28}y^{20} + 7681680x^{24}y^{24} + 3995376x^{20}y^{28} \\ &+ 535095x^{16}y^{32} + 17296x^{12}y^{36} + y^{48} \end{aligned}$$

EXERCISES

1. Verify (5.4.1).
2. Verify (5.4.2).
3. Prove Theorem 5.4.1.
4. Prove Theorem 5.4.2.
5. Verify that the group generated by A and I_2 in Example 5.4.1 is $G = \{I_2, -I_2, A, -A\}$.
6. Verify that applying a nonsingular matrix to a polynomial does not change its degree.
7. Prove that any invariant of G is a linear combination of monic homogeneous invariants of G.
8. Prove Theorem 5.4.5 in its full generality.
9. Show that the Molien series for the dihedral group $\mathfrak{D}_4$ and the Abelian group $\mathbb{Z}_2 \times \mathbb{Z}_4$ are both equal to $1/[(1-t^2)(1-t^4)]$.
10. Let $W(x,y)$ be the weight enumerator of an *even* binary self-dual code. Show that $W(x,y) = W(x,iy)$.

CHAPTER 6

Some Linear Codes

6.1 Hamming and Golay Codes

In this chapter, we take a look at three of the most well-known families of codes.

HAMMING CODES

The Hamming codes $\mathcal{H}_q(r)$ are probably the most famous of all error-correcting codes. These codes were discovered independently by Marcel Golay in 1949 and Richard Hamming in 1950. They are perfect, linear codes that decode in a very elegant manner. In addition, all *binary* Hamming codes are equivalent to cyclic codes, and some, but not all, non-binary Hamming codes are equivalent to cyclic codes.

According to Theorem 5.1.4, the minimum distance of a linear [n,k]-code with parity check matrix H is the smallest integer d for which there exists d linearly dependent columns in H. Hence, the parity check matrix of an [n,k,3]-code has the property that no two of its columns are linearly dependent, that is, no column is a scalar multiple of another column, but some set of three columns is linearly dependent.

For a given code alphabet F_q, we can construct a parity check matrix with these properties, and with the maximum possible number of columns, as follows. First, pick any nonzero column $\mathbf{c}_1$ in $V_1 = V(r,q)$. Then pick any nonzero column $\mathbf{c}_2$ in

$$V_2 = V_1 - \{\alpha \mathbf{c}_1 \mid \alpha \neq 0\}$$

We continue to pick nonzero columns and then discard all nonzero scalar multiples of the chosen column until all columns have been discarded. Since

$$|\{\alpha c_i \mid \alpha \neq 0\}| = q-1$$

the result is a parity check matrix with $(q^r-1)/(q-1)$ columns, for which no two columns are linearly independent, but for which some set of three columns is linearly dependent. (The matrices H_1 [for $q=2$] and H_2 [for $q=3$] defined below are examples of such matrices.)

The resulting matrix, known as a **Hamming matrix of order** r, is the parity check matrix of a q-ary linear [n,k,3]-code with parameters

$$n = \frac{q^r-1}{q-1}, \quad k = n-r, \quad d = 3 \tag{6.1.1}$$

that is known as a **q-ary Hamming code of order** r and is denoted by $\mathcal{H}_q(r)$. Notice that, while the size of a Hamming code is quite large, and the rate tends to 1 as $r \to \infty$, these codes are only single-error-correcting.

Notice also that the choice of columns is not unique, and so there are many different Hamming matrices, and Hamming codes, with a given set of parameters. However, any Hamming matrix can be obtained from any other (with the same parameters) by permuting the columns and multiplying some columns by nonzero scalars. Hence, any two Hamming codes of the same size are scalar multiple equivalent. In fact, up to scalar multiple equivalence, Hamming codes are uniquely defined by their parameters and the fact that they are linear.

The binary case is by far the most common, where $\mathcal{H}_2(r)$ is a binary linear [n,k,3]-code, with

$$n = 2^r - 1, \quad k = 2^r - 1 - r, \quad d = 3$$

In this case, the columns of the Hamming matrix of order r are simply the binary representations of the first $2^r - 1$ positive integers. (See the matrix H_1 below.)

It is not hard to see that a code with parameters (6.1.1) is indeed perfect, and we leave the details as an exercise.

Decoding with a Hamming Code

Some forms of the Hamming matrix H provide more elegant decoding than other forms. In the binary case, we choose the columns of H in increasing order, thought of as binary numbers (read from the top down). Thus, for instance, the parity check matrix for the binary Hamming code $\mathcal{H}_2(3)$ is

$$H_1 = \begin{bmatrix} 0 & 0 & 0 & 1 & 1 & 1 & 1 \\ 0 & 1 & 1 & 0 & 0 & 1 & 1 \\ 1 & 0 & 1 & 0 & 1 & 0 & 1 \end{bmatrix}$$

Now, if a single error occurs in transmission in the i-th position, resulting in the error vector e_i, the syndrome of the received word is $e_i H^T$, which is just the i-th column of H, written as a row. Moreover, this row, thought of as a binary number, is none other than the binary representation of the position of the error!

To illustrate, for the matrix H_1 above, if the error is $e_3 = 0010000$, then the syndrome is

$$e_3 H_1^T = \begin{bmatrix} 0 & 0 & 1 & 0 & 0 & 0 & 0 \end{bmatrix} \begin{bmatrix} 0 & 0 & 1 \\ 0 & 1 & 0 \\ 0 & 1 & 1 \\ 1 & 0 & 0 \\ 1 & 0 & 1 \\ 1 & 1 & 0 \\ 1 & 1 & 1 \end{bmatrix} = \begin{bmatrix} 0 & 1 & 1 \end{bmatrix} = 011_2 = 3_{10}$$

In the non-binary case, we can do almost as well by choosing the columns of the parity check matrix in increasing size as ternary numbers, but for which the first nonzero entry in each column is a 1. For instance, the parity check matrix for $\mathcal{H}_3(3)$ is

$$H_2 = \begin{bmatrix} 0 & 0 & 0 & 0 & 1 & 1 & 1 & 1 & 1 & 1 & 1 & 1 & 1 \\ 0 & 1 & 1 & 1 & 0 & 0 & 0 & 1 & 1 & 1 & 2 & 2 & 2 \\ 1 & 0 & 1 & 2 & 0 & 1 & 2 & 0 & 1 & 2 & 0 & 1 & 2 \end{bmatrix}$$

Now, if an error occurs in the i-th position, the error vector will have the form αe_i, for some nonzero scalar α. Hence, the syndrome is

$$\alpha e_i H^T$$

which is α times the i-th column of H (written as a row). Thus, because of the way H was constructed, we see that α is the first nonzero entry in the syndrome. Furthermore, multiplying the syndrome by α^{-1} will give us the i-th column of H, telling us the position of the error.

Example 6.1.1 Using the matrix H_2 above, the syndrome of the received word $\mathbf{y} = 1101112211201$ is

$$[1101112211201]\, H_2^T = [201] = 2\,[102] = 2 \times (\text{7th column of } H_2)$$

Hence, we subtract 2 from the 7th position of $\mathbf{y}$ to get the codeword

$$\mathbf{c} = 1101110211201$$

□

A Nonlinear Code with the Hamming Parameters

As we have seen, any *linear* code with the Hamming parameters must be equivalent to a Hamming code. However, there are nonlinear codes with the Hamming parameters. The following construction is due to Vasil'ev (1962).

Let $\lambda:\mathcal{H}_2(r)\to\mathbb{Z}_2$ be a *nonlinear* map from the Hamming code $\mathcal{H}_2(r)$ to $\mathbb{Z}_2$, for which $\lambda(0)=0$. Hence, there exist $\mathbf{c},\mathbf{d}\in\mathcal{H}_2(r)$ for which $\lambda(\mathbf{c}+\mathbf{d})\neq\lambda(\mathbf{c})+\lambda(\mathbf{d})$. For any $\mathbf{x}\in V(n,2)$, let $\pi(\mathbf{x})$ equal 0 if $\mathbf{x}$ has even weight and 1 if $\mathbf{x}$ has odd weight.

Now consider the code

$$\mathcal{V}=\{(\mathbf{x},\mathbf{x}+\mathbf{c},\pi(\mathbf{x})+\lambda(\mathbf{c})),\text{ for all }\mathbf{x}\in V(n,2),\ \mathbf{c}\in\mathcal{H}_2(r)\}$$

where $n=2^r-1$. We leave it as an exercise to show that $\mathcal{V}$ is a perfect nonlinear binary code with the Hamming parameters

$$(2^{r+1}-1,2^{2n-r},3)$$

Hamming Codes and Designs

We have already seen (Section 4.4) that the codewords of weight 3 in $\mathcal{H}_2(r)$ hold a Steiner system $S(2,3,2^r-1)$ and that the number of codewords of weight 3 is

$$A_3=\frac{(2^r-1)(2^{r-1}-1)}{3}$$

Example 6.1.2 The number of codewords of weight 3 in $\mathcal{H}_2(3)$ is $A_3=7$, and these 7 codewords hold an $S(2,3,7)$ design. We leave it as an exercise to show that $\mathcal{H}_2(3)$ holds the projective plane of order 2 (Figure 4.4.1). □

SIMPLEX CODES

The dual codes to the binary Hamming codes $\mathcal{H}_2(r)$ are the **simplex codes** Σ_r. The reason for this terminology will become apparent as we discover the properties of these codes.

Since $\mathcal{H}_2(r)$ is a $[2^r,2^r-1-r]$-code, the simplex code Σ_r has parameters $[2^r-1,r]$. Furthermore, a parity check matrix for $\mathcal{H}_2(r)$ is a generator matrix G_r for Σ_r. Let us look at the codes Σ_2 and Σ_3. For Σ_2, we have

$$G_2=\begin{bmatrix}0&1&1\\1&0&1\end{bmatrix}\qquad \Sigma_2=\begin{matrix}0&0&0\\0&1&1\\1&0&1\\1&1&0\end{matrix}$$

For Σ_3, we have (the lines are included to help see the pattern)

$$G_3 = \left[\begin{array}{c|c|c} 0\ 0\ 0 & 1 & 1\ 1\ 1 \\ \hline 0\ 1\ 1 & 0 & 0\ 1\ 1 \\ 1\ 0\ 1 & 0 & 1\ 0\ 1 \end{array}\right] = \left[\begin{array}{c|c|c} 0\ 0\ 0 & 1 & 1\ 1\ 1 \\ \hline G_2 & \mathbf{0} & G_2 \end{array}\right]$$

and so

$$\Sigma_3 = \begin{array}{c|c|c} 0\ 0\ 0 & 0 & 0\ 0\ 0 \\ 0\ 1\ 1 & 0 & 0\ 1\ 1 \\ 1\ 0\ 1 & 0 & 1\ 0\ 1 \\ 1\ 1\ 0 & 0 & 1\ 1\ 0 \\ \hline 0\ 0\ 0 & 1 & 1\ 1\ 1 \\ 0\ 1\ 1 & 1 & 1\ 0\ 0 \\ 1\ 0\ 1 & 1 & 0\ 1\ 0 \\ 1\ 1\ 0 & 1 & 0\ 0\ 1 \end{array} = \begin{array}{c|c|c} & 0 & \\ \Sigma_2 & 0 & \Sigma_2 \\ & 0 & \\ & 0 & \\ \hline & 1 & \\ \Sigma_2 & 1 & \Sigma_2^c \\ & 1 & \\ & 1 & \end{array}$$

where Σ_2^c is the set of complements of codewords in Σ_2.

We can make several conjectures from this information that can be proved by induction.

Theorem 6.1.1 The $[2^r-1,r]$-simplex code Σ_r has the following properties.

1) Every nonzero codeword in Σ_r has weight 2^{r-1}. Hence, Σ_r is a $[2^r-1,r,2^{r-1}]$-code.
2) The distance between any two codewords in Σ_r is 2^{r-1}.
3) If $\mathbf{c} = c_1c_2\cdots c_n$ is in Σ_r, then so is the word $\mathbf{d} = c_nc_1c_2\cdots c_{n-1}$, formed by cyclically permuting the coordinate positions in $\mathbf{c}$. In other words, the simplex code Σ_r is cyclic.

Proof. We leave the inductive proofs as exercises. As a suggestion for part 3), it is useful to think about what happens when the codewords in Σ_3 shown above are permuted cyclically. Consider the cases where the third position is a 0 or a 1 separately. ∎

Part 2) of Theorem 6.1.1 explains why the codes Σ_r are referred to as simplex codes. Figure 6.1.1 shows a picture of Σ_2. Notice that the line segments connecting the codewords form a regular simplex (a tetrahedron in this case).

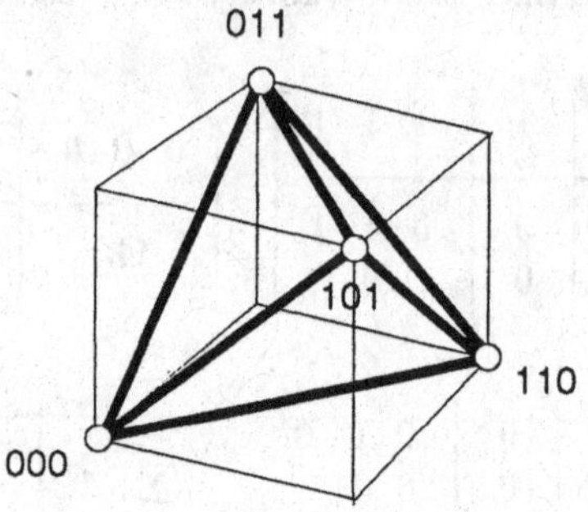

Figure 6.1.1

GOLAY CODES

There are a total of four Golay codes – two binary codes and two ternary codes. We will define these codes by giving generating matrices, as did Golay in 1949. As we will see in a subsequent chapter, two of the Golay codes can be defined as cyclic codes.

The Binary Golay Code $\mathcal{G}_{24}$

The binary Golay code $\mathcal{G}_{24}$ is a [24,12,8]-code whose generator matrix has the form $G = [\mathbf{I}_{12} \mid A]$, where

$$A = \begin{bmatrix} \cdot & 1 & 1 & 1 & 1 & 1 & 1 & 1 & 1 & 1 & 1 & 1 \\ 1 & 1 & 1 & \cdot & 1 & 1 & 1 & \cdot & \cdot & \cdot & 1 & \cdot \\ 1 & 1 & \cdot & 1 & 1 & 1 & \cdot & \cdot & \cdot & 1 & \cdot & 1 \\ 1 & \cdot & 1 & 1 & 1 & \cdot & \cdot & \cdot & 1 & \cdot & 1 & 1 \\ 1 & 1 & 1 & 1 & \cdot & \cdot & \cdot & 1 & \cdot & 1 & 1 & \cdot \\ 1 & 1 & 1 & \cdot & \cdot & \cdot & 1 & \cdot & 1 & 1 & \cdot & 1 \\ 1 & 1 & \cdot & \cdot & \cdot & 1 & \cdot & 1 & 1 & \cdot & 1 & 1 \\ 1 & \cdot & \cdot & \cdot & 1 & \cdot & 1 & 1 & \cdot & 1 & 1 & 1 \\ 1 & \cdot & \cdot & 1 & \cdot & 1 & 1 & \cdot & 1 & 1 & 1 & \cdot \\ 1 & \cdot & 1 & \cdot & 1 & 1 & \cdot & 1 & 1 & 1 & \cdot & \cdot \\ 1 & 1 & \cdot & 1 & 1 & \cdot & 1 & 1 & 1 & \cdot & \cdot & \cdot \\ 1 & \cdot & 1 & 1 & \cdot & 1 & 1 & 1 & \cdot & \cdot & \cdot & 1 \end{bmatrix}$$

(each dot represents a 0). We will show that $\mathcal{G}_{24}$ has minimum weight 8 through a series of simple facts about $\mathcal{G}_{24}$.

Lemma 6.1.2 The Golay code $\mathcal{G}_{24}$ is self-dual, that is, $\mathcal{G}_{24}^{\perp} = \mathcal{G}_{24}$.
Proof. It is straightforward to check that, if $\mathbf{r}$ and $\mathbf{s}$ are rows of G, then $\mathbf{r} \cdot \mathbf{s} = 0$. Hence, $\mathcal{G}_{24} \subset \mathcal{G}_{24}^{\perp}$. Furthermore, $\mathcal{G}_{24}$ and $\mathcal{G}_{24}^{\perp}$ have the

same dimension, and so they must be equal. ∎

Lemma 6.1.3 The Golay code $\mathcal{G}_{24}$ is also generated by the matrix $[A \mid I_{12}]$.
Proof. This follows from Lemma 6.1.2 and the fact that $A^T = A$. ∎

Lemma 6.1.4 The weight of every codeword in $\mathcal{G}_{24}$ is divisible by 4.
Proof. It is easy to see that the weight of every row of G is divisible by 4. If **r** and **s** are rows of G, then

$$w(\mathbf{r}+\mathbf{s}) = w(\mathbf{r}) + w(\mathbf{s}) - 2w(\mathbf{r} \cap \mathbf{s})$$

But $w(\mathbf{r} \cap \mathbf{s}) \equiv \mathbf{r} \cdot \mathbf{s} = 0 \bmod 2$, and so $w(\mathbf{r}+\mathbf{s})$ is also divisible by 4. We leave it as an exercise to show by induction that the weight of the sum of any number of rows of G is divisible by 4, which will complete the proof. ∎

Lemma 6.1.5 The code $\mathcal{G}_{24}$ has no codewords of weight 4.
Proof. We take advantage of the two generating matrices $G_1 = [I_{12} \mid A]$ and $G_2 = [A \mid I_{12}]$ for $\mathcal{G}_{24}$. Suppose that **c** is a codeword in $\mathcal{G}_{24}$ of weight 4, and consider the left half **L** and the right half **R** of **c**. Since any nontrivial linear combination of the rows of G_1 has a left half of weight at least 1, we see that **L** must have weight at least 1. Similarly, using G_2, we deduce that **R** must have weight at least 1. Furthermore, if $w(\mathbf{L}) = 1$ then **c** must be a row of G_1, none of which has weight 4. Hence, $w(\mathbf{L}) \geq 2$. Similarly, $w(\mathbf{R}) \geq 2$. This leaves only the possibility $w(\mathbf{L}) = w(\mathbf{R}) = 2$, which can be ruled out by checking that no sum of any two rows of G_1 has weight 4. ∎

We can now state the following.

Theorem 6.1.6 The binary Golay code $\mathcal{G}_{24}$ is a [24,12,8]-code. □

Let A_i denote the number of words of weight i in $\mathcal{G}_{24}$. According to Lemmas 6.1.4 and 6.1.5, the only possible nonzero A_i's are A_0, A_8, A_{12}, A_{16}, A_{20}, and A_{24}. However, since $\mathbf{1} \in \mathcal{G}_{24}$, we have $A_{20} = A_4 = 0$, as well as $A_8 = A_{16}$. In fact, we can state the following

$$A_0 = A_{24} = 1,\quad A_8 = A_{16} = ?,\quad A_{12} = ?,\quad A_i = 0 \text{ for all other } i$$

A counting argument (MacWilliams and Sloane (1977), Chapter 2, §6) or direct computer computation can then be used to show that the weight distribution of $\mathcal{G}_{24}$ is

$$A_0 = A_{24} = 1,\ A_8 = A_{16} = 759,\quad A_{12} = 2576$$

Now, let us consider the set W_8 of all codewords in $\mathcal{G}_{24}$ that have weight 8. Recall that a binary word $\mathbf{x}$ is said to **cover** a binary word $\mathbf{y}$ of the same length if whenever $\mathbf{y}$ has a 1 in the i-th position, so does $\mathbf{x}$. It is not hard to see that the codewords of minimum weight in $\mathcal{G}_{24}$ hold a design.

First, we observe that no binary word of weight 5 can be covered by more than one codeword in W_8, for two such codewords would then have distance at most 6 from each other, which is not possible in $\mathcal{G}_{24}$. Thus, if $\mathbf{c}$ and $\mathbf{d}$ are distinct codewords in W_8, the two sets consisting of $\binom{8}{5}$ words of length 5 that are covered by $\mathbf{c}$ and $\mathbf{d}$ are disjoint. But

$$|W_8|\binom{8}{5} = 759\binom{8}{5} = \binom{24}{5}$$

and so every word of weight 5 is in one of these sets. In other words, every word of weight 5 is covered by exactly one codeword in W_8. We have proved the following result.

Theorem 6.1.7 The codewords of minimum weight 8 in the Golay code $\mathcal{G}_{24}$ hold a Steiner system $S(5,8,24)$. ▯

In 1968, Pless showed that any binary *linear* code with the same parameters as $\mathcal{G}_{24}$ must be scalar multiple equivalent to $\mathcal{G}_{24}$. In 1975, Delsarte and Goethals extended this result, by showing that *any* binary code (linear or nonlinear) with the same parameters as $\mathcal{G}_{24}$ is equivalent to $\mathcal{G}_{24}$. (Actually, they showed that if such a code contains the zero codeword, then it must be linear, and so by the result of Pless, it must be equivalent to $\mathcal{G}_{24}$.) Thus, we have the following theorem.

Theorem 6.1.8 Any binary $(24,2^{12},8)$-code is equivalent to the Golay code $\mathcal{G}_{24}$. ▯

Decoding the Binary Golay Code $\mathcal{G}_{24}$

Since $\mathcal{G}_{24}$ is a $[24,12,8]$-code, syndrome decoding would require that we construct

$$\frac{2^{24}}{2^{12}} = 2^{12} = 4096$$

syndromes. On the other hand, using the structure of $\mathcal{G}_{24}$, we can considerably reduce the work involved in decoding.

Since $\mathcal{G}_{24}$ is self-dual, the matrices $G_1 = [\mathbf{I} \mid A]$ and $G_2 = [A \mid \mathbf{I}]$ are both parity check matrices. (Here $\mathbf{I}$ is $\mathbf{I}_{12}$.) Suppose that 3 or fewer errors occur in the transmission of a codeword, and let $\mathbf{x}$ be the received word and $\mathbf{e}$ be the error vector. Thus, $w(\mathbf{e}) \le 3$. Let us

write $\mathbf{e} = \mathbf{e}_1\mathbf{e}_2$, where $\mathbf{e}_i$ has length 12. We can compute the syndromes of the received word using both parity check matrices as follows

$$S_1 = \mathbf{e}G_1^T = [\mathbf{e}_1 \mid \mathbf{e}_2]\left[\frac{I}{A}\right] = \mathbf{e}_1 + \mathbf{e}_2A$$

(we are mixing matrix and vector notation here) and similarly

$$S_2 = \mathbf{e}G_2^T = [\mathbf{e}_1 \mid \mathbf{e}_2]\left[\frac{A}{I}\right] = \mathbf{e}_1A + \mathbf{e}_2$$

Now let us examine the possibilities.

1) If $w(\mathbf{e}_1) = 0$, then $\mathbf{e} = \mathbf{0}S_2$ and $w(S_2) \leq 3$.

2) If $w(\mathbf{e}_2) = 0$ then $\mathbf{e} = S_1\mathbf{0}$ and $w(S_1) \leq 3$.

3) If $w(\mathbf{e}_1) > 0$ and $w(\mathbf{e}_2) > 0$, then $w(S_1) \geq 5$ and $w(S_2) \geq 5$.

Thus, if either syndrome has weight at most 3, we can easily recover the error vector $\mathbf{e}$. If $w(S_1)$ and $w(S_2)$ are both greater than 3, we know that one of the following holds

3a) $w(\mathbf{e}_1) = 1$ and $w(\mathbf{e}_2) = 1$ or 2.

3b) $w(\mathbf{e}_1) = 2$ and $w(\mathbf{e}_2) = 1$.

Consider case 3a). Let ϵ_j be the vector of length 12 with a 1 in the j-th position and zeros elsewhere. Hence, if $\mathbf{e}_1$ has a zero in the i-th position, then $\mathbf{e}_1 = \epsilon_i$ and $\mathbf{e} = \epsilon_i\mathbf{e}_2$. For $j = 1,\ldots,12$, we compute

$$(\mathbf{x} + \epsilon_j\mathbf{0})G_2^T = (\mathbf{e} + \epsilon_j\mathbf{0})G_2^T = (\epsilon_i\mathbf{e}_2 + \epsilon_j\mathbf{0})G_2^T = \epsilon_iA + \mathbf{e}_2 + \epsilon_jA$$

Now, if $j = i$, this vector is just $\mathbf{e}_2$, which has weight at most 2. But if $j \neq i$, then this vector has weight at least 4. Hence, by checking the 12 syndromes

$$(\mathbf{x} + \epsilon_1\mathbf{0})G_2^T, \ldots, (\mathbf{x} + \epsilon_{12}\mathbf{0})G_2^T$$

we can determine the error position i.

Now consider case 3b). If $\mathbf{e}_2 = \epsilon_i$, then $\mathbf{e} = \mathbf{e}_1\epsilon_i$. In this case, we compute

$$(\mathbf{x} + \mathbf{0}\epsilon_j)G_1^T = (\mathbf{e} + \mathbf{0}\epsilon_j)G_1^T = (\mathbf{e}_1\epsilon_i + \mathbf{0}\epsilon_j)G_1^T = \mathbf{e}_1 + \epsilon_iA + \epsilon_jA$$

As before, this has weight at most 2 if $j = i$ and weight at least 4 if $j \neq i$. Hence, the weight will determine the value of i.

In summary, if at most 3 errors occur, then we can decode correctly by computing at most the 26 syndromes

$$\mathbf{x}G_1^T, \mathbf{x}G_2^T, (\mathbf{x} + \epsilon_1\mathbf{0})G_2^T, \ldots, (\mathbf{x} + \epsilon_{12}\mathbf{0})G_2^T, (\mathbf{x} + \mathbf{0}\epsilon_1)G_1^T, \ldots, (\mathbf{x} + \mathbf{0}\epsilon_{12})G_1^T$$

The Binary Golay Code $\mathcal{G}_{23}$

Although we will not prove it here, puncturing the code $\mathcal{G}_{24}$ in any of its coordinate positions will lead to an equivalent code, denoted by $\mathcal{G}_{23}$. This code is a [23,12,7]-code, and therefore is perfect. Note that we may obtain $\mathcal{G}_{24}$ from $\mathcal{G}_{23}$ by adding an overall parity check.

Since $\mathcal{G}_{23}$ is perfect, we may apply Theorem 4.4.5.

Theorem 6.1.9 The codewords in $\mathcal{G}_{23}$ of minimum weight 7 hold a Steiner system S(4,7,23). □

Also, Corollary 4.4.6 tell us that $\mathcal{G}_{23}$ has

$$A_7 = \binom{23}{4} \Big/ \binom{7}{4} = 253$$

codewords of minimum weight 7. Although we will not prove it, the complete weight distribution of $\mathcal{G}_{23}$ is

$$A_0 = A_{23} = 1, \;\; A_7 = A_{16} = 253, \;\; A_8 = A_{15} = 508, \;\; A_{11} = A_{12} = 1288$$

Pless (1968) and Delsarte and Goethals (1975) also established the uniqueness of $\mathcal{G}_{23}$.

Theorem 6.1.10 Any binary $(23,2^{12},7)$-code is equivalent to the Golay code $\mathcal{G}_{23}$. □

We will see that the code $\mathcal{G}_{23}$ can also be defined, perhaps more naturally, as a cyclic code, and this leads to efficient decoding procedures for $\mathcal{G}_{23}$.

The Ternary Golay Codes

The ternary Golay code $\mathcal{G}_{12}$ is the code with generating matrix $G = [I_6 \mid B]$, where

$$B = \begin{bmatrix} 0 & 1 & 1 & 1 & 1 & 1 \\ 1 & 0 & 1 & 2 & 2 & 1 \\ 1 & 1 & 0 & 1 & 2 & 2 \\ 1 & 2 & 1 & 0 & 1 & 2 \\ 1 & 2 & 2 & 1 & 0 & 1 \\ 1 & 1 & 2 & 2 & 1 & 0 \end{bmatrix}$$

We will leave proof of the following as an exercise.

Theorem 6.1.11
1) The ternary Golay code $\mathcal{G}_{12}$ is self-dual, that is, $\mathcal{G}_{12}^{\perp} = \mathcal{G}_{12}$.
2) The matrix B defined above is symmetric, that is, $B^{T} = B$.
3) $\mathcal{G}_{12}$ is a [12,6,6]-code.
4) The ternary code $\mathcal{G}_{11}$, obtained by puncturing $\mathcal{G}_{12}$ in its last coordinate position, is a perfect [11,6,5]-code. □

As with the binary Golay code, we get an equivalent code by puncturing $\mathcal{G}_{12}$ in any coordinate position, and we may recover $\mathcal{G}_{12}$ from $\mathcal{G}_{11}$ by adding a parity check. The code $\mathcal{G}_{11}$ can also be defined as a cyclic code.

Pless (1968) and Delsarte and Goethals (1975) also showed that the ternary Golay codes are unique.

Theorem 6.1.12 Any ternary $(12,3^6,6)$-code is equivalent to the Golay code $\mathcal{G}_{12}$, and any ternary $(11,3^6,5)$-code is equivalent to the Golay code $\mathcal{G}_{11}$. □

PERFECT CODES

Now that we have seen the Vasil'ev construction of nonlinear perfect codes with the same parameters as the Hamming codes, let us repeat Theorem 4.3.6.

Theorem 6.1.13 A nontrivial perfect q-ary code C, where q is a prime power, must have the same parameters as either a Hamming code $\mathcal{H}_q(r)$ or one of the Golay codes $\mathcal{G}_{23}$ or $\mathcal{G}_{11}$. Furthermore,
1) if C has the parameters of one of the Golay codes, then it is equivalent to that Golay code,
2) if C is *linear* and has the parameters of one of the Hamming codes, then it is equivalent to that Hamming code. □

THE NORDSTROM-ROBINSON CODE

One of the virtues of the so-called *Nordstrom-Robinson code* $\mathcal{N}$ is that it provides an example of a nonlinear code that is larger than any linear code with the same length and minimum distance.

There are many ways to construct the Nordstrom-Robinson code. In fact, this code was first constructed by Alan Nordstrom using trial and error, when he was but a high school student! Our approach will be through the binary Golay code $\mathcal{G}_{24}$.

Recall that $\mathcal{G}_{24}$ has generator matrix $G = [I_{12} \mid A]$, where A is given earlier in this section. We leave it as an exercise to show that, by permuting columns and using elementary row operations, the matrix G

can be brought to the form

$$G' = \begin{bmatrix} I_7 & * \\ 0_{5,7} & * \end{bmatrix}$$

where the 8th column is the sum of the first 7 columns. Thus, the code C with generating matrix G' is equivalent to $\mathcal{G}_{24}$.

Now, it is clear from the matrix G' that there are $2^5 = 32$ distinct codewords in C that have 0's in their first 7 positions, and that each of these codewords also has a 0 in its 8th position. Furthermore, for any binary string e of length 7 with exactly one 1, there are also 32 distinct codewords in C that begin with this string, and each of these codewords has a 1 in its 8th position. Hence, there are $8 \times 32 = 256$ codewords in C that begin as follows

$$\begin{array}{l} 0000\,0000\cdots \\ 1000\,0001\cdots \\ 0100\,0001\cdots \\ 0010\,0001\cdots \\ 0001\,0001\cdots \\ 0000\,1001\cdots \\ 0000\,0101\cdots \\ 0000\,0011\cdots \end{array}$$

The Nordstrom-Robinson code $\mathcal{N}$ is the code whose codewords are obtained from these 256 words by deleting the first 8 coordinate positions. Hence, $\mathcal{N}$ has length 16 and size 256. We leave it to the reader to show that $\mathcal{N}$ has minimum distance 6, and so is a (16,256,6)-code. We also leave it to the reader to show that $\mathcal{N}$ is not linear.

The linear programming bound can be used to show that $A_2(13,6) \le 32$. (MacWilliams and Sloane (1977), p. 538*ff.*) Hence, Theorem 4.5.2 gives

$$A_2(13,6) \le 32, \quad A_2(14,6) \le 64, \quad A_2(15,6) \le 128, \quad A_2(16,6) \le 256$$

which shows that the Nordstrom-Robinson code is optimal and that $A_2(16,6) = 256$.

In addition, by shortening the Nordstrom-Robinson code, we obtain a (15,M,6)-code C_1, where $M \ge 128$. In view of the above, we see that $M = 128$ and that this code is optimal. Similarly, successive shortening gives (14,64,6) and (13,32,6)-codes, each of which is optimal. In particular, note that $A_2(12,5) = A_2(13,6) = 32$.

Now, it is possible to show that no *linear* [12,5,5]-code exists. Hence, we have a situation in which there exists a *nonlinear* code that is

larger than any linear code with the same length and minimum distance.

EXERCISES

1. Write parity check for $\mathcal{H}_2(4)$, $\mathcal{H}_3(4)$.
2. If we extend the *binary* [n,k,3] Hamming code $\mathcal{H}_2(r)$ by adding an overall parity check, we get an [n+1,k,4]-code. How do the error correcting capabilities of this code compare with those of the unextended code? What about the error detecting capabilities?
3. Assuming a binary symmetric channel, show that the probability of correcting an error, using syndrome decoding, is the same for the code $\mathcal{H}_2(r)$ as it is for its extension, obtained by adding an overall parity check.
4. Prove that Hamming codes are perfect.
5. Use the fact that Hamming codes are perfect to determine the number of codewords of weight 3 in $\mathcal{H}_2(r)$. *Hint.* Think about words of weight 2.
6. Prove Theorem 6.1.1.
7. Complete the details of the proof of Lemma 6.1.3.
8. Finish the proof of Lemma 6.1.4.
9. Show that $\mathbf{1} \in \mathcal{G}_{24}$.
10. Show that, for the Golay code $\mathcal{G}_{24}$, we have $A_{20} = A_4 = 0$ and $A_8 = A_{16}$.
11. Let $G = [I_{12} \mid A]$ be the generator matrix of $\mathcal{G}_{24}$ discussed in the text. Prove that the sum of any two rows of the matrix A has weight at least 6.
12. Referring to the discussion of decoding $\mathcal{G}_{24}$, show that if both $w(\mathbf{e}_1)$ and $w(\mathbf{e}_2)$ are positive, then $w(S_1) \geq 5$ and $w(S_2) \geq 5$.
13. Fill in the details related to the decoding of the Golay code $\mathcal{G}_{24}$.
14. Does the decoding process described for the Golay code $\mathcal{G}_{24}$ apply in more generality? Explain.
15. Prove Theorem 6.1.11.
16. Show that $\mathcal{G}_{11}$ has 132 words of weight 5. *Hint.* Let A_5 be the number of words of weight 5. Count the number of pairs $(\mathbf{x},\mathbf{c})$, where $w(\mathbf{x}) = 3$, $w(\mathbf{c}) = 5$, $\mathbf{c} \in \mathcal{G}_{11}$, and $\mathbf{c}$ covers $\mathbf{x}$, in two ways. Use the fact that $\mathcal{G}_{11}$ is perfect.
17. Show that the Vasil'ev code $\mathcal{V}$ is a perfect nonlinear binary code with Hamming parameters.
18. Show that, by permuting columns and using elementary row operations, the generating matrix G for $\mathcal{G}_{24}$ can be brought to the form

$$G' = \begin{bmatrix} \mathbf{I}_7 & * \\ \mathbf{0}_5 & * \end{bmatrix}$$

where the 8th column is the sum of the first 7 columns. *Hint.* Use Theorem 5.1.4.

19. Show that the Nordstrom-Robinson code $\mathcal{N}$ has minimum distance 6.
20. Show that the Nordstrom-Robinson code $\mathcal{N}$ is not linear.
21. Show that the Hamming code $\mathcal{H}_2(3)$ holds the projective plane of order 2, pictured in Figure 4.4.1. (Some renumbering of coordinates may be necessary.)

6.2 Reed-Muller Codes

The Reed-Muller codes are among the oldest known codes. They are relatively easy to decode, but except for the first order codes, their minimum distance is not that good. In order to introduce these codes, we begin with a discussion of Boolean functions and Boolean polynomials.

BOOLEAN FUNCTIONS AND BOOLEAN POLYNOMIALS

Boolean Functions

Definition A **Boolean function** of m variables $x_1,\ldots,x_m$ is a function $f(x_1,\ldots,x_m)$ from $\mathbb{Z}_2^m$ to $\mathbb{Z}_2$, where $\mathbb{Z}_2 = \{0,1\}$. □

A Boolean function is often described by giving its *truth table*, which is simply a table of values of the function. For example, Table 6.2.1 describes a Boolean function of three variables.

Table 6.2.1

x_1	x_2	x_3	f
0	0	0	0
0	0	1	0
0	1	0	1
0	1	1	0
1	0	0	0
1	0	1	0
1	1	0	1
1	1	1	0

By reading the last column of this table from the top down, we obtain the binary string $\mathbf{a}_f = 00100010$. Clearly, if we always agree to list the variables and rows of a truth table in the same order, we obtain a one-to-one correspondence between Boolean functions f of m variables and binary strings $\mathbf{a}_f$ of length 2^m. (Note that many authors reverse the order of the variables in Table 6.2.1.)

The two *constant Boolean functions* are defined by

$$0(x_1,\ldots,x_m) = 0 \text{ and } 1(x_1,\ldots,x_m) = 1$$

and are associated with the strings **0** and **1**, respectively.

There are various logical operations that can be performed on Boolean functions. Our interest will center on the operations of

conjunction, defined by

$$(fg)(\mathbf{x}) = 1 \text{ if and only if } f(\mathbf{x}) = 1 \text{ and } g(\mathbf{x}) = 1$$

and **exclusive OR**, defined by

$$(f+g)(\mathbf{x}) = 1 \text{ if and only if } f(\mathbf{x}) = 1 \text{ or } g(\mathbf{x}) = 1, \text{ but not both}$$

Notice that conjunction corresponds to multiplication in $\mathbb{Z}_2$ and exclusive OR corresponds to addition, that is,

$$(fg)(\mathbf{x}) = f(\mathbf{x})g(\mathbf{x})$$

and

$$(f+g)(\mathbf{x}) = [f(\mathbf{x}) + g(\mathbf{x})] \bmod 2$$

If follows that the string versions of conjunction and exclusive OR are given by

$$\mathbf{a}_{fg} = \mathbf{a}_f \cap \mathbf{a}_g$$

and

$$\mathbf{a}_{f+g} = \mathbf{a}_f + \mathbf{a}_g$$

Under the operation of exclusive OR, the set $\mathcal{B}_m$ of all Boolean functions of m variables forms a vector space of size

$$|\mathcal{B}_m| = 2^{2^m}$$

over $\mathbb{Z}_2$.

Boolean Polynomials

It is customary in coding theory circles to adopt the following definition of Boolean polynomial. Note that this differs from the definition given in algebraic circles.

Definition A **Boolean monomial** in the variables $x_1,\ldots,x_m$ is an expression of the form

$$p = x_{i_1}x_{i_2}\cdots x_{i_s}$$

The **reduced form** of p is obtained by applying the rules

$$x_ix_j = x_jx_i \text{ and } x_i^2 = x_i$$

until the factors are distinct. The **degree** of a monomial p is the degree of its reduced form p', and this is the number of variables in p'.

A **Boolean polynomial** in the variables $x_1,\ldots,x_m$ is a linear combination of Boolean monomials in these variables, with coefficients in $\mathbb{Z}_2$. A Boolean polynomial is in **reduced form** if each monomial is in

reduced form, and if all like monomials have been canceled. The **degree** of a Boolean polynomial q is the degree of its reduced form q', and this is the largest of the degrees of the monomials that form q'. □

The set B_m of all Boolean polynomials in m variables is a vector space over $\mathbb{Z}_2$, as is the set $B_{m,r}$ of all Boolean polynomials in m variables of degree at most r.

Since there are $\binom{m}{k}$ distinct Boolean monomials of degree k in m variables, the total number of distinct Boolean monomials is

$$1+\binom{m}{1}+\cdots+\binom{m}{m}=2^m$$

and the total number of distinct Boolean polynomials in m variables is

$$2^{2^m}$$

This also happens to be the total number of Boolean functions in m variables, which is no mere a coincidence.

The Vector Spaces $\mathfrak{B}_m$ ***and*** B_m

Any Boolean polynomial p can be regarded as a Boolean function by allowing substitution of the elements of $\mathbb{Z}_2$ for the variables in p. As we now see, all Boolean functions are obtained in this way.

Theorem 6.2.1 For every Boolean function $f(x_1,\ldots,x_m)$ in $\mathfrak{B}_m$, there is a unique Boolean polynomial $p(x_1,\ldots,x_m)$ in B_m for which

$$p(x_1,\ldots,x_m)=f(x_1,\ldots,x_m)$$

as Boolean functions. Furthermore, this correspondence between Boolean functions and Boolean polynomials is a vector space isomorphism from $\mathfrak{B}_m$ to B_m.

Proof. For the first statement, we proceed by induction on the number m of variables. There are precisely four Boolean functions in 1 variable, which, in string form, are 00, 11, 01, and 10. These correspond to the Boolean polynomials 0, 1, x, and 1+x, respectively.

Assume the result is true for $m-1$ variables, and let $f(x_1,\ldots,x_m)$ be a Boolean function in m variables. Observe that

$$f(x_1,\ldots,x_m)=f(0,x_2,\ldots,x_m)+\pi_1[f(1,x_2,\ldots,x_m)-f(0,x_2,\ldots,x_m)]$$

where $\pi_1=\pi_1(x_1,\ldots,x_m)=x_1$ is a *projection function.* If we define Boolean functions of $m-1$ variables by

$$f_0(x_2,\ldots,x_m) = f(0,x_2,\ldots,x_m) \text{ and } f_1(x_2,\ldots,x_m) = f(1,x_2,\ldots,x_m)$$

then the induction hypothesis implies that there are Boolean polynomials p_0 and p_1 for which, as Boolean functions,

$$p_0(x_2,\ldots,x_m) = f_0(x_2,\ldots,x_m) = f(0,x_2,\ldots,x_m)$$

and

$$p_1(x_2,\ldots,x_m) = f_1(x_2,\ldots,x_m) = f(1,x_2,\ldots,x_m)$$

Furthermore, the Boolean polynomial $q(x_1,\ldots,x_m) = x_1$ is the same Boolean function as the projection π_1. Therefore, the polynomial

$$p(x_1,\ldots,x_m) = p_0(x_2,\ldots,x_m) + x_1[p_1(x_2,\ldots,x_m) - p_0(x_2,\ldots,x_m)]$$

is equal, as a Boolean function, to $f(x_1,\ldots,x_m)$.

Thus, we see that for each Boolean function f, there is a Boolean polynomial p for which $p = f$. Since no two distinct Boolean functions can be associated in this way to the same polynomial, the fact that there are the same number of Boolean functions as Boolean polynomials implies that this association is a one-to-one correspondence. We leave proof of the remaining portion to the reader. ∎

The method used in the proof of Theorem 6.2.1 can be used to find the Boolean polynomial corresponding to any Boolean function.

Example 6.2.1 Let $\mathbf{a} = 0110\,0011$ be the string representation of a Boolean function. Using a convenient abuse of notation, and writing a binary string in place of the corresponding polynomial, we have

$$\begin{aligned} 0110\,0011 &= 0110 + x_1(0011 - 0110) \\ &= 0110 + x_1(0101) \\ &= 01 + x_2(10 - 01) + x_1(01 + x_2(01 - 01)) \\ &= 01 + x_2(11) + x_1(01 + x_2 00) \\ &= x_3 + x_2 1 + x_1(x_3 + x_2 0) \\ &= x_3 + x_2 + x_1 x_3 \end{aligned}$$

□

REED-MULLER CODES

We can now define the Reed-Muller codes.

Definition Let $0 \le r \le m$. The **r-th order Reed-Muller code** $\mathcal{R}(r,m)$ is the set of all binary strings $\mathbf{a}_p$ of length $n = 2^m$ associated with the Boolean polynomials $p(x_1,\ldots,x_m)$ of degree at most r. □

Example 6.2.2 The 0th order Reed-Muller code $\mathcal{R}(0,m)$ consists of the binary strings associated with the constant polynomials 0 and 1, that is,

$$\mathcal{R}(0,m) = \{\mathbf{0},\mathbf{1}\} = Rep(2^m)$$

Thus, $\mathcal{R}(0,m)$ is just the repetition code of length 2^m.

On the other extreme, the m-th order Reed-Muller code $\mathcal{R}(m,m)$ consists of all binary strings of length 2^m, that is, $\mathcal{R}(m,m) = \mathbb{Z}_2^n$, where $n = 2^m$. □

Example 6.2.3 The first order Reed-Muller code of length $n = 8 = 2^3$ is the set of all binary strings associated with the Boolean polynomials $p(x_1,x_2,x_3)$ of degree at most 1. These polynomials have the form

$$\alpha_0 + \alpha_1 x_1 + \alpha_2 x_2 + \alpha_3 x_3$$

where $\alpha_i = 0$ or 1. The binary string corresponding to this polynomial is

$$\alpha_0(1111\ 1111) + \alpha_1(0000\ 1111) + \alpha_2(0011\ 0011) + \alpha_3(0101\ 0101)$$

Thus, we can list the codewords in $\mathcal{R}(1,3)$ as follows

Polynomial	*Codeword*
0	0000 0000
x_1	0000 1111
x_2	0011 0011
x_3	0101 0101
x_1+x_2	0011 1100
x_1+x_3	0101 1010
x_2+x_3	0110 0110
$x_1+x_2+x_3$	0110 1001
1	1111 1111
$1+x_1$	1111 0000
$1+x_2$	1100 1100
$1+x_3$	1010 1010
$1+x_1+x_2$	1100 0011
$1+x_1+x_3$	1010 0101
$1+x_2+x_3$	1001 1001
$1+x_1+x_2+x_3$	1001 0110

□

It is an easy exercise to show that any Boolean polynomial of the form

$$x_m + p(x_1,\ldots,x_{m-1})$$

where $p(x_1,\ldots,x_{m-1})$ is a Boolean polynomial, takes on the values 0 and 1 equally often. Since all nonconstant linear polynomials have this form, we get the following (compare with the simplex code).

Theorem 6.2.2 All codewords in $\mathcal{R}(1,m)$ except **0** and **1** have weight 2^{m-1}. □

According to the definition, the r-th order Reed-Muller code $\mathcal{R}(r,m)$ consists of all binary strings corresponding to linear combinations of monomials of degree at most r. Since there are

$$k = 1 + \binom{m}{1} + \binom{m}{2} + \cdots + \binom{m}{r}$$

such monomials, there are 2^k such linear combinations.

Theorem 6.2.3 The Reed-Muller code $\mathcal{R}(r,m)$ has length 2^m and dimension

$$k = 1 + \binom{m}{1} + \binom{m}{2} + \cdots + \binom{m}{r}$$

Thus, the rate of $\mathcal{R}(r,m)$ is

$$R = \frac{1 + \binom{m}{1} + \binom{m}{2} + \cdots + \binom{m}{r}}{2^m}$$ □

THE REED-MULLER CODES AS (u,u+v)-CONSTRUCTIONS

The Reed-Muller codes can be obtained using the **(u,u+v)**-construction. Recall from Section 4.3 that if C_i is a binary (n,M_i,d_i)-code, for $i = 1$ and 2, then the **(u,u+v)**-construction yields a code defined by

$$C_1 \oplus C_2 = \{\mathbf{c}(\mathbf{c}+\mathbf{d}) \mid \mathbf{c} \in C_1,\ \mathbf{d} \in C_2\}$$

which, according to Theorem 4.3.12 is a $(2n,M_1M_2,d')$-code, where

(6.2.1) $$d' = \min\{2d_1, d_2\}$$

Suppose that $0 < r < m$, and consider a codeword **a** in $\mathcal{R}(r,m)$. This word comes from a polynomial $p(x_1,\ldots,x_m)$ of degree at most r. We can factor the variable x_1 from those terms in which it appears, and write p in the form

$$p(x_1,\ldots,x_m) = x_1 g(x_2,\ldots,x_m) + h(x_2,\ldots,x_m)$$

where $g(x_2,\ldots,x_m)$ and $h(x_2,\ldots,x_m)$ are polynomials in $m-1$ variables, with $\deg(g) \le r-1$ and $\deg(h) \le r$.

Let $\mathbf{a}_g$ and $\mathbf{a}_h$ be the binary strings corresponding to the polynomials g and h, respectively, where each is a polynomial in $m-1$ variables. Thus, $\mathbf{a}_g \in \mathcal{R}(r-1,m-1)$ and $\mathbf{a}_h \in \mathcal{R}(r,m-1)$.

The string corresponding to $x_1 g(x_2,\ldots,x_m)$ is $\mathbf{0a}_g$ and, if we think of h as a polynomial in m variables $x_1,\ldots,x_m$, then the string

corresponding to h is $\mathbf{a}_h\mathbf{a}_h$. Hence, the string corresponding to $p(x_1,\ldots,x_m)$ is

$$\mathbf{a}_p = \mathbf{0a}_g + \mathbf{a}_h\mathbf{a}_h = \mathbf{a}_h(\mathbf{a}_g + \mathbf{a}_h)$$

and since $\mathbf{a}_h \in \mathcal{R}(r,m-1)$ and $\mathbf{a}_g \in \mathcal{R}(r-1,m-1)$, we see that

$$\mathbf{a}_p \in \mathcal{R}(r,m-1) \oplus \mathcal{R}(r-1,m-1)$$

Thus,

$$\mathcal{R}(r,m) \subset \mathcal{R}(r,m-1) \oplus \mathcal{R}(r-1,m-1)$$

We leave it to the reader to show that the codes $\mathcal{R}(r,m)$ and $\mathcal{R}(r,m-1) \oplus \mathcal{R}(r-1,m-1)$ have the same size, and so equality holds. Let us summarize.

Theorem 6.2.4 For the Reed-Muller codes $\mathcal{R}(r,m)$, we have
1) $\mathcal{R}(0,m) = \{\mathbf{0},\mathbf{1}\} = Rep(2^m)$,
2) $\mathcal{R}(m,m) = \mathbb{Z}_2^n$, $n = 2^m$
3) for $0 < r < m$,

$$\mathcal{R}(r,m) = \mathcal{R}(r,m-1) \oplus \mathcal{R}(r-1,m-1)$$

where $\oplus$ denotes the $(\mathbf{u},\mathbf{u}+\mathbf{v})$-construction. □

Corollary 6.2.5 The Reed-Muller code $\mathcal{R}(m-1,m)$ consists of all binary words of length 2^m that have even weight. Therefore, if $r < m$, $\mathcal{R}(r,m)$ contains codewords of even weight only.
Proof. Certainly, the code $\mathcal{R}(0,1) = \{00,11\}$ contains words of even weight only. Assuming that $\mathcal{R}(m-2,m-1)$ contains words of even weight only, part 3) of Theorem 6.2.4, gives

$$\begin{aligned}\mathcal{R}(m-1,m) &= \mathcal{R}(m-1,m-1) \oplus \mathcal{R}(m-2,m-1)\\ &= \mathbb{Z}_2^n \oplus \mathcal{R}(m-2,m-1)\end{aligned}$$

where $n = 2^{m-1}$. A typical word in $\mathcal{R}(m-1,m)$ then has the form

$$\mathbf{a} = \mathbf{b}(\mathbf{b}+\mathbf{c}) = \mathbf{bb} + \mathbf{0c}$$

where $\mathbf{b} \in \mathbb{Z}_2^n$ and $\mathbf{c} \in \mathcal{R}(m-2,m-1)$. Since $\mathbf{bb}$ has even weight for all $\mathbf{b}$, and since $\mathbf{0c}$ has even weight by hypothesis, we deduce that

$$w(\mathbf{bb} + \mathbf{0c}) = w(\mathbf{bb}) + w(\mathbf{0c}) - 2w(\mathbf{bb} \cap \mathbf{0c})$$

is also even. Thus, $\mathcal{R}(m-1,m)$ contains words of even weight only.

Since $dim(\mathcal{R}(m-1,m)) = 2^m - 1$, it must contain all even weight words of length 2^m. Finally, if $r < m$, then $\mathcal{R}(r,m) \subset \mathcal{R}(m-1,m)$ from which the second statement follows. ∎

Theorem 6.2.4 can be used to determine the minimum distance of the Reed-Muller codes. Clearly, we have $d(\mathcal{R}(0,m)) = 2^m$ and $d(\mathcal{R}(m,m)) = 1$. Now let us assume that $0 < r < m$. A few computations using (6.2.1) gives

$$d(\mathcal{R}(1,2)) = d(\mathcal{R}(1,1) \oplus \mathcal{R}(0,1)) = \min\{2 \cdot 1,2\} = 2 = 2^{2-1}$$

$$d(\mathcal{R}(1,3)) = d(\mathcal{R}(1,2) \oplus \mathcal{R}(0,2)) = \min\{2 \cdot 2,4\} = 4 = 2^{3-1}$$

$$d(\mathcal{R}(2,3)) = d(\mathcal{R}(2,2) \oplus \mathcal{R}(1,2)) = \min\{2 \cdot 1,2\} = 2 = 2^{3-2}$$

From this, we deduce the following, whose inductive proof is left as an exercise.

Theorem 6.2.6 The Reed-Muller code $\mathcal{R}(r,m)$ has minimum distance 2^{m-r}, and hence has parameters

$$[2^m,\ 1+\binom{m}{1}+\cdots+\binom{m}{m},\ 2^{m-r}]$$

□

THE DUAL OF $\mathcal{R}$**(r,m)**

Suppose that $p(x_1,\ldots,x_m)$ is a Boolean polynomial of degree at most r, and $q(x_1,\ldots,x_m)$ is a Boolean polynomial of degree at most $m-r-1$. Let $\mathbf{a}_p \in \mathcal{R}(r,m)$ and $\mathbf{a}_q \in \mathcal{R}(m-r-1,m)$ be the codewords corresponding to p and q. Observe that

$$\mathbf{a}_p \cdot \mathbf{a}_q \equiv w(\mathbf{a}_p \cap \mathbf{a}_q) \equiv w(\mathbf{a}_{pg}) \bmod 2$$

where $\mathbf{a}_{pq}$ is the codeword corresponding to the product polynomial

$$pq(x_1,\ldots,x_m) = p(x_1,\ldots,x_m)q(x_1,\ldots,x_m)$$

But since

$$\deg(pq) \le \deg(p) + \deg(q) \le m-1$$

we have $\mathbf{a}_{pq} \in \mathcal{R}(m-1,m)$, and so, according to Corollary 6.2.5, $w(\mathbf{a}_{pq})$ is even, which implies that $\mathbf{a}_p \cdot \mathbf{a}_q = 0$. Hence,

$$(6.2.2) \qquad \mathcal{R}(m-r-1,m) \subset \mathcal{R}(r,m)^{\perp}$$

However,

$$\begin{aligned} dim(\mathcal{R}(r,m)^{\perp}) &= 2^m - \left[1+\binom{m}{1}+\cdots+\binom{m}{r}\right] \\ &= \binom{m}{r+1}+\binom{m}{r+2}\cdots+\binom{m}{m} \\ &= \binom{m}{m-r-1}+\binom{m}{m-r-2}+\cdots+1 \\ &= dim(\mathcal{R}(m-r-1,m)) \end{aligned}$$

and so equality holds in (6.2.2).

Theorem 6.2.7 For $0 < r < m-1$, we have

$$\mathcal{R}(r,m)^{\perp} = \mathcal{R}(m-r-1,m) \qquad \square$$

EUCLIDEAN GEOMETRY

We can get additional insight into the Reed-Muller codes by taking a more geometric look at the vector space $\mathbb{Z}_2^m$ of all binary strings of length m. Our current view is to consider functions from $\mathbb{Z}_2^m$ to $\mathbb{Z}_2$, as shown in Table 6.2.1, where the entries in each row (excluding the last column) are the elements of $\mathbb{Z}_2^m$. However, we can endow $\mathbb{Z}_2^m$ with some additional geometric structure as follows.

Let S be a k-dimensional subspace of $\mathbb{Z}_2^m$. Since S is a linear code, it has a parity check matrix H, which has size $(m-k)\times m$. Thus, the vectors in S can be described by writing

$$\mathbf{x} \in S \text{ if and only if } \mathbf{x}H^T = \mathbf{0}$$

Recall that the cosets of S in $\mathbb{Z}_2^m$ are the subsets of the form $\mathbf{b}+S$, where $\mathbf{b} \in \mathbb{Z}_2^m$. Since $\mathbf{x} \in \mathbf{b}+S$ if and only if $\mathbf{x}-\mathbf{b} \in S$, which happens if and only if $(\mathbf{x}-\mathbf{b})H^T = \mathbf{0}$, we see that

$$\text{(6.2.3)} \qquad \mathbf{x} \in \mathbf{b}+S \text{ if and only if } \mathbf{x}H^T = \mathbf{b}H^T$$

Thus, while the vectors in S are precisely the solutions to a *homogeneous* system of $m-k$ linear equations in m variables with rank $m-k$, the vectors in the coset $\mathbf{b}+S$ (for $\mathbf{b} \neq \mathbf{0}$) are precisely the solutions to a *nonhomogeneous* system of $m-k$ linear equations in m variables with rank $m-k$.

In geometric terms, the vector space $\mathbb{Z}_2^m$ is referred to as a *Euclidean geometry* EG(m,2) of dimension m over $\mathbb{Z}_2$. If $dim(S) = k$, then the coset $\mathbf{b}+S$ is referred to as a **k-flat.** Thus, 0-flats have the form $\mathbf{b}+\{\mathbf{0}\} = \{\mathbf{b}\}$, and so are the *points* of EG(m,2). Similarly, 1-flats are the *lines* in EG(m,2) and have the form

$$\mathbf{b}+\{\lambda\mathbf{c}\} = \{\mathbf{b},\mathbf{b}+\mathbf{c}\}$$

where $\mathbf{c} \neq \mathbf{0}$. Thus, the lines are just the two-point sets. The (m−1)-flats are referred to as **hyperplanes** in EG(m,2).

Notice that, according to (6.2.3), a subset of EG(m,2) is a k-flat if and only if it is the solution set of a system of $m-k$ linear equations in m variables of rank $m-k$.

Example 6.2.4 The vector space $\mathbb{Z}_2^3$ is the Euclidean geometry EG(3,2). We have already seen that the 0-flats (points) are the one-point sets and the 1-flats (lines) are the two-point sets. Accordingly, there are $2^3 = 8$ points and $\binom{8}{2} = 28$ lines in EG(3,2).

The 2-flats (planes), which are the hyperplanes of EG(3,2), are sets of the form

$$\mathbf{b}+\{\lambda\mathbf{c}_1+\mu\mathbf{c}_2\}=\{\mathbf{b},\mathbf{b}+\mathbf{c}_1,\mathbf{b}+\mathbf{c}_2,\mathbf{b}+\mathbf{c}_1+\mathbf{c}_2\}$$

Notice that the first three vectors in this set are arbitrary and the fourth vector is the sum of the first three. Hence, there are

$$\binom{8}{3}\Big/\binom{4}{3}=14$$

planes in EG(3,2). For instance, the set {000,001,010,011} is a plane in EG(3,2), but the set {000,001,010,111} is not. ▯

A GEOMETRIC LOOK AT THE REED-MULLER CODES

Now let us take another look at Table 6.2.1. The rows of this table (excluding the last column) are the points in EG(3,2). Hence, the binary string $\mathbf{a}_p$, formed from the last column of the table, can be viewed as the *characteristic function*, or *characteristic vector*, of the subset

$$\{010,110\}$$

of EG(m,2), since these are the points of EG(3,2) associated with the positions in $\mathbf{a}_p$ that are equal to 1. Notice that, in this case, the subset of EG(m,2) associated with $\mathbf{a}_p$ is a 1-flat (line) in EG(m,2).

More generally, we can associate with each binary string $\mathbf{a}$ of length 2^m, not only a Boolean polynomial $p_\mathbf{a}$ in m variables, but also a subset $F_\mathbf{a}$ of the Euclidean geometry EG(m,2). Furthermore, given a Boolean polynomial $p(x_1,\ldots,x_m)$, the associated subset of EG(m,2) is

$$F=\{(x_1,\ldots,x_m)\mid p(x_1,\ldots,x_m)=1\}$$

Let us establish our notation.

1) Given a string $\mathbf{a}=a_1\cdots a_n$ $(n=2^m)$, we denote the Boolean polynomial whose truth table has $\mathbf{a}$ as its last column by $p_\mathbf{a}$, and the subset of EG(m,2) for which $\mathbf{a}$ is the characteristic vector by $F_\mathbf{a}$. We remark that $F_\mathbf{a}$ need not be a flat in EG(m,2).
2) Given a Boolean polynomial $p(x_1,\ldots,x_m)$, we denote the string that is the last column of the truth table of p by $\mathbf{a}_p$ and the subset of EG(m,2) for which $\mathbf{a}_p$ is the characteristic vector by F_p. Thus, F_p is, by definition, the same as $F_{\mathbf{a}_p}$.
3) Given a subset F of EG(m,2), which we will always take to be a flat, we denote the characteristic vector of F by $\mathbf{a}_F$ and the corresponding Boolean polynomial by p_F.

As we have seen, the Reed-Muller code $\mathcal{R}(r,m)$ is defined as the set of binary strings associated with the Boolean polynomials of degree at most r. We would also like to know how to describe this code in terms of subsets of EG(m,2). To this end, we explore the correspondence between Boolean polynomials of various degrees and subsets of EG(m,2).

As we have observed, a subset of EG(m,2) is a hyperplane if and only if it is the solution set of a single linear equation in the variables $x_1,\ldots,x_m$, that is, the solution set of an equation of the form

$$a_1x_1+\cdots+a_mx_m=e$$

or

$$a_1x_1+\cdots+a_mx_m-e+1=1$$

where $e=0$ or 1. Thus, hyperplanes are associated with nonconstant linear polynomials

$$p(x_1,\ldots,x_m)=a_1x_1+\cdots+a_mx_m-e+1$$

and conversely. More generally, we have the following result.

Theorem 6.2.8 If F is an (m−k)-flat in EG(m,2), then the corresponding Boolean polynomial p_F has degree k.

Proof. Let F be an (m−k)-flat. Then we have seen that $\mathbf{x}\in F$ if and only if $\mathbf{x}$ satisfies a system of k linear equations in m variables, with rank k, which can be written in the form

$$\begin{gathered}\ell_1(x_1,\ldots,x_m)=1\\ \ell_2(x_1,\ldots,x_m)=1\\ \vdots\\ \ell_k(x_1,\ldots,x_m)=1\end{gathered}$$

But $\mathbf{x}$ is a solution to this system if and only if it is a solution to the single polynomial equation

$$\prod_{i=1}^{k}\ell_i(x_1,\ldots,x_m)=1$$

We leave it to the reader to verify that, in this case, the polynomial on the left has degree k. ∎

We leave it as an exercise to show that the converse of Theorem 6.2.8 does not hold in general. That is, there are Boolean polynomials $p(x_1,\ldots,x_m)$ whose corresponding subsets F_p of EG(m,2) are not flats (of any dimension).

Nevertheless, it is true that the subset F_p associated with a *monomial* $p=x_{i_1}\cdots x_{i_j}$ of degree k is an (m−k)-flat, for it is precisely

the intersection of the hyperplanes

$$x_{i_1} = 1,\ x_{i_2} = 1, \ldots, x_{i_k} = 1$$

Moreover, if p and q are Boolean polynomials, then

$$\mathbf{a}_{p+q} = \mathbf{a}_p + \mathbf{a}_q$$

and so, if $\deg(f) = k$ and if f is the sum of monomials

$$f = \sum_i p_i$$

where if $\deg(p_i) \le k$, then

$$\mathbf{a}_f = \sum_i \mathbf{a}_{p_i}$$

But $\mathbf{a}_{p_i}$ is the characteristic vector of a flat of dimension $m - \deg(p_i) \ge m - k$. Hence, $\mathbf{a}_f$ is the sum of characteristic vectors of flats of dimension at least $m - \deg(f)$. Thus, we have the following.

Theorem 6.2.9 The Reed-Muller code $\mathcal{R}(r,m)$ is spanned by the characteristic vectors of all flats of dimension at least $m - r$. ☐

It is possible to strengthen Theorem 6.2.9. We will omit the proof of the following result, referring the interested reader to MacWilliams and Sloane (1977), p. 380.

Theorem 6.2.10

1) The codewords in $\mathcal{R}(r,m)$ of minimum weight are precisely the characteristic vectors of the $(m-r)$-flats in $EG(m,2)$.
2) The codewords in $\mathcal{R}(r,m)$ of minimum weight span $\mathcal{R}(r,m)$. ☐

DECODING THE REED-MULLER CODES

One of the virtues of the Reed-Muller codes is that they are easy to decode. Let us describe one (of several) decoding methods, called **Reed decoding**, that uses a form of majority logic.

According to the definition, a codeword in $\mathcal{R}(r,m)$ comes from a Boolean polynomial

$$p(x_1, \ldots, x_m) = \sum_{s=0}^{r} \sum_{i_1, \ldots, i_s} a_{i_1, \ldots, i_s} x_{i_1} \cdots x_{i_s}$$

of degree at most r. Thus, we can write

$$\mathbf{a}_p = \sum_{s=0}^{r} \sum_{i_1, \ldots, i_s} a_{i_1, \ldots, i_s} \mathbf{a}_{x_{i_1} \cdots x_{i_s}}$$

where $\mathbf{a}_{x_{i_1}\cdots x_{i_s}}$ is the binary string corresponding to the monomial $x_{i_1}\cdots x_{i_s}$.

Now, we are interested in ways to compute the coefficient

(6.2.4) $$a_{k_1,\ldots,k_r}$$

of $\mathbf{a}_p$. If we can compute this coefficient in several different ways, then we can employ a majority logic decision rule. As we will see, this can be done by taking dot products of $\mathbf{a}_p$ with various other vectors.

Let us begin with a few simple flat facts.

Lemma 6.2.11 If F and G are flats in EG(m,2), then

$$\mathbf{a}_F \cdot \mathbf{a}_G \equiv |F \cap G| \mod 2$$

Proof. This follows from

$$\mathbf{a}_F \cdot \mathbf{a}_G \equiv w(\mathbf{a}_F \cap \mathbf{a}_G) = w(\mathbf{a}_{F\cap G}) = |F \cap G| \quad \blacksquare$$

Lemma 6.2.12 Let $F = \mathbf{b} + S$ and $G = \mathbf{c} + T$ be flats in EG(m,2). Then either $F \cap G = \emptyset$ or else $F \cap G = \mathbf{x} + (S \cap T)$ for any nonzero $\mathbf{x}$ in $F \cap G$.

Proof. Left to the reader. ∎

Lemma 6.2.13 All flats in EG(m,2) except one-point flats have an even number of vectors.

Proof. Left to the reader. ∎

Now let

$$\{j_1,\ldots,j_{m-r}\} = \{k_1,\ldots,k_r\}^c$$

where the complement is taken with respect to $\{1,\ldots,m\}$. To get a feel for the situation, let us take the dot product

(6.2.5) $$\mathbf{a}_p \cdot \mathbf{a}_{x_{j_1}\cdots x_{j_{m-r}}}$$

Since the dot product is linear, we can compute the dot product with each term of $\mathbf{a}_p$ separately. Lemma 6.2.11 gives

$$\mathbf{a}_{x_{i_1}\cdots x_{i_s}} \cdot \mathbf{a}_{x_{j_1}\cdots x_{j_{m-r}}} \equiv \left| F_{x_{i_1}\cdots x_{i_s}} \cap F_{x_{j_1}\cdots x_{j_{m-r}}} \right|$$

$$\equiv \left| F_{x_{i_1}\cdots x_{i_s} x_{j_1}\cdots x_{j_{m-r}}} \right| \mod 2$$

Now, by Lemma 6.2.13, this will be 0 unless

$$F_{x_{i_1}\cdots x_{i_s} x_{j_1}\cdots x_{j_{m-r}}}$$

consists of a single vector, which is the case if and only if

$$\{i_1,\ldots,i_s,j_1,\ldots,j_{m-r}\} = \{1,\ldots,m\}$$

that is, if and only if $s = r$ and

$$\{i_1,\ldots,i_r\} = \{j_1,\ldots,j_{m-r}\}^c = \{k_1,\ldots,k_r\}$$

Hence, taking the dot product (6.2.5) singles out the desired coefficient, that is,

$$\mathbf{a}_p \cdot \mathbf{a}_{x_{j_1}\cdots x_{j_{m-r}}} = a_{k_1,\ldots,k_r}$$

Of course, this is only one computation of the coefficient (6.2.4). But we can generalize by taking the dot product

$$\mathbf{a}_p \cdot \mathbf{a}_{\mathbf{b}+F_{x_{j_1}\cdots x_{j_{m-r}}}}$$

where

$$\mathbf{b} + F_{x_{j_1}\cdots x_{j_{m-r}}}$$

is a translate of the flat $F_{x_{j_1}\cdots x_{j_{m-r}}}$. In this case, we have

$$\mathbf{a}_{x_{i_1}\cdots x_{i_s}} \cdot \mathbf{a}_{\mathbf{b}+F_{x_{j_1}\cdots x_{j_{m-r}}}} \equiv \left| F_{x_{i_1}\cdots x_{i_s}} \cap (\mathbf{b} + F_{x_{j_1}\cdots x_{j_{m-r}}}) \right| \mod 2$$

Let us denote the intersection on the right by N. Then this dot product will be 0 unless $|N| = 1$. Now, if $\mathbf{x} \in N$, then according to Lemma 6.2.12,

$$N = \mathbf{x} + (F_{x_{i_1}\cdots x_{i_s}} \cap F_{x_{j_1}\cdots x_{j_{m-r}}}) = \mathbf{x} + F_{x_{i_1}\cdots x_{i_s} x_{j_1}\cdots x_{j_{m-r}}}$$

and this is a one-point flat if and only if

$$F_{x_{i_1}\cdots x_{i_s} x_{j_1}\cdots x_{j_{m-r}}}$$

consists of a single vector, just as before. Hence,

$$\mathbf{a}_p \cdot \mathbf{a}_{\mathbf{b}+F_{x_{j_1}\cdots x_{j_{m-r}}}} = a_{k_1,\ldots,k_r}$$

where $\{j_1,\ldots,j_{m-r}\} = \{k_1,\ldots,k_r\}^c$ and where $\mathbf{b}$ is any vector in EG(m,2).

Since there are 2^{m-r} distinct translates of the r-flat

$$F_{x_{j_1}\cdots x_{j_{m-r}}} \tag{6.2.6}$$

we get 2^{m-r} different expressions for the coefficient

$$a_{k_1,\ldots,k_r}$$

each of which involves different components of $\mathbf{a}_p$. In this way, we can determine the correct value of this coefficient by majority logic, provided that at most $\frac{1}{2}(2^{m-r}-1)$ errors have occurred. An example should help clarify the situation.

Example 6.2.5 Table 6.2.2 shows the binary strings associated with the monomials used to construct the code $\mathcal{R}(2,4)$. (The columns are read from top to bottom.) A polynomial of degree at most 2 has the form

$$\begin{aligned} p = {} & a_0 + a_1x_1 + a_2x_2 + a_3x_3 + a_4x_4 \\ & + a_{12}x_1x_2 + a_{13}x_1x_3 + a_{14}x_1x_4 + a_{23}x_2x_3 + a_{24}x_2x_4 + a_{34}x_3x_4 \end{aligned}$$

where the coefficients are equal to 0 or 1. Let

$$\mathbf{a}_p = c_1 \cdots c_{16}$$

be the corresponding codeword. Let us determine the coefficient a_{13}. First, we observe that $\{1,3\}^c = \{2,4\}$, and so the flat (6.2.6) is

$$F_{x_2x_4} = \{0101, 0111, 1101, 1111\}$$

The $2^{4-2} = 4$ translates of this flat are

$$\begin{gathered} \{0101,0111,1101,1111\} \\ \{0000,0010,1000,1010\} \\ \{0100,0110,1100,1110\} \\ \{0001,0011,1001,1011\} \end{gathered}$$

and the 4 characteristic vectors corresponding to these flats are

0000 0101 0000 0101
1010 0000 1010 0000
0000 1010 0000 1010
0101 0000 0101 0000

Table 6.2.2

				1	x_1	x_2	x_3	x_4	x_1x_2	x_1x_3	x_1x_4	x_2x_3	x_2x_4	x_3x_4
0	0	0	0	1	0	0	0	0	0	0	0	0	0	0
0	0	0	1	1	0	0	0	1	0	0	0	0	0	0
0	0	1	0	1	0	0	1	0	0	0	0	0	0	0
0	0	1	1	1	0	0	1	1	0	0	0	0	0	1
0	1	0	0	1	0	1	0	0	0	0	0	0	0	0
0	1	0	1	1	0	1	0	1	0	0	0	0	1	0
0	1	1	0	1	0	1	1	0	0	0	0	1	0	0
0	1	1	1	1	0	1	1	1	0	0	0	1	1	1
1	0	0	0	1	1	0	0	0	0	0	0	0	0	0
1	0	0	1	1	1	0	0	1	0	0	1	0	0	0
1	0	1	0	1	1	0	1	0	0	1	0	0	0	0
1	0	1	1	1	1	0	1	1	0	1	1	0	0	1
1	1	0	0	1	1	1	0	0	1	0	0	0	0	0
1	1	0	1	1	1	1	0	1	1	0	1	0	1	0
1	1	1	0	1	1	1	1	0	1	1	0	1	0	0
1	1	1	1	1	1	1	1	1	1	1	1	1	1	1

Taking the dot product of each of these vectors with the codeword $\mathbf{a}_p = c_1 \cdots c_{16}$ gives the desired equations for the coefficient a_{13}

$$a_{13} = c_6 + c_8 + c_{14} + c_{16}$$

$$a_{13} = c_1 + c_3 + c_9 + c_{11}$$

$$a_{13} = c_5 + c_7 + c_{13} + c_{15}$$

$$a_{13} = c_2 + c_4 + c_{10} + c_{12}$$

Thus, if no more than one error occurs in the c_i, we can recover the value of a_{13} by taking the majority value obtained from the right-hand sides of these equations.

The other coefficients a_{ij} of monomials of degree 2 can be corrected in a similar way. Once we have the correct values for the coefficients a_{ij}, we can subtract the polynomial

$$a_{12}x_1x_2 + a_{13}x_1x_3 + a_{14}x_1x_4 + a_{23}x_2x_3 + a_{24}x_2x_4 + a_{34}x_3x_4$$

from p, to get

$$q = a_0 + a_1x_1 + a_2x_2 + a_3x_3 + a_4x_4$$

which is the polynomial associated with a codeword in $\mathcal{R}(1,4)$. Hence, the same procedure can be repeated in $\mathcal{R}(1,4)$ to determine the coefficients a_i. □

EXERCISES

1. Define the **negation** by
$$(f')(\mathbf{x}) = 1 \text{ if and only if } f(\mathbf{x}) = 0$$
Show that $f' = 1 + f$. What is the string version of negation?
2. Define **disjunction** by
$$(f \vee g)(\mathbf{x}) = 1 \text{ if and only if } f(\mathbf{x}) = 1 \text{ or } g(\mathbf{x}) = 1, \text{ or both}$$
Show that $f \vee g = f + g + fg$. What is the string version of disjunction?
3. Prove that $\mathcal{B}_m$ is a vector space.
4. Prove the remaining statement in Theorem 6.2.1.
5. Find Boolean polynomials corresponding to each of the binary strings
a) 1101 1001 b) 0000 1111
c) 1010 0101 d) 1101 1110 0001 1001
6. Show that the set of all monomials of degree at most m forms a basis for the vector space of all Boolean functions in m variables.
7. What is the weight of the codeword in $\mathcal{R}(r,m)$ corresponding to the monomial $x_{i_1} \cdots x_{i_j}$?
8. Show that, for any Boolean function $f(x_1,\ldots,x_{m-1})$, the function $x_m + f(x_1,\ldots,x_{m-1})$ takes on the values 0 and 1 equally often.
9. Show that the codes $\mathcal{R}(r,m)$ and $\mathcal{R}(r,m-1) \oplus \mathcal{R}(r-1,m-1)$ have the same size.
10. Use Theorem 6.2.4 to find an expression for a generating matrix for $\mathcal{R}(r,m)$ in terms of generating matrices for $\mathcal{R}(r,m-1)$ and $\mathcal{R}(r-1,m-1)$.
11. What is the relationship between $\mathcal{R}(r,m)$ and $\mathcal{R}(s,m)$?
12. What is the dual of $\mathcal{R}(r,m)$?
13. Which Reed-Muller codes are self-dual? Which are self-orthogonal?
14. Why does the polynomial mentioned at the end of the proof of Theorem 6.2.8 have degree $m-k$? (Note that the product of two linear polynomials may be linear, for example, $x_1x_1 = x_1$.)
15. Show that there are $2(2^m - 1)$ hyperplanes in $EG(m,2)$. *Hint.* How many one-dimensional subspaces are there in $EG(m,2)$?
16. Show that the converse of Theorem 6.2.8 does not hold. That is, there are Boolean polynomials $p(x_1,\ldots,x_m)$ of degree $m-k$

whose corresponding subsets F_p of EG(m,2) are not flats. *Hint.* Consider quadratic polynomials in three variables. Recall that, in EG(3,2), the planes (hyperplanes) correspond to the linear polynomials, so that if a quadratic polynomial does correspond to a flat, it must be a line (1-flat).

17. Let $F = \mathbf{b} + S$ and $G = \mathbf{c} + T$ be flats in EG(m,2). Prove that either $F \cap G = \emptyset$ or else $F \cap G = \mathbf{x} + (S \cap T)$ for any nonzero $\mathbf{x}$ in $F \cap G$.
18. Prove that all flats in EG(m,2) except one-point flats have an even number of vectors.

CHAPTER 7

Finite Fields and Cyclic Codes

7.1 Basic Properties of Finite Fields

Finite fields play a major role in coding theory, and so it is important to gain a solid understanding of the structure of such fields. Equally important is an understanding of the structure of polynomials whose coefficients lie in a finite field. We will be interested, for instance, in how to find the minimal polynomial of an element over a finite field, as well as how to factor the polynomial $x^n - 1$ over a finite field.

Throughout this chapter, we will rely on certain well-known facts from algebra. The reader may wish to refer to the appendix as needed for a brief review of these facts.

Let K and F be fields. If K is an extension of F, we write $F < K$. In this case, K is also a vector space over F. If the dimension of K over F is finite, we say that K is a *finite extension* of F and denote this dimension by $[K:F]$.

Lemma 7.1.1 Suppose that F is a finite field and that K is a finite extension of F, with $d = [K:F]$. Then $|K| = |F|^d$.

Proof. Let $\{\alpha_1, \ldots, \alpha_d\}$ be a basis for K over F. Then each element of K has a *unique* representation of the form

$$a_1\alpha_1 + \cdots + a_d\alpha_d$$

where $a_i \in F$. Since there are $|F|$ possibilities for each coefficient a_i, and since each possibility gives rise to a different element of K, we deduce that $|K| = |F|^d$. ∎

Theorem 7.1.2 If F is a finite field, then F has prime characteristic. Furthermore, if $char(F) = p$, then F has p^n elements, for some positive integer n.

Proof. In a field of characteristic 0, the positive integers 1, 2, ... are all distinct, and so a finite field must have nonzero characteristic. Hence, the characteristic of a finite field is the smallest positive integer n for which $n1 = 0$. Now suppose that $char(F) = n$. If $n = pq$, where $p,q < n$, then $pq1 = 0$. Hence, $(p1)(q1) = 0$, implying that $p1 = 0$ or $q1 = 0$. In either case, we have a contradiction to the fact that n is the *smallest* positive integer such that $n1 = 0$. Thus, n must be prime.

We have shown that the prime subfield of F is $\mathbb{Z}_p$. Since F is finite, it must be a finite extension of Z_p, and so if $n = [F:\mathbb{Z}_p]$, we deduce from Lemma 7.1.1 that $|F| = |\mathbb{Z}_p|^n = p^n$. ∎

From now on, unless otherwise stated, p will represent a prime number, and q will represent a prime power.

The following result, proved in the appendix for rings of prime characteristic, is a key reason why the theory of finite fields has its characteristic flavor.

Lemma 7.1.3 If F is a finite field of characteristic p, then

$$(\alpha \pm \beta)^{p^n} = \alpha^{p^n} + \beta^{p^n}$$

for any positive integer n, and for all $\alpha, \beta \in F$. □

A CHARACTERIZATION OF FINITE FIELDS

According to the definition, the set F^* of nonzero elements of a field F forms a group under multiplication. If $|F| = q$, then $|F^*| = q - 1$, and since the order of every element in a group divides the order of the group, we have

$$\alpha \in F^* \Rightarrow \alpha^{q-1} = 1$$

or, equivalently,

$$\alpha \in F \Rightarrow \alpha^q = \alpha$$

In other words, every element of F is a root of the polynomial $f_q(x) = x^q - x$. But since this polynomial has at most q roots, we see that F is the set of *all* roots of $f_q(x)$, and therefore is also the splitting field for $f_q(x)$.

Since polynomials of the form $x^q - x$ will occur frequently in our discussion, let us agree to denote them by $f_q(x)$. We have proved the following key theorem.

Theorem 7.1.4 If $|F| = q$, then F is both the set of all roots of the polynomial $f_q(x) = x^q - x$, and the splitting field for $f_q(x)$. □

Theorem 7.1.4 tells us that any finite field of size q is a splitting field for $f_q(x)$, and since any two splitting fields for the same polynomial are isomorphic, we see that any two finite fields of the same size are isomorphic.

It remains now to determine whether or not there is a finite field of size q for *every* prime power q. Suppose that $q = p^n$, let K be the splitting field for $f_q(x) = x^q - x$, and let R be the set of roots of $f_q(x)$. If α and β are in R, we have

$$\alpha^q = \alpha \quad \text{and} \quad \beta^q = \beta$$

Hence, by Lemma 7.1.3,

$$(\alpha \pm \beta)^q = \alpha^q \pm \beta^q = \alpha \pm \beta \quad \text{and} \quad (\alpha\beta^{-1})^q = \alpha^q(\beta^q)^{-1} = \alpha\beta^{-1}$$

which implies that $\alpha \pm \beta$ and $\alpha\beta^{-1}$ are also in R. Thus, R is a subfield of K, which implies that $R = K$. Furthermore, since

$$Df_q(x) = qx^{q-1} - 1 = -1 \quad \text{in } \mathbb{Z}_p$$

we see that $Df_q(x)$ has no common nonconstant factors with $f_q(x)$. Hence, $f_q(x)$ has no multiple roots, and so R is indeed a field of size q. Let us summarize our results.

Theorem 7.1.5
1) All finite fields have size $q = p^n$, for some prime p.
2) For every $q = p^n$, there is a unique (up to isomorphism) field of size q, which is both the set of all roots of the polynomial $f_q(x) = x^q - x$ and the splitting field for $f_q(x)$. □

It is customary to denote a finite field of size q by F_q, or $GF(q)$. (The symbol GF stand for *Galois field*, in honor of Evariste Galois.) According to Theorem 7.1.5, there is a unique (up to isomorphism) finite field F_q for each prime power q.

THE SUBFIELDS OF A FINITE FIELD

Our goal here is to describe the subfields of a finite field. In particular, we will show that any field of size p^n has exactly one subfield of size p^d, for all $d \mid n$, and furthermore, that these are all of the subfields of F.

Lemma 7.1.1 can be rephrased in terms of subfields as follows.

Theorem 7.1.6 Suppose that K is a finite field with $|K| = p^n$ and that F is a subfield of K. Then $|F| = p^d$, where $d \mid n$. ▯

As it happens, if $d \mid n$, then K has a *unique* subfield of size p^d. To see this, suppose that K is a field of size p^n, and let $d \mid n$. We know that K is the splitting field for $f_{p^n}(x) = x^{p^n} - x$. Now, it is not hard to show that

$$d \mid n \Rightarrow p^d - 1 \mid p^n - 1 \Rightarrow x^{p^d} - x \mid x^{p^n} - x \Rightarrow f_{p^d}(x) \mid f_{p^n}(x)$$

Hence, since $f_{p^n}(x)$ splits into linear factors over K, so does $f_{p^d}(x)$. In other words, K contains a splitting field for $f_{p^d}(x)$, that is, K contains a subfield of size p^d. Certainly, K cannot contain more than one such subfield, for then there would be more than p^d roots for the polynomial $f_{p^d}(x)$, which has degree p^d. Let us summarize.

Theorem 7.1.7 Let K be a finite field of size p^n. Then K has exactly one subfield of size p^d, for each $d \mid n$. Furthermore, this accounts for all of the subfields of K. ▯

THE MULTIPLICATIVE STRUCTURE OF A FINITE FIELD

The set F^* of all nonzero elements of a finite field F forms a finite group under multiplication. This group could not have a simpler structure – it is cyclic. Our plan is to prove this by showing that if $|F^*| = q-1$, then F^* has exactly $\phi(d)$ elements of order d, for each $d \mid q-1$, where ϕ is Euler's phi function. In particular, F^* must have at least one element of order $q-1$, and so it must be cyclic.

Let us recall a fact about cyclic groups, whose proof can be found in the appendix. If G is a cyclic group of order n, then G contains exactly $\phi(d)$ elements of each order d dividing n. This gives the formula

$$\sum_{d \mid n} \phi(d) = n$$

Now suppose that $|F^*| = q-1$ and that α is an element of F^* of order d. Thus, $d \mid q-1$. Consider the cyclic subgroup

$$\langle\alpha\rangle = \{1, \alpha, \alpha^2, \ldots, \alpha^{d-1}\}$$

generated by α. Every element of $\langle\alpha\rangle$ has order dividing d and so is a root of the polynomial $x^d - 1$. But this polynomial has at most d distinct roots in F (this is where we use the fact that F is a field), and so $\langle\alpha\rangle$ is the set of *all* roots of $x^d - 1$. In particular, all of the elements of F of order d must lie in $\langle\alpha\rangle$.

But the cyclic group $\langle\alpha\rangle$ has exactly $\phi(d)$ elements of order d, and so there are exactly $\phi(d)$ elements in F of order d. Thus, if F has one element of order $d \mid q-1$, it has exactly $\phi(d)$ such elements. Therefore, letting $\psi(d)$ denote the number of elements of F of order d, then $\psi(d) = 0$ or $\phi(d)$, we have

$$q-1 = \sum_{d \mid q-1} \psi(d) \le \sum_{d \mid q-1} \phi(d) = q-1$$

This implies that $\psi(d) = \phi(d)$ for all $d \mid q-1$. We have proven the following result.

Theorem 7.1.8 If F is a finite field with q elements, then F contains exactly $\phi(d)$ elements of order d, for each $d \mid q-1$. ▯

Corollary 7.1.9 The group F^* of nonzero elements of a finite field F is cyclic. ▯

Definition Any element of F_q that generates the cyclic group F_q^* is called a **primitive element** of F_q. ▯

We will generally reserve the Greek letter β for primitive elements of a finite field.

DESCRIBING THE ELEMENTS OF A FINITE FIELD

In general, there are several ways in which to represent the elements of a finite field. One way is to use a factor ring of the form $F_q[x]/\langle p(x)\rangle$, where p(x) is irreducible. Another is to use the fact that F_q^* is cyclic, and so its elements are all powers of a primitive element. As we will see, the first representation is ideally suited to performing addition, but not so convenient when it comes to multiplication. On the other hand, the second representation is well suited to performing multiplication but not addition. Fortunately, however, we can combine the two methods.

It is well known that if p(x) is an irreducible polynomial over F_q, then the factor ring

$$K = \frac{F_q[x]}{\langle p(x)\rangle}$$

is a field. (See the appendix.) In fact, if $\deg(p(x)) = d$, then K has degree d over F_q, and so $K = F_{q^d}$. This observation provides one way to describe the elements of a field. Since

$$F_{q^d} = \frac{F_q[x]}{\langle p(x)\rangle} = \{r(x)+\langle p(x)\rangle \mid \deg(r(x)) < d\}$$

we may identify F_{q^d} with the set of all polynomials of degree less than d, with addition and multiplication performed modulo p(x).

If α is a root of p(x) in a splitting field, then we can also describe the field F_{q^d} as the set of all polynomials in α of degree less than d, where addition and multiplication are performed modulo $p(\alpha)$. This is an informal way of saying that $F_q[x]/\langle p(x)\rangle$ is isomorphic to $F_q(\alpha)$.

Example 7.1.1 The polynomial $p(x) = x^4+x^3+x^2+x+1$ is irreducible over F_2. To see this, note that in order for p(x) to be reducible, it must have either a linear or a quadratic factor. But since $p(0) \neq 0$ and $p(1) \neq 0$, it has no linear factors. To see that p(x) has no quadratic factors, we note that there are precisely four quadratic polynomials over F_2, namely,

$$x^2,\ x^2+1,\ x^2+1,\ x^2+x+1$$

and it is easy to check that no product of any two of these polynomials equals p(x).

Now, since $q = 2$ and $d = 4$, we have

$$\frac{F_2[x]}{\langle x^4+x^3+x^2+x+1\rangle} = F_{16}$$

Thus, letting α be a root of p(x), we can represent the elements of F_{16} as the 16 binary polynomials of degree 3 or less in α

Constant: $0, 1,$

Linear: $\alpha,\ \alpha+1$

Quadratic: $\alpha^2,\ \alpha^2+1,\ \alpha^2+\alpha,\ \alpha^2+\alpha+1$

Cubic: $\alpha^3,\ \alpha^3+1,\ \alpha^3+\alpha,\ \alpha^3+\alpha^2,\ \alpha^3+\alpha+1,$
$\alpha^3+\alpha^2+1,\ \alpha^3+\alpha^2+\alpha,\ \alpha^3+\alpha^2+\alpha+1$

Any polynomial $f(\alpha) \in F_2[\alpha]$ can be reduced modulo $p(\alpha)$ simply by noting that $p(\alpha) = 0$ is equivalent to

$$\alpha^4 = \alpha^3+\alpha^2+\alpha+1$$

In this case, computations are made somewhat easier by the fact that α has a relatively small multiplicative order

$$\alpha^5 = \alpha\cdot\alpha^4 = \alpha\cdot(\alpha^3+\alpha^2+\alpha+1) = \alpha^4+\alpha^3+\alpha^2+\alpha = 1$$

Hence, $\alpha^6 = \alpha$, $\alpha^7 = \alpha^2$, and so on.

Addition in this representation is quite simple – it is just addition of polynomials of degree at most 3. For instance, since $2 = 0$ in F_2,

$$(\alpha^3 + \alpha + 1) + (\alpha^3 + \alpha^2 + 1) = \alpha^2 + \alpha$$

However, multiplication is not as simple, since we must reduce modulo $p(\alpha)$. Here is an example

$$\begin{aligned}(\alpha^3 + \alpha + 1)(\alpha^3 + \alpha^2 + 1) &= \alpha^6 + \alpha^5 + \alpha^3 + \alpha^4 + \alpha^3 + \alpha + \alpha^3 + \alpha^2 + 1 \\ &= \alpha^6 + \alpha^5 + \alpha^4 + \alpha^3 + 1 \\ &= \alpha + 1 + \alpha^4 + \alpha^3 + 1 \\ &= \alpha^4 + \alpha^3 + \alpha\end{aligned}$$

For larger fields, polynomial multiplication becomes impractical, especially compared to the alternative representation, which we consider next. ▯

If β is a primitive element of F_q, then

$$F_q = \{0, 1, \beta, \ldots, \beta^{q-2}\}$$

In this representation, multiplication is all but trivial

$$\beta^i \cdot \beta^j = \beta^{(i+j) \bmod (q-1)}$$

But addition is not at all clear. On the other hand, if we had the minimal polynomial $p(x)$ for β, then we could represent the elements of F_{16} as polynomials in β, for the purposes of addition, and as powers of β, for the purposes of multiplication. Let us consider an example.

Example 7.1.2 The polynomial $q(x) = x^4 + x + 1$ is irreducible over F_2. Suppose that β is a root of this polynomial. Then, as in the previous example, we can represent F_{16} as the set of polynomials in β of degree at most 3, where $\beta^4 = \beta + 1$. On the other hand,

$$\begin{aligned}\beta^{15} &= (\beta^5)^3 = (\beta \cdot \beta^4)^3 = (\beta \cdot (\beta+1))^3 = \beta^3(\beta+1)^3 \\ &= \beta^3 \cdot (\beta^3 + \beta^2 + \beta + 1) \\ &= \beta^6 + \beta^5 + \beta^4 + \beta^3 \\ &= (\beta^3 + \beta^2\ ) + (\beta^2 + \beta) + (\beta + 1) + \beta^3 \\ &= (\beta^3 + \beta^2) + (\beta^2 + \beta) + (\beta + 1) + \beta^3 \\ &= 1\end{aligned}$$

This shows that the order of β must divide 15. But $\beta^3 \neq 1$ and $\beta^5 \neq 1$, and so β has order 15, which means that it is primitive.

Since β is a primitive element of F_{16}, we know that every nonzero element of F_{16} has the form β^k, for some $k = 0,1,\ldots,14$. Now we can link the two representations, by computing a table showing how each element β^k can be represented as a polynomial in β of degree at most 3. Using the fact that $\beta^4 = 1 + \beta$, we have

$$\begin{aligned}
\beta^4 &= \beta + 1 \\
\beta^5 &= \beta \cdot \beta^4 = \beta(\beta+1) = \beta^2 + \beta \\
\beta^6 &= \beta \cdot \beta^5 = \beta^3 + \beta^2 \\
\beta^7 &= \beta \cdot \beta^6 = \beta^4 + \beta^3 = \beta^3 + \beta + 1
\end{aligned}$$

and so on. The complete list is given Table 7.1.1. As is customary, we write only the exponent k for β^k, and $a_3a_2a_1a_0$ for the polynomial $a_3\beta^3 + a_2\beta^2 + a_1\beta + a_0$.

Table 7.1.1

0	0001
1	0010
2	0100
3	1000
4	0011
5	0110
6	1100
7	1011
8	0101
9	1010
10	0111
11	1110
12	1111
13	1101
14	1001

Computations using this table are quite straightforward. Here is an example

$$\begin{aligned}
(\beta^8 + \beta^4 + 1)(\beta^3 + \beta) &= \beta^{11} + \beta^9 + \beta^7 + \beta^5 + \beta^3 + \beta \\
&= 1110 + 1010 + 1011 + 0110 + 1000 + 0010 \\
&= 0011 \\
&= \beta^4 \\
&= \beta + 1
\end{aligned}$$

□

We can see from the previous example that the key to doing arithmetic in a finite field is having a primitive element, along with its minimal polynomial. This motivates the following definition.

Definition Let β be a primitive element of F_{q^n}. The minimal polynomial of β over F_q is called a **primitive polynomial** for F_{q^n} over F_q. □

We will see in the next section that all of the roots of an irreducible polynomial over a finite field have the same multiplicative order. Hence, a primitive polynomial for F_{q^n} is a monic irreducible polynomial over F_q, *all* of whose roots are primitive elements of F_{q^n}. We will also see that all primitive polynomials for F_{q^n} over F_q have degree n.

In general, the task of finding primitive polynomials is not easy. There are various methods that do achieve some measure of success in certain cases, but we will not go into those methods here. Fortunately, extensive tables of primitive polynomials and field tables, such as Table 7.1.1, have been constructed. We give some tables in the back of this book. For more extensive tables, see Lidl and Niederreiter (1986).

Both of the representations of a finite field that we have just discussed can lead to a matrix representation. To see this, we begin with an idea from linear algebra. The **companion matrix** of a *monic* polynomial $p(x) = a_0 + a_1x + \cdots + a_{n-1}x^{n-1} + x^n$ is the matrix

$$C = \begin{bmatrix} 0 & 0 & \cdots & 0 & -a_0 \\ 1 & 0 & \cdots & 0 & -a_1 \\ 0 & 1 & \cdots & 0 & -a_2 \\ \vdots & \vdots & \vdots & \vdots & \vdots \\ 0 & 0 & 0 & 1 & -a_{n-1} \end{bmatrix}$$

It is possible to prove that the companion matrix of a polynomial satisfies the polynomial, that is, $p(C) = 0$. Hence, we can use C as a root of $p(x)$.

Thus, if $p(x)$ is a monic irreducible polynomial over F_q of degree d, with companion matrix C, then the field F_{q^d} can be represented as the set of polynomials in C of degree less than d, where addition and multiplication are modulo $p(C)$. But polynomials in C are just matrices, and so we get a representation of the elements of F_q as matrices over F_q. Let us illustrate.

Example 7.1.3 Consider the irreducible polynomial $p(x) = x^2 + 1$ over F_3. The elements of F_9 can be represented as polynomials of degree less than 2 in the companion matrix of $p(x)$

$$C = \begin{bmatrix} 0 & 2 \\ 1 & 0 \end{bmatrix}$$

In particular, the elements of F_9 are

$$0 = \begin{bmatrix} 0 & 0 \\ 0 & 0 \end{bmatrix}, \quad I = \begin{bmatrix} 1 & 0 \\ 0 & 1 \end{bmatrix}, \quad 2I = \begin{bmatrix} 2 & 0 \\ 0 & 2 \end{bmatrix}, \quad C = \begin{bmatrix} 0 & 2 \\ 1 & 0 \end{bmatrix}$$

$$I + C = \begin{bmatrix} 1 & 2 \\ 1 & 1 \end{bmatrix}, \quad 2I + C = \begin{bmatrix} 2 & 2 \\ 1 & 2 \end{bmatrix}, \quad 2C = \begin{bmatrix} 0 & 1 \\ 2 & 0 \end{bmatrix}$$

$$I+2C = \begin{bmatrix} 1 & 1 \\ 2 & 1 \end{bmatrix}, \quad 2I + 2C = \begin{bmatrix} 2 & 1 \\ 2 & 2 \end{bmatrix}$$

An alternative is to use the primitive polynomial $q(x) = x^2 + x + 2$ over F_3, for then we know that all elements of F_9 are powers of a root of $q(x)$. Therefore, since the companion matrix for $q(x)$ is

$$A = \begin{bmatrix} 0 & 1 \\ 1 & 2 \end{bmatrix}$$

the elements of F_9 are

$$0 = \begin{bmatrix} 0 & 0 \\ 0 & 0 \end{bmatrix}, \quad A = \begin{bmatrix} 0 & 1 \\ 1 & 2 \end{bmatrix}, \quad A^2 = \begin{bmatrix} 1 & 2 \\ 2 & 2 \end{bmatrix}, \quad A^3 = \begin{bmatrix} 2 & 2 \\ 2 & 0 \end{bmatrix}$$

$$A^4 = \begin{bmatrix} 2 & 0 \\ 0 & 2 \end{bmatrix}, \quad A^5 = \begin{bmatrix} 0 & 2 \\ 2 & 1 \end{bmatrix}, \quad A^6 = \begin{bmatrix} 2 & 1 \\ 1 & 1 \end{bmatrix}$$

$$A^7 = \begin{bmatrix} 1 & 1 \\ 1 & 0 \end{bmatrix}, \quad A^8 = \begin{bmatrix} 1 & 0 \\ 0 & 1 \end{bmatrix}$$

In either case, addition and multiplication are performed using the ordinary rules of matrix algebra. The advantage of these representations is that reductions are done automatically, for instance,

$$(I + C)(I + 2C) = \begin{bmatrix} 1 & 2 \\ 1 & 1 \end{bmatrix} \begin{bmatrix} 1 & 1 \\ 2 & 1 \end{bmatrix} = \begin{bmatrix} 2 & 0 \\ 0 & 2 \end{bmatrix} = 2I$$

On the other hand, if $\deg(p(x))$ is large, the matrices involved become difficult to handle. ☐

EXERCISES

1. Show that
$$d \mid n \Rightarrow p^d - 1 \mid p^n - 1 \Rightarrow f_{p^d}(x) \mid f_{p^n}(x)$$
2. Is $\mathbb{Z}_2 \approx F_2$? Is $\mathbb{Z}_4 \approx F_4$? When is $\mathbb{Z}_{q^n} \approx F_{q^n}$?
3. Determine the number of subfields of F_{1024}. Determine the number of subfields of F_{729}.
4. If $F < K$ show that F and K must have the same characteristic.
5. Show that, except for the case of F_2, the sum of all of the elements in a finite field is equal to 0.
6. Find all primitive elements of F_7, of F_9.
7. Finish computing the powers of β in Example 7.1.2.
8. Use Table 1 to compute $(\beta^{10} + \beta^5)(\beta^2 + \beta^4)$.
9. Find all subfields of F_{16}. Describe them in terms of the primitive element β in Example 7.1.2.
10. Verify that the polynomial $p(x)$ in Example 7.1.1 is irreducible.
11. Verify that the polynomial $q(x)$ in Example 7.1.2 is irreducible.
12. Referring to Example 7.1.1, show that $\alpha + 1$ is a primitive element of F_{16}.
13. Show that the polynomials $p(x) = x^2 + 1$ and $q(x) = x^2 + x + 4$ are irreducible over F_{11}. Are the fields $F_{11}[x]/\langle p(x)\rangle$ and $F_{11}[x]/\langle q(x)\rangle$ isomorphic? Explain.
14. Use the primitive polynomial $x^3 + x + 1$ to give matrix representations for F_8 over F_2, following Example 7.1.3.
15. Show that the polynomial $p(x) = x^4 + x^3 + 1$ is primitive over F_2, and compute a field table, similar to Table 1, using this polynomial.
16. If F is an arbitrary field, prove that if F^* is cyclic, then F must be a *finite* field.
17. Let $N_q(d)$ be the number of monic irreducible polynomials of degree d over F_q. Find $N_q(2)$.
18. Referring to the previous exercise, find an expression for $N_q(3)$ in terms of $N_q(2)$.
19. Find all irreducible polynomials of degree 2, 3, and 4 over F_2.
20. Find all irreducible polynomials of degree 2 and 3 over F_3.
21. Show that $p(x) = x^3 + x^2 + 1$ is primitive over F_2. Construct the field F_8 in both representations, as we did in Examples 7.1.1 and 7.1.2. Construct a table similar to Table 1 for this field.
22. Verify that $p(x) = x^2 + x + 2$ is primitive over F_3, and repeat the previous exercise for the field F_9.

7.2 Irreducible Polynomials Over Finite Fields

In this section, we discuss the basic properties of irreducible polynomials over finite fields. Let us denote the splitting field of any polynomial f(x) by *Split*(f(x)) or *Split*(f). The following lemma is very useful.

Lemma 7.2.1 Let f(x) be an irreducible polynomial over F_q, and let α be a root of f(x) in some extension field. Then if $g(x) \in F_q[x]$, we have $g(\alpha) = 0$ if and only if $f(x) \mid g(x)$.
Proof. This follows from the fact that f(x) is a nonzero scalar multiple of the minimal polynomial of α over F_q. ∎

We begin our discussion with the most fundamental question – that of existence.

Theorem 7.2.2 For every finite field F_q, and every positive integer d, there exists an irreducible polynomial f(x) of degree d over F_q.
Proof. Let β be a primitive element of F_{q^d}. Then $F_q(\beta) = F_{q^d}$, and since $[F_q(\beta):F_q] = [F_{q^d}:F_q] = d$, the irreducible polynomial for β over F_q must have degree d. ∎

THE SPLITTING FIELD OF AN IRREDUCIBLE POLYNOMIAL

In the case of nonfinite fields F, if α is a root of an irreducible polynomial $f(x) \in F[x]$ in some extension of F, then the field $F(\alpha)$, obtained by adjoining α to F is, in general, *not* the complete splitting field of f(x). In other words, adjoining one root α of f(x) does not, in general, pick up the other roots of f(x). In fact, the most we can say in general about the dimension of the splitting field of an irreducible polynomial of degree d is that it lies somewhere between d and $d! = 1 \cdot 2 \cdots d$.

However, for *finite* fields, adjoining a single root of an *irreducible* polynomial always gives the splitting field for that polynomial. Hence, the splitting field has degree equal to the degree of the polynomial itself.

To see this, suppose that f(x) is irreducible over F_q, with degree d. Let α be a root of f(x) in *Split*(f), and consider the fields

$$F_q < F_q(\alpha) < \mathit{Split}(f)$$

Since $F_q(\alpha)$ has degree d over F_q, it has size q^d, and so according to Theorem 7.1.5, it is the set of all roots of the polynomial $f_{q^d}(x) = x^{q^d} - x$. In particular, α is a root of $f_{q^d}(x)$. Hence, according to Lemma 7.2.1, we have $f(x) \mid f_{q^d}(x)$, which implies that *all* of the roots

of $f(x)$ are roots of $f_{q^d}(x)$ and so lie in $F_q(\alpha)$. Hence, $Split(f) < F_q(\alpha)$, and so $F_q(\alpha) = Split(f)$. This proves the following.

Theorem 7.2.3 Let $f(x) \in F_q[x]$ be an irreducible polynomial of degree d, and let α be any root of $f(x)$. Then the splitting field of $f(x)$ is

$$Split(f) = F_q(\alpha) = F_{q^d}$$

In particular, the splitting field of $f(x)$ has degree d over F_q. □

Corollary 7.2.4 Let $f(x) \in F_q[x]$ be an irreducible polynomial of degree d. Then $f(x) \mid x^{q^n} - x$ if and only if $d \mid n$.

Proof. First, we note that

$$Split(f) = F_{q^d} \quad \text{and} \quad Split(x^{q^n} - x) = F_{q^n}$$

and so, according to Theorem 7.1.7,

$$Split(f) < Split(x^{q^n} - x) \Leftrightarrow d \mid n$$

Now, if $f(x) \mid x^{q^n} - x$, then every root of $f(x)$ is a root of $x^{q^n} - x$. This implies that $Split(f) < Split(x^{q^n} - x)$, and so $d \mid n$. Conversely, if $d \mid n$, then $Split(f) < Split(x^{q^n} - x)$, and so any root α of $f(x)$ lies in $Split(x^{q^n} - x)$. But, $Split(x^{q^n} - x)$ is the set of all roots of $x^{q^n} - x$, and so α must actually be a root of $x^{q^n} - x$. Therefore, by Lemma 7.2.1, $f(x)$ divides $x^{q^n} - x$. ∎

THE NATURE OF THE ROOTS OF AN IRREDUCIBLE POLYNOMIAL

Let us take a closer look at the nature of the roots of an irreducible polynomial $f(x)$ over a finite field F_q. We know that the roots of $f(x)$ lie in F_{q^d}, where $d = \deg(f(x))$. Suppose that

$$f(x) = a_0 + a_1 x + \cdots + a_d x^d$$

where $a_i \in F_q$. If α is a root of $f(x)$, then

$$f(\alpha) = a_0 + a_1 \alpha + \cdots + a_d \alpha^d = 0$$

We can take advantage of the fact that we are working in a field of characteristic p as follows. Since $a_i \in F_q$, we know that $a_i^q = a_i$, and so

$$\begin{aligned} f(\alpha^q) &= a_0 + a_1 \alpha^q + \cdots + a_d \alpha^{qd} \\ &= a_0^q + a_1^q \alpha^q + \cdots + a_d^q \alpha^{qd} \\ &= a_0^q + (a_1 \alpha)^q + \cdots + (a_d \alpha^d)^q \\ &= (a_0 + a_1 \alpha + \cdots + a_d \alpha^d)^q = (f(\alpha))^q = 0 \end{aligned}$$

Thus, if α is a root of f(x), so is α^q. Accordingly, we see that

$$(7.2.1) \qquad \alpha, \alpha^q, \alpha^{q^2}, \ldots, \alpha^{q^{d-1}}$$

are roots of f(x). If we can show that these roots are distinct, we will know that this is a complete list of all roots of f(x).

If $\alpha^{q^i} = \alpha^{q^j}$, where say $i < j$, then taking q^{d-j} powers of both sides, we get

$$\alpha^{q^{d+i-j}} = \alpha^{q^d} = \alpha$$

and so α is a root of the polynomial $x^{q^{d+i-j}} - x$. Lemma 7.2.1 then implies that $f(x) \mid x^{q^{d+i-j}} - x$, which is a contradiction to Corollary 7.2.4. Hence, the roots (7.2.1) are distinct.

We note also that d is the smallest positive integer for which $\alpha^{q^d} = \alpha$. Let us summarize.

Theorem 7.2.5 Let $f(x) \in F_q[x]$ be an irreducible polynomial of degree d. If α is a root of f(x) in $Split(f) = F_{q^d}$, then all of the roots of f(x) are given by

$$\alpha, \alpha^q, \alpha^{q^2}, \ldots, \alpha^{q^{d-1}}$$

Furthermore, d is the smallest positive integer for which $\alpha^{q^d} = \alpha$. □

Theorem 7.2.5 has some very important consequences.

Corollary 7.2.6 Let $f(x) \in F_q[x]$ be irreducible. Then all of the roots of f(x) in *Split*(f) have the same multiplicative order.

Proof. This follows from the fact that $[Split(f)]^*$ has order $q^d - 1$, and q^d-1 and q^i are relatively prime for all i. ∎

Corollary 7.2.6 allows us to make the following definition.

Definition The multiplicative order of any root of an irreducible polynomial $f(x) \in F_q[x]$ in its splitting field is called the **order** of f(x), and will be denoted by $o(f(x))$ or $o(f)$. □

In view of the fact that every element of F_q is the root of exactly one monic irreducible polynomial over F_q (its minimal polynomial), we can make the following definition.

Definition The elements $\alpha, \beta \in F_{q^n}$ are said to be **conjugates over** $\mathbf{F_q}$ if they are roots of the same monic irreducible polynomial over F_q, that is, if they have the same minimal polynomial over F_q. □

Corollary 7.2.7 The conjugates of $\alpha \in F_{q^n}$ over F_q are

$$\alpha, \alpha^q, \alpha^{q^2}, \ldots, \alpha^{q^{d-1}}$$

where d is the smallest positive integer for which $\alpha^{q^d} = \alpha$. □

COMPUTING MINIMAL POLYNOMIALS

Theorem 7.2.5 also gives us the minimal polynomial for any $\alpha \in F_q$.

Corollary 7.2.8 Let $\alpha \in F_{q^n}$. Then the minimal polynomial for α over F_q is

$$\text{irr}(\alpha, F_q) = (x-\alpha)(x-\alpha^q)\cdots(x-\alpha^{q^{d-1}})$$

where d is the smallest positive integer for which $\alpha^{q^d} = \alpha$. In particular, the polynomial on the right (when multiplied out and simplified) must have coefficients in F_q. □

Example 7.2.1 In the previous section, we saw that the polynomial $p(x) = x^4 + x + 1$ is primitive for F_{16} over F_2. Hence, any root β of $p(x)$ generates F_{16}^*. Let us compute the minimal polynomials for the elements of F_{16}.

We begin by computing sets of conjugates

Conjugates of β: $\beta, \beta^2, \beta^4, \beta^8$ (done since $\beta^{16} = \beta$)

Conjugates of β^3: $\beta^3, \beta^6, \beta^{12}, \beta^{24} = \beta^9$ (done since $\beta^{48} = \beta^3$)

Conjugates of β^5: β^5, β^{10} (done since $\beta^{20} = \beta^5$)

Conjugates of β^7: $\beta^7, \beta^{14}, \beta^{28} = \beta^{13}, \beta^{56} = \beta^{11}$ (done since $\beta^{112} = \beta^7$)

Letting $m_k(x)$ be the minimal polynomial for β^k, we have, for example

$$m_5(x) = m_{10}(x) = (x-\beta^5)(x-\beta^{10}) = x^2 - (\beta^5 + \beta^{10})x + \beta^{15}$$

But, according to Table 7.1.1

$$\beta^5 + \beta^{10} = (0110) + (0111) = (0001) = \beta^0 = 1$$

and since $\beta^{15} = 1$, we have

$$m_5(x) = m_{10}(x) = x^2 + x + 1$$

The other minimal polynomials are computed similarly. The complete list is

$$m_0(x) = x+1$$
$$m_1(x) = m_2(x) = m_4(x) = m_8(x) = x^4 + x + 1$$
$$m_3(x) = m_6(x) = m_9(x) = m_{12}(x) = x^4 + x^3 + x^2 + x + 1$$
$$m_5(x) = m_{10}(x) = x^2 + x + 1$$
$$m_7(x) = m_{11}(x) = m_{13}(x) = m_{14}(x) = x^4 + x^3 + 1$$ □

Another method for computing minimal polynomials is discussed in Section A.4 of the appendix.

THE AUTOMORPHISM GROUP OF F_{q^n}

Corollary 7.2.7 allows us to characterize the automorphisms of an extension field. Let us first have the definition of automorphism.

Definition An **automorphism** of F_{q^n} over F_q is a bijective map $\sigma: F_{q^n} \to F_{q^n}$ for which
1) $\sigma(\alpha + \beta) = \sigma(\alpha) + \sigma(\beta)$
2) $\sigma(\alpha\beta) = \sigma(\alpha)\sigma(\beta)$
3) If $a \in F_q$, then $\sigma(a) = a$

The last condition says that σ **fixes** all elements of the field F_q. □

Theorem 7.2.9 The set of automorphisms of F_{q^n} over F_q forms a cyclic group (under composition) of order n, generated by the map $\sigma_q(\alpha) = \alpha^q$.

Proof. Let β be a primitive element of F_{q^n}. Thus, β has multiplicative order $q^n - 1$, and its minimal polynomial $m(x)$ has roots

$$\beta, \beta^q, \beta^{q^2}, \ldots, \beta^{q^{n-1}}$$

Now, let $f(x)$ be a polynomial over F_q. Since an automorphism τ of F_{q^n} over F_q fixes the coefficients of $f(x)$, we see that $f(\alpha) = 0$ if and only if $f(\tau(\alpha)) = 0$. In other words, τ permutes the roots of $f(x)$ that lie in F_{q^n}. In particular, τ must send the root β of $m(x)$ to another root, say

$$\tau(\beta) = \beta^{q^i}$$

But since β is a primitive element of F_{q^n}, τ is completely determined by its value on β, and since

$$\sigma_q^i(\beta) = \beta^{q^i} = \tau(\beta)$$

we deduce that $\tau = \sigma_q^i$. Hence, all automorphisms of F_{q^n} over F_q have the form σ_q^i for some i. ∎

**NORMAL BASES*

As we know, if α is an element of F_{q^n} of degree n over F_q, then the powers $\{1,\alpha,\alpha^2,\ldots,\alpha^{n-1}\}$ form a basis for F_{q^n} over F_q. Such a basis is called a **polynomial basis.** There is another type of basis that plays an important role in the theory of finite fields.

Definition A basis for F_{q^n} over F_q of the form $\{\alpha,\alpha^q,\alpha^{q^2},\ldots,\alpha^{q^{n-1}}\}$ is called a **normal basis** for F_{q^n} over F_q. □

Our goal here is to show that every finite field F_{q^n} has a normal basis over F_q. We begin with a well known result from algebra, which says that any set of distinct homomorphisms from a group G to the multiplicative group F^* of a field F must be linearly independent.

Theorem 7.2.10 **(Artin's Lemma)** Let $\psi_1,\ldots,\psi_s$ be distinct homomorphisms from a group G into the multiplicative group F^* of a field F. Then the ψ_i are linearly independent over F, that is, for any $c_1,\ldots,c_s$ in F, not all 0, there exists a $g \in G$ for which

$$c_1\psi_1(g) + \cdots + c_s\psi_s(x) \neq 0$$

Proof. Suppose to the contrary that the ψ_i's are linearly dependent. We can, by renumbering if necessary, assume that

$$c_i\psi_i(g) + c_{i+1}\psi_{i+1}(g) + \cdots + c_s\psi_s(g) = 0 \qquad (7.2.2)$$

for all $g \in G$, and that this is the *shortest* such nontrivial linear combination of the ψ_i's that is equal to 0 (for all $g \in G$). In particular, $c_i,\ldots,c_s$ are all nonzero. Since $\psi_i \neq \psi_s$, there must exist an element $h \in G$ for which $\psi_i(h) \neq \psi_s(h)$. Replacing g by gh in (7.2.2), we see that

$$c_i\psi_i(h)\psi_i(g) + c_{i+1}\psi_{i+1}(h)\psi_{i+1}(g) + \cdots + c_s\psi_s(h)\psi_s(g) = 0$$

for all $g \in G$. Multiplying (7.2.2) by $\psi_i(h)$ and subtracting the result from this equation gives

$$c_{i+1}[\psi_{i+1}(h)-\psi_i(h)]\psi_{i+1}(g) + \cdots + c_s[\psi_s(h)-\psi_i(h)]\psi_s(g) = 0$$

for all $g \in G$. But $c_s[\psi_s(h) - \psi_i(h)] \neq 0$, and so not all of the coefficients are zero. This gives a shorter nontrivial linear combination of the ψ_i's that is equal to 0, which is a contradiction. Hence, the ψ_i's are linearly independent. ∎

We will also need a result from linear algebra, which we will not prove here. Recall that, if $T:V\to V$ is a linear operator on an

n-dimensional vector space V over a field F, then the *minimal polynomial* $m_T(x)$ for T is the unique monic polynomial over F of smallest degree for which $m_T(T) = 0$. Recall also that since T satisfies its *characteristic polynomial* $c_T(x) = \det(xI - \mathrm{Mat}(T))$, we have $m_T(x) \mid c_T(x)$. Finally, a vector $v \in V$ is said to be **cyclic** for T if the vectors $\{v, Tv, T^2v, \ldots, T^{n-1}v\}$ form a basis for V.

Theorem 7.2.11 Let $T:V \to V$ be a linear operator on a finite-dimensional vector space V over a field F. Then V contains a cyclic vector for T if and only if the minimum polynomial $m_T(x)$ and the characteristic polynomial $c_T(x)$ are the same. □

Now we can establish the existence of normal bases.

Theorem 7.2.12 (The Normal Basis Theorem) There exists a normal basis for F_{q^n} over F_q.

Proof. If $n = 1$, there is nothing to prove, so assume that $n > 1$. According to Theorem 7.2.9, the distinct automorphisms of F_{q^n} over F_q are given by

(7.2.3) $$\epsilon, \sigma_q, \sigma_q^2, \ldots, \sigma_q^{n-1}$$

where ϵ is the identity map, and $\sigma_q(\alpha) = \alpha^q$, for all $\alpha \in F_{q^n}$. Of course, σ_q may also be thought of as a linear operator on the n-dimensional vector space F_{q^n} over F_q. As such, the minimal polynomial of σ_q is $x^n - 1$. To see this, note that σ_q does satisfy the polynomial $x^n - 1$. Furthermore, the maps in (7.2.3) are distinct homomorphisms from the group $F_{q^n}^*$ to itself, and so Artin's lemma implies that no polynomial in σ_q of degree less than n can be equal to 0. Hence, $x^n - 1$ is the minimal polynomial of σ_q.

On the other hand, the characteristic polynomial of σ_q has degree n, is monic, and is divisible by $x^n - 1$, and so it must be $x^n - 1$. Thus, we may apply Theorem 7.2.11, to deduce the existence of a cyclic vector α for σ_q. Hence

$$\alpha, \sigma_q(\alpha), \ldots, \sigma_q^{n-1}(\alpha)$$

form a normal basis for F_{q^n} over F_q. ∎

*LINEARIZED POLYNOMIALS

We now consider briefly a special type of polynomial that has applications in coding theory.

Definition A polynomial of the form

(7.2.4) $$L(x) = \sum_{i=0}^{m} \alpha_i x^{q^i}$$

with coefficients $\alpha_i \in F_{q^n}$ is called a **linearized polynomial.** □

The term linearized polynomial comes from the following theorem, whose proof we leave as an exercise.

Theorem 7.2.13 Let $L(x)$ be a linearized polynomial over F_{q^n}. If $\alpha, \beta \in F_{q^n}$ and $c \in F_q$, then

$$L(\alpha + \beta) = L(\alpha) + L(\beta)$$

and

$$L(c\alpha) = cL(\alpha)$$ □

The roots of a linearized polynomial, in a splitting field F_{q^s}, have some rather special properties, which we give in the next two theorems.

Theorem 7.2.14 Let $L(x)$ be a nonzero linearized polynomial over F_{q^n}, with splitting field F_{q^s}. Then each root of $L(x)$ has the same multiplicity, which must be either 1 or else a power of q. Furthermore, the roots of $L(x)$ form a vector subspace of F_{q^s} over F_q.
Proof. If $L(x)$ has the form (7.2.4), then $L'(x) = \alpha_0$. Hence, if $\alpha_0 \neq 0$, we deduce that all roots of $L(x)$ are simple. On the other hand, suppose that $\alpha_0 = \alpha_1 = \cdots = \alpha_{k-1} = 0$ but $\alpha_k \neq 0$. Then since $\alpha_i \in F_{q^n}$, we have

$$\alpha_i^{q^{nk}} = \alpha_i$$

and so

$$L(x) = \sum_{i=k}^{m} \alpha_i x^{q^i} = \sum_{i=k}^{m} \alpha_i^{q^{nk}} x^{q^i} = \left(\sum_{i=k}^{m} \alpha_i^{q^{(n-1)k}} x^{q^{i-k}} \right)^{q^k}$$

which is the q^k-th power of a linearized polynomial with the property that its constant term is nonzero, and so it has only simple roots. Hence, each root of $L(x)$ has multiplicity q^k. We leave proof of the fact that the roots form a vector subspace of F_{q^s} as an exercise. ∎

The following is a partial converse of Theorem 7.2.15. We will omit the proof, however, and refer the interested reader to Lidl and Niederreiter (1986).

Theorem 7.2.15 Let U be a vector subspace of F_{q^n} over F_q. Then for any nonnegative integer k, the polynomial

$$L(x) = \prod_{\alpha \in U} (x-\alpha)^{q^k}$$

is a linearized polynomial over F_{q^n}. □

*THE NUMBER OF IRREDUCIBLE POLYNOMIALS

We can get an explicit formula for the number of irreducible polynomials of a given degree over F_q by using Möbius inversion. (See the appendix for a discussion of Möbius inversion.) First, we need the following result.

Theorem 7.2.16 Let F_q be a finite field, and let n be a positive integer. Then the product of all monic irreducible polynomials in $F_q[x]$ whose degree divides n is the polynomial $f_{q^n}(x) = x^{q^n} - x$.

Proof. We know that $f_{q^n}(x)$ can be written as a product of monic irreducible polynomials in $F_q[x]$. On the other hand, according to Corollary 7.2.4, *every* monic irreducible polynomial in $F_q[x]$ whose degrees divide n is a factor of $f_{q^n}(x)$. Finally, since

$$Df_{q^n}(x) = q^n x^{q^n-1} - 1 = -1$$

we know that $f_{q^n}(x)$ has no multiple roots in any splitting field, and so each monic irreducible polynomial appears only once as a factor of $f_{q^n}(x)$. ∎

Let us denote the number of monic irreducible polynomials of degree d over F_q by $N_q(d)$. Theorem 7.2.16 gives us the following.

Corollary 7.2.17 For all positive integers d and n, we have

$$q^n = \sum_{d \mid n} d\, N_q(d) \qquad \square$$

Now we can apply Möbius inversion to get an explicit formula for $N_q(d)$. Classical Möbius inversion is

$$g(n) = \sum_{d \mid n} f(d) \;\Rightarrow\; f(n) = \sum_{d \mid n} g(d)\mu\left(\frac{n}{d}\right) \tag{7.2.5}$$

where the Möbius function μ is defined by

$$\mu(m) = \begin{cases} 1 & \text{if } m = 1 \\ (-1)^k & \text{if } m = p_1 p_2 \cdots p_k \text{ for } \textit{distinct} \text{ primes } p_i \\ 0 & \text{otherwise} \end{cases}$$

Corollary 7.2.18 The number $N_q(n)$ of monic irreducible polynomials of degree n over F_q is given by

$$N_q(n) = \frac{1}{n} \sum_{d \mid n} \mu(\tfrac{n}{d}) q^d = \frac{1}{n} \sum_{d \mid n} \mu(d) q^{n/d}$$

Proof. Letting $g(n) = q^n$ and $f(d) = dN_q(d)$ in (7.2.5), we get the desired result. ∎

Example 7.2.2 The number of monic irreducible polynomials of degree 12 over F_q is

$$\begin{aligned} N_q(12) &= \frac{1}{12}\Big(\mu(1)q^{12} + \mu(2)q^6 + \mu(3)q^4 + \mu(4)q^3 + \mu(6)q^2 + \mu(12)q\Big) \\ &= \frac{1}{12}\big(q^{12} - q^6 - q^4 + q^2\big) \end{aligned}$$

The number of monic irreducible polynomials of degree 4 over F_2 is

$$N_2(4) = \frac{1}{4}\Big(\mu(1)2^4 + \mu(2)2^2 + \mu(4)2^1\Big) = 3$$

just as we would expect from the results of Example 7.2.1. □

Mobius inversion can be used to find not only the *number* of monic irreducible polynomials of degree d over F_q but also the *product* of all such polynomials. Let us denote this product by $I(q,d;x)$. Then Theorem 7.2.16 is equivalent to

$$x^{q^n} - x = \prod_{d \mid n} I(q,d;x)$$

Applying the multiplicative version of Mobius inversion gives the following.

Corollary 7.2.19 The product $I(q,n;x)$ of all monic irreducible polynomials of degree n over F_q is given by

$$I(q,n;x) = \prod_{d \mid n} \left(x^{q^d} - x\right)^{\mu(n/d)} = \prod_{d \mid n} \left(x^{q^{n/d}} - x\right)^{\mu(d)} \qquad \square$$

Example 7.2.3 For $q = 2$ and $n = 4$, we get

$$I(2,4;x) = (x^{16} - x)^{\mu(1)}(x^4 - x)^{\mu(2)}(x^2 - x)^{\mu(4)}$$

$$= \frac{x^{16} - x}{x^4 - x} = \frac{x^{15} - 1}{x^3 - 1} = x^{12} + x^9 + x^6 + x^3 + 1 \qquad \square$$

EXERCISES

1. Consider the irreducible polynomial $p(x) = x^4 - 2$ over $\mathbb{Q}$. Show that adjoining one root of $p(x)$ to $\mathbb{Q}$ does not produce the splitting field for $p(x)$. What is the degree of the splitting field for $p(x)$ over $\mathbb{Q}$?
2. Show that the order $o(f)$ of an irreducible polynomial $f(x)$ is the smallest positive integer e for which $f(x) \mid x^e - 1$.
3. Show that the relation α is a conjugate of β over F_q is an equivalence relation.
4. Use Theorem 7.2.9 to show that every element in F_{q^n} has a unique q^i-th root, for $i = 1,\ldots,n-1$.
5. If $2 \nmid q$, show that exactly one-half of the nonzero elements of F_q have square roots. *Hint.* Let β be a primitive element of F_q. If $\beta = \alpha^2$, then $\alpha^{2k} = \alpha$ for some k.
6. Compute the minimal polynomials over F_2 for all elements of F_4.
7. Compute the minimal polynomials over F_2 for all elements of F_8.
8. Compute the minimal polynomials over F_3 for all elements of F_9.
9. Find $N_q(2)$ directly, without using Corollary 7.2.18. Then use that corollary to verify your answer.
10. Calculate $N_q(20)$.
11. Use Corollary 7.2.18 to show that $N_q(d) > 0$. What does this say in words? *Hint.* Use the fact that $\mu(d) \geq -1$ for $d \mid n$, $d > 1$.
12. Suppose that $\alpha \in F_{q^n}$ and that $f(x)$ is the minimal polynomial of α over F_q. If $\deg(f(x)) = d$, show that $f(x)$ divides $\gcd(x^{q^d} - x, x^{q^n} - x)$.
13. Suppose that $f(x), g(x) \in F_q[x]$ and that both of these polynomials split in F_{q^n}. Suppose also that $f(x)$ has no multiple roots in F_{q^n}. If all of the roots of $f(x)$ in F_{q^n} are also roots of $g(x)$, then prove that $f(x)$ divides $g(x)$ in F_q.
14. Use the results of the previous exercise to show that if $a \in F_q$ and n is a positive integer, then $x^q - x + a$ divides $x^{q^n} - x + na$.
15. Let $\alpha \in F_{q^n}$ have minimal polynomial $f(x)$ over F_q, with $\deg(f(x)) = d$. Consider the elements $\alpha, \alpha^q, \ldots, \alpha^{q^r}$.
   (a) If $r = d$, are these elements distinct?
   (b) If $r = kd$ for $k > 1$, what can you say about the distinctness

of these elements?

16. Find a normal basis for F_8 over F_2. *Hint.* Let α be a root of the irreducible polynomial x^3+x^2+1.
17. Prove Theorem 7.2.13.
18. Finish the proof of Theorem 7.2.14.
19. Use Corollary 7.2.17 to show that $q+nN_q(n)\le q^n$, and so
$$N_q(n)\le \tfrac{1}{n}(q^n-q)$$
20. Use Corollary 7.2.17 and the results of the previous exercise to show that
$$q^n=\sum_{d\mid n} dN_q(d)\le nN_q(n)+\sum_{k=0}^{\lfloor n/2\rfloor} q^k\le nN_q(n)+q^{1+n/2}$$
Hence, $N_q(n)\ge \frac{1}{n}(q^n-q^{1+n/2})$.
21. Use the results of the previous two exercises to show that $N_q(n)\approx q^n/n$.

If $f(x)$ is a polynomial of degree d, we define the **reciprocal polynomial** by $f_R(x)=x^d f(x^{-1})$. Thus, if
$$f(x)=a_nx^n+a_{n-1}x^{n-1}+\cdots+a_1x+a_0$$
then
$$f_R(x)=a_0x^n+a_1x^{n-1}+\cdots+a_{n-1}x+a_n$$

If a polynomial satisfies $f(x)=f_R(x)$, we say that $f(x)$ is **self-reciprocal.**

22. Show that $\alpha\ne 0$ is a root of $f(x)$ if and only if $\alpha^{-1}\ne 0$ is a root of $f_R(x)$.
23. Show that the reciprocal of an irreducible polynomial $f(x)\ne x$ is also irreducible.
24. Show that if a polynomial $f(x)$ is self-reciprocal and irreducible, then $\deg(f(x))$ must be even.
25. Suppose that $f(x)=p(x)q(x)$, where $p(x)$ and $q(x)$ are irreducible, and $f(x)$ is self–reciprocal. Show that either
(i) $p(x)=\delta p_R(x)$ and $q(x)=\delta q_R(x)$ with $\delta=\pm 1$, or
(ii) $p(x)=\alpha q_R(x)$ and $q(x)=\alpha^{-1}p_R(x)$ for some $\alpha\in F_q$.
What can you say about this if $\deg(p(x))$ is odd?

7.3 The Roots of Unity

In this section, we address the question of how to factor the polynomial $x^n - 1$ over a finite field F_q. This will be an important question when we discuss cyclic codes later in the chapter.

The first observation we should make is that if n and q are not relatively prime, then we can write $n = mp^k$, where $(m,q) = 1$ and p is the characteristic of F_q. In this case,

$$x^n - 1 = x^{mp^k} - 1 = (x^m - 1)^{p^k}$$

and so, from now on, *we will assume that* n *and* q *are relatively prime.*

ROOTS OF UNITY

Let us denote the splitting field of $x^n - 1$ over F_q by F_{q^s}. The polynomial $x^n - 1$ has no multiple roots in any extension of F_q, since $D[x^n - 1] = nx^{n-1}$ has no common factors with $x^n - 1$. Hence, $x^n - 1$ has n distinct roots in its splitting field F_{q^s}. (Note that this requires $(n,q) = 1$.)

Definition The roots of $x^n - 1$ in the splitting field F_{q^s} are called **n-th roots of unity over F_q**. The set of all n-th roots of unity will be denoted by $E^{(n)}$. □

We remark that the n-th roots of unity are defined even when n and q are not relatively prime. However, in that case, there are fewer than n distinct n-th roots of unity. When $(n,q) = 1$, the set $E^{(n)}$ has a particularly nice structure.

Theorem 7.3.1 The set $E^{(n)}$ of n-th roots of unity is a cyclic subgroup of size n of the multiplicative group $F^*_{q^s}$.

Proof. We have already seen that $|E^{(n)}| = n$. Suppose that $\alpha,\beta \in E^{(n)}$. Then $(\alpha\beta^{-1})^n = \alpha^n(\beta^n)^{-1} = 1$, and so $\alpha\beta^{-1} \in E^{(n)}$. Hence, $E^{(n)}$ is a subgroup of the cyclic group $F^*_{q^s}$ and so is also cyclic. ∎

Definition An n-th root of unity over F_q that generates the cyclic group $E^{(n)}$, that is, an n-th root of unity of order n, is called a **primitive n-th root of unity over F_q**. □

We will generally reserve the Greek letter ω for primitive roots of unity.

Because $E^{(n)}$ is cyclic, we have the following result.

Corollary 7.3.2 There are precisely $\phi(n)$ primitive n-th roots of unity over F_q. In particular, since $\phi(n) > 0$, there exists a primitive n-th root of unity over F_q for all positive integers n that are relatively prime to q. □

Corollary 7.3.2 allows us to find a simple formula for s (in the splitting field F_{q^s}) in terms of n. For if ω is a primitive n-th root of unity, then ω has order n, and since $\omega \neq 0$, we have

$$\omega \in F_{q^r} \Leftrightarrow \omega^{q^r} = \omega \Leftrightarrow \omega^{q^r-1} = 1 \Leftrightarrow n \mid q^r - 1$$

Since s is the smallest r for which $\omega \in F_{q^r}$, we have proved the following.

Corollary 7.3.3 If F_{q^s} is the splitting field for $x^n - 1$ over F_q, then s is the smallest positive integer for which $n \mid q^s - 1$, that is, s is the smallest positive integer for which $q^s \equiv 1 \bmod n$. Put another way, s is the *order* of q mod n, which we will denote by $o_n(q)$. □

Primitive Field Elements and Primitive Roots of Unity

It is important to keep in mind the distinction between a primitive field element of the splitting field F_{q^s} of $x^n - 1$, and a primitive n-th root of unity. By definition, β is a primitive element of F_{q^s} if it generates the cyclic group $F^*_{q^s}$. On the other hand, by definition, ω is a primitive n-th root of unity if it generates the cyclic subgroup $E^{(n)}$. In symbols,

$$\beta \text{ is a primitive element of } F_{q^s} \Leftrightarrow \{1,\beta,\beta^2,\ldots\} = F^*_{q^s}$$

and

$$\omega \text{ is a primitive n-th root of unity } \Leftrightarrow \{1,\omega,\omega^2,\ldots\} = E^{(n)}$$

Notice, however, that a primitive n-th root of unity ω generates F_{q^s} over F_q, as a *field* element (which allows for addition as well as multiplication), that is, $F_q(\omega) = F_{q^s}$. This follows from the fact that

$$F_q(\omega) = F_q(E^{(n)}) = F_{q^s}$$

If β is a primitive element of F_{q^s}, then β has order q^s-1. Since $n \mid q^s-1$, we may write $q^s - 1 = nr$, and so

$$o(\beta^k) = \frac{q^s-1}{(k,q^s-1)} = \frac{nr}{(k,nr)}$$

Hence, β^k is a primitive n-th root of unity if and only if

$$\frac{nr}{(k,nr)} = n$$

or, equivalently,

$$(k,nr) = r$$

But this holds if and only if $k = ru$, where u is relatively prime to n. Since there are precisely $\phi(n)$ primitive n-th roots of unity, we have proved the following result.

Theorem 7.3.4 Let β be a primitive element of the splitting field F_{q^s} of $x^n - 1$. Then the $\phi(n)$ primitive n-th roots of unity over F_q are precisely the elements

$$\{\beta^k \mid k = \frac{q^s-1}{n} u,\ u < n,\ (u,n) = 1\}$$

In particular, $\beta^{(q^s-1)/n}$ is a primitive n-th root of unity. □

We leave proof of the following as an exercise.

Theorem 7.3.5 Let F_q be a finite field. Let Ω be the set of primitive n-th roots of unity over F_q and let $\mathcal{F}$ be the set of primitive field elements of the splitting field F_{q^s} of $x^n - 1$. Then either $\Omega \cap \mathcal{F} = \emptyset$, or else $\Omega = \mathcal{F}$, the latter happening if and only if $n = q^s - 1$. □

A METHOD FOR FACTORING $X^n - 1$

For $(n,q) = 1$, we can factor the polynomial $x^n - 1$ over F_q, by using the simple fact that $x^n - 1$ has n distinct roots, and so it is just the product of the *distinct* minimal polynomials for these roots. These minimal polynomials can be computed using the methods of the previous section. The main drawback of this method is that we must work in the (larger) splitting field F_{q^s}.

We begin with a primitive element β of F_{q^s}, where $s = o_n(q)$. From β, using Theorem 7.3.4, we obtain a primitive n-th root of unity

$$\omega = \beta^{(q^s-1)/n}$$

Hence, the roots of $x^n - 1$ are given by

$$(7.3.1) \qquad 1, \omega, \omega^2, \ldots, \omega^{n-1}$$

All we need to do now is compute the minimal polynomials for these roots and take the product of the distinct polynomials so obtained.

For $i = 0, \ldots, n-1$, the conjugates of ω^i are

$$(7.3.2) \qquad \omega^i, \omega^{iq}, \omega^{iq^2}, \ldots, \omega^{iq^{d-1}}$$

where d is the smallest positive integer for which $\omega^{iq^d} = \omega^i$. But,

$$\omega^{iq^d} = \omega^i \Leftrightarrow \omega^{iq^d - i} = 1 \Leftrightarrow n \mid iq^d - i \Leftrightarrow iq^d \equiv i \bmod n$$

and so we may use the condition $iq^d \equiv i \bmod n$ in determining when we have all of the conjugates.

Thus, the minimal polynomial for the roots (7.3.2) is the product

$$m_i(x) = (x-\omega^i)(x-\omega^{iq})(x-\omega^{iq^2}) \cdots (x-\omega^{iq^{d-1}})$$

where d is the smallest positive integer for which $iq^d \equiv i \bmod n$.

The set of exponents in (7.3.2)

$$C_i = \{i, iq, \ldots, iq^{d-1}\}$$

where d is the smallest positive integer for which $iq^d \equiv i \bmod n$, is referred to as the **i-th cyclotomic coset** for q modulo n. These sets can be defined without specific reference to a primitive n-th root of unity.

Let us illustrate the factoring process with some examples.

Example 7.3.1 Consider the polynomial

$$h_{15}(x) = x^{15} - 1$$

over F_2. Here $n = 15$ and $q = 2$. Since $s = o_{15}(2) = 4$, the splitting field for $h_n(x)$ is $F_{q^s} = F_{16}$.

We saw in Example 7.1.2 that the polynomial $x^4 + x + 1$ is primitive for F_{16} over F_2, and so any root β of this polynomial is a primitive element of F_{16}, and gives a primitive 15th root of unity

$$\omega = \beta^{(q^s-1)/n} = \beta$$

(In this case, β is both a primitive field element and a primitive root of unity.) Hence, the roots of $x^{15} - 1$ are

(7.3.3) $$1, \beta, \beta^2, \ldots, \beta^{14}$$

The minimal polynomials for these roots were computed in Example 7.2.1. Referring to that example, we get the factorization

$$x^{15} - 1 = (x+1)(x^4+x+1)(x^4+x^3+x^2+x+1)(x^2+x+1)(x^4+x^3+1)$$ □

Example 7.3.2 Sometimes knowledge of the cyclotomic cosets can show us that a particular factorization of $x^n - 1$ is actually a factorization into irreducible polynomials. For instance, consider the polynomial $x^9 - 1$ over F_2. Since $q = 2$ and $n = 9$, the cyclotomic cosets in this case are

$$C_0 = \{0\}, \quad C_1 = \{1,2,4,8,7,5\}, \quad C_3 = \{3,6\}$$

Hence, $x^9 - 1$ factors into an irreducible linear factor, an irreducible quadratic factor, and an irreducible factor of degree 6. However, we easily see that, over F_2,

$$x^9 - 1 = (x^3)^3 - 1 = (x^3 - 1)(x^6 + x^3 + 1)$$
$$= (x - 1)(x^2 + x + 1)(x^6 + x^3 + 1)$$

and so this must be the desired factorization. □

THE ORDER OF AN IRREDUCIBLE POLYNOMIAL

Recall that the order $o(f)$ of an irreducible polynomial $f(x)$ is the multiplicative order of any of its roots, in the splitting field F_{q^d} of $f(x)$. We begin with some basic facts.

Theorem 7.3.6 Let $f(x)$ be an irreducible polynomial over F_q.
1) If $\deg(f(x)) = d$ then $o(f) \mid q^d - 1$.
2) $f(x) \mid x^{o(f)} - 1$.
3) $o(f) \mid n$ if and only if $f(x) \mid x^n - 1$.
4) $o(f)$ is the smallest positive integer e such that $f(x) \mid x^e - 1$.

Proof.
1) Since $Split(f) = F_{q^d}$, every root of $f(x)$ is a root of
$$x^{q^d-1} - 1$$
and so has order dividing $q^d - 1$. Hence, $o(f) \mid q^d - 1$.
2) Every root of $f(x)$ in $Split(f)$ has order $o(f)$, and so is a root of $x^{o(f)} - 1$. Hence, $f(x) \mid x^{o(f)} - 1$.
3) If $f(x) \mid x^n - 1$ then any root of $f(x)$ is a root of $x^n - 1$, and so has order a divisor of n. That is, $o(f) \mid n$. Conversely, if $n = k o(f)$, then
$$x^{o(f)} - 1 \mid x^{k o(f)} - 1 = x^n - 1$$
and the rest follows from part 2).
4) This follows immediately from part 3). ∎

We can determine the degree d of an irreducible polynomial of order e as follows. If α is a root of $f(x)$ in F_{q^d}, then according to Theorem 7.2.3,
$$F_q(\alpha) = F_{q^d}$$

But α is also a *primitive* e-th root of unity, and so

$$F_q(\alpha) = F_q(E^{(e)}) = Split(x^e - 1) = F_{q^s}$$

where $s = o_e(q)$ by Corollary 7.3.3. Hence, $d = s = o_e(q)$.

Theorem 7.3.7 Let f(x) be an irreducible polynomial over F_q of order $o(f)$. Then the degree of f(x) is the order of q mod $o(f)$, in symbols,

$$\deg(f(x)) = o_{o(f)}(q) \qquad \square$$

Corollary 7.3.8 If f(x) is a primitive polynomial for F_{q^n} over F_q, then $\deg(f(x)) = n$. $\square$

Notice that while, according to the previous theorem, the order of an irreducible polynomial uniquely determines its degree, the converse does not hold. As an example, the irreducible polynomials $q(x) = x^4 + x + 1$ and $p(x) + x^4 + x^3 + x^2 + x + 1$ both have degree 4, but $o(q(x)) = 15$ and $o(p(x)) = 5$. (See Examples 7.1.1 and 7.1.2.)

Theorem 7.3.7 can be used to determine when an irreducible polynomial f(x) over F_q is still irreducible, when thought of as a polynomial over an extension field F_{q^n}.

Theorem 7.3.9 Let f(x) be an irreducible polynomial of degree d over F_q. Then, over F_{q^n}, the polynomial f(x) factors into (n,d) irreducible polynomials, each of which has degree d/(n,d). In particular, f(x) is irreducible over F_{q^n} if and only if $(n,d) = 1$.
Proof. Suppose that g(x) is an irreducible factor of f(x) over F_{q^n}. Then $f(x) = q(x)g(x)$ over F_{q^n}. If $\deg(g(x)) = e$, then the splitting field for g(x) is $F_{q^{ne}}$. If $\alpha \in F_{q^{ne}}$ is a root of g(x), then it is also a root of f(x), which is an irreducible polynomial over F_q. Therefore, since F_{q^d} is the splitting field for f(x) over F_q, we deduce that $F_{q^d} = F_q(\alpha) < F_{q^{ne}}$.

At this point, we can follow two lines – one using brute force and the other using the concept of the order of an irreducible polynomial.

As to the brute force approach, since $F_{q^d} < F_{q^{ne}}$, we have $d \mid ne$. Furthermore, if $d \mid na$ for some a, then $F_{q^d} < F_{q^{na}}$ implies that f(x) splits over $F_{q^{na}}$. Thinking of f(x) as being in $F_{q^n}[x]$, where g(x) is a factor, we deduce that g(x) also splits over $F_{q^{na}}$, and so $F_{q^{ne}} < F_{q^{na}}$, from which we get $e \mid a$.

Thus, we have shown that $d \mid ne$ and that if $d \mid na$ then $e \mid a$. From this we can prove that $e = d/(n,d)$. Let $b = d/(n,d)$, and observe that $d \mid nb$. Hence $e \mid b$. On the other hand, since $d \mid ne$, we have $ne = kd$ for some k. Now, $(n,d) = sn+td$ for some s and t, and so

$$e(n,d) = esn+etd = skd+etd = d(sn+td)$$

which shows that $d \mid e(n,d)$ or, equivalently, that $b \mid e$. Hence, $e = b$, which is what we wanted to prove.

Now let us proceed more elegantly. We know that $F_{q^d} = F_q(\alpha) < F_{q^{ne}}$. Suppose that $o(f(x)) = u$. Then, since α is a root

of f(x), it has multiplicative order u, as an element of F_{q^d} and also as an element of $F_{q^{ne}}$. Thus, α is a root of the *irreducible* polynomial f(x) over F_q and a root of the *irreducible* polynomial g(x) over F_{q^n}, and in either case, it has the same order u. Hence, according to Theorem 7.3.7,

$$d = o_u(q) \quad \text{and} \quad e = o_u(q^n)$$

Now the set $G = \{q^i \mid i \in \mathbb{N}\}$ is a cyclic group under multiplication modulo u, with order $o_u(q) = d$. As a cyclic group, we know that $e = o_u(q^n) = d/(n,d)$. ∎

COMPUTING THE ORDER OF AN IRREDUCIBLE POLYNOMIAL

Theorem 7.3.6 can be used to determine the order of an irreducible polynomial f(x) of degree d as follows. First, we factor $q^d - 1$ into its prime factorization

$$q^d - 1 = \prod_i p_i^{r_i}$$

Now, since $o(f) \mid q^d - 1$,

$$\begin{aligned} p_i^{r_i} \mid o(f) &\Leftrightarrow o(f) \nmid \frac{q^d-1}{p_i} \\ &\Leftrightarrow f(x) \nmid x^{(q^d-1)/p_i} - 1 \\ &\Leftrightarrow x^{(q^d-1)/p_i} \not\equiv 1 \bmod f(x) \end{aligned}$$

Hence, by calculating the residues of $x^{(q^d-1)/p_i} \bmod f(x)$, we can determine the highest power of each prime p_i that divides $o(f)$. Let us illustrate with an example.

Example 7.3.3 Consider the irreducible polynomial $f(x) = x^6 + x + 1$ over F_2. Since $q = 2$, we have

$$q^6 - 1 = 63 = 3^2 \cdot 7$$

We first check to see whether or not 3^2 divides $o(f)$. Since $63/3 = 21$, we have

$$3^2 \mid o(f) \Leftrightarrow x^{21} \not\equiv 1 \bmod(x^6+x+1)$$

A little bit of division gives

$$x^{21} \bmod(x^6+x+1) = x^2 + x \not\equiv 1$$

and so $3^2 \mid o(f)$. Next, we check the prime factor 7. Since $63/7 = 9$,

we have

$$7 \mid o(f) \iff x^9 \not\equiv 1 \bmod(x^6+x+1)$$

Another division shows that $x^9 \bmod(x^6+x+1) = x^4+x^3 \not\equiv 1$, and so $7 \mid o(f)$. Hence, $o(f) = 63$. This shows, incidentally, that $f(x)$ is primitive over F_2.

As another example, consider the irreducible polynomial $g(x) = x^6+x^4+x^2+x+1$. In this case, we have

$$3^2 \mid o(g) \iff x^{21} \not\equiv 1 \bmod(x^6+x^4+x^2+x+1)$$

But $x^{21} \bmod(x^6+x^4+x^2+x+1) = 1$ and so $3^2 \nmid o(f)$. Hence, we must check the divisor 3

$$3 \mid o(g) \iff x^7 \not\equiv 1 \bmod(x^6+x^4+x^2+x+1)$$

But $x^7 \bmod(x^6+x^4+x^2+x+1) = x^5+x^3+x^2+x \neq 1$, and so $3 \mid o(g)$. Finally, we check the factor 7

$$7 \mid o(g) \iff x^9 \not\equiv 1 \bmod(x^6+x^4+x^2+x+1)$$

Since $x^9 \bmod(x^6+x^4+x^2+x+1) = x^4+x^2+1 \neq 1$, we deduce that $7 \mid o(g)$, and so $o(g) = 21$. □

THE CYCLOTOMIC POLYNOMIALS

As mentioned before, the factorization method used in Example 7.3.1 requires that we work in the splitting field F_{q^s}. Let us see how we might obtain a partial factorization of x^n-1, without requiring knowledge of this splitting field.

First, we observe that in the factorization of x^n-1,

$$x^n - 1 = \prod_i m_i(x) \tag{7.3.4}$$

over F_q, the minimal polynomials $m_i(x)$ need not have distinct orders. For instance, referring to Example 7.2.1, the minimal polynomials

$$m_1(x) = x^4+x+1 \quad \text{and} \quad m_7(x) = x^4+x^3+1$$

both have order 15, since $o(\beta) = o(\beta^7) = 15$.

Next, we observe that, if $k \mid n$, then each primitive k-th root of unity is also an n-th root of unity and hence, is a root of one of the minimal polynomials $m_i(x)$ of order k. Conversely, any root of a minimal polynomial of order k in (7.3.4) is a primitive k-th root of unity. In other words, the roots of the minimal polynomials in (7.3.4) of order k are *precisely* the primitive k-th roots of unity. This leads us to make the following definition.

Definition If F_q is a finite field and k is a positive integer that is relatively prime to q, then the **k-th cyclotomic polynomial over F_q** is the monic polynomial $Q_k(x)$ whose roots are precisely the *primitive* k-th roots of unity over F_q, that is, the elements of F_q of order k. □

Since the roots of $Q_k(x)$ lie in the splitting field of the polynomial $x^k - 1$, the definition tells us only that $Q_k(x)$ is a polynomial over that splitting field. However, since the primitive k-th roots of unity are precisely the roots of the minimal polynomials of order k in (7.3.4), we deduce that $Q_k(x)$ is the product of these minimal polynomials and, therefore, must lie in $F_q[x]$.

In terms of cyclotomic polynomials, we can factor $x^n - 1$ over F_q as follows

(7.3.5) $$x^n - 1 = \prod_{k \mid n} Q_k(x)$$

where the roots of $Q_k(x)$ are precisely the n-th roots of unity that have order k.

Notice also that, since there are $\phi(k)$ primitive k-th roots of unity, we have $\deg(Q_k(x)) = \phi(k)$.

Theorem 7.3.10 Let F_q be a finite field, and let $(n,q) = 1$. Then the n-th cyclotomic polynomial over F_q has coefficients in F_q and has degree $\phi(n)$. □

According to Theorem 7.3.7, the degree of an irreducible polynomial of order e is $o_e(q)$. Hence, each minimal polynomial that is a factor of $Q_k(x)$ has degree $o_e(q)$, and so the *number* of monic irreducible polynomials of order e is $\phi(e)/o_e(q)$. We have proved the following.

Theorem 7.3.11 The *number* of monic irreducible polynomials over F_q of order e is $\phi(e)/o_e(q)$, where $o_e(q)$ is the order of q mod e. □

Equation (7.3.5) is a prime candidate for Möbius inversion. Applying the multiplicative version gives

(7.3.6) $$Q_n(x) = \prod_{d \mid n} (x^d - 1)^{\mu(n/d)} = \prod_{d \mid n} (x^{n/d} - 1)^{\mu(d)}$$

Of course, some of the exponents $\mu(d)$ may be equal to -1, and so a little additional algebraic manipulation may be required to obtain $Q_n(x)$ as a product of polynomials.

Example 7.3.4 Let us look again at Example 7.3.1. Equation (7.3.5) gives

$$x^{15} - 1 = Q_1(x)Q_3(x)Q_5(x)Q_{15}(x)$$

We can compute the cyclotomic polynomials using (7.3.6)

$$Q_1(x) = x - 1 = x + 1$$

$$Q_3(x) = (x^3 - 1)(x - 1)^{-1} = \frac{x^3 - 1}{x - 1} = x^2 + x + 1$$

$$Q_5(x) = (x^5 - 1)(x - 1)^{-1} = \frac{x^5 - 1}{x - 1} = x^4 + x^3 + x^2 + x + 1$$

$$Q_{15}(x) = (x^{15} - 1)(x^5 - 1)^{-1}(x^3 - 1)^{-1}(x - 1)$$

$$= \frac{(x^{15} - 1)(x - 1)}{(x^5 - 1)(x^3 - 1)} = x^8 + x^7 + x^5 + x^4 + x^3 + x + 1$$

and so we get the following factorization of $x^{15} - 1$ into cyclotomic polynomials

$$x^{15}-1 = (x+1)(x^2+x+1)(x^4+x^3+x^2+x+1)(x^8+x^7+x^5+x^4+x^3+x+1)$$

Note also that, since $o_{15}(2) = 4$, the *number* of monic irreducible factors of $Q_{15}(x)$ is $\phi(15)/4 = 2$. □

We conclude this section by mentioning that, according to the definition, if $e = q^n - 1$, then the roots of the e-th cyclotomic polynomial $Q_e(x)$ over F_q are the primitive e-th roots of unity over F_q. Hence, they are the primitive elements of F_{q^n}. In other words, the monic irreducible factors of $Q_e(x)$ are precisely the primitive polynomials of F_{q^n} over F_q. Thus, one way to find primitive polynomials is to factor this cyclotomic polynomial.

In the appendix (Section A.4), we discuss an algorithm for factoring polynomials over finite fields. Then we find all primitive polynomials of degree 4 over F_2 by factoring the cyclotomic polynomial $Q_{15}(x)$.

EXERCISES

All cyclotomic polynomials are assumed to be over fields for which they are defined.

1. Prove that, for an irreducible polynomial $f(x)$, we have $f(x) \mid x^k - 1$ if and only if $o(f) \mid k$.
2. Evaluate the sum $1 + \alpha + \alpha^2 + \cdots + \alpha^{n-1}$, where α is an n-th root of unity.

3. Find $Q_2(x)$, $Q_4(x)$, $Q_8(x)$, and $Q_{12}(x)$.
4. Is a primitive n-th root of unity also a primitive element in some field?
5. If $(n,q) \neq 1$, how many n-th roots of unity are there over F_q?
6. Prove Corollary 7.3.8.
7. Factor $x^5 - 1$ over
   (a) F_2 (b) F_3
8. Factor $x^7 - 1$ over
   (a) F_2 (b) F_3 (c) F_5
9. Factor $x^8 - 1$ over
   (a) F_2 (b) F_3 (c) F_4 (d) F_5
10. Factor $x^{10} - 1$ over
    (a) F_2 (b) F_3
11. Factor $x^{13} - 1$ over
    (a) F_2 (b) F_3
12. What is the splitting field for $x^4 - 1$ over F_3? Find the primitive 4th roots of unity in this splitting field. Do the same for the 8th roots of unity over F_3.

In Exercises 13–20, compute the order of each of the following irreducible polynomials. Which are primitive?

13. $x^4 + x^3 + x^2 + x + 1$ over F_2.
14. $x^4 + x + 1$ over F_2.
15. $x^8 + x^4 + x^3 + x^2 + 1$ over F_2.
16. $x^8 + x^5 + x^4 + x^3 + 1$ over F_2.
17. $x^8 + x^7 + x^5 + x + 1$ over F_2.
18. $x^4 + x + 2$ over F_3.
19. $x^4 + x^3 + x^2 + 1$ over F_3.
20. $x^5 - x + 1$ over F_3.
21. Suppose that $\alpha_1, \ldots, \alpha_n$ are the n-th roots of unity over F_q. Show that for $1 < k < n$ we have $\alpha_1^k + \alpha_2^k + \cdots + \alpha_n^k = 0$.
22. Show that $Q_n(x) \in F_q[x]$ is irreducible if and only if $o_n(q) = \phi(n)$.
23. If n and q are relatively prime, prove that the polynomial $x^{n-1} + x^{n-2} + \cdots + x + 1$ is irreducible over F_q if and only if n is prime and $Q_n(x)$ is irreducible.
24. Show that $d \mid n$ implies that $Q_n(x)$ divides $(x^n - 1)(x^d - 1)$.
25. Show that if r is a prime, then $Q_{r^n}(x) = (x^{r^n} - 1)/(x^{r^{n-1}} - 1)$.
26. Evaluate $Q_n(1)$.
27. Evaluate $Q_n(-1)$.

Verify the following properties of the cyclotomic polynomials. As usual, p is a prime number.

28. $Q_{np}(x) = Q_n(x^p)/Q_n(x)$ for $p \nmid n$.

29. $Q_{np}(x) = Q_n(x^p)$ for all $p \mid n$.

30. $Q_{np^k}(x) = Q_{np}(x^{p^{k-1}})$

31. $Q_n(0) = 1$ for $n \geq 2$.

32. $Q_n(x^{-1})x^{\phi(n)} = Q_n(x)$ for $n \geq 2$.

7.4 Cyclic Codes

Cyclic codes are important for many reasons. Besides being of great theoretical interest by virtue of possessing a very rich mathematical structure, they are also important from a practical standpoint, since their encoding and decoding can be implemented quite efficiently using linear switching circuits.

We assume throughout our discussion of cyclic codes that n *and* q *are relatively prime. In particular, if* $q = 2$ *then* n *must be odd.*

Let us review the definition of a cyclic code. If L is a q-ary linear code, we may associate to each codeword $\mathbf{c} = c_0c_1\cdots c_{n-1}$ in L a polynomial in $F_q[x]$ as follows

(7.4.1) $$\phi: c_0c_1\cdots c_{n-1} \to c_0 + c_1x + \cdots + c_{n-1}x^{n-1}$$

The map ϕ is a vector space isomorphism from L onto the subspace $\phi(L)$ of $F_q[x]$. As is customary, we ignore this map and think of the codewords in L as polynomials or vice-versa.

Definition A linear code $L \subset V(n,q)$ is **cyclic** if

$$c_0c_1\cdots c_{n-1} \in L \text{ implies that } c_{n-1}c_0c_1\cdots c_{n-2} \in L$$

When codewords are thought of as polynomials using (7.4.1), a code L is cyclic if it is an ideal of

$$R_n = \frac{F_q[x]}{\langle x^n - 1\rangle}$$ □

Recall that R_n is the set of all polynomials over F_q of degree less than n. Addition in R_n is the usual addition of polynomials and multiplication is ordinary multiplication of polynomials, followed by reduction using the fact that $x^n = 1$ in R_n. Since all polynomial congruences will be modulo $x^n - 1$, we will use the notation $f(x) \equiv g(x)$ to denote congruence modulo $x^n - 1$.

Note that, according to the definition, a linear code C is cyclic if it is closed under the *cyclic shift*

$$c_0c_1\cdots c_{n-1} \to c_{n-1}c_0c_1\cdots c_{n-2}$$

If this is the case, then C is also closed under all cyclic shifts

$$c_0c_1\cdots c_{n-1} \to c_k\cdots c_{n-1}c_0c_1\cdots c_{k-1}$$

If $p(x) \in R_n$, then the **ideal generated by** $p(x)$, denoted by $\langle p(x)\rangle$, is the smallest ideal in R_n containing $p(x)$. We leave it as an exercise to show that

$$\langle p(x)\rangle = \{f(x)p(x) \mid f(x) \in R_n\}$$

where all polynomials are taken modulo $x^n - 1$.

THE GENERATOR POLYNOMIAL OF A CYCLIC CODE

The following theorem contains some basic facts about cyclic codes. It says, among other things, that the ring R_n is a principal ideal domain.

Theorem 7.4.1 Let C be an ideal in R_n, that is, a cyclic code of length n.

1) There is a unique monic polynomial $g(x)$ of minimum degree in C. This polynomial generates C, that is, $C = \langle g(x) \rangle$, and it is called the **generator polynomial** for C. (We remark that the generator polynomial is usually not the only polynomial that generates C.)
2) The generator polynomial $g(x)$ divides $x^n - 1$.
3) If $\deg(g(x)) = r$, then C has dimension $n - r$. In fact,
$$C = \langle g(x) \rangle = \{r(x)g(x) \mid \deg(r(x)) < n-r\}$$
4) If $g(x) = g_0 + g_1x + \cdots + g_rx^r$, then $g_0 \neq 0$ and C has generator matrix
$$G = \begin{bmatrix} g_0 & g_1 & g_2 & \cdots & & g_r & 0 & 0 & \cdots & 0 \\ 0 & g_0 & g_1 & g_2 & \cdots & & g_r & 0 & \cdots & 0 \\ 0 & 0 & g_0 & g_1 & g_2 & \cdots & & g_r & & \vdots \\ \vdots & \vdots & & \ddots & \ddots & \ddots & & & \ddots & 0 \\ 0 & 0 & \cdots & 0 & g_0 & g_1 & g_2 & \cdots & & g_r \end{bmatrix}$$
where each row of G is a cyclic shift of the previous row.

Proof.

1) Suppose that C contains two distinct monic polynomials $g_1(x)$ and $g_2(x)$ of minimum degree r. Then their difference $g_1(x) - g_2(x)$ would be a nonzero polynomial in C of degree less than r, which is not possible. Hence, there is a unique monic polynomial $g(x)$ of degree r in C.

   Since $g(x) \in C$ and C is an ideal, we have $\langle g(x) \rangle \subset C$. On the other hand, suppose that $p(x) \in C$, and let
$$p(x) = q(x)g(x) + r(x)$$
where $\deg(r(x)) < r$. Then $r(x) = p(x) - q(x)g(x) \in C$ has degree less than r, which is possible only if $r(x) = 0$. Hence, $p(x) = q(x)g(x) \in \langle g(x) \rangle$, and so $C \subset \langle g(x) \rangle$. Thus, $C = \langle g(x) \rangle$.
2) Dividing $x^n - 1$ by $g(x)$ gives
$$x^n - 1 = q(x)g(x) + r(x)$$
where $\deg(r(x)) < r$. Since in R_n, $x^n - 1 = 0 \in C$, we see that $r(x) \in C$, and so $r(x) = 0$, which shows that $g(x) \mid x^n - 1$.

3) The ideal generated by $g(x)$ is

$$\langle g(x)\rangle = \{f(x)g(x) \mid f(x) \in R_n\}$$

with the usual reduction modulo $x^n - 1$, and we must show that it is sufficient to restrict $f(x)$ to polynomials of degree less than $n-r$. We have seen that $g(x) \mid x^n - 1$, and so $x^n - 1 = h(x)g(x)$, for some polynomial $h(x)$ of degree $n-r$. Let us divide $f(x)$ by $h(x)$

$$f(x) = q(x)h(x) + r(x)$$

where $\deg(r(x)) < n-r$. Then

$$f(x)g(x) = q(x)h(x)g(x) + r(x)g(x) = q(x)(x^n - 1) + r(x)g(x)$$

and so $f(x)g(x) = r(x)g(x)$ in R_n, which is what we wanted to show. This also shows that the set

$$\{g(x), xg(x), \ldots, x^{n-r-1}g(x)\}$$

spans C, and since it is linearly independent, it forms a basis for C. Hence $dim(C) = n-r$.

4) If $g_0 = 0$, then $g(x) = xg_1(x)$, where $\deg(g_1(x)) < r$. But then we would have

$$g_1(x) = 1 \cdot g_1(x) \equiv x^n g_1(x) = x^{n-1}g(x)$$

Hence $g_1(x) \in C$, which contradicts the fact that no nonzero polynomial in C has degree less than r. Thus, $g_0 \neq 0$. Finally, G is a generating matrix of C since $\{g(x), xg(x), \ldots, x^{n-r-1}g(x)\}$ is a basis for C. ∎

It is important to note that a cyclic code C may be generated by polynomials other than the generating polynomial. The next example illustrates this point.

Example 7.4.1 Consider the cyclic code $C = \langle 1+x \rangle$ in

$$R_3 = \frac{F_2[x]}{\langle x^3 - 1\rangle}$$

The previous theorem tells us that $dim(C) = 3 - 1 = 2$ and that C contains the codewords

$$0, \quad 1+x, \quad x(1+x) = x+x^2, \quad (1+x)(1+x) = 1+x^2$$

Thus,

$$C = \{0,\ 1+x,\ 1+x^2,\ x+x^2\} = \{000, 110, 101, 011\}$$

We leave it to the reader to verify that

$$\langle 1+x^2\rangle = \{f(x)(1+x^2) \mid f(x) \in R_3\} = C$$

and so C is generated by the polynomial $1+x^2$ as well. □

In the previous example, we saw that $C = \langle 1+x \rangle = \langle 1+x^2 \rangle$. It will be convenient to adopt the notation $C = \langle\!\langle p(x) \rangle\!\rangle$ to denote the fact that C is the ideal generated by $p(x)$ *and* that $p(x)$ is the generator polynomial for C.

Let $p(x)$ be a monic polynomial in R_n, and let $C = \langle p(x) \rangle$ be the cyclic code generated by that polynomial. If $p(x)$ does not divide $x^n - 1$, then according to the previous theorem, it cannot be the generator polynomial for C. On the other hand, the next theorem shows that if $p(x)$ does divide $x^n - 1$, then it is the generator polynomial for C.

Theorem 7.4.2 A monic polynomial $p(x)$ in R_n is the generator polynomial for a cyclic code if and only if $p(x) \mid x^n - 1$.
Proof. We have already established one implication. As to the converse, suppose that $p(x) \mid x^n - 1$, and let $g(x)$ be the generator polynomial for $C = \langle p(x) \rangle$. Assume that $p(x) \neq g(x)$. Since $p(x)$ and $g(x)$ are both monic, we must have $\deg(p(x)) > \deg(g(x))$.

By assumption,

$$x^n - 1 = p(x)f(x) \tag{7.4.2}$$

for some polynomial $f(x)$. Furthermore, since $g(x) \in \langle p(x) \rangle$, we have

$$g(x) \equiv a(x)p(x)$$

for some $a(x) \in R_n$. Multiplying both sides of this by $f(x)$ and using (7.4.2) gives

$$g(x)f(x) \equiv a(x)p(x)f(x) \equiv a(x) \cdot (x^n - 1) \equiv 0$$

But $\deg(g(x)f(x)) < \deg(p(x)f(x)) = n$, and so $g(x)f(x) = 0$, which is not possible. Hence $p(x) = g(x)$. ∎

Theorems 7.4.1 and 7.4.2 tell us that the map $\phi: g(x) \to \langle\!\langle g(x) \rangle\!\rangle$, which sends each monic divisor $g(x)$ of $x^n - 1$ to the cyclic code $\langle\!\langle g(x) \rangle\!\rangle$, is a one-to-one correspondence between the set $\mathfrak{D}_n$ of all monic divisors of $x^n - 1$ and the set $\mathfrak{I}_n$ of all cyclic codes in R_n. This explains why is it important to be able to factor the polynomial $x^n - 1$ over a finite field.

Example 7.4.2 In Example 7.3.2, we saw that $x^9 - 1$ factors over F_2 into irreducible factors as follows

$$x^9 - 1 = (x-1)(x^2+x+1)(x^6+x^3+1)$$

Hence, there are $2^3 = 8$ cyclic codes in R_9. For instance, the cyclic

code $C_1 = \langle\langle x^6 + x^3 + 1\rangle\rangle$ has dimension $9 - 6 = 3$ and generator matrix

$$G_1 = \begin{bmatrix} 1 & 0 & 0 & 1 & 0 & 0 & 1 & 0 & 0 \\ 0 & 1 & 0 & 0 & 1 & 0 & 0 & 1 & 0 \\ 0 & 0 & 1 & 0 & 0 & 1 & 0 & 0 & 1 \end{bmatrix}$$ □

Example 7.4.3 It is not hard to show that $x^{23} - 1$ factors over F_2 as follows

$$x^{23} - 1$$
$$= (x+1)(x^{11}+x^9+x^7+x^6+x^5+x+1)(x^{11}+x^{10}+x^6+x^5+x^4+x^2+1)$$

It is also possible to show that the binary cyclic code

$$C_1 = \langle\langle x^{11} + x^{10} + x^6 + x^5 + x^4 + x^2 + 1\rangle\rangle$$

has the same parameters as the Golay code $\mathcal{G}_{23}$. The details are straightforward, and we refer the interested reader to Hill (1986). Hence, as a result of the uniqueness theorem for Golay codes, $\mathcal{G}_{23}$ is equivalent to the cyclic code C_1.

In a similar manner, we have over F_3,

$$x^{11} - 1 = (x - 1)(x^5 + x^4 - x^3 + x^2 - 1)(x^5 - x^3 + x^2 - x - 1)$$

and the ternary cyclic code

$$C_2 = \langle\langle x^5 + x^4 - x^3 + x^2 - 1\rangle\rangle$$

has the same parameters as the ternary Golay code $\mathcal{G}_{11}$, which is thus equivalent to the cyclic code C_2. □

The map $\phi: g(x) \to \langle\langle g(x)\rangle\rangle$ has some additional properties related to a natural partial ordering on $\mathfrak{D}_n$ and $\mathfrak{I}_n$. Note that, if C_1 and C_2 are cyclic codes in R_n, then the *sum*

$$C_1 + C_2 = \{c_1 + c_2 \mid c_1 \in C_1, c_2 \in C_2\}$$

is the smallest cyclic code in R_n containing C_1 and C_2. Readers who are familiar with the terminology of lattices will recognize the next theorem as implying that ϕ is an *order anti-isomorphism* from the lattice $(\mathfrak{D}_n, \mid)$ to the lattice $(\mathfrak{I}_n, \subset)$. We leave the proof as an exercise.

Theorem 7.4.3 Let $C_1 = \langle\langle g_1(x)\rangle\rangle$ and $C_2 = \langle\langle g_2(x)\rangle\rangle$ be cyclic codes in R_n. Then

1) $C_1 \subset C_2$ if and only if $g_2(x) \mid g_1(x)$,
2) $C_1 \cap C_2 = \langle\langle \operatorname{lcm}(g_1(x), g_2(x))\rangle\rangle$,
3) $C_1 + C_2 = \langle\langle \gcd(g_1(x), g_2(x))\rangle\rangle$. □

THE CHECK POLYNOMIAL OF A CYCLIC CODE

Since the generator polynomial $g(x)$ of a cyclic $[n,n-r]$-code in R_n divides $x^n - 1$, we have

$$x^n - 1 = g(x)h(x)$$

where $h(x)$ is a polynomial of degree $n-r$, called the **check polynomial** of C. This terminology is explained by the next theorem.

Theorem 7.4.4 Let $h(x)$ be the check polynomial for a cyclic code C in R_n.

1) The code C can be described by

$$C = \{p(x) \in R_n \mid p(x)h(x) \equiv 0\}$$

2) If $h(x) = h_0 + h_1 x + \cdots + h_{n-r}x^{n-r}$, then a parity check matrix for C is given by

$$H = \begin{bmatrix} h_{n-r} & \cdots & & h_0 & 0 & 0 & \cdots & 0 \\ 0 & h_{n-r} & \cdots & & h_0 & 0 & \cdots & 0 \\ 0 & 0 & h_{n-r} & \cdots & & h_0 & & \vdots \\ \vdots & \vdots & \ddots & \ddots & \cdots & & \ddots & 0 \\ 0 & 0 & \cdots & 0 & h_{n-r} & \cdots & & h_0 \end{bmatrix}$$

3) The dual code $C^{\perp}$ is the cyclic code of dimension r with generator polynomial

$$h^{\perp}(x) = h_0^{-1}x^{n-r}h(x^{-1}) = h_0^{-1}(h_0x^{n-r}+h_1x^{n-r-1}+\cdots+h_{n-r})$$

where the last polynomial in parentheses is the *reverse polynomial* of the check polynomial $h(x)$. (Note that $C^{\perp}$ is *not* generated by $h(x)$.)

Proof.

1) Let $g(x)$ be the generator polynomial of C. If $p(x) \in C$, then $p(x) = f(x)g(x)$ for some polynomial $f(x) \in R_n$. Hence,

$$p(x)h(x) = f(x)g(x)h(x) = f(x)(x^n - 1) \equiv 0$$

On the other hand, if $p(x) \in R_n$ and $p(x)h(x) \equiv 0$, then we write

$$p(x) = q(x)g(x) + r(x)$$

where $\deg(r(x)) < r$. Multiplying by $h(x)$ gives

$$p(x)h(x) = q(x)g(x)h(x) + r(x)h(x)$$

which tells us that $r(x)h(x) \equiv 0$. However, $\deg(r(x)h(x)) < r + (n-r) = n$, and so we deduce that $r(x)h(x) = 0$. Hence $r(x) = 0$, and $p(x) = q(x)g(x) \in C$.

2) If $c(x) \in C$, then $c(x)h(x) \equiv 0$. Now, $\deg(c(x)h(x)) < 2n - r$, and

from this we deduce that the coefficients of $x^{n-r}, x^{n-r+1}, \ldots, x^{n-1}$ in the product $c(x)h(x)$ must be 0, that is,

$$\begin{aligned} c_0 h_{n-r} + c_1 h_{n-r-1} + \cdots + c_{n-r} h_0 &= 0 \\ c_1 h_{n-r} + c_2 h_{n-r-1} + \cdots + c_{n-r+1} h_0 &= 0 \\ &\vdots \\ c_{r-1} h_{n-r} + c_r h_{n-r-1} + \cdots + c_{n-1} h_0 &= 0 \end{aligned}$$

But this is equivalent to $(c_0 c_1 \ldots c_{n-1})H^T = 0$, and so H generates a code C' that is orthogonal to C, that is, $C' \subset C^{\perp}$. However, since $h_{n-r} \neq 0$, it follows that $dim(C') = r$, and so $C' = C^{\perp}$.

3) If we show that $h^{\perp}(x)$ divides $x^n - 1$, then we will know that it is the generator polynomial for a cyclic code $\langle h^{\perp}(x) \rangle$ that has generator matrix H, and so $\langle h^{\perp}(x) \rangle = C^{\perp}$. But

$$h(x)g(x) = x^n - 1$$

implies

$$h(x^{-1})g(x^{-1}) = x^{-n} - 1$$

or

$$x^{n-r}h(x^{-1})x^r g(x^{-1}) = 1 - x^n$$

which shows that $h^{\perp}(x) \mid x^n - 1$. ∎

Example 7.4.4 Referring to Example 7.4.2, the code $C_1 = \langle\langle x^6 + x^3 + 1 \rangle\rangle$ has check polynomial

$$h(x) = (x - 1)(x^2 + x + 1) = x^3 - 1$$

and since $h^{\perp}(x) = x^3(x^{-3} - 1) = x^3 + 1$, the code C_1 has parity check matrix

$$H = \begin{bmatrix} 1 & 0 & 0 & 1 & 0 & 0 & 0 & 0 & 0 \\ 0 & 1 & 0 & 0 & 1 & 0 & 0 & 0 & 0 \\ 0 & 0 & 1 & 0 & 0 & 1 & 0 & 0 & 0 \\ 0 & 0 & 0 & 1 & 0 & 0 & 1 & 0 & 0 \\ 0 & 0 & 0 & 0 & 1 & 0 & 0 & 1 & 0 \\ 0 & 0 & 0 & 0 & 0 & 1 & 0 & 0 & 1 \end{bmatrix}$$ □

The following theorem provides an example of what we can learn about a cyclic code from its generator polynomial.

Theorem 7.4.5 Let E_n be the binary even weight code of length n, that is, the code consisting of all even weight words in V(n,2). Let C be a binary cyclic code of length n. Then

1) $E_n = \langle\langle x - 1 \rangle\rangle$,
2) $C = \langle\langle g(x) \rangle\rangle \subset E_n$ if and only if $x - 1 \mid g(x)$.

Proof. To prove part 1), we observe that, for the cyclic code $\langle\langle x-1\rangle\rangle$,

$$h(x) = \frac{x^n - 1}{x - 1} = x^{n-1} + x^{n-2} + \cdots + 1 = h^{\perp}(x)$$

and so

$$\langle\langle x-1\rangle\rangle^{\perp} = \langle\langle x^{n-1} + x^{n-2} + \cdots + 1\rangle\rangle = \{0,1\}$$

Hence, $\langle\langle x-1\rangle\rangle = \langle\langle x-1\rangle\rangle^{\perp\perp} = \{0,1\}^{\perp} = E_n$. Part 2) follows from part 1) and Theorem 7.4.3. ∎

Example 7.4.5 The code $C_1 = \langle\langle x^6 + x^3 + 1\rangle\rangle$ of Example 7.4.2 has codewords of odd weight, since $x - 1 \nmid x^6 + x^3 + 1$. However, the code $C_2 = \langle\langle (x-1)(x^6 + x^3 + 1)\rangle\rangle$ is a subcode of C_1 and has only even weight codewords. □

THE ZEROS OF A CYCLIC CODE

If we have convenient access to the roots of the polynomial $x^n - 1$ (that is, to the n-th roots of unity), then it is possible to characterize the cyclic codes in R_n in a slightly different way than through generator polynomials. Let

$$x^n - 1 = \prod_i m_i(x)$$

be the factorization of $x^n - 1$ into monic irreducible factors over F_q. If α is a root of $m_i(x)$ in some extension field of F_q, then $m_i(x)$ is the minimal polynomial of α over F_q. Hence, for any polynomial $f(x) \in F_q[x]$, we have $f(\alpha) = 0$ if and only if $f(x) = a(x)m_i(x)$ for some polynomial $a(x)$. In particular, if $f(x) \in R_n$, then $f(\alpha) = 0$ if and only if $f(x) \in \langle\langle m_i(x)\rangle\rangle$. Let us generalize.

Theorem 7.4.6 Let $g(x) = q_1(x)\cdots q_t(x)$ be a product of irreducible factors of $x^n - 1$, and let $\{\alpha_1, \ldots, \alpha_u\}$ be the roots of $g(x)$ in the splitting field of $x^n - 1$ over F_q. Then

$$\langle\langle g(x)\rangle\rangle = \{f(x) \in R_n \mid f(\alpha_1) = 0, \ldots, f(\alpha_u) = 0\}$$

Furthermore, it is sufficient to take a *single* root of each irreducible factor of $g(x)$. That is, if β_i is a root of $q_i(x)$ for $i = 1, \ldots, t$, then

$$\langle\langle g(x)\rangle\rangle = \{f(x) \in R_n \mid f(\beta_1) = 0, \ldots, f(\beta_t) = 0\}$$ □

Definition The roots of the generator polynomial of a cyclic code are called the **zeros** of the code. All other *roots of unity* are called **nonzeros** of the code. □

Notice that if $\{\alpha_1,\ldots,\alpha_u\}$ is *any* set of roots of $x^n - 1$, then the generator polynomial of the code

$$\{f(x) \in R_n \mid f(\alpha_1) = 0,\ldots,f(\alpha_u) = 0\}$$

is the least common multiple of the minimal polynomials for the roots $\alpha_1,\ldots,\alpha_u$.

The representation of a cyclic code through its zeros can be used to obtain a parity check matrix for the code as well. Let $\{\alpha_1,\ldots,\alpha_u\}$ be any set of roots of $x^n - 1$, and suppose they lie in the extension field F_{q^d}. If $f(x) = \sum f_j x^j$ is a polynomial in R_n then $f(\alpha_i) = 0$ if and only if

(7.4.3)
$$\sum_j f_j \alpha_i^j = 0$$

Since F_{q^d} can also be thought of as the vector space F_q^d of dimension d over F_q, we may write each of the powers α_i^j as a column vector $[\alpha_i^j]$ of length d over F_q. Furthermore, since $f_j \in F_q$, we have

$$[f_j \alpha_i^j] = f_j [\alpha_i^j]$$

and so (7.4.3) is equivalent to

$$\sum_j f_j [\alpha_i^j] = \left[\sum_j f_j \alpha_i^j\right] = \mathbf{0}$$

Thus, if we define the $ud \times n$ matrix H by

$$H = \begin{bmatrix} [\alpha_1^0] & [\alpha_1^1] & \cdots & [\alpha_1^{n-1}] \\ [\alpha_2^0] & [\alpha_2^1] & \cdots & [\alpha_2^{n-1}] \\ \vdots & \vdots & \vdots & \vdots \\ [\alpha_u^0] & [\alpha_u^1] & \cdots & [\alpha_u^{n-1}] \end{bmatrix}$$

and let $\mathbf{f} = (f_0,\ldots,f_{n-1})$, then

$$f(\alpha_i) = 0 \text{ for all } i = 1,\ldots,u \text{ if and only if } \mathbf{f}H^T = \mathbf{0}$$

Of course, the rows of H may not be linearly independent, but by removing any dependent rows, we obtain a parity check matrix for the code with zeros $\{\alpha_1,\ldots,\alpha_u\}$.

HAMMING CODES AS CYCLIC CODES

The representation of a cyclic code by its zeros can be used to show that some Hamming codes are cyclic codes. The binary case is the easiest, so let us consider it first.

Theorem 7.4.7 The binary Hamming code $\mathcal{H}_2(r)$ is equivalent to a cyclic code.
Proof. Recall that the binary Hamming code $\mathcal{H}_2(r)$ has parameters $[2^r-1,2^r-1-r]$ and parity check matrix whose columns consist of all 2^r-1 nonzero binary vectors of length r.

Now, let $n=2^r-1$, and let ω be a primitive n-th root of unity over F_2. Since $o_n(2)=r$, the splitting field of x^n-1 is F_{2^r}, and since ω has order $n=2^r-1$, it is a primitive field element of F_{2^r}. As a result, the powers of ω form all nonzero elements of F_{2^r}, and so the columns of the matrix

$$H=[[\omega^0],[\omega^1],\ldots,[\omega^{n-1}]]$$

consist of all nonzero binary vectors of length r. This shows that the Hamming code $\mathcal{H}_2(r)$ with this particular parity check matrix is the cyclic code whose *only* zeros consist of a primitive (2^r-1)-st root of unity ω and (of necessity) all the other roots of the minimal polynomial of ω. ∎

Example 7.4.6 Consider the Hamming code $\mathcal{H}_2(4)$. In this case, $n=2^4-1=15$, and the splitting field for x^n-1 is F_{16}. If β is a primitive element of F_{16}, then it is also a primitive 15th root of unity. The conjugates of β are

$$\beta,\beta^2,\beta^4,\beta^8$$

since $\beta^{16}=\beta$. Hence, the minimal polynomial for β is

$$(x-\beta)(x-\beta^2)(x-\beta^4)(x-\beta^8)$$

which we found in Example 7.2.1 to be the polynomial x^4+x+1. Hence, $\mathcal{H}_2(4)$ is the cyclic code generated in R_{15}, by $g(x)=x^4+x+1$. □

For q-ary $[n=\frac{q^r-1}{q-1},n-r]$ Hamming codes $\mathcal{H}_q(r)$, we have the following theorem.

Theorem 7.4.8 Let $n=(q^r-1)/(q-1)$, and assume that $(r,q-1)=1$. Then the q-ary Hamming code $\mathcal{H}_q(r)$ is equivalent to a cyclic code.
Proof. If F_{q^s} is the splitting field for x^n-1, then $s=o_n(q)$. In other words, s is the smallest positive integer for which

$$\frac{q^r-1}{q-1}\,\Big|\, q^s-1$$

But this holds if $s=r$, and since $(q^r-1)/(q-1)>q^{r-1}-1$, we see that r is the smallest such integer. Hence, $s=r$, and the splitting field for x^n-1 over F_q is F_{q^r}. Now let β be a primitive element of F_{q^r}.

Then

$$\omega = \beta^{(q^s-1)/n} = \beta^{q-1}$$

is a primitive n-th root of unity over F_q. Let H be the matrix whose columns are the vector representations of the powers of ω,

$$H = [[\omega^0],[\omega^1],\ldots,[\omega^{n-1}]]$$

We wish to show that any two distinct columns of H are linearly independent. Once this has been done, we will know that the cyclic code C with parity check matrix H has length n, dimension $k \geq n-r$ (we do not yet know that the rows of H are independent), and minimum distance $d \geq 3$. Then the sphere-packing bound can be used to show that $k = n-r$ and $d = 3$, and so the linear code C has the same parameters as $\mathcal{H}_q(r)$ and is, therefore, equivalent to $\mathcal{H}_q(r)$.

Now, the columns $[\omega^i]$ and $[\omega^j]$ of H are linearly dependent over F_q if and only if one is a scalar multiple of the other, that is, if and only if $\omega^{i-j} \in F_q$. But $\alpha \in F_q^*$ if and only if $\alpha^{q-1} = 1$, and so $[\omega^i]$ and $[\omega^j]$ are linearly dependent if and only if

$$\omega^{(i-j)(q-1)} = 1$$

Since ω is a primitive n-th root of unity, this happens if and only if

$$(i-j)(q-1) \equiv 0 \bmod n \tag{7.4.4}$$

Finally, if we show that $(r,q-1) = 1$ is equivalent to $(n,q-1) = 1$, then since we are assuming that $(r,q-1) = 1$, we can deduce from (7.4.4) that $i = j$, which will complete the proof. To this end, write

$$n = \frac{q^r - 1}{q-1} = 1 + q + \cdots + q^{r-1}$$

Since $(q-1) \mid q^i - 1$, we have

$$q^i = (q-1)u_i + 1$$

for some constant u_i. Summing from $i = 0$ to $r-1$ gives

$$n = (q-1)\sum u_i + r$$

which shows that $(r,q-1) = 1$ if and only if $(n,q-1) = 1$, and concludes the proof. ∎

In the exercises, we ask the reader to find all ternary cyclic codes of length 4. Since none of these codes has minimum distance 3, we see that the ternary Hamming code $\mathcal{H}_3(2)$ is not equivalent to a cyclic code.

THE IDEMPOTENT GENERATOR OF A CYCLIC CODE

We have seen that a complete list of all cyclic codes in R_n can be obtained from a factorization of $x^n - 1$ into monic irreducible factors over F_q. However, factoring $x^n - 1$ is not an easy task in general. One way to do this is to first obtain a primitive n-th root of unity, but this requires working in the splitting field F_{q^s} of $x^n - 1$.

In this subsection, we explore another approach to describing cyclic codes, involving a different type of generating polynomial than *the* generator polynomial.

Definition A polynomial $e(x) \in R_n$ is said to be **idempotent** in R_n if $e^2(x) \equiv e(x)$. □

For example, the polynomial $x^3 + x^5 + x^6$ is an idempotent in R_7 since $(x^3 + x^5 + x^6)^2 \equiv x^3 + x^5 + x^6$.

Theorem 7.4.9 Let C be a cyclic code in R_n, with generator polynomial $g(x)$ and check polynomial $h(x)$. Then $g(x)$ and $h(x)$ are relatively prime, and so there exist polynomials $a(x)$ and $b(x)$ for which

$$a(x)g(x) + b(x)h(x) = 1 \tag{7.4.5}$$

The polynomial $e(x) = a(x)g(x) \bmod (x^n - 1)$ has the following properties.

1) $e(x)$ is the unique identity in C, that is,

$$p(x)e(x) \equiv p(x) \quad \text{for all } p(x) \in C$$

2) $e(x)$ is the unique polynomial in C that is both idempotent and generates C, that is, $C = \langle e(x) \rangle$.

Proof. If $e_1(x)$ and $e_2(x)$ are both identities in R_n, then

$$e_1(x) \equiv e_1(x)e_2(x) \equiv e_2(x)$$

and so $e_1(x) = e_2(x)$. Thus, if an identity exists, its is unique. Since $g(x)h(x) = x^n - 1$ has no multiple roots in any extension field, $g(x)$ and $h(x)$ are relatively prime, and so (7.4.5) holds. If $p(x) \in C$, then $p(x)h(x) \equiv 0$, and so (7.4.5) gives

$$a(x)g(x)p(x) \equiv p(x)$$

which says that $e(x) = a(x)g(x) \bmod (x^n - 1)$ is indeed the identity in C and also that $e(x)$ generates C since any polynomial in C is a multiple of $e(x)$.

Multiplying (7.4.5) by $a(x)g(x)$ gives

$$[a(x)g(x)]^2 + a(x)b(x)g(x)h(x) = a(x)g(x)$$

and so

$$[a(x)g(x)]^2 \equiv a(x)g(x)$$

Thus, e(x) is idempotent.

To complete the proof, we need only show that an idempotent f(x) that also generates C must be equal to e(x). Since f(x) generates C, there exists a $q(x) \in R_n$ for which $e(x) \equiv q(x)f(x)$. Hence,

$$f(x) \equiv e(x)f(x) \equiv q(x)f^2(x) \equiv q(x)f(x) \equiv e(x)$$

which implies that $f(x) = e(x)$. This completes the proof. ∎

We will refer to the polynomial e(x) of the previous theorem as the **generating idempotent** of C, and write $C = [\![e(x)]\!]$ to denote the fact that e(x) is this generating idempotent. Theorem 7.4.9 shows that we can compute e(x) from g(x) using the Euclidean algorithm. The next result shows how to compute g(x) from e(x).

Theorem 7.4.10 The generator polynomial of the code $[\![e(x)]\!]$ is

$$g(x) = \gcd(e(x), x^n - 1)$$

Proof. Referring to (7.4.5), since $x^n - 1 = g(x)h(x)$ and $e(x) \equiv a(x)g(x)$, we have

$$\gcd(e(x), x^n - 1) = \gcd(a(x)g(x), h(x)g(x))$$

But, according to (7.4.5), a(x) and h(x) are relatively prime, and so this is equal to g(x). ∎

The following result is the analog of Theorem 7.4.3. We again leave the proof to the reader.

Theorem 7.4.11 Let $C_1 = [\![e_1(x)]\!]$ and $C_2 = [\![e_2(x)]\!]$ be cyclic codes in R_n. Then

1) $C_1 \subset C_2$ if and only if $e_1(x)e_2(x) \equiv e_1(x)$,
2) $C_1 \cap C_2 = [\![e_1(x)e_2(x)]\!]$,
3) $C_1 + C_2 = [\![e_1(x) + e_2(x) - e_1(x)e_2(x)]\!]$,

where all polynomials are taken modulo $x^n - 1$. □

The next theorem describes an interesting relationship between the generator polynomial and the generating idempotent of a cyclic code.

Theorem 7.4.12 Let C be a cyclic code in R_n, with generator polynomial g(x) and generating idempotent e(x). Then g(x) and e(x) have exactly the same roots, in the splitting field for $x^n - 1$, from among the n-th roots of unity.

Furthermore, if $f(x)$ is an idempotent in R_n that has exactly the same roots as $g(x)$ from among the n-th roots of unity, then $f(x)$ is the generating idempotent of $\langle\langle g(x)\rangle\rangle$.

Proof. Let ω be a primitive n-th root of unity. Since $e(x) \equiv a(x)g(x)$, we deduce that $g(\omega^i) = 0$ implies $e(\omega^i) = 0$. On the other hand, if $h(x)$ is the check polynomial for C, then $g(\omega^i) = 0$ if and only if $h(\omega^i) \neq 0$. Hence, by (7.4.5), $e(\omega^i) = 0$ implies $h(\omega^i) \neq 0$, which implies that $g(\omega^i) = 0$.

For the second part of the theorem, we observe that since every root of $g(x)$ is a root of $f(x)$, and since $g(x)$ has no multiple roots in any extension, we must have $g(x) \mid f(x)$. Furthermore, since the roots of the check polynomial $h(x)$ are precisely the *non-roots* of $g(x)$, from among the n-th roots of unity, we see that $h(x)$ and $f(x)$ have no common roots in any extension. Hence, they are relatively prime. But if D is the cyclic code in R_n with generating idempotent $f(x)$, then by Theorem 7.4.10, the generator polynomial for D is

$$\gcd(f(x),x^n-1) = \gcd(f(x),h(x)g(x)) = g(x)$$

and so $D = C$. Thus, $f(x)$ is the generating idempotent of C. ∎

We also have the following result that relates the generating idempotents of a cyclic code and its dual.

Theorem 7.4.13 Let $C = [\![e(x)]\!]$ be a cyclic code with check polynomial $h(x)$. Then the cyclic code $\langle\langle h(x)\rangle\rangle$ has generating idempotent $1-e(x)$ and $C^{\perp} = \langle\langle h^{\perp}(x)\rangle\rangle$ has generating idempotent

$$[1-e(x^{n-1})] \bmod (x^n-1)$$

Proof. Since

$$h(x)(1-e(x)) \equiv h(x)(1-a(x)g(x)) \equiv h(x)$$

we see that $1-e(x)$ is the identity in $\langle\langle h(x)\rangle\rangle$. Similarly, since $h^{\perp}(x) = h_0^{-1}x^k h(x^{-1}) \equiv h_0^{-1}x^k h(x^{n-1})$, we have

$$\begin{aligned} h^{\perp}(x)(1-e(x^{n-1})) &\equiv h^{\perp}(x) - h_0^{-1}x^k h(x^{n-1})e(x^{n-1}) \\ &\equiv h^{\perp}(x) - h_0^{-1}x^k h(x^{n-1})a(x^{n-1})g(x^{n-1}) \\ &\equiv h^{\perp}(x) \end{aligned}$$

and so $[1-e(x^{n-1})] \bmod (x^n-1)$ is the identity in $C^{\perp} = \langle\langle h^{\perp}(x)\rangle\rangle$. ∎

MINIMAL CYCLIC CODES

Among all cyclic codes in R_n, two special types deserve mention. Suppose that

$$x^n - 1 = \prod_i m_i(x)$$

is the factorization of $x^n - 1$ into monic irreducible polynomials over F_q. In view of Theorem 7.4.3, the cyclic codes $M_i = \langle\!\langle m_i(x) \rangle\!\rangle$ are *maximal*, since the only cyclic code that properly contains M_i is R_n itself. Similarly, if we let

$$\widehat{m}_i(x) = \frac{x^n - 1}{m_i(x)}$$

then the cyclic codes $\widehat{M}_i = \langle\!\langle \widehat{m}_i(x) \rangle\!\rangle$ are *minimal*, since the only cyclic code in R_n that is properly contained in $\widehat{M}_i$ is the zero code $\{\mathbf{0}\}$. Minimal cyclic codes are also called **irreducible codes.**

As the next result shows, the minimal cyclic codes in R_n are actually fields.

Theorem 7.4.14 Let $\widehat{M}_i = \langle\!\langle \widehat{m}_i(x) \rangle\!\rangle$ be a minimal cyclic code in R_n, where $\deg(m_i(x)) = d$. Then $\widehat{M}_i$ is a field isomorphic to F_{q^d}.
Proof. Let $f(x)$ be a nonzero polynomial in $\widehat{M}_i$. Then the ideal $\langle f(x) \rangle$ is contained in the minimal $\widehat{M}_i$, and so we must have $\widehat{M}_i = \langle f(x) \rangle$. In particular, if $e(x)$ is the idempotent generator for $\widehat{M}_i$, then $e(x) \equiv a(x)f(x)$, for some polynomial $a(x) \in R_n$. If $a(x) \in \widehat{M}_i$, it is the inverse of $f(x)$. In any case, since $e(x) \equiv e^2(x) \equiv a(x)e(x)f(x)$, and since $a(x)e(x) \bmod (x^n-1)$ is in $\widehat{M}_i$, we see that $a(x)e(x) \bmod (x^n-1)$ is the inverse of $f(x)$. Hence, $\widehat{M}_i$ is a field. Finally, since $\deg(\widehat{m}_i(x)) = n - d$, we have $\mathit{dim}(\widehat{M}_i) = d$ over F_q. ∎

It is evident that the generator polynomial of a cyclic code can be obtained, by multiplication, from the generator polynomials of the maximal cyclic codes $M_i = \langle\!\langle m_i(x) \rangle\!\rangle$. Similarly, the generating idempotent of a cyclic code can be obtained, by addition, from the generating idempotents for the minimal codes $\widehat{M}_i = \langle\!\langle \widehat{m}_i(x) \rangle\!\rangle$.

Definition The generating idempotent of a minimal cyclic code $\widehat{M}_i = \langle\!\langle \widehat{m}_i(x) \rangle\!\rangle$ is called a **primitive idempotent** and is denoted by $\theta_i(x)$. □

Theorem 7.4.15 Let $\theta_1(x), \ldots, \theta_s(x)$ be the primitive idempotents in R_n. Then

1) $\theta_i(x)\theta_j(x) \equiv 0$ for $i \neq j$,
2) $\theta_1(x) + \cdots + \theta_s(x) = 1$,
3) The generating idempotent of the code $\langle\!\langle m_{i_1}(x) m_{i_2}(x) \cdots m_{i_k}(x) \rangle\!\rangle$ is

$$1 - \theta_{i_1}(x) - \theta_{i_2}(x) - \cdots - \theta_{i_k}(x)$$

Proof.

1) According to Theorem 7.4.11, $\theta_i(x)\theta_j(x) \bmod (x^n - 1)$ is the identity for the code $\widehat{M}_i \cap \widehat{M}_j$. But since $\widehat{M}_i$ and $\widehat{M}_j$ are

minimal, their intersection consists of just the zero polynomial.

2) Using Theorem 7.4.11 and part 1), we deduce that

$$\widehat{M}_1 + \widehat{M}_2 \cdots + \widehat{M}_s = [\![\theta_1(x) + \cdots + \theta_s(x)]\!]$$

But $\widehat{M}_1 + \widehat{M}_2 \cdots + \widehat{M}_s = R_n$, which has identity element 1.

3) Let

$$\widehat{m}_{j_1 j_2 \cdots j_r}(x) = \frac{x^n - 1}{m_{j_1}(x) m_{j_2}(x) \cdots m_{j_r}(x)}$$

Then since

$$\widehat{m}_{j_1 j_2 \cdots j_r}(x) = \gcd(\widehat{m}_{j_1}(x) \widehat{m}_{j_2}(x) \cdots \widehat{m}_{j_r}(x))$$

we deduce from Theorems 7.4.3 and 7.4.11 and part 1) that

$$\begin{aligned}\langle\!\langle \widehat{m}_{j_1 j_2 \cdots j_r}(x) \rangle\!\rangle &= \widehat{M}_{j_1} + \widehat{M}_{j_2} + \cdots + \widehat{M}_{j_r} \\ &= [\![\theta_{j_1}(x) + \theta_{j_2}(x) + \cdots + \theta_{j_r}(x)]\!]\end{aligned}$$

If $\{i_1, \ldots, i_k\} = \{1, \ldots, s\} - \{j_1, \ldots, j_r\}$, then part 2 gives

$$\langle\!\langle m_{i_1}(x) m_{i_2}(x) \cdots m_{i_k}(x) \rangle\!\rangle = [\![1 - \theta_{i_1}(x) - \theta_{i_2}(x) - \cdots - \theta_{i_k}(x)]\!] \quad \blacksquare$$

FINDING GENERATING IDEMPOTENTS

Now let us turn to the question of how we might find idempotents without factoring $x^n - 1$. We restrict attention in this subsection to the case $q = 2$.

In $F_2[x]$, we have $f^2(x) = f(x^2)$, and so a polynomial $e(x) = e_0 + e_1 x + \cdots + e_{n-1} x^{n-1}$ in R_n is idempotent if and only if

$$e(x^2) \equiv e(x)$$

In R_n, this holds if and only if whenever $e_i \neq 0$, then $e_{2i (\mathrm{mod}\ n)} \neq 0$. It follows that $e(x)$ must be a sum of polynomials of the form

$$x^i + x^{2i} + \cdots + x^{2^{d-1} i}$$

where the exponents form a cyclotomic coset. We have established the following.

Theorem 7.4.16 Let $q = 2$. The idempotents in R_n are precisely the sums of polynomials of the form

$$x^i + x^{2i} + \cdots + x^{2^{d-1} i}$$

where $C_i = \{i, 2i, \ldots, 2^{d-1} i\}$ is a cyclotomic coset for 2 modulo n. $\square$

Example 7.4.7 Let $n = 9$. The cyclotomic cosets for 2 modulo 9 are

$$C_0 = \{0\}, \quad C_1 = \{1,2,4,8,7,5\}, \quad C_3 = \{3,6\}$$

and so there are $2^3 = 8$ idempotents

$$\begin{aligned} e_1(x) &= 0 \\ e_2(x) &= x^0 = 1 \\ e_3(x) &= x + x^2 + x^4 + x^5 + x^7 + x^8 \\ e_4(x) &= x^3 + x^6 \\ e_5(x) &= e_2(x) + e_3(x) \\ e_6(x) &= e_2(x) + e_4(x) \\ e_7(x) &= e_3(x) + e_4(x) \\ e_8(x) &= e_2(x) + e_3(x) + e_4(x) \end{aligned}$$

The corresponding generator polynomials can be computed using Theorem 7.4.10. For example, the Euclidean algorithm (see Chapter 8) gives

$$g_3(x) = \gcd(e_3(x), x^9 - 1) = x + 1$$ □

A FORMULA FOR PRIMITIVE IDEMPOTENTS

When we have access to the n-th roots of unity over F_q, we can obtain an explicit formula for the primitive idempotents. Consider again the factorization

$$x^n - 1 = \prod_i m_i(x)$$

of $x^n - 1$ into irreducible factors over F_q, where

(7.4.6) $$m_i(x) = \prod_{j \in C_i} (x - \omega^j)$$

and $C_i = \{i, qi, \ldots, q^{d-1}i\}$. We need a few preliminary lemmas.

Lemma 7.4.17 Let ω be a primitive n-th root of unity over F_q. Then for $i = 0, \ldots, n-1$,

$$\sum_{j=0}^{n-1} (\omega^i)^j = n\delta_{i,0}$$

Proof. If $i = 0$, the result is clear. If $i > 0$, then $\omega^i \neq 1$, and so

$$\sum_{j=0}^{n-1} (\omega^i)^j = \frac{1 - (\omega^i)^n}{1 - \omega^i} = 0$$ ∎

The next lemma shows how to recover a polynomial of degree n−1 or less from its values on the n-th roots of unity.

Lemma 7.4.18 Let $p(x) = p_0 + p_1x + \cdots + p_{n-1}x^{n-1}$ be a polynomial over F_q, and let ω be a primitive n-th root of unity over F_q. Then

$$p_i = \frac{1}{n}\sum_{j=0}^{n-1} p(\omega^j)\omega^{-ij}$$

where the computations on the right take place in the splitting field for $x^n - 1$ over F_q.

Proof. We have

$$\begin{aligned}\frac{1}{n}\sum_{j=0}^{n-1} p(\omega^j)\omega^{-ij} &= \frac{1}{n}\sum_{j=0}^{n-1}\sum_{k=0}^{n-1} p_k\omega^{jk}\omega^{-ij}\\ &= \frac{1}{n}\sum_{k=0}^{n-1} p_k\sum_{j=0}^{n-1}\omega^{j(k-i)}\\ &= \frac{1}{n}\sum_{k=0}^{n-1} p_k n\delta_{k,i} = p_i \qquad \blacksquare\end{aligned}$$

Now suppose that $\theta_k(x) \in F_q[x]$ is the primitive idempotent for the minimal code $\widehat{M}_k = \langle\langle \widehat{m}_k(x)\rangle\rangle$ in R_n. Since $\theta_k(x)$ is idempotent, we have $\theta_k^2(x) \equiv \theta_k(x) \bmod (x^n-1)$, and so

$$\theta_k^2(x) = \theta_k(x) + f(x)(x^n-1)$$

for some polynomial $f(x)$. Therefore,

$$\theta_k^2(\omega^j) = \theta_k(\omega^j)$$

and so $\theta_k(\omega^j) = 0$ or 1. But Theorem 7.4.10 tells us that $\theta_k(x)$ and $\widehat{m}_k(x)$ have the same zeros among the n-th roots of unity, and so

$$\theta_k(\omega^j) = \begin{cases} 0 & \text{if } j \notin C_k \\ 1 & \text{if } j \in C_k \end{cases}$$

If $\theta_k(x) = \sum_{i=0}^{n-1} \epsilon_i x^i$, we may apply Lemma 7.4.18 to get

$$\epsilon_i = \frac{1}{n}\sum_{j=0}^{n-1}\theta_k(\omega^j)\omega^{-ij} = \frac{1}{n}\sum_{j\in C_k}\omega^{-ij}$$

This proves the following.

Theorem 7.4.19 Let $\widehat{M}_k = \langle\langle \widehat{m}_k(x) \rangle\rangle$ be a minimal cyclic code in R_n, and suppose that

$$m_k(x) = \prod_{j \in C_k} (x - \omega^j)$$

where C_k is a cyclotomic coset. Then the primitive idempotent for $\widehat{M}_k$ is

$$\theta_k(x) = \frac{1}{n} \sum_{i=0}^{n-1} \left(\sum_{j \in C_k} \omega^{-ij} \right) x^i \qquad \square$$

Note that the inside sums in the above expression for $\theta_k(x)$, while computed in the splitting field for $x^n - 1$ over F_q, must actually lie in F_q.

Example 7.4.8 Let ω be a primitive 9th root of unity over F_2. Since the cyclotomic cosets for 2 modulo 9 are

$$C_0 = \{0\}, \quad C_1 = \{1,2,4,8,7,5\}, \quad C_3 = \{3,6\}$$

we have

$$\theta_0(x) = 1 + x + \cdots + x^8$$

$$\theta_1(x) = \sum_{i=0}^{8} (\omega^{-i} + \omega^{-2i} + \omega^{-4i} + \omega^{-5i} + \omega^{-7i} + \omega^{-8i}) x^i$$

$$\theta_3(x) = \sum_{i=0}^{8} (\omega^{-3i} + \omega^{-6i}) x^i$$

Let us compute $\theta_3(x)$. The splitting field for $x^9 - 1$ over F_2 is F_{q^s}, where $s = o_9(2) = 6$, that is, $F_{q^s} = F_{64}$. Furthermore, if β is a primitive element of F_{64}, then $\omega = \beta^7$ is a primitive n-th root of unity. Referring to the field table for F_{64} in the appendix, we see that the coefficients of $\theta_3(x)$ are

coef of $x^0 : 1 + 1 = 0$
coef of $x^1 : \omega^{-3} + \omega^{-6} = \omega^6 + \omega^3 = \beta^{42} + \beta^{21} = 111010 + 111011 = 1$
coef of $x^2 : \omega^{-6} + \omega^{-12} = \omega^3 + \omega^6 = 1$
coef of $x^3 : \omega^{-9} + \omega^{-18} = 1 + 1 = 0$
coef of $x^4 : \omega^{-12} + \omega^{-24} = \omega^6 + \omega^3 = 1$
coef of $x^5 : \omega^{-15} + \omega^{-30} = \omega^3 + \omega^6 = 1$
coef of $x^6 : \omega^{-18} + \omega^{-36} = 1 + 1 = 0$
coef of $x^7 : \omega^{-21} + \omega^{-42} = \omega^6 + \omega^3 = 1$
coef of $x^8 : \omega^{-24} + \omega^{-48} = \omega^3 + \omega^6 = 1$

Hence,

$$\theta_3(x) = x + x^2 + x^4 + x^5 + x^7 + x^8$$

which is the idempotent $e_3(x)$ in Example 7.4.7. □

The following simple result is sometimes useful. Its proof follows from Lemma 7.4.18, and we leave it as an exercise.

Theorem 7.4.20 Let $q = 2$, and let ω be a primitive n-th root of unity over F_2. A polynomial $f(x)$ in R_n is idempotent if and only if $f(\omega^i) = 0$ or 1 for all $i = 0,\ldots,n-1$.

EXERCISES

In all exercises, we assume that $(n,q) = 1$. *In particular, a binary cyclic code has odd length.*

1. If $p(x) \in R_n$, show that $\langle p(x)\rangle = \{f(x)p(x) \mid f(x) \in R_n\}$.
2. Prove that $f(x) \in R_n$ has a multiplicative inverse if and only if $f(x)$ is relatively prime to $x^n - 1$. Use this to prove that R_n is not a field.
3. Referring to Example 7.4.1, verify that $1 + x^2$ also generates C.
4. Find all binary cyclic codes of length 3. Write down a generating matrix and parity check matrix for each code.
5. Find all ternary cyclic codes of length 4. Write down a generating matrix and parity check matrix for each code.
6. Describe the fact that a linear code C is cyclic in terms of the automorphism group *Aut*(C).
7. Show that any set of k consecutive coordinate positions in a cyclic [n,k]-code is an information set.
8. Prove Theorem 7.4.3.
9. Describe the smallest cyclic code containing the codeword 0011010.
10. If $h(x)$ is the check polynomial for C, write $h(x^{-1})$ as a polynomial in $C^\perp$.
11. Show that the binary [7,4]-code $\langle\langle x^3 + x + 1\rangle\rangle$ and the binary [7,3]-code $\langle\langle x^4 + x^3 + x^2 + 1\rangle\rangle$ are dual codes.
12. Let $g(x)$ be the generator matrix of a binary cyclic code C. Suppose that C contains at least one codeword of odd weight. Show that the set of codewords in C of even weight is a cyclic subcode of C. What is the generator polynomial of C?
13. If C is a cyclic code and if L is a linear code that is scalar multiple equivalent to C, must L be cyclic as well?
14. Let $h(x)$ be the check polynomial of a cyclic code C. What is the relationship between C and the code $\langle h(x)\rangle$? (Are they equal? Are they equivalent?)
15. Prove that a binary cyclic code $C = \langle\langle g(x)\rangle\rangle$ contains the codeword $\mathbf{1}$ if and only if $g(1) \neq 0$.

16. Let C be a binary cyclic code. Prove that C contains a codeword of odd weight if and only if $1 \in C$.
17. Find the number of cyclic codes in R_n in terms of the number of cyclotomic cosets of q modulo n.
18. If $C = \langle\langle g(x) \rangle\rangle$ is a self-orthogonal binary cyclic code, show that $x - 1 \mid g(x)$.
19. If $C = \langle\langle g(x) \rangle\rangle$ is a cyclic code, prove that C is self-orthogonal if and only if $h^{\perp}(x) \mid g(x)$, where $h(x)$ is the check polynomial.
20. Let C_1 and C_2 be cyclic codes in R_n. Show that $C_1 + C_2$ is the smallest cyclic code containing C_1 and C_2.
21. Prove Theorem 7.4.11.
22. Find all binary cyclic codes of length 7. Give the generator polynomial, generator matrix, check polynomial, and parity check matrix for each code. Identify all dual codes. Find the idempotent generator for each code.
23. Let C be a cyclic code. Show that the zeros of the dual code $C^{\perp}$ are the inverses of the nonzeros of the original code C.
24. Let $C = \langle\langle g(x) \rangle\rangle$ be a cyclic code in R_n, with check polynomial $h(x)$. Suppose that $p(x)$ and $h(x)$ are relatively prime. Show that $C = \langle\langle p(x)g(x) \rangle\rangle$.
25. A code C is **reversible** if $c_0c_1\cdots c_{n-1} \in C$ implies $c_{n-1}\cdots c_1c_0 \in C$. Show that a cyclic code $C = \langle\langle g(x) \rangle\rangle$ is reversible if and only if $g(\alpha) = 0$ implies $g(\alpha^{-1}) = 0$.
26. Show that a cyclic code of length n is reversible (see the previous exercise) if -1 is a power of q modulo n.
27. Let ω be a primitive n-th root of unity. Let $C = \langle\langle g(x) \rangle\rangle = [\![e(x)]\!]$ be a cyclic code in R_n.
    a) Show that $e(\omega^i) = 0$ or 1.
    b) Show that $dim(C) =$ number of ω^i such that $g(\omega^i) \neq 0$.
    c) Show that $dim(C) =$ number of ω^i such that $e(\omega^i) = 1$.
28. Prove Theorem 7.4.20. *Hint.* First show that $f(\omega^{2j}) = f(\omega)$, and so f is constant on cyclotomic cosets. Then use Lemma 7.4.18 to show that $f_i = f_{2i}$.
29. Find all primitive idempotents in R_7, where $q = 2$.
30. Let $C = \langle\langle g(x) \rangle\rangle$ be a cyclic code, with check polynomial $h(x)$ and generating idempotent $e(x)$. Show that $e(x) \equiv \frac{1}{n}xg(x)h'(x)$, where $h'(x)$ is the formal derivative of $h(x)$. *Hint.* Differentiate $x^n - 1 = g(x)h(x)$. Try this formula out for $n = 9$, $q = 2$. (See Example 7.4.7.)
31. A cyclic code C of length n is **degenerate** if there exists an $r \mid n$ such that each codeword $\mathbf{c}$ has the form $\mathbf{c} = \mathbf{c}'\mathbf{c}'\cdots\mathbf{c}'$, where $\mathbf{c}'$ is a string of length r. (Thus, each codeword consists of n/r copies of a string of length r.) Prove that C is degenerate if and only if its check polynomial $h(x)$ divides $x^r - 1$. *Hint.* Show that the

generator polynomial $g(x)$ has the form

$$g(x) = a(x)(1+x^r+x^{2r}+\cdots+x^{n-r})$$

and deduce from this that $h(x) \mid x^r - 1$.

32. Prove that a q-ary cyclic code C of length n is invariant under the permutation π for which $\pi(i) = qi \bmod n$. *Hint*. - If $p(x) \in C$, then show that $p(x^q) \bmod (x^n - 1) \in C$.

7.5 More on Cyclic Codes

In this section, we continue our study of cyclic codes.

MATTSON-SOLOMON POLYNOMIALS

Let ω be a primitive n-th root of unity over F_q. For each polynomial $p(x) \in R_n$, we associate another polynomial $p_{ms}(x)$, called the **Mattson-Solomon polynomial** for $p(x)$, defined by

$$p_{ms}(x) = \sum_{i=0}^{n-1} p(\omega^{-i})x^i$$

We will not study the Mattson-Solomon polynomials in detail, but we do want to present some of the more basic properties of these polynomials.

The Mattson-Solomon polynomial $p_{ms}(x)$ can also be written

$$p_{ms}(x) = \sum_{i=0}^{n-1} p(\omega^{n-i})x^i = \sum_{i=1}^{n} p(\omega^i)x^{n-i}$$

If $p(x) = \sum p_j x^j$ and $p_{ms}(x) = \sum P_i x^i$, we have

$$\sum_{i=0}^{n-1} P_i x^i = \sum_{i=0}^{n-1}\left(\sum_{j=0}^{n-1} p_j \omega^{-ij}\right) x^i$$

which leads to the matrix form (note the order of the elements in the first matrix)

$$\begin{bmatrix} P_{n-1} \\ P_{n-2} \\ P_{n-3} \\ \vdots \\ P_0 \end{bmatrix} = \begin{bmatrix} 1 & 1 & 1 & \cdots & 1 \\ 1 & \omega^1 & \omega^2 & \cdots & \omega^{(n-1)} \\ 1 & \omega^2 & \omega^4 & \cdots & \omega^{2(n-1)} \\ \vdots & \vdots & \vdots & \vdots & \vdots \\ 1 & \omega^{n-1} & \omega^{2(n-1)} & \cdots & \omega^{(n-1)^2} \end{bmatrix} \begin{bmatrix} p_0 \\ p_1 \\ p_2 \\ \vdots \\ p_{n-1} \end{bmatrix}$$

Given the Mattson-Solomon polynomial $p_{ms}(x)$, it is a simple matter to recover the original polynomial $p(x)$.

Theorem 7.5.1 For any polynomial $p(x)$ in R_n, we have

$$p(x) = \frac{1}{n} \sum_{i=0}^{n-1} p_{ms}(\omega^i)x^i$$

Proof. If $p(x) = \sum p_i x^i$, then Lemma 7.4.18, with ω replaced by ω^{-1}, gives

$$p_i = \frac{1}{n}\sum_{j=0}^{n-1} p(\omega^{-j})\omega^{ij} = \frac{1}{n}p_{ms}(\omega^i)$$

from which the result follows. ∎

The following corollaries describe one possible use for the Mattson-Solomon polynomials.

Corollary 7.5.2 The weight of $p(x) \in R_n$ is equal to $n-s$, where s is the number of zeros of $p_{ms}(x)$ among the n-th roots of unity. □

Corollary 7.5.3 The weight of $p(x) \in R_n$ is at least $n - \deg(p_{ms}(x))$. □

The following corollary is referred to as the BCH bound. We will use this bound in the next chapter as motivation for defining the so-called BCH codes.

Corollary 7.5.4 (The BCH bound) Let ω be a primitive n-th root of unity. Let C be a cyclic code in R_n, with generator polynomial $g(x)$. Assume that $g(x)$ has the numbers

$$\omega^b, \omega^{b+1}, \ldots, \omega^{b+d-1}$$

among its roots, where the exponents are d consecutive integers. Then the minimum weight of C is at least $d+1$.
Proof. Let $c(x) \in C$. Since $g(x) \mid c(x)$, we have $c(\omega^i) = 0$ for $i = b, \ldots, b+d-1$. Hence,

$$c_{ms}(x) = c(\omega)x^{n-1} + \cdots + c(\omega^{b-1})x^{n-b+1} + c(\omega^{b+d})x^{n-b-d} + \cdots + c(\omega^n)$$

Now, let us multiply this by x^{b-1} and rearrange

$$\begin{aligned} x^{b-1}c_{ms}(x) &= c(\omega)x^{n+b-2} + \cdots + c(\omega^{b-1})x^n + c(\omega^{b+d})x^{n-d-1} + \cdots + c(\omega^n)x^{b-1} \\ &= x^n[c(\omega)x^{b-2} + \cdots + c(\omega^{b-1})] + \{c(\omega^{b+d})x^{n-d-1} + \cdots + c(\omega^n)x^{b-1}\} \end{aligned}$$

Letting $p(x)$ be the polynomial in the square brackets and $q(x)$ be the polynomial in the curly brackets, we have

$$x^{b-1}c_{ms}(x) = x^n p(x) + q(x) = (x^n - 1)p(x) + p(x) + q(x)$$

This equation shows that ω^i is a root of $c_{ms}(x)$ if and only if it is a root of the polynomial $p(x) + q(x)$, which has degree $n-d-1$. Hence,

$c_{ms}(x)$ has at most $n-d-1$ roots from among the n-th roots of unity, and Corollary 7.5.2 tells us that the weight of $c(x)$ is at least $n-(n-d-1)=d+1$. ∎

For a proof of the following result, see MacWilliams and Sloane (1977).

Theorem 7.5.5 Suppose that $q=2$, and let $p(x)\in R_n$. Let

$$f_1(x)=\gcd[p_{ms}(x),x^n-1]$$
$$f_2(x)=\gcd[p_{ms}(x),xp'_{ms}(x)]$$

where $p'_{ms}(x)$ is the formal derivative of $p_{ms}(x)$. Then the weight of $p(x)$ is given by

$$w(p(x))=n-\deg(f_1(x))=n-\deg(c_{ms}(x))+\deg(f_2(x)) \qquad \square$$

ENCODING WITH A CYCLIC CODE

There are two rather straightforward ways to encode message strings using a cyclic code – one systematic method and one nonsystematic method. We consider the nonsystematic method first. Let $C=\langle\langle g(x)\rangle\rangle$ be a q-ary cyclic $[n,n-r]$-code, where $\deg(g(x))=r$. Thus, C is capable of encoding q-ary messages of length $n-r$ and requires r redundancy symbols.

A Nonsystematic Method

Given a source string $a_0a_1\cdots a_{n-r-1}$, we form the *message polynomial*

$$a(x)=a_0+a_1x+\cdots+a_{n-r-1}x^{n-r-1}$$

This polynomial is encoded as the product $c(x)=a(x)g(x)\in C$.

A Systematic Method

To obtain a systematic encoder, we form the message polynomial

$$\bar{a}(x)=a_0x^{n-1}+a_1x^{n-2}+\cdots+a_{n-r-1}x^r$$

Notice that $\bar{a}(x)$ has no terms of degree less than r. Next, we divide $\bar{a}(x)$ by $g(x)$,

$$\bar{a}(x)=q(x)g(x)+r(x), \text{ where } \deg(r(x))<r$$

and send the codeword $c(x)=\bar{a}(x)-r(x)=q(x)g(x)$. Since $\bar{a}(x)$ and $r(x)$ have no terms of the same degree, this encoder is systematic. In fact, reading the terms of a polynomial from highest degree to lowest degree, we see that the first $n-r$ positions are information symbols and the remaining r positions are check symbols.

Example 7.5.1 Consider the binary cyclic [7,4]-code generated by the polynomial $g(x) = x^3 + x + 1$. Since

$$h(x) = \frac{x^7 - 1}{x^3 + x + 1} = x^4 + x^2 + x + 1$$

we get the parity check matrix

$$H = \begin{bmatrix} 1 & 0 & 1 & 1 & 1 & 0 & 0 \\ 0 & 1 & 0 & 1 & 1 & 1 & 0 \\ 0 & 0 & 1 & 0 & 1 & 1 & 1 \end{bmatrix}$$

whose columns consist of all nonzero binary strings of length 3. Hence, this code is the Hamming code $\mathcal{H}_2(3)$, which is single-error-correcting.

Now consider the message 1001. Using the systematic encoder, we have

$$\bar{a}(x) = x^6 + x^3$$

and since

$$x^6 + x^3 = (x^3 + x)(x^3 + x + 1) + (x^2 + x)$$

the encoded message is

$$c(x) = \bar{a}(x) - r(x) = x^6 + x^3 + x^2 + x$$ □

DECODING WITH A CYCLIC CODE

Since a cyclic code is a linear code, we can decode using the polynomial form of syndrome decoding. Let C be a cyclic code. If $c(x) \in C$ is the codeword sent and $u(x)$ is the *received polynomial*, then $err(x) = u(x) - c(x)$ is the **error polynomial**. The **weight** of a polynomial is the number of nonzero coefficients.

Definition Let $C = \langle\!\langle g(x) \rangle\!\rangle$ be a cyclic [n,n−r]-code. The **syndrome** of a polynomial $u(x)$, denoted by $syn(u(x))$, is the remainder upon dividing $u(x)$ by $g(x)$, that is,

$$u(x) = q(x)g(x) + syn(u(x)), \quad \deg(syn(u(x))) < r$$ □

In the exercises, we ask the reader to show that this definition of syndrome coincides with the definition of syndrome given for linear codes in Chapter 5.

As expected, a received polynomial $u(x)$ is a codeword if and only if its syndrome is the zero polynomial. Also, two polynomials have the same syndrome if and only if they lie in the same coset of C. Thus, the polynomial form of syndrome decoding is analogous to the vector form. Let us continue Example 7.5.1.

Example 7.5.2 Since the code $\mathcal{H}_2(3)$ is single-error-correcting, it is capable of correcting all error polynomials of weight at most 1. Hence, the coset leaders and corresponding syndromes are

Table 7.5.1

coset leader	*syndrome*
0	0
1	1
x	x
x^2	x^2
x^3	$x+1$
x^4	x^2+x
x^5	x^2+x+1
x^6	x^2+1

If, for example, the polynomial $u(x) = x^6 + x + 1$ is received, we compute its syndrome

$$x^6 + x + 1 = (x^3 + x + 1)(x^3 + x + 1) + (x^2 + x)$$

Since $syn(u(x)) = x^2 + x$, Table 7.5.1 shows that the coset leader is $a(x) = x^4$, and so we decode $u(x)$ as

$$c(x) = u(x) - a(x) = x^6 + x^4 + x + 1$$ □

The main practical difficulty with syndrome decoding is that coset leader-syndrome decoding tables become quite long. However, we can take advantage of the fact that the code in question is cyclic as follows.

Suppose we have a method for decoding the leading coefficient of any received word, which for the purposes of this discussion is the coefficient of x^{n-1}, *regardless* of whether or not that coefficient is nonzero. Then, if $u(x)$ is received, we can decode its leading coefficient, perform a cyclic shift modulo $x^n - 1$, and decode the new leading coefficient, which is the coefficient of x^{n-2} in $u(x)$. Repeating this process will decode the entire received word. The reason that this method may save time is that we only need those rows of the coset leader-syndrome table that contain coset leaders of degree $n-1$.

Example 7.5.3 Referring to Example 7.5.2, the only coset leader of weight 1 and degree $n - 1 = 6$ is x^6, and so we need only a one-row table

coset leader	*syndrome*
x^6	x^2+1

As in Example 7.5.2, suppose we receive $u(x) = x^6 + x + 1$. Since $syn(u(x)) = x^2 + x$ is not in the table, we assume that the leading coefficient of $u(x)$ is correct.

Then we shift $u(x)$

$$x(x^6 + x + 1) \bmod (x^7 - 1) = x^2 + x + 1$$

and compute its syndrome, which is just $x^2 + x + 1$. Again this is not in the table, and so the coefficient of x^5 in $u(x)$ is assumed to be correct. Shifting again and computing the syndrome gives

$$x(x^2 + x + 1) = x^3 + x^2 + x = (x^3 + x + 1) + x^2 + 1$$

Since the syndrome $x^2 + 1$ is in the table, we deduce that the coefficient of x^4 in $u(x)$ is incorrect. Continuing in this manner, we decode $u(x)$ as $x^6 + x^4 + x + 1$, just as before. ▯

The encoding and decoding of cyclic codes can be achieved using linear switching circuits, which can perform polynomial multiplication and division over finite fields quite efficiently. We will not discuss such circuits in this book, but refer the interested reader to Peterson and Weldon (1972). Suffice it to say that the following theorem, whose proof we leave as an exercise, can sometimes be used to advantage when implementing syndrome decoding using linear circuits.

Theorem 7.5.6 (Meggitt (1960)) Let $C = \langle\langle g(x) \rangle\rangle$ be a cyclic code in R_n. For any polynomial $u(x) \in R_n$,

$$syn[xu(x) \bmod (x^n - 1)] = syn(x\,syn(u(x)))$$

In words, the cyclic shift of $u(x)$, taken modulo $x^n - 1$, and the cyclic shift of the syndrome of $u(x)$ both have the same syndrome. ▯

ERROR TRAPPING

Let us take a closer look at the relationship between the unknown error polynomial $err(x)$, defined by

$$u(x) = c(x) + err(x)$$

where $c(x)$ is the codeword sent, and the known syndrome $syn(u(x))$, defined by

$$u(x) = q(x)g(x) + syn(u(x)), \quad \deg(syn(u(x)) < r$$

Of course, these polynomials need not be the same and, in fact, often have different degrees. However, we can say the following.

Lemma 7.5.7 Let C be a t-error-correcting cyclic code, and suppose that at most t errors have occurred in the transmission of a codeword c(x). If the syndrome of the received word u(x) has weight at most t, then $err(x) = syn(u(x))$.

Proof. If $w(syn(u(x))) \le t$, then since $w(err(x)) \le t$, the codeword

$$c(x) - q(x)g(x) = syn(u(x)) - err(x)$$

has weight at most 2t and so must be the zero codeword. Hence, $err(x) = syn(u(x))$. ∎

Of course, we may not be lucky enough to encounter syndromes of weight at most t. However, if the syndrome of a *cyclic shift* of u(x) has weight at most t, then it is almost as easy to obtain the error polynomial from this syndrome.

Let us denote the polynomial obtained from p(x) by performing k cyclic shifts by $p^{(k)}(x)$. Suppose that the syndrome of the cyclic shift $u^{(k)}(x)$ of u(x) has weight at most t. Then since

$$u^{(k)}(x) = c^{(k)}(x) + err^{(k)}(x)$$

Lemma 7.5.7 gives

$$err^{(k)}(x) = syn(u^{(k)}(x))$$

and so $err(x)$ can be easily recovered from $syn(u^{(k)}(x))$ by shifting an additional $n-k$ places.

Thus, we may define a decoding strategy as follows. Perform successive cyclic shifting of the received word u(x) until a word $u^{(k)}(x)$ is encountered, whose syndrome $s(x) = syn(u^{(k)}(x))$ has weight at most t. Then

$$err(x) = s^{(n-k)}(x)$$

This strategy is known as **error trapping.**

The next result says that error trapping can correct any t errors that happen to fall within r consecutive positions, including wrap around. (It does *not* say that any burst of length r or less can be corrected, and we know that this is not possible from Theorem 5.1.17.)

Theorem 7.5.8 Let C be a t-error-correcting cyclic [n,n−r]-code. If t or fewer errors occur, and if they are confined to r consecutive positions (including wrap-around), then the syndrome of some cyclic shift $u^{(k)}(x)$ of the received polynomial u(x) has weight at most t.

Proof. There must exist some k for which the cyclic shift $u^{(k)}(x)$ of u(x) has its errors confined to the r coefficients of $x^0, \ldots, x^{r-1}$. Thus,

$$u^{(k)}(x) = c^{(k)}(x) + err^{(k)}(x)$$

where $\deg(err^{(k)}(x)) < r$. Hence, since $c^{(k)}(x)$ is a codeword, we have

$$syn(u^{(k)}(x)) = syn(err^{(k)}(x)) = err^{(k)}(x)$$

and so the syndrome of $u^{(k)}(x)$ has weight at most t. ∎

BURST ERROR DETECTION AND CORRECTION WITH CYCLIC CODES

The subject of burst error detection and correction by cyclic codes is covered quite thoroughly by Peterson and Weldon (1972), who state that "from the engineering viewpoint, at least, the problem of designing burst-correcting codes and decoders appears to be solved." We give only a sampling of results here.

It is convenient (and desirable) to make a slight alteration in the definition of a burst, to allow wrap-around.

Definition A **burst** of length b in R_n is a polynomial in R_n whose nonzero coefficients are confined to b consecutive positions, allowing for wrap-around, the first and last of which must be nonzero. □

For example, the polynomial $p(x) = 1 + x + x^5$ is a burst of length 3 in R_6, starting at x^5 and ending at x.

We leave the proof of the following result on error detection as an exercise.

Theorem 7.5.9 A cyclic [n,n−r]-code C contains no bursts of length r or less. Hence, it can detect any burst error of length r or less. □

According to Theorem 5.1.17, if a cyclic [n,n−r]-code C can correct all burst errors of length b or less (not including wrap around), then we must have $2b \leq r$. Hence, the degree r of the generator polynomial g(x) of C provides an upper bound on the quantity 2b. As we will see, this upper bound is easily achieved.

Interleaving

Suppose that $C = \langle\langle g(x)\rangle\rangle$ is a cyclic [n,n−r]-code that is capable of correcting burst errors of length b or less. Let $\mathbf{c}_0, \ldots, \mathbf{c}_{k-1}$ be codewords in C. For convenience, we use string notation (rather than polynomial notation), and let

$$\mathbf{c}_i = c_{i,0}c_{i,1}\cdots c_{i,n-1}$$

Then we may *interleave* these codewords by juxtaposing the first

positions in each codeword, followed by the second positions in each codeword, and so on, to obtain the word

$$(c_{0,0}c_{1,0}\cdots c_{k-1,0})(c_{0,1}c_{1,1}\cdots c_{k-1,1})\cdots(c_{0,n-1}c_{1,n-1}\cdots c_{k-1,n-1})$$

(The parentheses are included to aid readability only.)

Let us denote by $C^{(k)}$ the set of all words formed in this way from all possible choices of k codewords $\mathbf{c}_0,\ldots,\mathbf{c}_{k-1}$ (taken in all possible orders). The code $C^{(k)}$ is said to be formed from C by **interleaving** (or **symbol interleaving**), and k is called the **interleaving degree.** The key properties of $C^{(k)}$ are given in the next theorem, whose proof is left for the reader.

Theorem 7.5.10 Let $C = \langle\langle g(x)\rangle\rangle$ be a cyclic [n,s]-code. Then the code $C^{(k)}$, obtained from C by symbol interleaving, is a cyclic [nk,sk]-code with generator polynomial $g(x^k)$. Furthermore, if C is capable of correcting burst errors of length b or less, then $C^{(k)}$ is capable of correcting burst errors of length bk or less. □

Theorem 7.5.10 tells us that if we can find short cyclic codes with good burst error correcting capabilities, then through the process of symbol interleaving, we can also find long cyclic codes with proportionally good burst error correcting capabilities.

Computer searches have been done to find short cyclic codes with good burst error correcting capabilities, to which interleaving can apply. For a list of such codes, see Peterson and Weldon (1972). Here is one example.

Example 7.5.4 The binary cyclic [7,3]-code $C = \langle\langle x^4+x^3+x^2+1\rangle\rangle$ is capable of correcting burst errors of length 2 or less, which is the best possible since, in this case, $r = 2b$. Hence, according to Theorem 7.5.10, the interleaved code $C^{(k)} = \langle\langle x^{4k}+x^{3k}+x^{2k}+1\rangle\rangle$ is a binary cyclic [7k,3k]-code that can correct burst errors of length $2k$ or less. Note that this is again best possible. □

EXERCISES

1. How many rows are there in the error polynomial-syndrome decoding table of a binary t-error-correcting cyclic code? How many are there if we use Meggitt decoding?
2. Prove Theorem 7.5.6.
3. To see that the syndrome as we have defined it here is a syndrome in the sense of Chapter 4, proceed as follows. Let $C = \langle\langle g(x)\rangle\rangle$ be a cyclic [n,n−r]-code.

a) Let $s_i(x)$ be the remainder obtained by dividing x^{r+i} by $g(x)$,

$$x^{r+i} = a_i(x)g(x) + s_i(x)$$

Show that $x^{r+i} - s_i(x)$, for $i = 0,\ldots,n-r-1$, is a basis for C.

b) Find the generator matrix for the basis in part a, and a corresponding parity check matrix H.

c) If $u(x)$ is a received polynomial, write

$$u(x) = \sum_{j=0}^{r-1} u_j x^j + \sum_{i=0}^{n-r-1} u_{r+i} x^{r+i}$$

and substitute for x^{r+i} using the results of part a.

d) Use the computations in part c) to compute the coefficient of x^j in the syndrome of $u(x)$. How is this related to $(u_0 \cdots u_{n-1})H^T$?

4. Prove Theorem 7.5.9. *Hint.* If $b(x) \in C$ is a burst of length r or less, what about the cyclic shifts of $b(x)$?
5. Prove Theorem 7.5.10.
6. If C is best possible at burst error correction, in the sense that $r = 2b$, what about the interleaved code $C^{(k)}$?
7. Let $q = 2$, and let $p(x)$ be a polynomial in R_n. Show that the Mattson-Solomon polynomial $p_{ms}(x)$ is idempotent, that is, $p_{ms}^2(x) \equiv p_{ms}(x)$.

CHAPTER 8

Some Cyclic Codes

8.1 BCH Codes

THE BCH BOUND

Let us begin with a short review. We have seen that a cyclic code can be defined through its zeros. In particular, if $\alpha_1,\ldots,\alpha_u$ are n-th roots of unity over F_q, then the code

$$C = \{p(x) \in R_n \mid p(\alpha_1) = 0,\ldots,p(\alpha_u) = 0\}$$

is a cyclic code whose generator polynomial $g(x)$ is the product of the *distinct* minimal polynomials of $\alpha_1,\ldots,\alpha_u$ over F_q. This approach suggests that we may find some interesting codes by specifying certain special sets of n-th roots of unity as zeros of $g(x)$.

Recall that if

$$x^n - 1 = \prod_i m_i(x)$$

is the factorization of $x^n - 1$ into irreducible factors over F_q, and if ω is a primitive n-th root of unity over F_q, then the roots of the polynomial $m_i(x)$ are conjugates, that is, they have the form

$$\{\omega^i,\omega^{iq},\ldots,\omega^{iq^{d-1}}\}$$

where d is the smallest positive integer such that $iq^d \equiv i \bmod n$. Recall also that the set

$$C_i = \{i,qi,\ldots,q^{d-1}i\}$$

is called the i-th cyclotomic coset of q modulo n. Thus,

$$m_i(x) = \prod_{j \in C_i} (x - \omega^j)$$

Finally, let us recall the BCH bound of Section 7.5, phrased somewhat differently.

Theorem 8.1.1 (The BCH bound) Let ω be a primitive n-th root of unity over F_q. Let C be a cyclic code in R_n, whose generator polynomial g(x) is the monic polynomial of smallest degree over F_q that has the $\delta - 1$ numbers

$$\omega^b, \omega^{b+1}, \ldots, \omega^{b+\delta-2}$$

among its zeros, where $b \geq 0$. Then C has minimum distance at least δ. □

This theorem tells us that we can obtain a code with minimum distance at least a prescribed amount δ by specifying that its generator polynomial have $\delta - 1$ "consecutive" zeros. (That is, the exponents are consecutive.) Theorem 8.1.1 can be generalized as follows. (We leave the proof as an exercise.)

Theorem 8.1.2 Let ω be a primitive n-th root of unity. Let C be a cyclic code in R_n, whose generator polynomial g(x) is the monic polynomial of smallest degree that has the $\delta - 1$ numbers

$$\omega^b, \omega^{b+r}, \ldots, \omega^{b+(\delta-2)r}$$

among its zeros, where r and n are relatively prime and $b \geq 0$. Then C has minimum distance at least δ. □

BCH CODES

We are now ready to define the BCH codes. These codes were discovered independently by R.C. Bose and D.K. Ray-Chaudhuri (1960) and by A. Hocquenghem (1959). They form an extremely important class of codes for several reasons. For instance, they have good error-correcting properties when the length is not too great, they can be encoded and decoded relatively easily, and they provide a good foundation upon which to base other families of codes.

Definition Let ω be a primitive n-th root of unity over F_q, and let g(x) be the monic polynomial over F_q of smallest degree that has the $\delta - 1$ numbers

$$\omega^b, \omega^{b+1}, \ldots, \omega^{b+\delta-2}$$

among its zeros, where $b \geq 0$ and $\delta \geq 1$. Thus,

$$g(x) = \text{lcm}\{m_b(x), m_{b+1}(x), \ldots, m_{b+\delta-2}(x)\}$$

The q-ary cyclic code $\mathcal{B}_q(n,\delta,\omega,b)$ of length n, with generator polynomial g(x), is called a **BCH code** with **designed distance** δ. (Note that the designed distance is one greater than the number of zeros.)

When $b = 1$, the code $\mathcal{B}_q(n,\delta,\omega) = \mathcal{B}_q(n,\delta,\omega,1)$ is referred to as a **narrow sense** BCH code. When ω is a primitive *field element*, that is, when $n = q^s - 1$ for some $s \geq 1$, then $\mathcal{B}_q(n,\delta,\omega,b)$ is referred to as a **primitive** BCH code. ▯

The binary narrow sense BCH codes are generalizations of the Hamming codes. In particular, we have seen that the binary Hamming codes can be defined as cyclic codes whose generator polynomial is the monic polynomial of smallest degree over F_2 that has a primitive n-th root of unity ω as a zero. Hence, these Hamming codes are binary primitive narrow sense BCH codes of designed distance $\delta = 2$.

According to Theorem 8.1.1, the BCH code $\mathcal{B}_q(n,\delta,\omega,b)$ has minimum distance *at least* equal to its designed distance δ. Furthermore, since the splitting field of $x^n - 1$ over F_q has degree q^s over F_q, where $s = o_n(q)$, we deduce that

$$\deg(m_{b+i}(x)) \leq s$$

and so

$$dim(\mathcal{B}_q(n,\delta,\omega,b)) = n - \deg(g(x)) \geq n - s(\delta-1)$$

Let us summarize.

Theorem 8.1.3 The q-ary BCH code $\mathcal{B}_q(n,\delta,\omega,b)$ of length n and designed distance δ has parameters

$$dim(\mathcal{B}_q(n,\delta,\omega,b)) \geq n - (\delta-1)o_n(q)$$

and

$$d(\mathcal{B}_q(n,\delta,\omega,b)) \geq \delta$$

where $o_n(q)$ is the order of q modulo n. ▯

From the definition of BCH code, we see that

$$\mathcal{B}_q(n,\delta,\omega,b) = \{p(x) \in R_n \mid p(\omega^b) = p(\omega^{b+1}) = \cdots = p(\omega^{b+\delta-2}) = 0\}$$

and so, if $[\omega^i]$ denotes the column vector in F_q^s corresponding to the element ω^i in the splitting field F_{q^s} of $x^n - 1$, then the $s(\delta - 1)$ rows of the matrix

$$H = \begin{bmatrix} 1 & [\omega^b] & [\omega^{2b}] & \cdots & [\omega^{(n-1)b}] \\ 1 & [\omega^{b+1}] & [\omega^{2(b+1)}] & \cdots & [\omega^{(n-1)(b+1)}] \\ \vdots & \vdots & \vdots & \vdots & \vdots \\ 1 & [\omega^{b+\delta-2}] & [\omega^{2(b+\delta-2)}] & \cdots & [\omega^{(n-1)(b+\delta-2)}] \end{bmatrix}$$

form a complete set of parity checks, and once any dependent rows are removed, the remaining matrix is a parity check matrix for $\mathcal{B}_q(n,\delta,\omega,b)$.

BINARY BCH CODES

In the binary case, if C_k is a cyclotomic coset, then $u \in C_k$ if and only if $2u \bmod n \in C_k$. Hence, $m_i(x) = m_{2i}(x)$. This implies that for $\epsilon \geq 1$, the polynomials

$$g_1(x) = \mathrm{lcm}\{m_1(x), m_2(x), \ldots, m_{2\epsilon}(x)\}$$
$$g_2(x) = \mathrm{lcm}\{m_1(x), m_2(x), \ldots, m_{2\epsilon-1}(x)\}$$

and

$$g_3(x) = \mathrm{lcm}\{m_1(x), m_3(x), \ldots, m_{2\epsilon-1}(x)\}$$

are identical. (Note that the last set of polynomials $m_i(x)$ has odd indices only.) Since $g_2(x) = g_1(x)$, we have

$$\mathcal{B}_2(n,2\epsilon+1,\omega) = \mathcal{B}_2(n,2\epsilon,\omega)$$

Hence, we may restrict attention to binary narrow sense BCH codes of *odd* designed distance. The fact that $g_1(x) = g_3(x)$ leads to the following improvement on the lower bound on dimension.

Theorem 8.1.4 A binary narrow sense BCH code of length n and odd designed distance $\delta = 2\epsilon+1$ has dimension at least $n - \epsilon\, o_n(2)$, where $o_n(2)$ is the order of 2 modulo n. In symbols,

$$dim(\mathcal{B}_2(n,2\epsilon+1,\omega)) \geq n - \epsilon\, o_n(2)$$ □

Example 8.1.1 Let $q = 2$, $n = 2^5 - 1 = 31$. To determine the narrow sense (primitive) BCH codes $\mathcal{B}_2(31,\delta,\omega)$ with various designed distances, we first compute the cyclotomic cosets

$$C_0 = \{0\}$$
$$C_1 = \{1,2,4,8,16\}$$
$$C_3 = \{3,6,12,24,17\}$$
$$C_5 = \{5,10,20,9,18\}$$
$$C_7 = \{7,14,28,25,19\}$$
$$C_{11} = \{11,22,13,26\}$$
$$C_{15} = \{15,30,29,27,23\}$$

The generator matrix for the narrow sense BCH code $\mathcal{B}_2(31,9,\omega)$ with designed distance $\delta = 9$, for instance, is

$$g(x) = \mathrm{lcm}\{m_1(x),\ldots,m_8(x)\} = m_1(x)\cdot m_3(x)\cdot m_5(x)\cdot m_7(x)$$

Hence, the dimension of $\mathcal{B}_2(31,9,\omega)$ is $n - \deg(g(x)) = 31 - 20 = 11$. Notice that this is the same as the lower bound on the dimension given in Theorem 8.1.4.

Notice also that since $9 \in C_5$, we have $m_9(x) = m_5(x)$, and so

$$\mathcal{B}_2(31,11,\omega) = \mathcal{B}_2(31,9,\omega)$$

Thus, in this case, the lower bound on the dimension given by Theorem 8.1.4 is $31 - 25 = 6$, which is strictly less than the actual dimension.

Table 8.1.1 contains a complete list of all binary narrow sense BCH codes of length 31. In this table, we use the abbreviation

$$m_{i_1,i_2,\ldots,i_k}(x) = m_{i_1}(x)\cdots m_{i_k}(x)$$ □

Table 8.1.1

designed distance	*generator polynomial*	*dimension*	*actual distance*
1	1	31	1
3	$m_1(x)$	26	3
5	$m_{1,3}(x)$	21	5
7	$m_{1,3,5}(x)$	16	7
9 or 11	$m_{1,3,5,7}(x)$	11	11
13 or 15	$m_{1,3,5,7,9}(x)$	6	15
17,19,...,31	$x^{31}-1$	1	31

The previous example shows that a BCH code with odd designed distance δ may coincide with a BCH code with odd designed distance $\delta' > \delta$. This leads to the following definition.

Definition Let $\mathcal{B}$ be a BCH code. The largest designed distance δ for which $\mathcal{B}$ is the BCH code with designed distance δ is called the **Bose** distance of the code C. □

Example 8.1.2 Let $q = 2$ and $n = 23$. Since $s = o_{23}(2) = 11$, the splitting field for $x^{23} - 1$ is $F_{2^{11}}$. The cyclotomic cosets for 2 modulo 23 are

$$C_0 = \{0\}$$
$$C_1 = \{1,2,4,8,16,9,18,13,3,6,12\}$$
$$C_5 = \{5,10,20,17,11,22,21,19,15,7,14\}$$

and so the binary narrow sense BCH code $\mathcal{B}_2(23,5,\omega)$ with designed distance $\delta = 5$ has generator polynomial

$$g(x) = \text{lcm}\{m_1(x),m_2(x),m_3(x)\} = m_1(x)$$

In particular, $\mathcal{B}_2(23,5,\omega)$ is a binary [23,12]-code. It is not hard to see, using the results of Example 7.6.3, that $\mathcal{B}_2(23,5,\omega)$ is equivalent to the Golay code $\mathcal{G}_{23}$, which has minimum distance 7. Hence, in this case, the actual minimum distance is greater than the designed distance. ▯

THE AUTOMORPHISMS OF BINARY BCH CODES

According to Corollary 4.3.16, if we can show that an extended code $\widehat{C}$ is invariant under a transitive group of automorphisms, we have established that the code C has odd minimum weight. Let us do this for the binary primitive BCH codes.

Let $n = 2^m - 1$, and consider the field F_{2^m}. If ω is a primitive n-th roots of unity, then it is also a primitive field element of F_{2^m}, and so

$$F_{2^m} = \{0,\omega^0,\omega^1,\ldots,\omega^{n-1}\}$$

Let $u,v \in F_{2^m}$ with $u \neq 0$. The **affine transformation** $P_{u,v}:F_{2^m} \to F_{2^m}$ is defined by

$$P_{u,v}(\alpha) = u\alpha + v$$

The set of all affine transformations $P_{u,v}$ is referred to as the **affine group** on F_{2^m}.

Now, consider a binary primitive narrow sense BCH code $\mathcal{B} = \mathcal{B}(n{=}2^m{-}1,\delta,\omega)$. The extended code $\widehat{\mathcal{B}}$ has length 2^m, and any codeword

$$c(x) = c_0 + c_1x + \cdots + c_{n-1}x^{n-1}$$

in $\mathcal{B}$ gives rise to an extended codeword

$$\widehat{c}(x) = c_0 + c_1x + \cdots + c_{n-1}x^{n-1} + c_nx^n$$

in $\widehat{\mathcal{B}}$, where $\widehat{c}(1) = 0$. Thus, the coordinate positions of the extended code are $0,\ldots,n$, and the affine permutation $P_{u,v}$ can be used to define a permutation ρ on these positions as follows.

For $i = 0,\ldots,n-1$, we set

$$\rho(i) = \begin{cases} j & \text{if } u\omega^i + v = \omega^j \\ n & \text{if } u\omega^i + v = 0 \end{cases}$$

and for $i = n$,

$$\rho(n) = \begin{cases} j & \text{if } v = \omega^j \\ n & \text{if } v = 0 \end{cases}$$

We leave it to the reader to show that ρ is indeed a permutation of the coordinate positions $0,\ldots,n$ and that the group of all such permutations ρ, obtained from all affine transformations $P_{u,v}$, is **doubly transitive**, that is, for any $i \neq j$ and $k \neq \ell$, there exists a permutation ρ for which $\rho(i) = k$ and $\rho(j) = \ell$. We will now show that $\widehat{\mathfrak{B}}$ is invariant under ρ.

Let

$$\hat{c}(x) = \sum_{i=0}^{n} c_i x^i$$

be an extended codeword in $\widehat{\mathfrak{B}}$, where $\hat{c}(1) = 0$. Applying the permutation ρ to the coordinate positions gives the word

$$\hat{d}(x) = \sum_{i=0}^{n} c_i x^{\rho(i)}$$

We must show that the codeword $d(x)$ that comes from $\hat{d}(x)$ by dropping the term involving x^n is in the BCH code $\mathfrak{B}$. This will show that $\hat{d}(x)$ is in $\widehat{\mathfrak{B}}$. To show that $d(x)$ lies in $\mathfrak{B}$, we must show that $d(\omega^k) = 0$ for $k = 1,\ldots,\delta-1$.

Now, let us suppose first that $v \neq 0$, in which case $u\omega^r + v = 0$, for some $r = 0,\ldots,n-1$. Then we may write

$$\hat{d}(x) = \sum_{\substack{i=0 \\ i \neq r}}^{n} c_i x^{\rho(i)} + c_r x^n$$

and so, for $k = 1,\ldots,\delta - 1$, using the fact that $c(\omega^k) = 0$, we have

$$d(\omega^k) = \sum_{\substack{i=0 \\ i \neq r}}^{n} c_i \omega^{k\rho(i)} = \sum_{\substack{i=0 \\ i \neq r}}^{n-1} c_i \omega^{k\rho(i)} + c_n v^k$$

$$= \sum_{\substack{i=0 \\ i \neq r}}^{n-1} c_i [u\omega^i + v]^k + c_n v^k = \sum_{i=0}^{n-1} c_i [u\omega^i + v]^k + c_n v^k$$

$$= \sum_{i=0}^{n-1} c_i \sum_{j=0}^{k} \binom{k}{j} u^j \omega^{ij} v^{k-j} + c_n v^k = \sum_{j=0}^{k} \binom{k}{j} u^j v^{k-j} \left(\sum_{i=0}^{n-1} c_i \omega^{ij} \right) + c_n v^k$$

$$= \sum_{j=0}^{k} \binom{k}{j} u^j v^{k-j} \, c(\omega^j) + c_n v^k = c(1) v^k + c_n v^k = \hat{c}(1) v^k = 0$$

which shows that $d(x) \in \mathfrak{B}$.

There remains to consider the case $v = 0$, which we leave to the reader. Let us summarize our results.

Theorem 8.1.5 Let $\mathfrak{B}$ be a binary primitive narrow sense BCH code. The extended code $\widehat{\mathfrak{B}}$ is invariant under the (doubly) transitive group of permutations ρ, that come from the affine transformations $P_{u,v}$ defined above. As a result, the minimum distance of $\mathfrak{B}$ is odd. □

THE TRUE MINIMUM DISTANCE OF A BCH CODE

Much of the study of BCH codes is devoted to determining the true minimum distance of a BCH code with designed distance δ. Let us briefly discuss some of the results along these lines.

Theorem 8.1.6 Let $n = 2^s - 1$. If

(8.1.1)
$$2^{s\epsilon} < \sum_{i=0}^{\epsilon+1} \binom{n}{i}$$

then a (primitive) binary narrow sense BCH code of length n and designed distance $2\epsilon + 1$ has minimum distance $2\epsilon + 1$, that is,

$$d(\mathfrak{B}_2(2^s-1, 2\epsilon+1, \omega)) = 2\epsilon + 1$$

Proof. Let $\mathfrak{B} = \mathfrak{B}_2(2^s-1, 2\epsilon+1, \omega)$. According to Theorem 8.1.5, the minimum distance d of $\mathfrak{B}$ is odd. Thus, $d \neq 2\epsilon + 2$. Suppose that $d \geq 2\epsilon + 3$. Since the dimension of $\mathfrak{B}$ is at least $n - s\epsilon$, the sphere-packing bound then implies that

$$2^{n-s\epsilon} \sum_{i=0}^{\epsilon+1} \binom{n}{i} \leq 2^n$$

which is contrary to (8.1.1). Therefore, $\mathfrak{B}$ has minimum distance $d = 2\epsilon + 1$. ∎

Example 8.1.3 Let $s = 5$ and $n = 31$. Then we easily verify that

$$2^{5\epsilon} < \sum_{i=0}^{\epsilon+1} \binom{31}{i}$$

for $\epsilon = 1$, 2 or 3. Hence, the last column in the second, third, and fourth rows of Table 8.1.1 are verified. □

Corollary 8.1.7 Let $n = 2^s - 1$. If

$$s > 1 + \log_2(\epsilon+1)!$$

then a binary (primitive) narrow sense BCH code of length n and

designed distance $2\epsilon+1$ has minimum distance $2\epsilon+1$, that is

$$d(\mathcal{B}_2(2^s-1,2\epsilon+1,\omega)) = 2\epsilon+1$$

Proof. It can be shown that $s > 1+\log_2(\epsilon+1)!$ implies (8.1.1). However, we will omit the details. ∎

Example 8.1.4 For the binary primitive narrow sense BCH code $\mathcal{B}_2(2^s-1,5,\omega)$, we have $\epsilon = 2$, and since

$$1+\log_2(2+1)! < 4$$

Corollary 8.1.7 implies that if $s \geq 4$, the code $\mathcal{B}_2(2^s-1,5,\omega)$ has minimum distance 5. Hence $\mathcal{B}_2(2^s-1,5,\omega)$ is exactly double-error-correcting. We will discuss these codes again later in this section. □

Theorem 8.1.8 If $\delta \mid n$, then the binary narrow sense BCH code $\mathcal{B}_2(n,\delta,\omega)$ has minimum distance δ.
Proof. Let $n = \delta b$, and observe that

$$x^n - 1 = x^{\delta b} - 1 = (x^b-1)(1+x^b+x^{2b}+\cdots+x^{(\delta-1)b})$$

Since $\omega^{ib} \neq 1$ for $i = 1,\ldots,\delta-1$, we see that $\omega,\omega^2,\ldots,\omega^{\delta-1}$ are not zeros of x^b-1, and hence they must be zeros of

$$p(x) = 1+x^b+x^{2b}+\cdots+x^{(\delta-1)b}$$

which is therefore a codeword in $\mathcal{B}_2(n,\delta,\omega)$ of weight δ. Hence, the minimum distance of $\mathcal{B}_2(n,\delta,\omega)$ is δ. ∎

We state the following result without proof. The interested reader may consult MacWilliams and Sloane (1977).

Theorem 8.1.9 Any q-ary primitive narrow sense BCH code of the form $\mathcal{B}_q(q^s-1,q^k-1,\omega)$ for some k has minimum distance q^k-1. □

Example 8.1.5 Let $s = 5$ and $k = 4$. According to Theorem 8.1.9, the binary narrow sense BCH code $\mathcal{B}_2(31,15,\omega)$ has minimum distance 15. □

Theorem 8.1.10 The minimum distance of a q-ary *primitive* narrow sense BCH code $\mathcal{B}_q(n,\delta,\omega)$ of designed distance δ is at most $q\delta-1$, that is,

$$d(\mathcal{B}_q(q^s-1,\delta,\omega)) \leq q\delta-1$$

Proof. Let k be the positive integer with the property that

$$q^{k-1} \leq \delta \leq q^k-1$$

Let $\delta' = q^k-1$, and let $\mathcal{B}'$ be the q-ary primitive narrow sense BCH

code with length n and designed distance δ'. Then the zeros of the generator polynomial of $\mathcal{B}_q(n,\delta,\omega)$ are also zeros of the generator polynomial of $\mathcal{B}'$, and so $\mathcal{B}' \subset \mathcal{B}_q(n,\delta,\omega)$. But according to Theorem 8.1.9, $\mathcal{B}'$ has minimum distance δ', and so if d is the minimum distance of $\mathcal{B}_q(n,\delta,\omega)$, then

$$d \le \delta' = q^k - 1 \le q\delta - 1$$ ∎

THE QUALITY OF BCH CODES

We have already mentioned that BCH codes have good error correcting properties when the length is not too great. In fact, for lengths of a few thousand or less, the BCH codes are among the best codes known. On the other hand, for large lengths, the BCH codes are not good. Recall that, according to Corollary 4.2.7, we cannot expect asymptotically good error correcting properties from a family of codes for which the quantity d/n tends to 0 as n tends to infinity. In this regard, we have the following theorem, whose proof can be found in MacWilliams and Sloane (1977).

Theorem 8.1.11 There does not exist an infinite sequence of q-ary primitive BCH codes with the property that both d/n and k/n are bounded away from 0. □

DOUBLE-ERROR-CORRECTING BCH CODES

Let $\mathcal{B} = \mathcal{B}_2(2^s-1,5,\omega)$ be a binary primitive narrow sense BCH code of length $n = 2^s - 1$ $(s \ge 4)$ and designed distance $\delta = 5$. Then, according to Theorem 8.1.4, the dimension of $\mathcal{B}$ is at least $n - 2s$. Example 8.1.4 shows that $\mathcal{B}$ has minimum distance 5 and so is exactly double-error-correcting. In short, $\mathcal{B}$ has parameters

$$[n=2^s-1,\ k \ge n-2s,\ 5]$$

It is interesting to compare this with the single-error-correcting Hamming code $\mathcal{H}_2(s)$, which has parameters $[n=2^s-1,n-s,3]$. Notice that the rates

$$R(\mathcal{B}_2(2^s-1,5,\omega)) \ge 1 - \frac{2s}{2^s-1}, \qquad R(\mathcal{H}_2(s)) = 1 - \frac{s}{2^s-1}$$

are quite similar for even modest codeword lengths, and yet the BCH code corrects an additional error. We also mention that these BCH codes, while not perfect like the Hamming codes, are the next best thing – they are quasi-perfect. (See Section 4.2.)

DECODING BCH CODES

There are several efficient methods for the decoding of BCH codes. We will briefly discuss one method here, and then consider another method in detail in Section 8.3, where we can apply it to a broader class of codes that includes the BCH codes. To illustrate the present technique, we consider the case of a binary primitive narrow sense BCH code $\mathfrak{B} = \mathfrak{B}_2(n,5,\omega)$ with designed distance 5.

The defining zeros of such a code $\mathfrak{B}$ are $\omega,\omega^2,\omega^3,\omega^4$, where ω is a primitive n-th root of unity over F_2. We have already observed that ω and ω^3 will suffice as defining zeros, and so

$$\mathfrak{B} = \{p(x) \in R_n \mid p(\omega) = p(\omega^3) = 0\}$$

Now suppose that the codeword $c(x)$ was sent and that the received word $u(x)$ has at most 2 errors. Thus, we may write

$$u(x) = c(x) + e(x)$$

where the error polynomial $e(x)$ has weight at most 2. There are three possibilities for $e(x)$ to consider, namely, $e(x) = 0$, $e(x) = x^i$, or $e(x) = x^i + x^j$ with $0 \le i \ne j \le n-1$. If we let $u(\omega) = u_1$ and $u(\omega^3) = u_3$, then

$$e(\omega) = u_1 \quad \text{and} \quad e(\omega^3) = u_3 \tag{8.1.2}$$

No Errors

We leave it to the reader to show that no errors have occurred if and only if $u_1 = 0$, in which case $u_3 = 0$ as well.

Exactly One Error

If exactly one error has occurred, then $e(x) = x^i$, and so

$$u_3 = e(\omega^3) = \omega^{3i} = (\omega^i)^3 = [e(\omega)]^3 = u_1^3 \ne 0$$

Conversely, if $u_3 = u_1^3 \ne 0$, then $e(\omega^3) = [e(\omega)]^3$. Now, if we were to have $e(x) = x^i + x^j$ with $i \ne j$, then

$$\omega^{3i} + \omega^{3j} = (\omega^i + \omega^j)^3 = \omega^{3i} + \omega^{2i}\omega^j + \omega^i\omega^{2j} + \omega^{3j}$$

or

$$\omega^{2i}\omega^j + \omega^i\omega^{2j} = 0$$

Dividing by $\omega^i\omega^j$ gives

$$\omega^i + \omega^j = 0$$

which implies that $i = j$, a contradiction. Thus, we see that exactly one error has occurred if and only if $u_3 = u_1^3 \ne 0$.

Exactly Two Errors

Suppose now that exactly two errors have occurred, and so

$e(x) = x^i + x^j$ for $i \neq j$. Then (8.1.2) gives

$$\omega^i + \omega^j = u_1$$
$$\omega^{3i} + \omega^{3j} = u_3$$

Letting $X_1 = \omega^i$ and $X_2 = \omega^j$, these equations become

$$\begin{aligned} X_1 + X_2 &= u_1 \\ X_1^3 + X_2^3 &= u_3 \end{aligned} \tag{8.1.3}$$

The quantities X_1 and X_2 are called **error locators**, since they identify the location of the errors.

One way to solve equations (8.1.3) is to introduce a polynomial whose roots are X_1 and X_2. It is customary, however, to use the polynomial whose roots are the *inverses* of X_1 and X_2

$$\ell(x) = (1 - X_1x)(1 - X_2x) = 1 - (X_1 + X_2)x + X_1X_2x^2$$

This is known as an **error locator polynomial.** Now, $X_1 + X_2 = u_1$ and

$$u_3 = X_1^3 + X_2^3 = (X_1 + X_2)(X_1^2 + X_1X_2 + X_2^2) = u_1(u_1^2 + X_1X_2)$$

which gives $X_1X_2 = \frac{u_3}{u_1} - u_1^2$. Thus, the error locator polynomial is

$$\ell(x) = 1 - u_1x + \left(\frac{u_3}{u_1} - u_1^2\right)x^2$$

If n is not too large, we can find the roots of the error locator polynomial simply by trying all $2^n - 1$ nonzero values of F_{2^n}. (The quadratic formula does *not* apply over a field of characteristic 2!) Let us summarize.

Theorem 8.1.12 For $n \geq 15$, let $\mathcal{B}_2(n,5,\omega)$ be a double-error-correcting binary primitive narrow sense BCH code with designed distance 5. Assume that at most two errors have occurred in receiving the word $u(x)$, and let $u(\omega) = u_1$ and $u(\omega^3) = u_3$.

1) If $u_1 = 0$ then no errors have occurred.
2) If $u_3 = u_1^3 \neq 0$, then exactly one error has occurred, and if $u_1 = \omega^i$, the error occurred in the i-th position.
3) If $u_1 \neq 0$ and $u_3 \neq u_1^3$, then two errors have occurred. Furthermore, the locator polynomial

$$\ell(x) = 1 - u_1x + \left(\frac{u_3}{u_1} - u_1^2\right)x^2$$

has two distinct roots ω^{n-i} and ω^{n-j}, and the errors have occurred in the i-th and j-th positions. ▯

Example 8.1.6 Consider the binary narrow sense BCH code $\mathfrak{B} = \mathfrak{B}_2(15,5,\omega)$. The splitting field for $x^{15}-1$ is $F_{2^4} = F_{16}$, and so we work in this field. Recall from Example 7.2.1 that x^4+x+1 is a primitive polynomial over F_{16}. Let ω be a root of this polynomial. In Example 7.2.1, we saw that

$$m_1(x) = x^4+x+1 \quad \text{and} \quad m_3(x) = x^4+x^3+x^2+x+1$$

and so the generator polynomial for $\mathfrak{B}$ is

$$g(x) = m_1(x)m_3(x) = x^8+x^7+x^6+x^4+1$$

We can use this to obtain a codeword in $\mathfrak{B}$. In fact, let us take $g(x)$ itself.

By way of example, we introduce two errors – at the 4th and 12th positions – by taking

$$u(x) = x^{12}+x^8+x^7+x^6+1$$

Using the field table for F_{16}, we get

$$u_1 = u(\omega) = \omega^{12}+\omega^8+\omega^7+\omega^6+1 = \omega^6$$

and

$$u_3 = u(\omega^3) = \omega^{36}+\omega^{24}+\omega^{21}+\omega^{18}+1 = \omega^9+\omega^3+1 = \omega^4$$

Since $u_1 \neq 0$ and $u_1^3 = \omega^{18} = \omega^3 \neq u_3$, we deduce that two or more errors have been made (as we knew). The locator polynomial in this case is

$$\ell(x) = 1+\omega^6 x+(\omega^{13}+\omega^{12})x^2$$

By substituting $1,\omega,\ldots,\omega^{14}$ into this polynomial, we find that ω^3 and ω^{11} are roots, and so we discover that, if exactly two errors have occurred, then those errors have indeed occurred in the 12th and 4th positions, as arranged. ▯

The General Case

The decoding procedure described above can easily be generalized to all binary BCH codes. Let us describe this procedure in brief terms. For more details, we refer the reader to MacWilliams and Sloane (1977), Berlekamp (1984), Peterson and Weldon (1972) or Gallagher (1968).

If $u(x) = c(x)+e(x)$ is the received word, the quantities

$$u_1 = e(\omega),\ u_3 = e(\omega^3),\ldots,\ u_{\delta-2} = e(\omega^{\delta-2}) \tag{8.1.4}$$

are easily determined. (We assume that δ is odd.) Suppose that errors have occurred at locations $i_1,\ldots,i_w$ in $e(x)$, and so

$$e(x) = \sum_{j=1}^{w} x^{i_j}$$

Substituting this into (8.1.4) gives the system of equations

$$\sum_{j=1}^{w} \omega^{i_j} = u_1, \quad \sum_{j=1}^{w} \omega^{3i_j} = u_3, \cdots, \quad \sum_{j=1}^{w} \omega^{(\delta-2)i_j} = u_{\delta-2}$$

Next, we define the **error locators** by $X_j = \omega^{i_j}$. Observe that knowing the error locators tells us the positions of the errors. The above system can now be written in the more compact form

$$\sum_{j=1}^{w} X_j = u_1, \quad \sum_{j=1}^{w} X_j^3 = u_3, \cdots, \quad \sum_{j=1}^{w} X_j^{\delta-2} = u_{\delta-2}$$

The **error locator polynomial** is defined by

$$\ell(x) = \prod_{j=1}^{w} (1 - X_j x)$$

The roots of $\ell(x)$ are the inverses of the locators X_j. Of course, we do not know the values of the locators X_j ahead of time, and so we must determine the locator polynomial by some other means. In fact, if

$$\ell(x) = \sum_{i=0}^{w} \sigma_i x^i$$

then we want to determine the coefficients σ_i directly from the known quantities u_i in (8.1.4).

One way to do this (although not the most efficient way) is to use the fact that the σ_i's and u_i's are related by the matrix equation

$$\begin{bmatrix} 1 & 0 & 0 & 0 & 0 & \cdots & 0 \\ u_2 & u_1 & 1 & 0 & 0 & \cdots & 0 \\ u_4 & u_3 & u_2 & u_1 & 1 & \cdots & 0 \\ \vdots & \vdots & \vdots & \vdots & \vdots & \cdots & \vdots \\ u_{2w-4} & u_{2w-5} & & \cdots & & \cdots & u_{w-3} \\ u_{2w-2} & u_{2w-3} & & \cdots & & \cdots & u_{w-1} \end{bmatrix} \begin{bmatrix} \sigma_1 \\ \sigma_2 \\ \sigma_3 \\ \vdots \\ \sigma_{2w-3} \\ \sigma_{2w-1} \end{bmatrix} = \begin{bmatrix} u_1 \\ u_3 \\ u_5 \\ \vdots \\ u_{2w-3} \\ u_{2w-1} \end{bmatrix}$$

where we must also compute u_k when k is even, which can easily be done by observing that

$$u_{2j} = e(\omega^{2j}) = [e(\omega^j)]^2 = u_j^2$$

For instance, in our previous example of double-error-correcting

codes, we have $w = 2$ and so this matrix equation becomes

$$\begin{bmatrix} 1 & 0 \\ u_2 & u_1 \end{bmatrix}\begin{bmatrix} \sigma_1 \\ \sigma_2 \end{bmatrix} = \begin{bmatrix} u_1 \\ u_3 \end{bmatrix}$$

or, equivalently, since $u_2 = u_1^2$,

$$\begin{aligned} \sigma_1 &= u_1 \\ u_1^2\sigma_1 + u_1\sigma_2 &= u_3 \end{aligned}$$

whose solution is

$$\sigma_1 = u_1,\ \sigma_2 = \frac{u_3 - u_1^3}{u_1} = \frac{u_3}{u_1} - u_1^2$$

which agrees with our previous results.

Once the coefficients of the locator polynomial are found, the roots can be determined by trial and error, if the field is not too large.

EXERCISES

1. Does a narrow sense BCH code constructed using a primitive n-th root of unity ω depend on the choice of ω? Justify your answer. For a fixed primitive n-th root of unity ω, does the set of all narrow sense BCH codes of a given designed distance depend on the choice of ω?
2. Describe the code $\mathfrak{B}_2(7,5,\omega)$.
3. Determine all binary narrow sense BCH codes of length 15.
4. Determine all binary narrow sense BCH codes of length 23.
5. Show that the contents of Table 8.1.1 (except for the last column) are accurate.
6. What is the minimum distance of the binary narrow sense BCH $\mathfrak{B}_2(255,51,\omega)$?
7. Construct a ternary BCH code $\mathfrak{B}_3(26,5,\omega)$.
8. Prove Theorem 8.1.2.
9. Find the generator polynomial of a ternary BCH code of length 8 and dimension 5. What is the Bose distance of this code?
10. With reference to the decoding of BCH codes, show that $e(x) = 0$ if and only if $u_1 = 0$, in which case $u_3 = 0$ as well.
11. Referring to Example 8.1.6, decode the received word $u(x) = x^{10} + x^5 + x^4 + x + 1$.
12. Referring to Example 8.1.6, suppose we receive the word $u(x) = x^{11} + x^{10} + x^6 + x^5 + x^4 + x + 1$. What can you conclude about the number of errors?
13. Show that the dual of a BCH code need not be another BCH code.

14. Find a binary BCH code of length $n = 7$ that is self-orthogonal. *Hint.* What is the relationship between the zeros of a cyclic code and the zeros of its dual?
15. Show that the map ρ defined in the subsection on the automorphisms of a BCH code is indeed a permutation of the coordinate positions $0, \ldots, n$ and that the set of all such permutations ρ, obtained from all affine transformations $P_{u,v}$, is **doubly transitive**, that is, for any $i \neq j$ and $k \neq \ell$, there exists a permutation ρ for which $\rho(i) = k$ and $\rho(j) = \ell$.
16. Finish the proof of Theorem 8.1.5 by discussing the case $v = 0$.

8.2 Reed-Solomon and Justesen Codes

In this section, we discuss two special types of BCH codes, beginning with the very useful Reed-Solomon codes.

REED-SOLOMON CODES

Reed-Solomon codes are used quite often in practice. In fact, use of these codes, in conjunction with non-block codes, has been a standard procedure at NASA since the launch of the Voyager spacecraft in 1977. These codes have been used, for instance, on the Galileo, Magellan, and Ulysses missions.

As we will see, Reed-Solomon codes are maximum distance separable (MDS) and hence have the largest possible minimum distance for codes of their size and dimension. Also, they are very useful as building blocks for other codes and in correcting burst errors.

Definition Let $q \geq 3$. A q-ary **Reed-Solomon code** (or **RS code**) $\mathcal{S} = \mathcal{S}(n,\delta,\omega,b)$ is a q-ary BCH code $\mathcal{B}_q(q-1,\delta,\omega,b)$ with length $n = q-1$. □

Thus, the length of a q-ary Reed-Solomon code is equal to the number of nonzero elements in the base field F_q.

Since $n = q-1$, and since the roots of the polynomial $x^{q-1}-1$ are precisely the nonzero elements of F_q, we have

$$x^n - 1 = x^{q-1} - 1 = \prod_{\alpha \in F_q^*} (x-\alpha)$$

Thus, if ω is a primitive (q−1)-st root of unity over F_q (that is, a primitive element of F_q), then

$$m_i(x) = x - \omega^i$$

As a result, a Reed-Solomon code with designed distance δ has generator polynomial

$$g(x) = (x-\omega^b)(x-\omega^{b+1})\cdots(x-\omega^{b+\delta-1})$$

where $b \geq 0$.

Example 8.2.1 Let $q = 8$, and so $n = 7$. A root ω of the primitive polynomial x^3+x+1 over F_2 gives us a primitive element of F_8. To obtain an RS code of dimension 5, we want $\deg(g(x)) = 7-5 = 2$. For instance, we could take

$$\begin{aligned} g(x) &= (x - \omega)(x - \omega^2) \\ &= x^2 - (\omega + \omega^2)x + w^3 \\ &= x^2 + \omega^4 x + \omega^3 \end{aligned}$$ □

PROPERTIES OF THE REED-SOLOMON CODES

The Reed-Solomon Codes are MDS Codes

The generator polynomial of a Reed-Solomon code of designed distance δ has degree $\deg(g(x)) = \delta - 1$. Hence,

$$k = dim(\mathcal{S}(n,\delta,\omega,b)) = n - \deg(g(x)) = n - \delta + 1$$

and so, according to the BCH bound, the minimum distance of this code satisfies

$$d \geq \delta = n - k + 1$$

But the Singleton bound is $d \leq n - k + 1$, and so

$$d = \delta = n - k + 1$$

which tells us that $\mathcal{S}(n,\delta,\omega,b)$ is an MDS code and also that the minimum distance is equal to the designed distance.

Theorem 8.2.1 Reed-Solomon codes are maximum distance separable (MDS) codes. Furthermore, the minimum distance of a Reed-Solomon code is equal to its designed distance. □

Hence, RS codes have the largest possible minimum distance among all q-ary codes of length $n = q - 1$ and dimension $k = n - \delta + 1$.

The Dual of a Reed-Solomon Code

In general, the dual of a BCH code may not be another BCH code. For instance, let $n = 25$ and $q = 2$. The cyclotomic cosets of q modulo n are

$$\begin{aligned} C_0 &= \{0\} \\ C_1 &= \{1,2,4,8,16,7,14,3,6,12,24,23,21,17,9,18,11,22,19,13\} \\ C_5 &= \{5,10,20,15\} \end{aligned}$$

Let $\mathcal{B} = \mathcal{B}_2(25,1,\omega)$ be the BCH code with defining zero ω. The dual code $\mathcal{B}^\perp$ has zeros that are the reciprocals of the nonzeros of $\mathcal{B}$, and so $\mathcal{B}^\perp$ has zeros

$$(8.2.1) \qquad \{\omega^0,\omega^{-5},\omega^{-10},\omega^{-20},\omega^{-15}\} = \{\omega^0,\omega^{20},\omega^{15},\omega^5,\omega^{10}\}$$

Hence, $\mathfrak{B}^\perp$ is not a BCH code. On the other hand, for Reed-Solomon codes, we have the following, whose proof is left as an exercise.

Theorem 8.2.2 The dual of a Reed-Solomon code is a Reed-Solomon code. □

Extending a Reed-Solomon Code

Recall from Section 4.3 that if C is a q-ary (n,M,d)-code, the extended code $\widehat{C}$ is defined by adding an overall parity check c_n to each codeword $\mathbf{c} = c_0 \cdots c_{n-1}$ that satisfies

$$\sum_{k=0}^{n} c_k = 0$$

If $\widehat{C}$ is an $(\widehat{n},\widehat{M},\widehat{d})$-code, then

$$\widehat{n} = n+1, \quad \widehat{M} = M, \quad \widehat{d} = d \text{ or } d+1$$

In the exercises, we ask the reader to provide an example of a code C_1 for which $\widehat{d}_1 = d_1$ and a code C_2 for which $\widehat{d}_2 = d_2 + 1$. On the other hand, for Reed-Solomon codes, the minimum distance of the extended code is always larger than the minimum distance of the original.

Theorem 8.2.3 Let $\mathcal{S}$ be an [n,k,d] Reed-Solomon code with generator polynomial

$$g(x) = (x-\omega)(x-\omega^2)\cdots(x-\omega^{d-1})$$

Then the extended code $\widehat{\mathcal{S}}$ is an [n+1,k,d+1]-code.
Proof. Let $c(x) = c_0+c_1x+\cdots+c_{n-1}x^{n-1}$ be a codeword in $\mathcal{S}$ of weight d. Extending c(x) gives

$$\widehat{c}(x) = c_0 + c_1x + \cdots + c_{n-1}x^{n-1} + c_nx^n$$

where

$$c_n = -\sum_{i=0}^{n-1} c_i = -c(1)$$

We want to show that the weight of $\widehat{c}(x)$ is $d+1$, which is true if and only if $c(1) \neq 0$.

Since g(x) is the generator polynomial for $\mathcal{S}$, we have $c(x) = p(x)g(x)$ for some polynomial p(x), and so $c(1) = p(1)g(1)$. Clearly, $g(1) \neq 0$, since $\omega^i \neq 1$ for $i = 1,\ldots,d-1$. If $p(1) = 0$, then the polynomial $g_1(x) = (x-1)g(x)$ would divide c(x). Hence, c(x) would be in the cyclic code generated by

$$g_1(x) = (x-1)(x-\omega)(x-\omega^2)\cdots(x-\omega^{d-1})$$

which has d "consecutive" zeros $\omega^0,\ldots,\omega^{d-1}$, and so by the BCH bound, $c(x)$ would have weight at least $d+1$. This is a contradiction to the fact that $c(x)$ has weight d. Hence, $p(1) \neq 0$ and the proof is complete. ∎

OBTAINING A BINARY CODE FROM A 2^m-ARY CODE

A q-ary code with large q may not be practical in some situations. However, if $q = 2^m$ for some m, we may turn a q-ary $[n,k]$-code into a *binary* $[mn,mk]$-code in a very straightforward way.

Let $\{\alpha_1,\ldots,\alpha_m\}$ be an *ordered* basis for F_{2^m} over F_2. To each element of F_{2^m}

$$a = a_1\alpha_1 + \cdots + a_m\alpha_m$$

we can associate the string $a_1 \cdots a_m$. Let us denote this by writing $a \to a_1 \cdots a_m$.

If C is a 2^m-ary $[n,k]$-code, each codeword in C has the form $\mathbf{c} = c_0 \cdots c_{n-1}$, where $c_i \in F_{2^m}$. Therefore, if

$$c_i \to c_{i1} \cdots c_{im}$$

we can associate to $\mathbf{c} \in C$ the string

$$\mathbf{c} \to (c_{11} \cdots c_{1m})(c_{21} \cdots c_{2m}) \cdots (c_{n1} \cdots c_{nm})$$

of length mn. (The parentheses are for readability only.) The code $C^{\to}$ consisting of all such strings is a binary code of length mn.

Furthermore, $C^{\to}$ is a vector space over F_2 of size $|C^{\to}| = |C| = (2^m)^k = 2^{mk}$. Hence, it has dimension mk over F_2. Finally, if $a \neq b$ and

$$a \to a_1 \cdots a_m \quad \text{and} \quad b \to b_1 \cdots b_m$$

then for some i we must have $a_i \neq b_i$. Hence, the minimum distance of $C^{\to}$ is at least d. Let us summarize.

Theorem 8.2.4 Let C be a 2^m-ary $[n,k,d]$-code. Let $\{\alpha_1,\ldots,\alpha_m\}$ be an ordered basis for F_{2^m} over F_2. The code $C^{\to}$, obtained by replacing each code symbol $a = \sum a_i\alpha_i$ in each codeword of C by the corresponding m-string $a_1,\ldots,a_m$, is a binary $[n^{\to},k^{\to},d^{\to}]$-code, where

$$n^{\to} = mn, \quad k^{\to} = mk, \quad d^{\to} \geq d$$ □

In addition to replacing each codeword symbol in C by the corresponding m-string, we may also add a parity check symbol to the codeword for each m-string, as follows

$$\mathbf{c} \to (c_{11} \cdots c_{1m}x_1)(c_{21} \cdots c_{2m}x_2) \cdots (c_{n1} \cdots c_{nm}x_n)$$

where

$$c_{i1} + \cdots + c_{im} + x_i = 0$$

for $i = 1,\ldots,n$. The resulting code $C^{\uparrow}$ has parameters

$$n^{\uparrow} = (m+1)n, \quad k^{\uparrow} = mk, \quad d^{\uparrow} \geq 2d$$

To see that $d^{\uparrow} \geq 2d$, observe that at least d of the original m-strings in any codeword in $C^{\rightarrow}$ must be nonzero. But if any of these strings has weight 1, then the addition of a parity check will increase the weight to 2. In other words, the weight of each (m+1)-string is at least 2, and so the total weight of a codeword in $C^{\uparrow}$ is at least 2d. Notice that the rate of $C^{\uparrow}$ is

$$R(C^{\uparrow}) = \frac{mk}{(m+1)n} = \frac{m}{m+1} R(C)$$

which, for reasonably large m, is not much less than the rate of the original code C. On the other hand, its minimum distance is at least twice that of C. Thus, for a rather simple device, we can gain significant error correcting ability

We should mention that if C is a cyclic code, the codes $C^{\rightarrow}$ and $C^{\uparrow}$ need not be cyclic. Also, the minimum distance of the code $C^{\rightarrow}$ can depend on the choice of basis for F_{2^m} over F_2.

Example 8.2.2 Let $q = 4$, $n = 3$, and $k = 2$. The polynomial $x^2 + x + 1$ is primitive for F_4 over F_2 and so if ω is a root of this polynomial, we have $F_4 = \{0,1,\omega,\omega^2\}$.

A 4-ary [3,2,2] Reed-Solomon code $\mathcal{S}$ has generator polynomial of degree 1. Let us take $g(x) = x - \omega$. The code $\mathcal{S}$ is then

$$\mathcal{S} = \{p(x)(x-\omega) \mid \deg(p(x)) \leq 1\}$$

For instance, one codeword in $\mathcal{S}$ is $(x+\omega)(x-\omega) = x^2 - \omega^2$, which, in string notation, is $\omega^2 01$. We leave it for the reader to verify that the 16 codewords in $\mathcal{S}$ are

$$000,\ \omega 10,\ \omega^2\omega 0,\ 1\omega^2 0$$

$$0\omega 1,\ 0\omega^2\omega,\ 01\omega^2,\ \omega\omega^2 1$$

$$\omega^2 01,\ 111,\ \omega\omega\omega,\ \omega^2 1\omega$$

$$10\omega,\ \omega\omega\omega^2,\ \omega^2\omega^2\omega^2,\ 1\omega\omega^2$$

The set $\{1,\omega\}$ is an ordered basis for F_4 over F_2. Under this basis, we have $0 \rightarrow 00$, $1 \rightarrow 10$, $\omega \rightarrow 01$, and $\omega^2 \rightarrow 11$, and so the binary [6,4]-code $\mathcal{S}^{\rightarrow}$ is

$$000000,\ 011000,\ 110100,\ 101100$$
$$000110,\ 001101,\ 001011,\ 011110$$
$$110010,\ 101010,\ 010101,\ 111001$$
$$100001,\ 010111,\ 111111,\ 100111$$

We leave it to the reader to compute the code $\mathcal{S}^{\uparrow}$. □

BURST ERROR CORRECTION

The expanded binary code $C^{\rightarrow}$ can have good burst error correcting capabilities. If the original code C is t-error-correcting, then the expanded code $C^{\rightarrow}$ can correct bursts of length up to

$$b = (t-1)m + 1$$

since any burst of length b bits or less can effect at most t m-strings. (See Figure 8.2.1.)

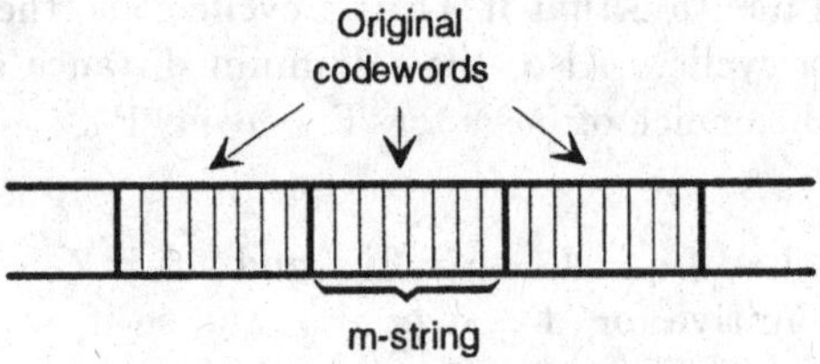

Figure 8.2.1

Example 8.2.3 Suppose we wish to find a binary code, based on an RS code, that can correct bursts of length $b = 13$ or less. Then we want

$$(t-1)m + 1 = 13$$

or

$$(t-1)m = 12$$

Assuming an odd minimum distance $d = 2t + 1$, this is equivalent to

$$(d-3)m = 24$$

Since $d \le n = 2^m - 1$, we may choose $m = 4$, $d = 9$. In this case, the Reed-Solomon code $\mathcal{S}$ has parameters [15,7,9], and is over F_{16}. The expanded binary code $\mathcal{S}^{\rightarrow}$ has parameters $[60,28,\ge 9]$ and can correct bursts of length 13 or less. □

IDEMPOTENTS OF REED-SOLOMON CODES

For $q = 2^m$, the minimal q-ary cyclic codes $\widehat{M}_i$ of length $n = 2^m - 1$ have check polynomial $h_i(x) = x - \omega^i$ and generating polynomial

$$g_i(x) = \frac{x^n - 1}{x - \omega^i}$$

It is easy to see that each codeword in $\widehat{M}_i$ has weight n.

Now suppose that $\theta_i(x)$ is the primitive idempotent for $\widehat{M}_i$, and let

$$\theta_i(x) = 1 + \epsilon_1 x + \cdots + \epsilon_{n-1} x^{n-1} \tag{8.2.2}$$

(Why is the constant term equal to 1?) Shifting this cyclically j times gives the codeword

$$p(x) = \epsilon_{n-j} + \epsilon_{n-j+1} x + \cdots + \epsilon_{n-j-1} x^{n-1}$$

Since $\theta_i(x)$ and $p(x)$ are in $\widehat{M}_i$, so is

$$\epsilon_{n-j+1}\theta_i(x) - \epsilon_1 p(x)$$

But this codeword has weight less than n, since the linear term has vanished, and so must be the zero polynomial. In particular, the constant term must be 0, giving

$$\epsilon_{n-j+1} = \epsilon_{n-j}\epsilon_1$$

or

$$\epsilon_{k+1} = \epsilon_k \epsilon_1$$

Setting $\epsilon_1 = \epsilon$, we thus have $\epsilon_k = \epsilon^k$, and so

$$\theta_i(x) = 1 + \epsilon x + (\epsilon x)^2 + \cdots + (\epsilon x)^{n-1}$$

$$= \frac{(\epsilon x)^n - 1}{\epsilon x - 1} = \frac{x^n - 1}{\epsilon x - 1} = \epsilon^{-1}\frac{x^n - 1}{x - \epsilon^{-1}} = \omega^j \frac{x^n - 1}{x - \omega^j}$$

since $\epsilon^{-1} = \omega^j$ for some j. But, according to Theorem 7.4.10,

$$g_i(x) = \gcd(\theta_i(x), x^n - 1) = \gcd\left(\omega^j \frac{x^n - 1}{x - \omega^j}, x^n - 1\right) = \frac{x^n - 1}{x - \omega^j}$$

(recall that $g_i(x)$ must be monic) and so $j = i$ and $\epsilon = \omega^{-i}$. We have established the following.

Theorem 8.2.5 The minimal q-ary cyclic codes $\widehat{M}_i$ of length $n = 2^m - 1$ have generating polynomials and (primitive) idempotents given by

$$g_i(x) = \frac{x^n - 1}{x - \omega^i}$$

and

$$\theta_i(x) = \omega^i g_i(x) = 1 + \omega^{-i}x + (\omega^{-i}x)^2 + \cdots + (\omega^{-i}x)^{n-1} \qquad \square$$

Once we have determined the primitive idempotents in R_n, we can determine the generating idempotent for any cyclic code in R_n, using Theorem 7.4.15.

Example 8.2.4 Let $q = 8$. Then $m = 3$ and $n = 7$. If ω is a primitive 7th root of unity, then the primitive idempotents in R_7 are (note the relationship between the coefficients of $\theta_i(x)$ and $\theta_{n-i}(x)$)

$$\theta_0(x) = 1 + x + x^2 + x^3 + x^4 + x^5 + x^6$$

$$\theta_1(x) = 1 + \omega^6x + \omega^5x^2 + \omega^4x^3 + \omega^3x^4 + \omega^2x^5 + \omega x^6$$

$$\theta_2(x) = 1 + \omega^5x + \omega^3x^2 + \omega x^3 + \omega^6x^4 + \omega^4x^5 + \omega^2x^6$$

$$\theta_3(x) = 1 + \omega^4x + \omega x^2 + \omega^5x^3 + \omega^2x^4 + \omega^6x^5 + \omega^3x^6$$

$$\theta_4(x) = 1 + \omega^3x + \omega^6x^2 + \omega^2x^3 + \omega^5x^4 + \omega x^5 + \omega^4x^6$$

$$\theta_5(x) = 1 + \omega^2x + \omega^4x^2 + \omega^6x^3 + \omega x^4 + \omega^3x^5 + \omega^5x^6$$

$$\theta_6(x) = 1 + \omega x + \omega^2x^2 + \omega^3x^3 + \omega^4x^4 + \omega^5x^5 + \omega^6x^6$$

Now, let us consider the [7,2,6] Reed-Solomon code $\mathcal{S}$ with generator polynomial $g(x) = (x - \omega)\cdots(x - \omega^5)$. According to Theorem 7.4.15, the generating idempotent of this code is

$$\begin{aligned} e(x) &= \theta_0(x) + \theta_6(x) \\ &= (\omega{+}1)x + (\omega^2{+}1)x^2 + (\omega^3{+}1)x^3 + (\omega^4{+}1)x^4 \\ &\quad + (\omega^5{+}1)x^5 + (\omega^6{+}1)x^6 \\ &= \omega^3x + \omega^6x^2 + \omega x^3 + \omega^5x^4 + \omega^4x^5 + \omega^2x^6 \end{aligned} \qquad \square$$

ENCODING REED-SOLOMON CODES

Let us briefly discuss the method of encoding RS codes that Reed and Solomon originally used. While it is not systematic, it does have some advantages over the methods discussed in Section 7.5.

Let $\mathcal{S}$ be an [n,k,d] Reed-Solomon code, where $n = q - 1$. Suppose that $a_0 \cdots a_{k-1}$ is a message string. We set

$$a(x) = \sum_{i=0}^{k-1} a_i x^i$$

and encode $a(x)$ as the polynomial

$$c(x) = \sum_{j=0}^{n-1} a(\omega^j)x^j$$

Of course, we must show that $c(x)$ is a codeword in $\mathcal{S}$, in other words, that

$$c(\omega) = \cdots = c(\omega^{d-1}) = 0$$

But, thinking of $a(x)$ as the polynomial

$$a(x) = \sum_{i=0}^{n-1} a_i x^i$$

with the additional coefficients a_i $(i \geq k)$ set equal to 0, Lemma 7.4.18 gives, for $i = 0,\ldots,n-1$,

$$a_i = \frac{1}{n}\sum_{j=0}^{n-1} a(\omega^j)\omega^{-ij} = \frac{1}{n}\,c(\omega^{-i}) = \frac{1}{n}\,c(\omega^{n-i})$$

and so

$$c(\omega^{n-i}) = na_i \tag{8.2.3}$$

In particular, for $0 \leq j \leq d-1 = n-k$,

$$c(\omega^j) = na_{n-j} = 0$$

which shows that $c(x)$ is indeed in $\mathcal{S}$. Furthermore, (8.2.3) shows that it is easy to recover the original message symbols from the codeword by evaluating $c(x)$ at various powers of ω.

DECODING REED-SOLOMON CODES

Since Reed-Solomon codes are BCH codes, they can be decoded by the method described in the previous section. Also, we will discuss another, more efficient, decoding procedure in Section 8.3. However, let us now briefly discuss the original method of Reed and Solomon, which is based on a form of majority logic. Unfortunately, this method, while interesting, is not very practical for [n,k,d]-codes in which $\binom{n}{k}$ is large.

As in the previous subsection, suppose that the original message is

$$a(x) = \sum_{i=0}^{k-1} a_i x^i$$

and that the codeword

$$c(x) = \sum_{i=0}^{n-1} a(\omega^i)x^i$$

is transmitted. In string notation, the message is

$$\mathbf{a} = a_0 a_1 \cdots a_{k-1}$$

and the codeword is

$$\mathbf{c} = c_0 c_1 \cdots c_{n-1} = a(1)a(\omega)\cdots a(\omega^{n-1})$$

Now, let $\mathbf{u} = u_0 \cdots u_{n-1}$ be the received word and $\mathbf{e} = e_0 \cdots e_{n-1}$ be the error. Thus, the quantities

$$u_i = e_i + c_i = e_i + a(\omega^i)$$

are known, and

$$\begin{aligned} u_0 &= e_0 + a_0 + a_1 + \cdots + a_{k-1} \\ u_1 &= e_1 + a_0 + \omega a_1 + \cdots + \omega^{k-1} a_{k-1} \\ u_2 &= e_2 + a_0 + \omega^2 a_1 + \cdots + \omega^{2(k-1)} a_{k-1} \\ &\vdots \\ u_{n-1} &= e_{n-1} + a_0 + \omega^{n-1} a_1 + \cdots + \omega^{(k-1)(n-1)} a_{k-1} \end{aligned} \tag{8.2.4}$$

If there are no errors in transmission, then the e_i's are all 0, and any k of these equations can be solved for the k unknowns $a_0, \ldots, a_{k-1}$. This follows from the fact that the matrix of coefficients of any k equations is a Vandermonde matrix. (See Lemma 5.3.14.)

On the other hand, suppose that t errors have occurred. Let us refer to the equations in (8.2.4) for which $e_i = 0$ as *good* equations and the equations for which $e_i \neq 0$ as the *bad* equations. Hence, there are t bad equations and $n-t$ good equations.

If we solve all subsystems of k equations in (8.2.4), then the $\binom{n-t}{k}$ subsystems consisting of all good equations will give the correct values of the a_i's. Moreover, a given *incorrect* solution cannot satisfy any set of k good equations, and so it is a solution to at most $t+k-1$ equations, that is, it can be obtained at most $\binom{t+k-1}{k}$ times from subsystems. Therefore, if

$$\binom{n-t}{k} > \binom{t+k-1}{k} \tag{8.2.5}$$

then the majority solution from among the $\binom{n}{k}$ solutions will give the true values of the a_i's.

Finally, we observe that (8.2.5) holds if and only if $n-t > t+k-1$, that is, if and only if $d = n-k+1 > 2t$.

Theorem 8.2.6 Let $\mathcal{S}$ be an $[n,k,d]$ Reed-Solomon code. The majority logic decoding method of this subsection will correct up to $t < \frac{1}{2}d$ errors in transmission, at the cost of having to solve $\binom{n}{k}$ systems of equations of size $k \times k$. □

ASYMPTOTICALLY GOOD CODES

We remarked in the previous section that BCH codes do not have good error correcting capabilities for large codeword lengths. More precisely, Theorem 8.1.11 says that there does not exist an infinite sequence of q-ary primitive BCH codes with the property that both d/n and k/n are bounded away from 0.

A family $\mathcal{F}$ of q-ary codes is said to be **asymptotically good** if $\mathcal{F}$ contains an infinite sequence of codes $C_1, C_2, \ldots$, where C_i has parameters $[n_i, k_i, d_i]$, with the property that the rates $R(C_i) = k_i/n_i$ and the numbers d_i/n_i are bounded away from 0. Otherwise, the family $\mathcal{F}$ is **asymptotically bad.** As we saw in Corollary 4.2.7, asymptotically bad families of codes either have their rates tending to 0 or their probabilities of error correction tending to 0.

Thus, BCH codes form an asymptotically bad family of codes. It can also be shown that the binary codes $\mathcal{S}^{\rightarrow}$ that come from Reed-Solomon codes, as described earlier in this section, are asymptotically bad.

FINDING GOOD FAMILIES OF CODES

There is an elegantly simple way to construct a good family of codes – at least in theory. Let $0 \neq \alpha \in F_{2^m}$. Thinking of the elements of F_{2^m} as m-tuples in F_2^m as well, let

$$C_\alpha = \{(\mathbf{a}, \alpha\mathbf{a}) \mid \mathbf{a} \in F_2^m\}$$

This is a binary code of length $n = 2m$, dimension $k = m$, and hence rate $R = 1/2$.

Notice that we can recover α from any nonzero element of C_α, since if $\mathbf{0} \neq (\mathbf{a}, \mathbf{b}) \in C_\alpha$, then $\alpha = \mathbf{b}/\mathbf{a}$ (division in F_{2^m}). Therefore, if $\alpha_1 \neq \alpha_2$, we have

$$C_{\alpha_1} \cap C_{\alpha_2} = \{\mathbf{0}\}$$

Given $\lambda = \lambda_m$, suppose we wish to find a $\beta = \beta_m$ for which C_β has minimum weight at least $2m\lambda$. Since a nonzero binary (2m)-tuple can appear in at most one C_α, we can find such a β provided that the number of nonzero binary (2m)-tuples of weight less than $2m\lambda$ is less than the number $2^m - 1$ of distinct codes C_α, that is, provided that

$$\sum_{i=1}^{2m\lambda-1} \binom{2m}{i} < 2^m - 1$$

or, a fortiori,

$$\sum_{i=1}^{2m\lambda} \binom{2m}{i} < 2^m - 1$$

Using Theorem 1.2.8, we may find such a code C_β provided that

$$2^{2mH(\lambda)} < 2^m - 1$$

where $H(x)$ is the binary entropy function. If we take, for instance,

$$\lambda = \lambda_m = H^{-1}(\tfrac{1}{2} - \tfrac{1}{\log m})$$

(for $\lambda < 1/2$) this becomes

$$2^{m-2m/\log m} < 2^m - 1$$

which certainly holds for m large.

Finally, since $\lambda_m \to H^{-1}(\frac{1}{2})$ as $m \to \infty$, we have found a sequence of codes C_{β_m} of length $n = 2m$, rate $R = 1/2$, and

$$\frac{d}{n} \geq \frac{2m\lambda_m}{2m} = \lambda_m \to H^{-1}(\tfrac{1}{2}) > 0$$

Thus, the family $\{C_{\beta_m}\}$ is good.

Unfortunately, as is often the case, the practice has not kept up with the theory, and we do not yet know how to choose the appropriate β_m's to obtain a concrete example of a good family of codes in this way. However, in 1972, Justesen devised a clever method to avoid having to find appropriate β_m's. His idea was to use *all* nonzero elements of F_{2^m} and employ a construction known as *concatenation of codes.*

CONCATENATION OF CODES

Concatenation of codes is a construction that is used quite often in practice. For example, NASA often uses a code constructed by concatenating a Reed-Solomon code with a convolutional (nonblock) code.

Consider a binary k_1k_2-tuple **a**. Referring to Figure 8.2.2, we group **a** into k_1 individual k_2-tuples. Each binary k_2-tuple can be thought of as an element of the field $GF(2^{k_2})$, and so **a** can be thought of as a k_1-tuple

$$\mathbf{a} = a_1 \cdots a_{k_1-1}$$

over $GF(2^{k_2})$.

Now let C_1 be an $[n_1,k_1,d_1]$-code over $GF(2^{k_2})$. We refer to C_1 as the **outer code.** The string **a** is encoded, using the outer code C_1, into an n_1-tuple

$$\mathbf{c} = c_1 \cdots c_{n_1-1}$$

also over $GF(2^{k_2})$. Next, each one of the n_1 individual 2^{k_2}-tuples is encoded, using an $[n_2,k_2,d_2]$-code C_2, known as the **inner code.** The

result is a string of length n_1n_2.

Thus, the two-step encoding process, referred to as **superencoding**, takes a k_1k_2-tuple as input and produces an n_1n_2-tuple as output. The resulting code is referred to as a **supercode** and has length n_1n_2 and dimension k_1k_2. We leave it as an exercise to show that the minimum distance of the supercode is at least d_1d_2.

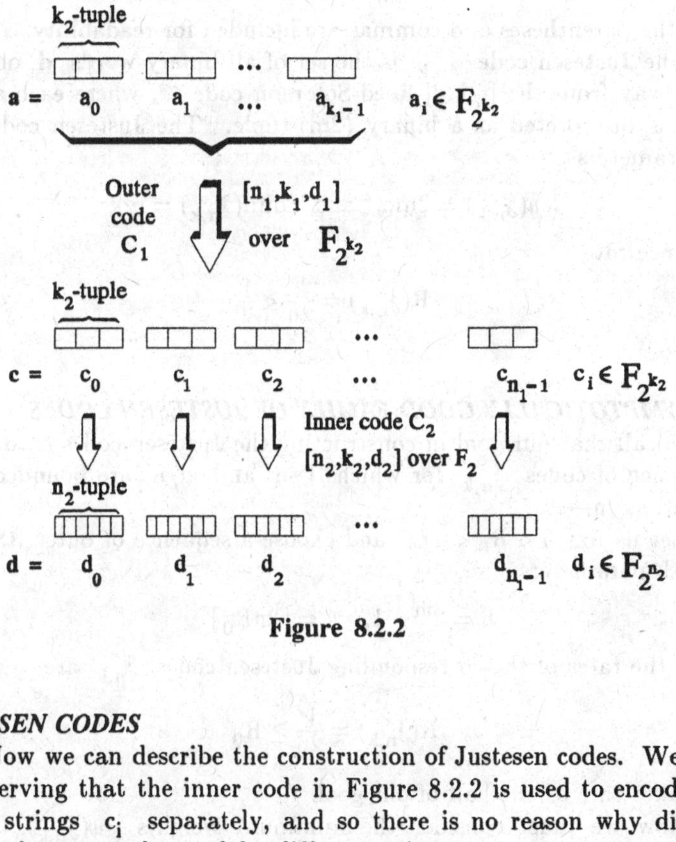

Figure 8.2.2

JUSTESEN CODES

Now we can describe the construction of Justesen codes. We begin by observing that the inner code in Figure 8.2.2 is used to encode each of the strings c_i separately, and so there is no reason why different inner codes cannot be used for different c_i's.

Let the outer code C_1 be an $[n=2^m-1,k,d]$ Reed-Solomon code $\mathcal{S}$ over F_{2^m}. Any binary $(k2^m)$-tuple **a** of message symbols can be written in the form

$$\mathbf{a} = a_0 \cdots a_{k-1}, \quad a_i \in F_{2^m}$$

Encoding **a** using $\mathcal{S}$ gives a codeword

$$\mathbf{c} = c_0 \cdots c_{n-1}, \quad c_i \in F_{2^m}$$

Let β be a primitive element of F_{2^m}. For each $i = 0,\ldots,n-1 =$

2^m-2, we consider the inner code

$$C_{\beta^i} = \{(\mathbf{c},\beta^i\mathbf{c}) \mid \mathbf{c} \in F_2^m\}$$

discussed earlier. Encoding **c** using these inner codes gives the supercodeword

$$\mathbf{d} = (c_0,c_0)(c_1,\beta c_1)(c_2\beta^2,c_2)\cdots(c_{n-1},\beta^{n-1}c_{n-1})$$

where the parentheses and commas are included for readability.

The **Justesen code** $\mathcal{J}_{n,k}$ is the set of all binary words **d** obtained in this way from the [n,k,d]-Reed-Solomon code $\mathcal{S}$, where each symbol in **d** is interpreted as a binary (2m)-tuple. The Justesen code $\mathcal{J}_{n,k}$ has parameters

$$len(\mathcal{J}_{n,k}) = 2mn \quad \text{and} \quad \dim(\mathcal{J}_{n,k}) = mk$$

and hence rate

$$R(\mathcal{J}_{n,k}) = \frac{k}{2n} < \frac{1}{2}$$

AN ASYMPTOTICALLY GOOD FAMILY OF JUSTESEN CODES

Recall that our goal in constructing the Justesen codes is to obtain a sequence of codes $\mathcal{J}_{n,k}$ for which k/n and d/n are bounded away from 0 as $n\to\infty$.

Let us fix $0 < R_0 < 1/2$ and choose a sequence of outer RS codes $\mathcal{S}_n$ with parameters

$$n = 2^m - 1, \quad k = \lceil 2nR_0 \rceil$$

Hence, the rates of the corresponding Justesen codes $\mathcal{J}_{n,k}$ are

$$R(\mathcal{J}_{n,k}) = \frac{k}{2n} \geq R_0$$

This takes care of one-half of our goal.

Now we must consider the minimum weights $w(\mathcal{J}_{n,k})$. Fix a value of n, and consider the code $\mathcal{J}_{n,k}$. If the outer Reed-Solomon code $\mathcal{S}_n$ has minimum weight d, any supercodeword

$$\mathbf{d} = (c_0,c_0)(c_1,\beta c_1)(c_2\beta^2,c_2)\cdots(c_{n-1},\beta^{n-1}c_{n-1})$$

in $\mathcal{J}_{n,k}$ will have at least d nonzero first components $c_0,\ldots,c_{n-1}$. Furthermore, recalling an earlier discussion, we know that any two of the inner codes

$$C_{\beta^0},C_{\beta^1},C_{\beta^2},\ldots,C_{\beta^{n-1}}$$

have only the zero string in common, and so the corresponding d

ordered pairs, being in different inner codes, must be distinct. In other words, **d** has at least d distinct nonzero binary (2m)-tuples.

Since the weight of the supercodeword $\mathbf{d} \in \mathcal{J}_{n,k}$ is at least equal to the sum of the weights of these d distinct nonzero binary (2m)-tuples, we need to determine a lower bound on such a sum. To this end, we observe that

$$d = n - k + 1 = n(1 - \frac{k-1}{n}) \geq n(1 - 2R_0) = nS$$

where $S = 1 - 2R_0$, and so a supercodeword **d** in $\mathcal{J}_{n,k}$ has at least nS distinct nonzero binary (2m)-tuples. The following lemma is just what we need.

Lemma 8.2.7 The sum of the weights of any nS distinct nonzero binary (2m)-tuples is at least

$$2mnS\left(H^{-1}(\tfrac{1}{2}) - o(n)\right)$$

where o(n) is a quantity that tends to 0 as $n \to \infty$.

Proof. Using Theorem 1.2.8, we deduce that the number of nonzero binary (2m)-tuples having weight at most $2m\lambda$ is at most

$$\sum_{i=1}^{2m\lambda} \binom{2m}{i} \leq 2^{2mH(\lambda)}$$

for any $0 \leq \lambda \leq \frac{1}{2}$. Hence, discounting these small (2m)-tuples, the total weight is at least

$$2m\lambda(nS - 2^{2mH(\lambda)}) = 2mn\lambda S\left(1 - \frac{2^{2mH(\lambda)}}{nS}\right)$$

Now, we pick $\lambda < \frac{1}{2}$ such that

$$\lambda = H^{-1}\left(\frac{1}{2} - \frac{1}{\log(2m)}\right)$$

(Restricting the domain of H(x) to [0,1/2] gives an invertible function and allows us to do this.) Then since $H^{-1}(x)$ is continuous, as $m \to \infty$, $\lambda \to H^{-1}(\frac{1}{2})$, and so we can write

$$\lambda = H^{-1}\left(\frac{1}{2} - \frac{1}{\log(2m)}\right) = H^{-1}(\tfrac{1}{2}) - o(m)$$

where o(m) (read "little oh of m") represents a quantity that tends to 0 as m tends to ∞.

Since $n = 2^m - 1$, we have

$$\frac{2^{2mH(\lambda)}}{nS} = \frac{1}{S}\,\frac{2^{m-2m/\log(2m)}}{2^m - 1} = \frac{1}{S}\,\frac{2^m}{2^m - 1}\,\frac{1}{2^{2m/\log(2m)}} \to 0 \text{ as } m \to \infty$$

and so the total weight of the nS distinct (2m)-tuples is at least

$$2mnS\left(H^{-1}(\tfrac{1}{2}) - o(m)\right)\left(1 - o(m)\right) = 2mnS\left(H^{-1}(\tfrac{1}{2}) - o(m)\right)$$

Since $m \to \infty$ if and only if $n \to \infty$, the result follows. ∎

Using Lemma 8.2.7, we conclude that the code $\mathcal{J}_{n,k}$ has minimum weight at least

$$w(\mathcal{J}_{n,k}) \geq 2mn(1 - 2R_0)\left(H^{-1}(\tfrac{1}{2}) - o(n)\right)$$

and so

$$\frac{w(\mathcal{J}_{n,k})}{len(\mathcal{J}_{n,k})} \geq (1 - 2R_0)\left(H^{-1}(\tfrac{1}{2}) - o(n)\right)$$

$$\to (1 - 2R_0)H^{-1}(\tfrac{1}{2}) \approx 0.1(1 - 2R_0) > 0$$

Thus, we see that the sequence of Justesen codes $\mathcal{J}_{n,k}$, based on Reed-Solomon codes $\mathcal{S}_n$ with parameters

$$n = 2^m - 1, \quad k = \lceil 2nR_0 \rceil$$

where $0 < R_0 < \frac{1}{2}$, has

$$R(\mathcal{J}_{n,k}) \geq R_0 > 0 \text{ and } \frac{w(\mathcal{J}_{n,k})}{len(\mathcal{J}_{n,k})} \to c(1 - 2R_0) > 0 \quad (c \approx 0.1)$$

and is therefore an asymptotically good family of codes.

The family of Justesen codes $\mathcal{J}_{n,k}$ have rate at most 1/2. However, by puncturing the Justesen codes $\mathcal{J}_{n,k}$ in appropriate places, it is possible to obtain an asymptotically good family of codes whose rate approaches any number R_0 with $0 < R_0 < 1$. The details of this procedure can be found in MacWilliams and Sloane (1977).

EXERCISES

1. Find the generator polynomial of an RS code of length 15 and dimension 11. What is the minimum distance of this code.
2. Find the generator polynomial of a [10,6] Reed-Solomon code. What is the minimum distance of this code?
3. Find a double-error-correcting Reed-Solomon code over F_{16}. What is the length and the dimension of this code?
4. Show that the code $C^{\perp}$ whose zeros are given by (8.2.1) is not a BCH code.

5. Prove that the dual of a Reed-Solomon code is also a Reed-Solomon code.
6. Find a code C_1 for which $\hat{d}_1 = d_1$ and a code C_2 for which $\hat{d}_2 = d_2 + 1$.
7. Verify the list of codewords for $\mathcal{I}$ in Example 8.2.2.
8. Using the codewords $\mathcal{I}$ in Example 8.2.2, compute the extended code $\hat{\mathcal{I}}$.
9. Referring to Example 8.2.2, compute the codewords in $\mathcal{I}^{\uparrow}$.
10. Verify that, if a 2^m-ary code C is t-error-correcting, then the expanded version $C^{\rightarrow}$ can correct bursts of length up to $(t-1)m + 1$.
11. Design a code, based on a Reed-Solomon code, that can correct burst errors of length up to 25.
12. Referring to the subsection on idempotents, show that each codeword in the minimal code $\hat{M}_i$ has weight n. *Hint.* Show that $g_i(x)$ has weight n. Start with the constant term.
13. Show that the primitive idempotent $\theta_i(x)$ given by (8.2.2) really does have constant terms equal to 1.
14. What do the Mattson-Solomon polynomials have to do with the encoding of RS codes as described in this section?
15. Show that the minimum distance of a supercode formed from an $[n_1,k_1,d_1]$ outer code and an $[n_2,k_2,d_2]$ outer code is at least d_1d_2.
16. Let **a** be a binary k_1k_2-tuple. Suppose we group **a** into k_2 individual k_1-tuples and encode each k_1-tuple using an $[n_1,k_1,d_1]$-code C_1, resulting in a string of k_2 individual n_1-tuples. Then we encode this string using an $[n_2,k_2,d_2]$-code over $GF(2^{n_1})$. How does this construction differ from concatenation? What can you say about the minimum distance of this type of "super" code?
17. Let the inner code be the [7,4,3] Hamming code $\mathcal{H}_2(3)$, and let the outer code be the [15,13,7] Reed-Solomon code C. Describe the superencoding procedure, using the nonsystematic encoding procedure of Section 7.5. Then perform the encoding on the 52-tuple

$$\mathbf{a} = 0001\ 0100\ \underbrace{0000 \cdots 0000}_{10\ 0000\text{'s}}\ 1000$$

8.3 Alternant Codes and Goppa Codes

We pause in our study of cyclic codes to discuss a generalization of BCH codes known as *alternant codes*, which include several important families of codes. While these codes are, in general, not cyclic, they are very powerful codes. This is exemplified by the fact that, unlike BCH codes, they do form an asymptotically good family in the sense of Section 8.2.

ALTERNANT CODES

We begin with a few remarks on matrices. Let

$$M = \begin{bmatrix} c_{11} & \cdots & c_{1n} \\ \vdots & \cdots & \vdots \\ c_{r1} & \cdots & c_{rn} \end{bmatrix}$$

be an $r \times n$ matrix over the field F_{q^m}. Thinking of the elements c in F_{q^m} as column vectors $[c]$ in F_q^m, we can write

$$M' = \begin{bmatrix} [c_{11}] & \cdots & [c_{1n}] \\ \vdots & \cdots & \vdots \\ [c_{r1}] & \cdots & [c_{rn}] \end{bmatrix}$$

where M' is an $mr \times n$ matrix with entries in the base field F_q.

If $a_i \in F_{q^m}$, then of course

$$\begin{bmatrix} c_{11} & \cdots & c_{1n} \\ \vdots & \cdots & \vdots \\ c_{r1} & \cdots & c_{rn} \end{bmatrix} \begin{bmatrix} a_1 \\ \vdots \\ a_n \end{bmatrix} = \begin{bmatrix} a_1c_{11}+a_2c_{12}+\cdots+a_nc_{1n} \\ \vdots \\ a_1c_{r1}+a_2c_{r2}+\cdots+a_nc_{rn} \end{bmatrix}$$

Furthermore, *provided that* a *lies in the base field* F_q, we have

$$a[c] = [ac]$$

as well as

$$[c+d] = [c]+[d]$$

for $c,d \in F_{q^m}$. Thus, if $a_1,\ldots,a_n \in F_q$,

$$\begin{bmatrix} [c_{11}] & \cdots & [c_{1n}] \\ \vdots & \cdots & \vdots \\ [c_{r1}] & \cdots & [c_{rn}] \end{bmatrix} \begin{bmatrix} a_1 \\ \vdots \\ a_n \end{bmatrix} = \begin{bmatrix} [a_1 c_{11} + a_2 c_{12} + \cdots + a_n c_{1n}] \\ \vdots \\ [a_1 c_{r1} + a_2 c_{r2} + \cdots + a_n c_{rn}] \end{bmatrix}$$

It follows that if the columns of M are linearly independent as r-vectors over F_{q^m}, they are also linearly independent when thought of as vectors over F_q, and so the columns of M' are linearly independent over F_q.

Now let us recall from Section 8.1 that if ω is a primitive n-th root of unity in F_{q^m}, the matrix

$$H_1 = \begin{bmatrix} 1 & [\omega] & [\omega^2] & \cdots & [\omega^{(n-1)}] \\ 1 & [\omega^2] & [\omega^4] & \cdots & [\omega^{2(n-1)}] \\ \vdots & \vdots & \vdots & \cdots & \vdots \\ 1 & [\omega^{\delta-1}] & [\omega^{2(\delta-1)}] & \cdots & [\omega^{(n-1)(\delta-1)}] \end{bmatrix}$$

is a parity check matrix for a narrow sense BCH code $\mathcal{B}$ (once all dependent rows have been purged).

We can generalize this by considering the matrices

$$H = \begin{bmatrix} h_1 & h_2 & h_3 & \cdots & h_n \\ h_1\alpha_1 & h_2\alpha_2 & h_3\alpha_3 & \cdots & h_n\alpha_n \\ \vdots & \vdots & \vdots & \cdots & \vdots \\ h_1\alpha_1^{r-1} & h_2\alpha_2^{r-1} & h_3\alpha_3^{r-1} & \cdots & h_n\alpha_n^{r-1} \end{bmatrix}$$

and

$$H' = \begin{bmatrix} [h_1] & [h_2] & [h_3] & \cdots & [h_n] \\ [h_1\alpha_1] & [h_2\alpha_2] & [h_3\alpha_3] & \cdots & [h_n\alpha_n] \\ \vdots & \vdots & \vdots & \cdots & \vdots \\ [h_1\alpha_1^{r-1}] & [h_2\alpha_2^{r-1}] & [h_3\alpha_3^{r-1}] & \cdots & [h_n\alpha_n^{r-1}] \end{bmatrix}$$

where $h_1,\ldots,h_n$ are *nonzero* elements of F_{q^m}, and $\alpha_1,\ldots,\alpha_n$ are *distinct* elements of F_{q^m}. Thus, H is an $r \times n$ matrix over F_{q^m}, whereas H' is an $rm \times n$ matrix over F_q. We will assume that $r < n$.

Observe that any r columns of H forms a square submatrix K that is similar to a Vandermonde matrix. We leave it to the reader to show that any such submatrix has nonzero determinant, and so any r columns of H are linearly independent over F_{q^m}. Therefore, according to our previous remarks, any r columns of H′ are linearly independent over F_q.

We can now define alternant codes.

Definition Let $\boldsymbol{\alpha} = (\alpha_1,\ldots,\alpha_n)$, where the α_i's are *distinct* elements of F_{q^m}, and let $\mathbf{h} = (h_1,\ldots,h_n)$, where the h_i's are *nonzero* elements of F_{q^m}. Let $r < n$. An **alternant code** $\mathcal{A}(\boldsymbol{\alpha},\mathbf{h})$ is a code over F_q, with parity check matrix of the form H′ (deleting any dependent rows). □

Theorem 8.3.1 The alternant code $\mathcal{A}(\boldsymbol{\alpha},\mathbf{h})$ with parity check matrix H′ (deleting any dependent rows) has length n, dimension k satisfying

$$n - mr \leq k \leq n - r$$

and minimum distance $d \geq r + 1$. □

Thus, we see that alternant codes are generalizations of BCH codes. Although we will not go into the details, it can be shown that there exist arbitrarily long alternant codes that meet the Gilbert-Varshamov bound (See Section 4.5 on asymptotic results). Hence, the family of alternant codes is asymptotically good.

Example 8.3.1 Let us take $q = 2$ and $m = 3$. If ω is a primitive element of F_8, let

$$\boldsymbol{\alpha} = (1,\omega,\omega^2,\ldots,\omega^6) \quad \text{and} \quad \mathbf{h} = (1,1,\ldots,1)$$

Then if $r = 2$, we have

$$H = \begin{bmatrix} 1 & 1 & 1 & 1 & 1 & 1 & 1 \\ 1 & \omega & \omega^2 & \omega^3 & \omega^4 & \omega^5 & \omega^6 \end{bmatrix}$$

Using the field table for F_8, we find that

$$H' = \begin{bmatrix} 1 & 1 & 1 & 1 & 1 & 1 & 1 \\ 0 & 0 & 0 & 0 & 0 & 0 & 0 \\ 0 & 0 & 0 & 0 & 0 & 0 & 0 \\ 1 & 0 & 0 & 1 & 0 & 1 & 1 \\ 0 & 1 & 0 & 1 & 1 & 1 & 0 \\ 0 & 0 & 1 & 0 & 1 & 1 & 1 \end{bmatrix}$$

Purging the second and third rows of this matrix gives us a parity check matrix H'' of size 4×7. Hence the binary alternant code $\mathcal{A}(\boldsymbol{\alpha},\mathbf{h})$ with parity check matrix H'' has parameters [7,3,4]. Note that, in this case, the minimum distance exceeds the lower bound $r+1=3$. We leave it as an exercise to show that this code is not equivalent to a cyclic code, showing that alternant codes are not always cyclic. □

There are several important subclasses of alternant codes, besides the BCH codes. One of the most important subclasses consists of the Goppa codes, which we study next.

GOPPA CODES

We will define Goppa codes without reference to alternant codes, and then show how these codes are, in fact, alternant codes.

Let $G(x)$ be a polynomial over F_{q^m}. We will let S_m denote the ring

$$S_m = \frac{F_{q^m}[x]}{\langle G(x) \rangle}$$

of polynomials over F_{q^m} modulo $G(x)$. Note that S_m is not a field unless $G(x)$ is irreducible.

However, if $G(\alpha) \neq 0$, then the polynomial $x - \alpha$ is invertible in S_m. To see this, we divide $G(x)$ by $x - \alpha$, to get

$$G(x) = q(x)(x-\alpha) + G(\alpha)$$

Hence, $q(x)(x-\alpha) \equiv -G(\alpha) \mod G(x)$, and so

$$[-G(\alpha)^{-1}q(x)](x-\alpha) \equiv 1 \mod G(x)$$

But

$$q(x) = \frac{G(x)-G(\alpha)}{x-\alpha}$$

and so

$$\text{(8.3.1)} \qquad \frac{1}{x-\alpha} = -\frac{G(x)-G(\alpha)}{x-\alpha}G(\alpha)^{-1} \quad \text{in } S_m$$

With this in mind, we can *define* the expression $\frac{1}{x-\alpha}$ to be the *polynomial* in S_m given by (8.3.1). Now we can define Goppa codes.

Definition Let $G(x)$ be a polynomial over F_{q^m}, and let $L = \{\alpha_1,\ldots,\alpha_n\}$ be a set of elements of F_{q^m} with the property that $G(\alpha_i) \neq 0$, and where $n > \deg(G(x))$. Finally, for any $\mathbf{a} = a_1 \cdots a_n$ with $a_i \in F_q$, let

$$R_{\mathbf{a}}(x) = \sum_{i=1}^{n} \frac{a_i}{x-\alpha_i} \in S_m$$

The q-ary **Goppa code** $\Gamma = \Gamma(L,G)$ is defined by

$$\Gamma(L,G) = \{\mathbf{a} \in F_q^n \mid R_{\mathbf{a}}(x) \equiv 0 \bmod G(x)\}$$

The polynomial $G(x)$ is referred to as the **Goppa polynomial** for $\Gamma(L,G)$. If the Goppa polynomial is irreducible, we refer to the Goppa code as **irreducible.** ▯

It is not hard to see that the Goppa code $\Gamma(L,G)$ is linear.

THE PARAMETERS OF $\Gamma(G,L)$

Let us find a parity check matrix for the Goppa code $\Gamma(L,G)$. According to the definition, $\mathbf{a} = a_1 \cdots a_n \in \Gamma(L,G)$ if and only if

$$\sum_{i=1}^{n} \frac{a_i}{x-\alpha_i} \equiv 0$$

or, equivalently, using (8.3.1),

$$\sum_{i=1}^{n} a_i \frac{G(x)-G(\alpha_i)}{x-\alpha_i} G(\alpha_i)^{-1} \equiv 0$$

But since

$$\deg\left(\frac{G(x)-G(\alpha_i)}{x-\alpha_i}\right) < \deg(G(x))$$

this is equivalent to

(8.3.2) $$\sum_{i=1}^{n} a_i \frac{G(x)-G(\alpha_i)}{x-\alpha_i} G(\alpha_i)^{-1} = 0$$

Now we need some good old-fashioned computation. Let

$$G(x) = \sum_{i=0}^{r} g_i x^i$$

with $g_r \neq 0$. Then

$$\frac{G(x)-G(\alpha_i)}{x-\alpha_i} = \sum_{j=0}^{r} g_j \frac{x^j - \alpha_i^j}{x-\alpha_i} = \sum_{i=0}^{r} g_j \sum_{u=0}^{j-1} \alpha_i^{j-i-u}$$

and so the left hand side of (8.3.2) is

$$\sum_{i=1}^{n} a_i \left(\sum_{j=0}^{r} g_j \sum_{u=0}^{j-1} \alpha_i^{j-i-u} \right) G(\alpha_i)^{-1}$$

$$= \sum_{i=1}^{n} a_i G(\alpha_i)^{-1} \sum_{u=0}^{r-1} \left(\sum_{j=u+1}^{r} g_j \alpha_i^{j-i-u} \right) x^u$$

$$= \sum_{u=0}^{r-1} \left(\sum_{i=1}^{n} a_i G(\alpha_i)^{-1} \sum_{j=u+1}^{r} g_j \alpha_i^{j-i-u} \right) x^u$$

Therefore, $\mathbf{a} \in \Gamma(L,G)$ if and only if

$$\text{(8.3.3)} \qquad \sum_{i=1}^{n} a_i G(\alpha_i)^{-1} \left(\sum_{j=u+1}^{r} g_j \alpha_i^{j-i-u} \right) = 0$$

for all $0 \le u \le r-1$.

We recognize the left-hand side of (8.3.3) as a product of matrices

$$\left[\sum_{j=u+1}^{r} g_j \alpha_1^{j-1-u}, \ldots, \sum_{j=u+1}^{r} g_j \alpha_n^{j-1-u} \right] \begin{bmatrix} a_1 G(\alpha_1)^{-1} \\ a_2 G(\alpha_2)^{-1} \\ \vdots \\ a_n G(\alpha_n)^{-1} \end{bmatrix}$$

$$= \left[\sum_{j=u+1}^{r} g_j \alpha_1^{j-1-u}, \ldots, \sum_{j=u+1}^{r} g_j \alpha_n^{j-1-u} \right] \begin{bmatrix} G(\alpha_1)^{-1} & 0 & \cdots & 0 \\ 0 & G(\alpha_2)^{-1} & \cdots & 0 \\ \vdots & \vdots & \ddots & 0 \\ 0 & 0 & 0 & G(\alpha_n)^{-1} \end{bmatrix} \begin{bmatrix} a_1 \\ a_2 \\ \vdots \\ a_n \end{bmatrix}$$

Let us write this as

$$\left[\sum_{j=u+1}^{r} g_j \alpha_1^{j-1-u}, \ldots, \sum_{j=u+1}^{r} g_j \alpha_n^{j-1-u} \right] \cdot G_1 \cdot A$$

Now,

$$\left[\sum_{j=u+1}^{r} g_j \alpha_1^{j-1-u}, \ldots, \sum_{j=u+1}^{r} g_j \alpha_n^{j-1-u} \right]$$

$$= \begin{bmatrix} g_{u+1} \cdots & g_r & 0 & \cdots & 0 \end{bmatrix} \begin{bmatrix} 1 & 1 & \cdots & 1 \\ \alpha_1 & \alpha_2 & \cdots & \alpha_n \\ \alpha_1^2 & \alpha_2^2 & \cdots & \alpha_n^2 \\ \vdots & \vdots & \cdots & \vdots \\ \alpha_1^{r-1} & \alpha_2^{r-1} & \cdots & \alpha_n^{r-1} \end{bmatrix}$$

which can be written, say

$$\begin{bmatrix} g_{u+1} \cdots & g_r & 0 & \cdots & 0 \end{bmatrix} \cdot V$$

and so (8.3.2) holds if and only if

$$\begin{bmatrix} g_{u+1} \cdots & g_r & 0 & \cdots & 0 \end{bmatrix} \cdot V \cdot G_1 \cdot A = \mathbf{0}$$

for all $0 \le u \le r-1$, that is, if and only if

$$\begin{bmatrix} g_r & 0 & 0 & \cdots & 0 \\ g_{r-1} & g_r & 0 & \cdots & 0 \\ g_{r-2} & g_{r-1} & g_r & \cdots & 0 \\ \vdots & \vdots & \vdots & \cdots & \vdots \\ g_1 & g_2 & g_3 & \cdots & g_r \end{bmatrix} \cdot V \cdot G_1 \cdot A = \mathbf{0}$$

But the matrix on the left is invertible, and so this is equivalent to

$$VG_1A = \mathbf{0}$$

Thus, by interpreting the elements of the matrix

$$H = VG_1 = \begin{bmatrix} 1 & 1 & \cdots & 1 \\ \alpha_1 & \alpha_2 & \cdots & \alpha_n \\ \alpha_1^2 & \alpha_2^2 & \cdots & \alpha_n^2 \\ \vdots & \vdots & \cdots & \vdots \\ \alpha_1^{r-1} & \alpha_2^{r-1} & \cdots & \alpha_n^{r-1} \end{bmatrix} \begin{bmatrix} G(\alpha_1)^{-1} & 0 & \cdots & 0 \\ 0 & G(\alpha_2)^{-1} & \cdots & 0 \\ \vdots & \vdots & \ddots & 0 \\ 0 & 0 & 0 & G(\alpha_n)^{-1} \end{bmatrix}$$

$$= \begin{bmatrix} G(\alpha_1)^{-1} & G(\alpha_2)^{-1} & \cdots & G(\alpha_n)^{-1} \\ \alpha_1 G(\alpha_1)^{-1} & \alpha_2 G(\alpha_2)^{-1} & \cdots & \alpha_n G(\alpha_n)^{-1} \\ \alpha_1^2 G(\alpha_1)^{-1} & \alpha_2^2 G(\alpha_2)^{-1} & \cdots & \alpha_n^2 G(\alpha_n)^{-1} \\ \vdots & \vdots & \cdots & \vdots \\ \alpha_1^{r-1} G(\alpha_1)^{-1} & \alpha_2^{r-1} G(\alpha_2)^{-1} & \cdots & \alpha_n^{r-1} G(\alpha_n)^{-1} \end{bmatrix}$$

as column m-vectors over F_q, the rows of H form a complete set of parity checks for $\Gamma(L,G)$.

This shows that the Goppa code $\Gamma(L,G)$ is indeed an alternant code $\mathcal{A}(\boldsymbol{\alpha},\mathbf{h})$, with

$$\boldsymbol{\alpha} = (\alpha_1,\alpha_2,\ldots,\alpha_n) \quad \text{and} \quad \mathbf{h} = (G(\alpha_1)^{-1},G(\alpha_2)^{-1},\ldots,G(\alpha_n)^{-1})$$

Accordingly, Theorem 8.3.1 gives the following.

Theorem 8.3.2 The Goppa code $\Gamma(L,G)$, where $\deg(G(x)) = r < n$, has length $n = |L|$, dimension k satisfying

$$n - mr \leq k \leq n - r$$

and minimum distance $d \geq r+1$. ▯

BINARY GOPPA CODES

In the case of binary Goppa codes, we can improve upon the lower bound on the minimum distance. Let $\mathbf{a} = a_1 \cdots a_n$ be a codeword in the binary Goppa code $\Gamma(L,G)$. Assume that $w(\mathbf{a}) = w$ and that

$$a_{i_1} = \cdots = a_{i_w} = 1$$

Let

$$f_{\mathbf{a}}(x) = \prod_{j=1}^{w} (x - \alpha_{i_j})$$

Then if $f'_{\mathbf{a}}(x)$ denotes the derivative of $f_{\mathbf{a}}(x)$, we have

$$R_{\mathbf{a}}(x) f_{\mathbf{a}}(x) = f'_{\mathbf{a}}(x)$$

Moreover, since the polynomials $f_{\mathbf{a}}(x)$ and $G(x)$ have no common roots in any extension, they are relatively prime, and so

$$\mathbf{a} \in \Gamma(L,G) \Leftrightarrow G(x) \mid R_{\mathbf{a}}(x) \Leftrightarrow G(x) \mid f'_{\mathbf{a}}(x)$$

Now, since $q = 2$, the polynomial $f'_{\mathbf{a}}(x)$ has only even powers of x, and so $f'_{\mathbf{a}}(x) = h(x^2) = [h(x)]^2$ for some polynomial $h(x)$. Thus, if $G^*(x)$ is the polynomial of smallest degree that is a perfect square and

is divisible by $G(x)$, we see that $G(x) \mid f'_{\mathbf{a}}(x)$ if and only if $G^*(x) \mid f'_{\mathbf{a}}(x)$. Hence,

$$\mathbf{a} \in \Gamma(L,G) \Leftrightarrow G^*(x) \mid f'_{\mathbf{a}}(x) \Leftrightarrow R_{\mathbf{a}}(x) \equiv 0 \bmod G^*(x)$$

Thus, we have the following result.

Theorem 8.3.3 Let $\Gamma(L,G)$ be a *binary* Goppa code. If $G^*(x)$ is the polynomial of smallest degree that is a perfect square and is divisible by $G(x)$, then $\Gamma(L,G) = \Gamma(L,G^*)$. In particular, $\Gamma(L,G)$ has minimum distance at least $\deg(G^*(x)) + 1$. ▯

The following special case is of particular interest.

Corollary 8.3.4 Let $\Gamma(L,G)$ be a *binary* Goppa code, and suppose that the Goppa polynomial has no multiple roots in any extension field. Then $\Gamma(L,G) = \Gamma(L,G^2)$. In particular, the minimum distance of $\Gamma(L,G)$ is at least $2\deg(G(x)) + 1$. Hence, $\Gamma(L,G)$ can correct at least $\deg(G(x))$ errors. ▯

A binary Goppa code whose Goppa polynomial has no multiple roots is called a **separable** Goppa code.

Example 8.3.2 Let $G(x) = x^2 + x + 1$ and $L = \{0,1,\omega,\ldots,\omega^6\} = F_8$. Thus, $q = 2$, $m = 3$, $r = 2$, $n = 8$, and ω is a primitive element of F_8. The zeros of $G(x)$ do not lie in F_8. One way to see this (aside from direct substitution) is to observe that the polynomial $G(x)$ is irreducible over F_2, and so according to Theorem 7.2.3, its splitting field is F_{2^2}. Hence, the roots of $G(x)$ lie in F_{2^i} if and only if i is a multiple of 2. (See Theorem 7.1.6.)

The parity check matrix of the Goppa code $\Gamma(L,G)$ is obtained from the matrix

$$H = \begin{bmatrix} \frac{1}{G(0)} & \frac{1}{G(1)} & \frac{1}{G(\omega)} & \cdots & \frac{1}{G(\omega^6)} \\ \frac{0}{G(0)} & \frac{1}{G(1)} & \frac{\omega}{G(\omega)} & \cdots & \frac{\omega^6}{G(\omega^6)} \end{bmatrix}$$

Using the field table for F_8, we get

$$H = \begin{bmatrix} 1 & 1 & \omega^2 & \omega^4 & \omega^2 & \omega & \omega & \omega^4 \\ 0 & 1 & \omega^3 & \omega^6 & \omega^5 & \omega^5 & \omega^6 & \omega^3 \end{bmatrix}$$

and so

$$H' = \begin{bmatrix} 1&1&0&0&0&0&0&0 \\ 0&0&0&1&0&1&1&1 \\ 0&0&1&1&1&0&0&1 \\ 0&1&1&1&1&1&1&1 \\ 0&0&1&0&1&1&0&1 \\ 0&0&0&1&1&1&1&0 \end{bmatrix}$$

Since this matrix has rank 6, it is a parity check matrix for $\Gamma(L,G)$, which therefore has dimension $8-6=2$. A little matrix manipulation gives

$$\Gamma(L,G) = \{00000000, 11110100, 11001011, 00111111\}$$

which has minimum distance 5. ▯

Example 8.3.3 Let $G(x) = x^2+1$, and $L = \{0, \omega, \omega^2, \omega^3, \omega^5, \omega^6\} \subset F_8$, where ω is a primitive element of F_8. Thus, $q=2$, $m=3$, $r=2$ and $n=6$. Since F_8 has characteristic 2, we have $G(x) = (x+1)^2$, and according to Corollary 8.3.4, $\Gamma = \Gamma(L,G) = \Gamma(L, x+1)$. We deduce that Γ has length 6, dimension $k \geq 6-3=3$, and minimum distance $d \geq 3$. Furthermore,

$$H = \begin{bmatrix} \frac{1}{0+1} & \frac{1}{\omega+1} & \frac{1}{\omega^2+1} & \frac{1}{\omega^3+1} & \frac{1}{\omega^5+1} & \frac{1}{\omega^6+1} \end{bmatrix}$$

$$= \begin{bmatrix} 1 & \omega^4 & \omega & \omega^6 & \omega^3 & \omega^5 \end{bmatrix}$$

and so a parity check matrix for Γ is

$$H' = \begin{bmatrix} 1&1&0&1&0&1 \\ 0&1&1&0&1&1 \\ 0&0&0&1&1&1 \end{bmatrix}$$

Hence, Γ has length 6, dimension 3, and minimum distance 3. ▯

FAST DECODING OF ALTERNANT CODES

We now wish to describe a decoding procedure, using the Euclidean algorithm, that applies to alternant codes and hence, in particular, to BCH codes and Goppa codes. Quoting from MacWilliams and Sloane (1977): "Nevertheless, decoding using the Euclidean algorithm is by far the simplest to understand, and is certainly at least comparable in speed with the other methods (for $n < 10^6$) and so it is the method we prefer."

Let us begin with a close look at the Euclidean algorithm.

The Euclidean Algorithm

The Euclidean algorithm is a procedure for computing the greatest common divisor of two integers or polynomials. We will confine our attention to polynomials, since that is our immediate concern. Let $f(x)$ and $g(x)$ be nonzero polynomials over the same field F (finite or infinite), and assume that $\deg(g(x)) \leq \deg(f(x))$. It will be convenient to let $f(x) = r_{-1}(x)$ and $g(x) = r_0(x)$.

The steps are as follows:

Divide $f(x) = r_{-1}(x)$ by $g(x) = r_0(x)$:

$$r_{-1}(x) = q_1(x)r_0(x) + r_1(x), \qquad \text{where } \deg r_1 < \deg r_0$$

Divide $g(x) = r_0(x)$ by $r_1(x)$:

$$r_0(x) = q_2(x)r_1(x) + r_2(x), \qquad \text{where } \deg r_2 < \deg r_1$$

Divide $r_1(x)$ by $r_2(x)$:

$$r_1(x) = q_3(x)r_2(x) + r_3(x), \qquad \text{where } \deg r_3 < \deg r_2$$

$$\vdots$$

Divide $r_{k-2}(x)$ by $r_{k-1}(x)$:

$$r_{k-2}(x) = q_k(x)r_{k-1}(x) + r_k(x), \qquad \text{where } \deg r_k < \deg r_{k-1}$$

$$\vdots$$

Continue to divide the previous remainder by the current remainder until the current remainder becomes 0:

$$\vdots$$

Divide $r_{s-2}(x)$ by $r_{s-1}(x)$:

$$r_{s-2}(x) = q_s(x)r_{s-1}(x) + r_s(x), \qquad \text{where } \deg r_s < \deg r_{s-1}$$

Divide $r_{s-1}(x)$ by $r_s(x)$:

$$r_{s-1}(x) = q_{s+1}(x)r_s(x)$$

Then

$$\gcd[f(x),g(x)] = r_s(x)$$

Notice that, in each step of the process, we can write the current remainder $r_k(x)$ in terms of the two previous remainders. Hence, it should be possible to write all remainders, including $r_s(x)$, in terms of $f(x) = r_{-1}(x)$ and $g(x) = r_0(x)$.

For inductive purposes, suppose that $a_i(x)$ and $b_i(x)$ are polynomials with the property that

$$r_k(x) = a_k(x)f(x) + b_k(x)g(x)$$

for $k \geq -1$. In particular,

$$a_{-1}(x) = 1, \quad b_{-1}(x) = 0$$

and

$$a_0(x) = 0, \quad b_0(x) = 1$$

and if $j = 1$, since

$$r_1(x) = r_{-1}(x) - q_1(x)r_0(x)$$

we have

$$a_1(x) = 1, \quad b_1(x) = -q_1(x)$$

More generally, from the k-th step in the Euclidean algorithm, we get

$$\begin{aligned} r_k(x) &= r_{k-2}(x) - q_k(x)r_{k-1}(x) \\ &= [a_{k-2}(x)f(x) + b_{k-2}(x)g(x)] \\ &\qquad - q_k(x)[a_{k-1}(x)f(x) + b_{k-1}(x)g(x)] \\ &= [-q_k(x)a_{k-1}(x) + a_{k-2}(x)]f(x) \\ &\qquad + [-q_k(x)b_{k-1}(x) + b_{k-2}(x)]g(x) \end{aligned}$$

and so

$$a_k(x) = -q_k(x)a_{k-1}(x) + a_{k-2}(x)$$

and

$$b_k(x) = -q_k(x)b_{k-1}(x) + b_{k-2}(x)$$

We can now state the following theorem.

Theorem 8.3.5 The remainders $r_k(x)$ in the Euclidean algorithm satisfy

$$r_k(x) = a_k(x)f(x) + b_k(x)g(x) \tag{8.3.4}$$

for $k \geq -1$, where

$$a_{-1}(x) = 1, \quad b_{-1}(x) = 0$$

$$a_0(x) = 0, \quad b_0(x) = 1$$

$$a_k(x) = -q_k(x)a_{k-1}(x) + a_{k-2}(x), \quad k \geq 1 \tag{8.3.5}$$

$$b_k(x) = -q_k(x)b_{k-1}(x) + b_{k-2}(x), \quad k \geq 1 \tag{8.3.6}$$

In particular, there exist polynomials $a(x) = a_s(x)$ and $b(x) = b_s(x)$ for which

$$\gcd(f(x),g(x)) = a(x)f(x) + b(x)g(x)$$

In addition, we have the following properties

1) $\deg a_k = \sum_{i=2}^{k} \deg q_i, \quad \deg b_k = \sum_{i=1}^{k} \deg q_i$

2) $\deg r_k = \deg f - \sum_{i=1}^{k+1} \deg q_k$

3) $\deg b_k = \deg f - \deg r_{k-1}$

4) $a_k(x)b_{k+1}(x) - a_{k+1}(x)b_k(x) = (-1)^{k+1}$

5) $a_k(x)$ and $b_k(x)$ are relatively prime

6) $r_k(x)b_{k+1}(x) - r_{k+1}(x)b_k(x) = (-1)^{k+1}f(x)$

7) $r_{k+1}(x)a_k(x) - r_k(x)a_{k+1}(x) = (-1)^{k+1}g(x)$

Proof. We leave the proofs of 1–3 as exercises. As to part 4), let

$$D_k = \begin{vmatrix} a_k & b_k \\ a_{k+1} & b_{k+1} \end{vmatrix}$$

(We omit the variable for compactness.) Using (8.3.5) and (8.3.6), we have

$$D_k = \begin{vmatrix} a_k & b_k \\ -q_{k+1}a_k + a_{k-1} & -q_{k+1}b_k + b_{k-1} \end{vmatrix}$$

$$= \begin{vmatrix} 1 & 0 \\ -q_{k+1} & 1 \end{vmatrix} \begin{vmatrix} a_k & b_k \\ a_{k-1} & b_{k-1} \end{vmatrix} = -D_{k-1}$$

and so $D_k = (-1)^{k+1}D_{-1} = (-1)^{k+1}$. Part 5) follows easily from part 4). Parts 6) and 7) are proved in a manner similar to that of part 4). We leave the details to the reader. ∎

Decoding of Alternant Codes – The Initial Setup

Let $\mathcal{A}(\boldsymbol{\alpha},\mathbf{h})$ be an alternant code over F_q, with parity check matrix H' derived from the matrix

$$H = \begin{bmatrix} h_1 & h_2 & h_3 & \cdots & h_n \\ h_1\alpha_1 & h_2\alpha_2 & h_3\alpha_3 & \cdots & h_n\alpha_n \\ \vdots & \vdots & \vdots & \cdots & \vdots \\ h_1\alpha_1^{r-1} & h_2\alpha_2^{r-1} & h_3\alpha_3^{r-1} & \cdots & h_n\alpha_n^{r-1} \end{bmatrix}$$

by replacing the entries by column vectors of length m and deleting any dependent rows. Recall that $h_1,\ldots,h_n$ are *nonzero* elements of F_{q^m} and that $\alpha_1,\ldots,\alpha_n$ are *distinct* elements of F_{q^m}.

Now suppose that a codeword $\mathbf{u} = \mathbf{c}+\mathbf{e}$ is received, where $\mathbf{c}$ is the correct codeword and $\mathbf{e}$ is the error. We will assume that r is even, so that $r+1$ is odd (see Theorem 8.3.1) and that $t \le r/2$ errors have occurred, say in positions

$$1 \le i_1,\ldots,i_t \le n$$

If the i_j-th component of $\mathbf{e}$ is e_{i_j}, then we define the **error locators** by

$$X_j = \alpha_{i_j}$$

and the **error values** by

$$Y_j = e_{i_j}$$

Observe that, since the α_i's are assumed to be distinct, knowing the error locators tells us the positions of the errors.

The first step in the decoding algorithm is to calculate the **syndrome**

$$\mathbf{u}H^{\mathsf{T}} = \mathbf{e}H^{\mathsf{T}} = \left[\, 0 \cdots e_{i_1} \cdots e_{i_t} \cdots 0 \,\right] H^{\mathsf{T}} = \left[\, S_0 \ \cdots S_{r-1} \right]$$

where

$$S_\ell = \sum_{k=1}^{t} h_{i_k}\alpha_{i_k}^{\ell} e_{i_k} = \sum_{k=1}^{t} h_{i_k} X_k^{\ell} Y_k$$

for $\ell = 0,\ldots,r-1$. Let us also define

$$S(x) = \sum_{\ell=0}^{r-1} S_\ell x^\ell$$

Since the syndrome is calculated directly from the matrix H and the received word $\mathbf{u}$, it is a known quantity.

Now let us define the **error locator polynomial** by

$$\sigma(x) = \prod_{i=1}^{t}(1 - X_i x) = \sum_{i=0}^{t}\sigma_i x^i$$

where the coefficients σ_i are defined by this formula. Note that $\sigma_0 = 1$. Note also that the roots of the error locator polynomial are the inverses of the error locators X_i. Hence, finding the roots of the locator polynomial is equivalent to finding the locations of the errors.

Next, we define the **error evaluator polynomial** by

$$\omega(x) = \sum_{k=1}^{t} h_{i_k} Y_k \prod_{\substack{j=1 \\ j \neq k}}^{t} (1 - X_j x)$$

This formidable looking polynomial will give us the error values Y_u and hence the magnitudes e_{i_u} of the errors.

Lemma 8.3.6

$$Y_u = \frac{\omega(X_u^{-1})}{h_{i_u} \prod_{\substack{j=1 \\ j \neq u}}^{t} (1 - X_j X_u^{-1})}$$

Proof. Left to the reader. ∎

Our goal is to determine the error locator polynomial and the error evaluator polynomial from the syndrome. To this end, we observe that $\omega(x)$, $\sigma(x)$, and $S(x)$ are related as described in the following lemma.

Lemma 8.3.7 The error evaluator polynomial $\omega(x)$, the error locator polynomial $\sigma(x)$, and the syndrome polynomial $S(x)$ are related by

$$\omega(x) \equiv \sigma(x) S(x) \bmod x^r$$

Proof. The computation is lengthy, but not difficult

$$\omega(x) - \sigma(x)S(x)$$

$$= \sum_{k=1}^{t} h_{i_k} Y_k \prod_{\substack{j=1 \\ j \neq k}}^{t} (1 - X_j x) - \sigma(x) \sum_{\ell=0}^{r-1} S_\ell x^\ell$$

$$= \sum_{k=1}^{t} h_{i_k} Y_k \prod_{\substack{j=1 \\ j \neq k}}^{t} (1 - X_j x) - \sigma(x) \sum_{\ell=0}^{r-1} \left[\sum_{k=1}^{t} h_{i_k} X_k^\ell Y_k \right] x^\ell$$

$$= \sum_{k=1}^{t} h_{i_k} Y_k \prod_{\substack{j=1 \\ j \neq k}}^{t} (1 - X_j x) - \sigma(x) \sum_{k=1}^{t} h_{i_k} Y_k \sum_{\ell=0}^{r-1} X_k^\ell x^\ell$$

$$= \sum_{k=1}^{t} h_{i_k} Y_k \left(\prod_{\substack{j=1 \\ j \neq k}}^{t} (1 - X_j x) - \sigma(x) \sum_{\ell=0}^{r-1} X_k^{\ell} x^{\ell} \right)$$

$$= \sum_{k=1}^{t} h_{i_k} Y_k \left(\prod_{\substack{j=1 \\ j \neq k}}^{t} (1 - X_j x) \right) \left(1 - (1 - X_k x) \sum_{\ell=0}^{r-1} X_k^{\ell} x^{\ell} \right)$$

$$= \sum_{k=1}^{t} h_{i_k} Y_k \left(\prod_{\substack{j=1 \\ j \neq k}}^{t} (1 - X_j x) \right) \left(1 - (1 - X_k x) \frac{1 - X_k^r x^r}{1 - X_k x} \right)$$

$$= \sum_{k=1}^{t} h_{i_k} Y_k \left(\prod_{\substack{j=1 \\ j \neq k}}^{t} (1 - X_j x) \right) (X_k^r x^r)$$

$$\equiv 0 \bmod x^r$$

Decoding of Alternant Codes – The Decoding Step

Let us assemble what we know about $\omega(x)$ and $\sigma(x)$. According to Lemma 8.3.7, there exists a polynomial $\theta(x)$ for which

(8.3.7) $$\omega(x) = \theta(x)x^r + \sigma(x)S(x)$$

Furthermore, we have

$$\deg \omega \leq t - 1 < \tfrac{1}{2}r \quad \text{and} \quad \deg \sigma = t \leq \tfrac{1}{2}r$$

We also know that $\omega(x)$ and $\sigma(x)$ are relatively prime, for they have no common roots in any extension.

We would like to be able to find $\omega(x)$ and $\sigma(x)$ using the Euclidean algorithm. So, suppose we apply the algorithm to the (known) polynomials

$$f(x) = x^r \quad \text{and} \quad g(x) = S(x)$$

According to Theorem 8.3.5, a typical step will produce a remainder of the form

(8.3.8) $$r_k(x) = a_k(x)x^r + b_k(x)S(x)$$

Of course, if we expect $r_k(x)$ and $b_k(x)$ to give $\omega(x)$ and $\sigma(x)$, respectively, their degrees must match. For this, we must at least have

(8.3.9) $$\deg r_k < \tfrac{1}{2}r \quad \text{and} \quad \deg b_k \leq \tfrac{1}{2}r$$

Thus, we want to carry out the Euclidean algorithm until

$$\deg r_k < \tfrac{1}{2}r \quad \text{but} \quad \deg r_{k-1} \geq \tfrac{1}{2}r$$

For then, according to part 3) of Theorem 8.3.5,

$$\deg b_k = \deg x^r - \deg r_{k-1} \leq r - \tfrac{1}{2}r = \tfrac{1}{2}r$$

Fortunately, this can be done, since the Euclidean algorithm can be continued until $r_k(x)$ is the greatest common divisor of x^r and $S(x)$, say $h(x)$. But by (8.3.7), $h(x)$ divides $\omega(x)$, and so $\deg h \leq \deg \omega < \frac{1}{2}r$. So let us assume that (8.3.9) holds.

For convenience, let us repeat the relevant equations

(8.3.7) $$\omega(x) = \theta(x)x^r + \sigma(x)S(x)$$

$$\deg \omega < \tfrac{1}{2}r \quad \text{and} \quad \deg \sigma \leq \tfrac{1}{2}r$$

$$(\omega(x),\sigma(x)) = 1$$

and

(8.3.8) $$r_k(x) = a_k(x)x^r + b_k(x)S(x)$$

(8.3.9) $$\deg r_k < \tfrac{1}{2}r \quad \text{and} \quad \deg b_k \leq \tfrac{1}{2}r$$

Our goal is to show that the polynomials $r_k(x)$ and $b_k(x)$, which we can easily obtain from the Euclidean algorithm, are just scalar multiples of the desired polynomials $\omega(x)$ and $\sigma(x)$, respectively.

To this end, we eliminate $S(x)$ from (8.3.7) and (8.3.8) to get

$$b_k(x)\omega(x) - r_k(x)\sigma(x) = [b_k(x)\theta(x) - a_k(x)\sigma(x)]x^r$$

But

$$\deg b_k\omega = \deg b_k + \deg \omega < \tfrac{1}{2}r + \tfrac{1}{2}r = r$$

and

$$\deg r_k\sigma = \deg r_k + \deg \sigma < \tfrac{1}{2}r + \tfrac{1}{2}r = r$$

and so the left-hand side has degree less than r. This implies that both sides must be the zero polynomial, that is

(8.3.10) $$\sigma(x)r_k(x) = b_k(x)\omega(x)$$

and

$$b_k(x)\theta(x) = a_k(x)\sigma(x)$$

Hence,

$$\sigma(x) \mid b_k(x)\omega(x) \quad \text{and} \quad b_k(x) \mid a_k(x)\sigma(x)$$

But $(\sigma(x),\omega(x)) = 1$ and $(a_k(x),b_k(x)) = 1$ [by part 5) of Theorem 8.3.5] and so

$$\sigma(x) \mid b_k(x) \quad \text{and} \quad b_k(x) \mid \sigma(x)$$

which means that $\sigma(x)$ is a scalar multiple of $b_k(x)$, say

$$\sigma(x) = \lambda b_k(x)$$

Then (8.3.10) gives

$$\omega(x) = \lambda r_k(x)$$

Finally, since $\sigma(0) = 1$, we see that $\lambda = b_k(0)^{-1}$.

We can at last state the complete decoding algorithm for alternant codes.

Theorem 8.3.8 (Decoding algorithm for alternant codes) Let $\mathcal{A}(\alpha, \mathbf{h})$ be an alternant code. Assume that r is even and that $t \leq r/2$ errors have occurred in the transmission of a codeword. These errors can be corrected as follows.

Step 1: Compute the syndrome

$$uH^T = \left[S_0 \ \cdots \ S_{r-1} \right]$$

of the received word **u**, and let

$$S(x) = \sum_{\ell=0}^{r-1} S_\ell x^\ell$$

Step 2: Apply the Euclidean algorithm to the polynomials

$$f(x) = x^r \quad \text{and} \quad g(x) = S(x)$$

to get

$$r_k(x) = a_k(x)x^r + b_k(x)S(x)$$

where

$$\deg r_k(x) < \tfrac{1}{2}r \quad \text{but} \quad \deg r_{k-1}(x) \geq \tfrac{1}{2}r$$

Step 3: Let

$$\sigma(x) = b_k(0)^{-1} b_k(x) \quad \text{and} \quad \omega(x) = b_k(0)^{-1} r_k(x)$$

Then $\sigma(x)$ is the error locator polynomial. The *inverses* of the roots of $\sigma(x)$ are the error locators

$$X_1 = \alpha_{i_1}, \ldots, X_t = \alpha_{i_t}$$

where $i_1, \ldots, i_t$ are the locations of the errors. The magnitudes Y_i of the errors can be found from the formula

$$Y_u = \frac{\omega(X_u^{-1})}{h_{i_u} \prod_{\substack{j=1 \\ j \neq u}}^{t} (1 - X_j X_u^{-1})}$$ □

Example 8.3.4 Let us consider the binary BCH code of length $n = 15$ and designed distance $\delta = 5$ $(r = 4)$ discussed in Example 8.1.6. Again taking the received word

$$u(x) = x^{12} + x^8 + x^7 + x^6 + 1$$

we see that the syndrome is

$$S_0 = u(\omega) = \omega^{12} + \omega^8 + \omega^7 + \omega^6 + 1 = \omega^6$$
$$S_1 = u(\omega^2) = \omega^{24} + \omega^{16} + \omega^{14} + \omega^{12} + 1 = \omega^{12}$$
$$S_2 = u(\omega^3) = \omega^4$$
$$S_3 = u(\omega^4) = \omega^9$$

and so

$$S(x) = \omega^6 + \omega^{12}x + \omega^4x^2 + \omega^9x^3$$

Now we apply the Euclidean algorithm to the polynomials

$$f(x) = x^4 \quad \text{and} \quad g(x) = S(x)$$

Divide $f(x) = x^4$ by $g(x) = S(x)$:

$$f(x) = q_1(x)g(x) + r_1(x)$$
$$x^4 = (\omega^6x)S(x) + (\omega^{10}x^3 + \omega^3x^2 + \omega^{12}x)$$

Divide $g(x) = S(x)$ by $r_1(x) = \omega^{10}x^3 + \omega^3x^2 + \omega^{12}x$:

$$g(x) = q_2(x)r_1(x) + r_2(x)$$
$$S(x) = \omega^{14}(\omega^{10}x^3 + \omega^3x^2 + \omega^{12}x) + (\omega^{10}x^2 + x + \omega^6)$$

Divide $r_1(x) = \omega^{10}x^3 + \omega^3x^2 + \omega^{12}x$ by $r_2(x) = \omega^{10}x^2 + x + \omega^6$:

$$r_1(x) = q_3(x)r_2(x) + r_3(x):$$
$$\omega^{10}x^3 + \omega^3x^2 + \omega^{12}x = (x + \omega^4)(\omega^{10}x^2 + x + \omega^6) + (\omega^{10})$$

At this point, we see that $\deg r_3 = 0 < \frac{1}{2}r = 2$ and that $\deg r_2 = 2 \geq \frac{1}{2}r$, and so we stop.

Next we compute $b_3(x)$, using (8.3.6). (The error evaluator polynomial is not needed since we are dealing with a binary code.)

$$b_{-1}(x) = 0$$
$$b_0(x) = 1$$
$$b_1(x) = q_1(x)b_0(x) + b_{-1}(x) = \omega^6x$$
$$b_2(x) = q_2(x)b_1(x) + b_0(x) = \omega^{20}x + 1 = \omega^5x + 1$$
$$b_3(x) = q_3(x)b_2(x) + b_1(x) = (x + \omega^4)(\omega^5x + 1) + \omega^6x$$
$$= \omega^5x^2 + \omega^{10}x + \omega^4$$

Therefore,

$$\sigma(x) = b_3(0)^{-1}b_3(x) = \omega^{11}(\omega^5x^2 + \omega^{10}x + \omega^4) = \omega x^2 + \omega^6 x + 1$$

Now we must find the zeros of $\sigma(x)$, which we do by trial and error, to get ω^3 and ω^{11}. The inverses of these values are ω^4 and ω^{12}, and so errors have occurred in positions 4 and 12. Thus, the correct codeword is

$$c(x) = u(x) + x^4 + x^{12} = x^8 + x^7 + x^6 + x^4 + 1 \qquad \square$$

EXERCISES

1. Verify the statement made in the first subsection. To wit: if the columns of M are linearly independent as vectors over F_{q^m}, they are also linearly independent as vectors over F_q, and so the columns of M′ are linearly independent over F_q.
2. Show that any submatrix K of H discussed in the first subsection has nonzero determinant.
3. Show that any BCH code (not necessarily narrow sense) is an alternant code.
4. Show that the alternant code in Example 8.3.1 is not equivalent to a cyclic code.
5. Let ω be a primitive element of F_8, and let $\boldsymbol{\alpha} = \mathbf{h} = (1,\omega,\omega^2,\ldots,\omega^6)$. Find a parity check matrix for the binary alternant code $\mathcal{A}(\boldsymbol{\alpha},\mathbf{h})$. What are the parameters of this code? Do you recognize this code? Is it cyclic?
6. Repeat the previous exercise, using instead $\mathbf{h} = (\omega,\omega,\ldots,\omega)$.
7. Show that the Goppa code $\Gamma(L,G)$ is linear.
8. Prove that if H is a parity check matrix for a code C, and if K is an invertible matrix, then KH is also a parity check matrix for C.
9. Prove Corollary 8.3.4.
10. Referring to Example 8.3.2, find the codewords in $\Gamma(L,G)$ directly from the matrix H′.
11. Let $G(x) = x^3 + x + 1$. Is this irreducible over F_2? Where are the zeros of $G(x)$? If $3 \nmid m$, then let $L = F_{2^m}$. What do Theorem 8.3.2 and Corollary 8.3.4 tell you about the parameters of the Goppa code $\Gamma(L,G)$?
12. Let $G(x)$ be a polynomial over F_{2^t} with degree r and with no multiple roots. Let F_{2^s} be the splitting field for $G(x)$. Show that we may choose $L = F_{2^m}$ provided that $t \mid m$ and that $(s,m) = 1$. What are the lower bounds on the parameters of such a code?

13. Let $L \subset F_{16}$ be the *primitive* 15th roots of unity. Let $G(x) = x^2 + 1$ over F_{16}. Discuss the Goppa code $\Gamma(L,G)$.
14. Let C be the binary cyclic code of length 15 with generator polynomial $g(x) = x^2 + x + 1$. Let ω be a primitive 15th root of unity.
    a) Show that ω^5 is a root of $g(x)$.
    b) Show that C is a BCH code.
    c) Show that the minimum distance of C is 2.
    d) Show that C cannot be a Goppa code. *Hint.* Consider the cases $\deg(G(x)) = 1$ and 2 separately.
15. Find the greatest common divisor of the binary polynomials $f(x) = 1 + x + x^2 + x^3 + x^4$ and $g(x) = 1 + x^3$.
16. Prove the unproven statements in Theorem 8.3.5.
17. Prove Lemma 8.3.6.
18. Suppose that we receive the word $u(x) = 1 + x^3 + x^9$, while using a binary BCH code of length 15 and designed distance $\delta = 5$. What is the correct codeword?

8.4 Quadratic Residue Codes

Let ω be a primitive n-th root of unity over F_q, and let

$$x^n - 1 = \prod_i m_i(x)$$

where the minimal polynomials $m_i(x)$ are given by

$$m_i(x) = \prod_{j \in C_i} (x - \omega^j)$$

and $C_i = \{i, qi, \ldots, q^{d-1}i\}$ is the i-th cyclotomic coset modulo q.

We have seen that a cyclic code can be defined by specifying the complete set of its zeros

$$\mathcal{Z} = \{\omega^i \mid i \in Z\}$$

where Z is a union of cyclotomic cosets modulo q. In the case of BCH codes, we specify a subset of the zeros, whose exponents are consecutive integers

$$S = \{\omega^i \mid i = b, b+1, \ldots, b+\delta-2\}$$

and then obtain the set of all zeros by taking the union of all cyclotomic cosets containing any of the numbers $b, b+1, \ldots, b+\delta-2$. In the Reed-Solomon case, the cyclotomic cosets are singleton sets, and so the set S is a complete set of zeros.

The point that we wish to make here is that the families of cyclic codes that we have studied so far are defined by choosing the zeros of their codes in a special way. As we shall see, this is also true of the quadratic residue codes.

Let us turn now to a brief discussion of quadratic residues. Proofs of the following results can be found in standard texts on number theory.

QUADRATIC RESIDUES

Definition Let p be an odd prime. If $(a,p) = 1$, then a is a **quadratic residue modulo p** if there is a number x for which

$$x^2 \equiv a \bmod p$$

If there is no number x satisfying this congruence, then a is a **quadratic nonresidue modulo p.** ▯

In other words, a is a quadratic residue modulo p if a has a *square root* modulo p.

The set of quadratic residues modulo p that lie in the set $\mathbb{Z}_p^* = \{1,2,\ldots,p-1\}$ will be denoted by QR. Similarly, the set of quadratic nonresidues that lie in $\mathbb{Z}_p^* = \{1,2,\ldots,p-1\}$ will be denoted by NQR. It is not hard to see that any quadratic residue modulo p is congruent to a member of QR.

Theorem 8.4.1 The set QR has size $(p-1)/2$, and is equal to

$$QR = \left\{1^2, 2^2, \ldots, \left(\tfrac{p-1}{2}\right)^2\right\}$$

where all numbers are taken modulo p. Therefore, the set NQR also has size $(p-1)/2$. □

Example 8.4.1 The quadratic residues modulo 11 in QR are

$$1^2 = 1, 2^2 = 4,\ 3^2 = 9,\ 4^2 = 16 \equiv 5,\ 5^2 = 25 \equiv 3 \qquad \square$$

Definition Let p be an odd prime, and let $(a,p) = 1$. The **Legendre symbol** $\left(\frac{a}{p}\right)$ is defined by

$$\left(\frac{a}{p}\right) = \begin{cases} 1 & \text{if } a \text{ is a quadratic residue mod } p \\ -1 & \text{if } a \text{ is a quadratic nonresidue mod } p \end{cases} \qquad \square$$

Theorem 8.4.2 (Properties of the Legendre symbols) Let p and q be distinct odd primes, and $(a,p) = (b,p) = 1$.

1) $\left(\frac{1}{p}\right) = \left(\frac{a^2}{p}\right) = 1.$

2) If $a \equiv b \bmod p$, then $\left(\frac{a}{p}\right) = \left(\frac{b}{p}\right)$.

3) (Euler's criterion) $\left(\frac{a}{p}\right) \equiv a^{(p-1)/2} \bmod p.$

4) $\left(\frac{ab}{p}\right) = \left(\frac{a}{p}\right)\left(\frac{b}{p}\right)$.

5) $\left(\frac{-1}{p}\right) = 1$ if and only if $p \equiv 1 \bmod 4$.

6) $\left(\frac{2}{p}\right) = 1$ if and only if $p \equiv \pm 1 \bmod 8$.

7) $\left(\frac{3}{p}\right) = 1$ if and only if $p \equiv \pm 1 \bmod 12$. □

We get the following directly from part 4) of Theorem 8.4.2.

Corollary 8.4.3 The product of two quadratic residues mod p is another quadratic residue mod p. The product of two quadratic nonresidues mod p is a quadratic residue mod p. The product of a quadratic residue and a quadratic nonresidue mod p is a quadratic nonresidue mod p. ▯

Corollary 8.4.4 If $s \in QR$ then $QR = \{sr \bmod p \mid r \in QR\}$. ▯

Example 8.4.3 Since

$$\left(\frac{-a^2}{p}\right) = \left(\frac{-1}{p}\right)\left(\frac{a^2}{p}\right) = \left(\frac{-1}{p}\right) = (-1)^{(p-1)/2}$$

we see that $x^2 \equiv -a^2 \bmod p$ has a solution if and only if $p \equiv 1 \bmod 4$. ▯

Example 8.4.4 Consider the congruence

$$x^2 \equiv 15 \bmod 19$$

We have

$$\left(\frac{15}{19}\right) = \left(\frac{-4}{19}\right) = \left(\frac{-1}{19}\right)\left(\frac{2}{19}\right)^2 = \left(\frac{-1}{19}\right) = -1$$

since $19 \not\equiv 1 \bmod 4$. Hence, this congruence has no solution. ▯

QUADRATIC RESIDUE CODES

We will discuss binary quadratic residue codes only. However, since it is just as easy to do so, we give the definition in the general case.

According to Theorem 8.4.1,

$$|QR| = |NQR| = \frac{p-1}{2}$$

and so the sets QR and NQR form a partition of $\{1,2,\ldots,p-1\}$ into equally sized blocks.

Now let ω be a primitive p-th root of unity over F_q. Hence, the p-th roots of unity are

$$\{1,\omega,\omega^2\ldots,\omega^{p-1}\}$$

In line with the discussion at the beginning of this section, we would like the set

$$\mathcal{Z} = \{\omega^i \mid i \in QR\}$$

to be a *complete* set of zeros of a q-ary cyclic code, for some prime q. To accomplish this, the set QR must be a union of cyclotomic cosets modulo q. That is, we must have

$$i \in QR \Rightarrow C_i = \{i, iq, iq^2, \ldots, iq^{d-1}\} \subset QR$$

But Corollary 8.4.3 tells us that, for $i \in QR$, we have $iq \in QR$ if and only if $q \in QR$. Thus, q must be a prime that is a quadratic residue modulo p. For any such q, the polynomials

$$q(x) = \prod_{r \in QR} (x - \omega^r) \qquad \text{and} \qquad n(x) = \prod_{u \in NQR} (x - \omega^u)$$

have coefficients in F_q, and

$$x^p - 1 = (x-1)q(x)n(x)$$

We can now define quadratic residue codes.

Definition Let p be an odd prime, and let q be a prime that is a quadratic residue modulo p. The q-ary cyclic codes

$$\mathcal{Q}(p) = \langle\langle q(x) \rangle\rangle, \quad \overline{\mathcal{Q}}(p) = \langle\langle (x-1)q(x) \rangle\rangle$$

$$\mathcal{N}(p) = \langle\langle n(x) \rangle\rangle, \quad \overline{\mathcal{N}}(p) = \langle\langle (x-1)n(x) \rangle\rangle$$

of length p in

$$R_p = \frac{F_q[x]}{\langle x^p - 1 \rangle}$$

are called **quadratic residue codes.** ▯

Since we will work only with binary quadratic residue codes, we require that $q = 2$ be a quadratic residue modulo p. In view of part 6) of Theorem 8.4.2, *from now on we assume that* p *is a prime of the form* $8m \pm 1$.

It is clear that

$$\overline{\mathcal{Q}}(p) \subset \mathcal{Q}(p) \quad \text{and} \quad \overline{\mathcal{N}}(p) \subset \mathcal{N}(p)$$

In fact, $\overline{\mathcal{Q}}(p)$ is the subset of $\mathcal{Q}(p)$ consisting of all even weight codewords in $\mathcal{Q}(p)$, and similarly for $\overline{\mathcal{N}}(p)$ and $\mathcal{N}(p)$.

In view of Theorem 8.4.1, we have

$$\mathit{dim}(\mathcal{Q}(p)) = p - \deg(q(x)) = p - |QR| = \frac{p+1}{2}$$

and

$$\mathit{dim}(\mathcal{N}(p)) = p - \deg(n(x)) = p - |NQR| = \frac{p+1}{2}$$

The codes $\mathcal{Q}(p)$ and $\mathcal{N}(p)$ are equivalent. To see this, let $v \in NQR$, and consider the permutation of the coordinate positions defined by

$$\pi_v(i) = vi \bmod p$$

Note that π_v can be implemented by replacing x by x^v and then reducing modulo $x^p - 1$, and further that

$$\pi_v(f(x)g(x)) = \pi_v(f(x))\pi_v(g(x))$$

To show that π_v sends $\mathcal{Q}(p)$ to $\mathcal{N}(p)$, we $p(x) = \pi_v(q(x))$. Then

$$p(x) = \pi_v(q(x)) \equiv q(x^v) \equiv \prod_{r \in QR} (x^v - \omega^r) \bmod (x^p - 1)$$

and so there exists a polynomial $a(x)$ for which

$$p(x) = \prod_{r \in QR} (x^v - \omega^r) + a(x)(x^p - 1)$$

Since the product of any two quadratic nonresidues is a quadratic residue, we deduce that for any $u \in NQR$, the product uv is in QR. Hence,

$$p(\omega^u) = \prod_{r \in QR} (\omega^{uv} - \omega^r) + a(\omega^u)(\omega^{up} - 1) = 0$$

This implies that $x - \omega^u$ divides $p(x)$, for all $u \in NQR$, and so $n(x) \mid p(x)$. Hence, $n(x)$ also divides $p(x) \bmod (x^p-1)$, which implies that

$$\pi_v(\mathcal{Q}(p)) = \langle q(x^v) \bmod (x^p-1) \rangle \subset \langle n(x) \rangle = \mathcal{N}(p)$$

But since π_v is injective and $\mathcal{Q}(p)$ and $\mathcal{N}(p)$ have the same size, we deduce that $\pi_v(\mathcal{Q}(p)) = \mathcal{N}(p)$, and so $\mathcal{N}(p)$ and $\mathcal{Q}(p)$ are equivalent.

Theorem 8.4.5 For the quadratic residue codes of length p, we have

$$\mathit{dim}(\mathcal{Q}(p)) = \mathit{dim}(\mathcal{N}(p)) = \frac{p+1}{2}$$

$$\mathit{dim}(\overline{\mathcal{Q}}(p)) = \mathit{dim}(\overline{\mathcal{N}}(p)) = \frac{p-1}{2}$$

Furthermore, the codes $\mathcal{Q}(p)$ and $\mathcal{N}(p)$ are equivalent, as are the codes $\overline{\mathcal{Q}}(p)$ and $\overline{\mathcal{N}}(p)$. □

The next example shows that the codes $\mathcal{Q}(p)$ and $\mathcal{N}(p)$ can be interchanged by replacing one primitive p-th root of unity ω by another.

Example 8.4.5 Let $p = 7$. Then according to Theorem 8.4.1,

$$QR = \{1,4,2\} \quad \text{and} \quad NQR = \{3,5,6\}$$

Hence, if ω is a primitive 7th root of unity over F_2, then

$$q(x) = (x-\omega)(x-\omega^2)(x-\omega^4) = x^3+x+1$$

and

$$n(x) = (x-\omega^3)(x-\omega^5)(x-\omega^6) = x^3+x^2+1$$

Thus $\mathcal{Q}(7) = \langle\langle x^3+x+1 \rangle\rangle$ is the binary [7,4,3] Hamming code $\mathcal{H}_2(3)$. (Recall that $\mathcal{H}_2(3)$ can be characterized as the cyclic code with ω as zero.)

Similarly, the code $\mathcal{N}(7) = \langle\langle x^3+x^2+1 \rangle\rangle$ is the cyclic code with zero ω^3 (which is also a primitive 7th root of unity) and so is also the Hamming code $\mathcal{H}_2(3)$. ☐

THE GOLAY CODES AS QUADRATIC RESIDUE CODES

With reference to Example 7.4.3, since we have the factorization

$$x^{23}-1$$
$$= (x+1)(x^{11}+x^9+x^7+x^6+x^5+x+1)(x^{11}+x^{10}+x^6+x^5+x^4+x^2+1)$$

into irreducible polynomials over F_2, and since 2 is a quadratic residue modulo 23, this must also be the factorization

$$x^{23}-1 = (x-1)q(x)n(x)$$

Therefore, the Golay code $\mathcal{G}_{23}$ is equivalent to the quadratic residue code $\mathcal{Q}(23)$.

In a similar manner, we have over F_3,

$$x^{11}-1 = (x-1)(x^5+x^4-x^3+x^2-1)(x^5-x^3+x^2-x-1)$$

where the factors are irreducible, and since 3 is a quadratic residue modulo 11, we deduce that the ternary Golay code $\mathcal{G}_{11}$ is equivalent to the ternary quadratic residue code $\mathcal{Q}(11)$.

THE SQUARE ROOT BOUND

Now let us consider the minimum distance of a binary quadratic residue code. It is possible to show, using Corollary 4.3.16, that the minimum distance of a quadratic residue code is odd. We then have the following result, known as the **square root bound**. Unfortunately, this bound is often not very good.

Theorem 8.4.6 (The square root bound) The minimum distance d of the binary codes $\mathcal{Q}(p)$ and $\mathcal{N}(p)$ satisfies

$$d^2 \geq p$$

Furthermore, if $p = 8m - 1$, then

$$d^2 - d + 1 \geq p$$

Proof. Let $c(x)$ be a codeword in $\mathcal{Q}(p)$ of minimum nonzero weight d. We leave it to the reader to show that if $u \in NQR$, then

$$\bar{c}(x) = c(x^u) \bmod (x^p - 1)$$

is a word of minimum weight d in $\mathcal{N}(p)$. Hence, the polynomial

$$p(x) = [c(x)\bar{c}(x)] \bmod (x^p - 1) = [c(x)c(x^u)] \bmod (x^p - 1)$$

is in the intersection $\mathcal{Q}(p) \cap \mathcal{N}(p)$ and so must be divisible by

$$q(x)n(x) = \frac{x^p - 1}{x - 1} = 1 + x + x^2 + \cdots + x^{p-1}$$

The fact that $\mathcal{Q}(p)$ has odd minimum weight implies that $p(1) = c(1)^2 \neq 0$, and so $p(x) \neq 0$. In summary, $p(x)$ is nonzero, is divisible by $q(x)n(x)$, and has degree at most $p - 1$, and so $p(x)$ must be a nonzero scalar multiple of $q(x)n(x)$. In particular, $p(x)$ must have weight p.

It follows that the product $c(x)c(x^u)$ has weight at least p. But $c(x)$ and $c(x^u)$ have d nonzero terms, and so this product can have no more than d^2 nonzero terms. Therefore, $d^2 \geq p$.

If $p = 8m - 1$, then according to part 5) of Theorem 8.4.2, we can take $u = -1$. But $c(x)c(x^{-1})$ has, among its at most d^2 nonzero terms, d terms of the form $a_i x^i x^{-i}$, which are constants, and which therefore collect into a single term. Hence, the product $c(x)c(x^{-1})$ has weight at most $d^2 - d + 1$, and so $d^2 - d + 1 \geq p$. ∎

Example 8.4.6 The binary quadratic residue codes of length $p = 7$ have minimum distance d satisfying $d^2 - d + 1 \geq 7$. Hence, $d \geq 3$. As we have seen, the binary [7,4] quadratic residue code has minimum distance $d = 3$. □

Table 8.4.1 lists the parameters of some binary quadratic codes, along with the lower bounds of the minimum distance guaranteed by Theorem 8.4.6.

Table 8.4.1 Quadratic Residue Codes

p	*k*	*d*	*lower bound on d*
7	4	3	3
17	9	5	4
23	12	7	6
31	16	7	6
41	21	9	6
47	24	11	8
71	36	11	9
73	37	13	9
79	40	15	10
89	45	17	9
97	49	15	9
103	52	19	11
127	64	19	12
151	76	19	13

THE IDEMPOTENTS OF A BINARY QUADRATIC RESIDUE CODE

Let us determine the generating idempotents of the binary codes $\mathcal{Q}(p)$ and $\mathcal{N}(p)$. Our plan is to employ Theorem 7.4.12. We begin by letting

$$e(x) = \sum_{r \in QR} x^r$$

According to Corollary 8.4.4, since 2 is a quadratic residue modulo p, we have

$$[e(x)]^2 = e(x^2) = \sum_{r \in QR} x^{2r} \equiv \sum_{r \in QR} x^r = e(x)$$

and so $e(x)$ is an idempotent. In order to determine the zeros of $e(x)$ from among the p-th roots of unity, we compute $e(\omega^i)$ for all $i = 0,1,\ldots,p-1$. Note that ω^i lies in the splitting field F_{2^s}, which has characteristic 2.

Since $e(x)$ is idempotent, we have $e(\omega^i) = 0$ or 1 for all i. Furthermore, if $s \in QR$, then

$$e(\omega^s) = \sum_{r \in QR} \omega^{sr} = \sum_{r \in QR} \omega^r = e(\omega)$$

and if $u \in NQR$, then

$$e(\omega^u) = \sum_{r \in QR} \omega^{ur} = \sum_{u \in NQR} \omega^u \neq e(\omega)$$

In other words, $e(x)$ is constant on the sets QR and NQR and is different on these two sets. Hence, we must have either

1) $e(\omega^s) = \begin{cases} 0 & \text{if } s \in QR \\ 1 & \text{if } s \in NQR \end{cases}$

or

2) $e(\omega^s) = \begin{cases} 1 & \text{if } s \in QR \\ 0 & \text{if } s \in NQR \end{cases}$

Should case 2) prevail, however, letting $\beta = \omega^v$ for some $v \in NQR$ implies that for all $s \in QR$

$$e(\beta^s) = e(\omega^{vs}) = 0$$

Hence, by replacing ω with a different primitive p-th root of unity, it is always possible to ensure case 1, which we assume holds from now on.

Next, we observe that, in the splitting field F_{2^s},

$$e(1) = \frac{p-1}{2} = \begin{cases} 1 & \text{if } p = 8m - 1 \\ 0 & \text{if } p = 8m + 1 \end{cases}$$

This leads to two possibilities

3) If $p = 8m - 1$, then

$$e(\omega^s) = \begin{cases} 0 & \text{if } s \in QR \\ 1 & \text{if } s \in NQR \cup \{0\} \end{cases}$$

4) If $p = 8m + 1$, then

$$e(\omega^s) = \begin{cases} 0 & \text{if } s \in QR \cup \{0\} \\ 1 & \text{if } s \in NQR \end{cases}$$

Now we may apply Theorem 7.4.12 to deduce that, under case 3), $e(x)$ is the generating idempotent for $\mathcal{Q}(p) = \langle\langle q(x) \rangle\rangle$, whereas under case 4), $e(x)$ is the generating idempotent for $\overline{\mathcal{Q}}(p) = \langle\langle (x-1)q(x) \rangle\rangle$. We fill in the other idempotents in the following theorem.

Theorem 8.4.7 Let

$$e(x) = \sum_{r \in QR} x^r \quad \text{and} \quad f(x) = \sum_{u \in NQR} x^u$$

There is a primitive p-th root of unity ω for which

$$e(\omega^s) = \begin{cases} 0 & \text{if } s \in QR \\ 1 & \text{if } s \in NQR \end{cases}$$

With this choice of ω, the following holds.

1) Assume $p = 8m - 1$. Then $e(1) = 1$, and
   a) the generating idempotent for $\mathcal{Q}(p)$ is $e(x)$,
   b) the generating idempotent for $\overline{\mathcal{Q}}(p)$ is $1 + f(x)$,
   c) the generating idempotent for $\mathcal{N}(p)$ is $f(x)$,
   d) the generating idempotent for $\overline{\mathcal{N}}(p)$ is $1 + e(x)$.
2) Assume $p = 8m + 1$. Then $e(1) = 0$, and
   a) the generating idempotent for $\mathcal{Q}(p)$ is $1 + f(x)$,
   b) the generating idempotent for $\overline{\mathcal{Q}}(p)$ is $e(x)$,
   c) the generating idempotent for $\mathcal{N}(p)$ is $1 + e(x)$,
   d) the generating idempotent for $\overline{\mathcal{N}}(p)$ is $f(x)$.

Proof. We leave it to the reader to show that $f(x)$ is idempotent. Since

$$f(x) + e(x) + 1 = \sum_{i=0}^{p-1} x^i = \frac{x^p - 1}{x - 1}$$

we deduce that $f(\omega^i) \neq e(\omega^i)$ for all $i \neq 0$. The value of

$$f(1) = \frac{p-1}{2}$$

depends on the value of p. If $p = 8m - 1$, then in F_{2^s}

$$f(1) = \frac{p-1}{2} = 1$$

Hence, the zeros are as follows.

For $e(x)$: QR For $f(x)$: NQR
For $1 + e(x)$: $NQR \cup \{0\}$ For $1 + f(x)$: $QR \cup \{0\}$

Part 1) of the theorem follows from Theorem 7.4.12.

On the other hand, if $p = 8m + 1$, then in F_{2^s}

$$f(1) = \frac{p-1}{2} = 0$$

and so the zeros are as follows.

For $e(x)$: $QR \cup \{0\}$ For $f(x)$: $NQR \cup \{0\}$
For $1 + e(x)$: NQR For $1 + f(x)$: QR

Part 2) of the theorem follows from Theorem 7.4.12. ∎

DUALS OF THE QUADRATIC RESIDUE CODES

We leave proof of the following result as an exercise.

Theorem 8.4.8

a) If $p = 8m - 1$, then

$$\mathcal{Q}(p)^{\perp} = \overline{\mathcal{Q}}(p) \quad \text{and} \quad \mathcal{N}(p)^{\perp} = \overline{\mathcal{N}}(p)$$

b) If $p = 8m + 1$, then

$$\mathcal{Q}(p)^{\perp} = \overline{\mathcal{N}}(p) \quad \text{and} \quad \mathcal{N}(p)^{\perp} = \overline{\mathcal{Q}}(p)$$

□

THE EXTENDED QUADRATIC RESIDUE CODES

We may extend the quadratic residue codes by adding an overall parity check in the usual way to get

$$\widehat{\mathcal{Q}}(p) = \{a_0 \cdots a_{p-1} a_p \mid a_0 \cdots a_{p-1} \in \mathcal{Q}(p), \ \sum_i a_i = 0\}$$

and

$$\widehat{\mathcal{N}}(p) = \{a_0 \cdots a_{p-1} a_p \mid a_0 \cdots a_{p-1} \in \mathcal{N}(p), \ \sum_i a_i = 0\}$$

Let us take a look at the duals of these extended codes. Let

$$E(x) = \sum_{i=0}^{p-1} E_i x^i$$

be the generating idempotent of the even weight code $\overline{\mathcal{Q}}(p)$, where $E(x)$ can be determined using Theorem 8.4.7. Then the rows of the $p \times p$ *circulant matrix*

$$\overline{G} = \begin{bmatrix} E_0 & E_1 & \cdots & E_{p-1} \\ E_{p-1} & E_0 & \cdots & E_{p-2} \\ \vdots & \vdots & \vdots & \vdots \\ E_1 & E_2 & \cdots & E_0 \end{bmatrix}$$

being all possible cyclic shifts of $E(x)$, span the code $\overline{\mathcal{Q}}(p)$. (Why?) Since the word $\mathbf{1} = 11 \cdots 1$ is in $\mathcal{Q}(p)$, but not in $\overline{\mathcal{Q}}(p)$, the rows of the matrix

$$G = \begin{bmatrix} \overline{G} \\ \mathbf{1} \end{bmatrix}$$

span $\mathcal{Q}(p)$. Finally, adding an overall parity check is accomplished by augmenting the matrix $\overline{G}$ as follows

$$\widehat{G} = \begin{bmatrix} & & & 0 \\ & \overline{\mathbf{G}} & & 0 \\ & & & \vdots \\ 1 & 1 & \cdots & 1 \end{bmatrix}$$

Thus, the rows of $\widehat{G}$ span $\widehat{Q}(p)$.

According to Theorem 8.4.8, if $p = 8m - 1$, then

$$\overline{Q}(p) \subset Q(p) = Q(p)^{\perp\perp} = [\overline{Q}(p)]^{\perp}$$

which implies that all rows of $\widehat{G}$, except the last, are orthogonal. However, since these rows have even weight, they are orthogonal to the last row. In other words, all rows of $\widehat{G}$ are orthogonal, and so

$$\widehat{Q}(p) \subset \widehat{Q}(p)^{\perp}$$

But

$$\begin{aligned} dim(\widehat{Q}(p)^{\perp}) &= (p+1) - dim(\widehat{Q}(p)) \\ &= (p+1) - dim(Q(p)) = \frac{p+1}{2} = dim(\widehat{Q}(p)) \end{aligned}$$

and so

$$\widehat{Q}(p) = \widehat{Q}(p)^{\perp}$$

Thus, $\widehat{Q}(p)$ is self-dual. Similarly, $\widehat{N}(p)$ is self-dual.

Theorem 8.4.9 If $p = 8m - 1$, then the extended quadratic residue codes $\widehat{Q}(p)$ and $\widehat{N}(p)$ are self-dual. ▯

The proof of the following result is left to the reader.

Theorem 8.4.10 Let $p = 8m - 1$.

1) All codewords in the extended code $\widehat{Q}(p)$ have weight divisible by 4.
2) All codewords in the code $Q(p)$ have weight congruent to 0 or 3 modulo 4. ▯

If $p = 8m + 1$, Theorem 8.4.10 does not hold – all we can say is that the codewords in $\widehat{Q}(p)$ have even weight.

EXERCISES

1. Prove Theorem 8.4.1.
2. Prove parts 1) and 2) of Theorem 8.4.2.

3. Use Euler's criterion to prove part 4) of Theorem 8.4.2.
4. Use Euler's criterion to prove part 5) of Theorem 8.4.2.
5. Prove Corollary 8.4.4.
6. If ρ is a primitive p-th root of unity, show that $QR = \{\rho^{2k} \mid k = 1,2,\ldots,\frac{p-1}{2}\}$.
7. Show that the permutation $\pi_v : i \to vi \bmod p$ can be implemented by replacing x by x^n and then reducing modulo $x^p - 1$. Show also that $\pi_v(f(x)g(x)) = \pi_v(f(x))\pi_v(g(x))$.
8. Prove that $\overline{\mathcal{N}}(p)$ and $\overline{\mathcal{Q}}(p)$ are equivalent.
9. Explain how Example 8.4.5 shows that the codes $\mathcal{Q}(p)$ and $\mathcal{N}(p)$ can be interchanged by replacing one primitive p-th root of unity ω by another.
10. Describe the binary quadratic residue codes of length 17.
11. Describe the ternary quadratic residue codes of length 11.
12. Let $c(x)$ be a codeword in $\mathcal{Q}(p)$ of minimum nonzero weight d. Show that if $u \in NQR$, then $\bar{c}(x) = c(x^u) \bmod (x^p - 1)$ is a word of minimum weight d in $\mathcal{N}(p)$.
13. Determine which single-error-correcting quadratic residue codes are perfect.
14. Show that
$$f(x) = \sum_{u \in NQR} x^u$$
is an idempotent.
15. Prove Theorem 8.4.8. *Hint.* First show that the zeros of the dual code are the inverses of the nonzeros of the original code.
16. Prove that the word $\mathbf{1} = 11\cdots1$ is in $\mathcal{Q}(p)$, but not in $\overline{\mathcal{Q}}(p)$.
17. Prove Theorem 8.4.10. *Hint.* Show that the weight of each row of $\widehat{G}$ is divisible by 4.

APPENDIX

Preliminaries

The purpose of this section is to review some basic facts that will be needed in the book. The discussion is not intended to be complete, nor are all proofs supplied.

A.1 Algebraic Preliminaries

GROUPS

Definition A **group** is a nonempty set G, together with a binary operation $*$ that satisfies the following properties.

1) **(associativity)** For all $a,b,c \in G$

$$(a*b)*c = a*(b*c)$$

2) **(identity)** There exists an element $e \in G$ for which

$$e*a = a*e = a$$

for all $a \in G$.

3) **(inverses)** For each $a \in G$, there is an element $a^{-1} \in G$ for which

$$a*a^{-1} = a^{-1}*a = e$$

□

Definition A group G is **abelian, or commutative,** if $a*b = b*a$, for all $a,b \in G$. □

We will often denote the action of $*$ by juxtaposition, thus writing $a*b$ simply as ab. When a group is abelian, it is customary to denote the operation by $+$, thus writing $a*b$ as $a+b$.

Definition A **subgroup** S of a group G is a subset of G that is a group in its own right, using the same operation as defined on G. ☐

Example A.1.1 The set $\mathbb{Z}_n = \{0,1,\ldots,n-1\}$ is an abelian group under addition modulo n

$$a \oplus b = (a+b) \bmod n$$ ☐

A group G is **finite** if it contains only a finite number of elements. The cardinality of a finite group G is called its **order** and is denoted by $o(G)$.

If $a \in G$, and if $a^k = e$ for some integer k, then we say that k is an **exponent** of a. The smallest positive exponent for $a \in G$ is called the **order** of a and is denoted by $o(a)$.

Theorem A.1.1 Let G be a group and let $a \in G$. Then k is an exponent of a if and only if k is a multiple of the order $o(a)$. ☐

Theorem A.1.2 Let G be a finite group.

1) The order of any subgroup of G divides the order of G.
2) The order of any element of G divides the order of G. ☐

EULER'S FORMULA

If a and b are integers, we denote their greatest common divisor by (a,b). If $(a,b) = 1$, we say that a and b are **relatively prime.** The **Euler phi function** ϕ is defined by letting $\phi(n)$ be the number of positive integers less than or equal to n that are relatively prime to n.

Two integers a and b are **congruent modulo n**, written

$$a \equiv b \bmod n$$

if $a-b$ is divisible by n. Theorem A.1.2 can be used to prove the following important theorem.

Theorem A.1.3 **(Euler's Theorem)** If $(a,n) = 1$, then

$$a^{\phi(n)} \equiv 1 \bmod n$$

Proof. The set $G = \{b \in \mathbb{Z}_n \mid (b,n) = 1\}$ is a group of order $\phi(n)$ under multiplication modulo n. For if $b, c \in G$, then so is bc. Also, if $b \in G$, then there exists p, q such that $pb + qn = 1$, and so

$pb \equiv 1 \bmod n$. That is, $b^{-1} = p \bmod n$. Thus, Theorem A.1.2 implies that $a^{\phi(n)} \equiv 1 \bmod n$, for all $a \in G$.

If $a \notin G$, then there exists an $a' \in G$ for which $a' \equiv a \bmod n$. In other words, $a' = a+kn$ for some integer k, and so $(a,n) = 1$ implies $(a',n) = 1$. Thus, we have $a^{\phi(n)} \equiv (a')^{\phi(n)} \equiv 1 \bmod n$. ∎

Corollary A.1.4 (Fermat's Theorem) If p is a prime not dividing a, then

$$a^{p-1} \equiv 1 \bmod p$$ □

CYCLIC GROUPS

If G is a group and $a \in G$, then the set of all powers of a

$$\langle a \rangle = \{a^n \mid n \in \mathbb{Z}\}$$

is a subgroup of G, called the **cyclic subgroup generated by** a. A group G is **cyclic** if it has the form $G = \langle a \rangle$, for some $a \in G$. In this case, we say that a **generates** G.

Theorem A.1.5 Every subgroup of a cyclic group is cyclic. □

The following theorem summarizes some key results about finite cyclic groups.

Theorem A.1.6 Let $G = \langle a \rangle$ be a cyclic group of order n.

1) For $1 \leq k < n$, we have
$$o(a^k) = \frac{n}{(n,k)}$$
In particular, a^k generates $G = \langle a \rangle$ if and only if $(n,k) = 1$.

2) If $d \mid n$, then
$$o(a^k) = d \Leftrightarrow k = r\tfrac{n}{d}, \text{ where } (r,d) = 1$$
In words, the elements of G of order d are the elements of the form $a^{rn/d}$, where r is relatively prime to d.

3) For each $d \mid n$, the group G has exactly one subgroup of order d and $\phi(d)$ elements of order d.

Proof. To prove part 1), we first observe that

$$(a^k)^{n/(n,k)} = (a^n)^{k/(n,k)} = e$$

and so $n/(n,k)$ is an exponent of a^k. On the other hand, if m is an exponent of a^k, that is, if $a^{km} = e$, then km is an exponent of a. Therefore, $n = o(a)$ divides km. But

$$n \mid km \text{ if and only if } \frac{n}{(n,k)} \,\Big|\, \frac{km}{(n,k)}$$

and since $n/(n,k)$ and $k/(n,k)$ are relatively prime, we must have

$$\frac{n}{(n,k)} \,\Big|\, m$$

which shows that the order of a^k is $n/(n,k)$.

To prove part 2), we let $d \mid n$ and determine for which k satisfying $1 \le k < n$ is it true that

$$\frac{n}{(n,k)} = d$$

that is, for which k is

$$(n,k) = \frac{n}{d}$$

Certainly, such a k must be a multiple of n/d, say $k = r\frac{n}{d}$. So we must determine for which r is it true that

$$(n, r\tfrac{n}{d}) = \frac{n}{d}$$

or, equivalently,

$$(d\tfrac{n}{d}, r\tfrac{n}{d}) = \frac{n}{d}$$

But this is equivalent to

$$(d,r) = 1$$

which proves part 2).

To prove part 3), we observe that if $d \mid n$, then according to part 2), an element of $\langle a \rangle$ has order d if and only if it is of the form $(a^{n/d})^r$, where $(r,d) = 1$. Hence, the subgroup $\langle a^{n/d} \rangle$ has order d and also contains all elements of order d. This implies that no other subgroup can have order d, for any other subgroup of order d, being cyclic, would have to contain an element of order d, which is impossible. ∎

Corollary A.1.6 For any positive integer n,

$$n = \sum_{d \mid n} \phi(d)$$

Proof. Since each element of $\mathbb{Z}_n$ has a unique order d dividing n, and since there are precisely $\phi(d)$ elements of order d, the right hand sum counts the total number of elements in $\mathbb{Z}_n$. Hence, it must equal $|\mathbb{Z}_n| = n$. ∎

RINGS AND FIELDS

Definition A **ring** is a nonempty set R, together with two binary operations, called *addition* (denoted by +), and *multiplication* (denoted by juxtaposition), satisfying the following properties.

1) R is an abelian group under the operation +.
2) (**associativity**) For all $a,b,c \in R$,

$$(ab)c = a(bc)$$

3) (**distributivity**) For all $a,b,c \in R$,

$$(a+b)c = ac + ab \quad \text{and} \quad c(a+b) = ca + cb \qquad \square$$

Definition
1) A ring R is called a **ring with identity** if there exists an element $1 \in R$ for which $a1 = 1a = a$, for all $a \in R$.
2) A ring R is called a **commutative ring** if multiplication is commutative, that is, if $ab = ba$ for all $a,b \in R$.
3) A ring R with identity is called an **integral domain** if it is commutative, and if there are no *zero divisors* in R, where a nonzero element $a \in R$ is a **zero divisor** if there exists a nonzero element $b \in R$ for which $ab = 0$.
4) A ring R with identity $1 \neq 0$ is called a **field** if the nonzero elements of R form an abelian group under multiplication. □

Theorem A.1.7 Every finite integral domain is a field. □

Definition A **subring** of a ring R is a subset S of R that is a ring in its own right, using the same operations as defined on R. □

HOMOMORPHISMS

Intuitively speaking, a *homomorphism* between two rings R and S is a function $\psi:R\to S$ that preserves the operations of addition and multiplication. More specifically, we have the following definitions.

Definition Let R and S be rings (or fields). A function $\psi:R\to S$ is a **homomorphism** if, for all $a,b \in R$,

$$\psi(a+b) = \psi(a) + \psi(b) \quad \text{and} \quad \psi(ab) = \psi(a)\psi(b)$$

A homomorphism that is also a bijection (one-to-one and onto) is called an **isomorphism**. If $\psi:R\to S$ is an isomorphism, we say that R and S are **isomorphic**. □

When two rings (or fields) are isomorphic, then from the point of view of their algebraic structure, they are essentially the same. Put another way, isomorphic rings (or fields) can be thought of merely as different concrete representations of the same abstract ring (or field).

IDEALS

Definition A subset I of a ring R is called an **ideal** if it satisfies
1) $a,b \in I$ implies $a-b \in I$.
2) $a \in R$, $i \in I \Rightarrow ai \in I$ and $ia \in I$. □

If S is a nonempty subset of a ring R, then the **ideal generated** by S is defined to be the smallest ideal I of R containing S. If R is a commutative ring with identity, and if $a \in R$, then the ideal generated by a is the set

$$\langle a\rangle = \{ra \mid r \in R\}$$

Any ideal of the form $\langle a\rangle$ is called a **principal ideal.**

Definition A ring R is called a **principal ideal domain** if every ideal of R is principal. □

FACTOR RINGS

Suppose that R is a ring and I is an ideal in R. Then for each $a \in R$, we can form the set

$$a+I = \{a+i \mid i \in I\}$$

which is called a **coset** of I. It can be shown that $a+I = b+I$ if and only if $a-b \in I$, and that any two cosets $a+I$ and $b+I$ are either disjoint or identical. Furthermore, the collection of all (distinct) cosets can be made into a ring itself with addition and multiplication defined by

$$(a+I)+(b+I) = (a+b)+I$$

and

$$(a+I)(b+I) = ab+I$$

The ring of all cosets of I is called a **factor ring** and is denoted by R/I, read " R mod I."

Example A.1.2 The factor ring $\mathbb{Z}/\langle n\rangle$ is isomorphic to the ring $\mathbb{Z}_n$, under addition and multiplication modulo n. The ring $\mathbb{Z}_n$ is a field if

and only if n is a prime. ▯

Definition An ideal I is **maximal** if $I \neq R$, and if whenever J is an ideal for which $I \subseteq J \subseteq R$, then $J = I$ or $J = R$. ▯

Theorem A.1.8 Let R be a commutative ring with identity. Then the factor ring R/I is a field if and only if I is a maximal ideal. ▯

If p(x) is a polynomial over F_q of degree d, then the factor ring

$$K = \frac{F_q[x]}{\langle p(x) \rangle}$$

is the set of all cosets of $I = \langle p(x) \rangle$,

$$f(x) + I$$

where $f(x) \in F_q[x]$.

However, it turns out that we need only consider polynomials f(x) whose degree is less than the degree of p(x). For if $\deg(f(x)) \geq d$, then dividing f(x) by p(x) gives

$$f(x) = q(x)p(x) + r(x)$$

where $\deg(r(x)) < d$. But $q(x)p(x) \in I$, and so $q(x)p(x) + I = 0 + I$. Hence,

$$f(x) + I = r(x) + I$$

and

$$K = \{r(x) + I \mid \deg(r(x)) < d\}$$

Thus, we may identify the factor ring K with the set of all polynomials of degree less than d, with addition and multiplication performed modulo p(x).

THE CHARACTERISTIC OF A RING

If R is a ring, and if there exists a positive integer n for which

$$na = \underbrace{a + a + \cdots + a}_{\text{n times}} = 0$$

for all $a \in R$, then the *smallest* such positive integer is called the **characteristic** of the ring and is denoted by char(R). If no such integer n exists, we say that R has characteristic 0. For instance, the ring $\mathbb{Z}_n$ has characteristic n. On the other hand, the ring of integers $\mathbb{Z}$ has characteristic 0.

Theorem A.1.9 Any integral domain has either characteristic 0 or a prime characteristic. In particular, a finite field has prime characteristic.

Proof. First we observe that the characteristic of a ring with identity is the smallest positive integer n for which $n1 = 0$. Now suppose that R is an integral domain of finite characteristic n. If $n = pq$, where $p,q < n$, then $pq1 = 0$. Hence, $(p1)(q1) = 0$, implying that $p1 = 0$ or $q1 = 0$. In either case, we have a contradiction to the fact that n is the *smallest* positive integer such that $n1 = 0$. Hence, n must be prime. Since a finite field cannot have characteristic 0 (why not?), the second part follows from the first. ∎

The following result will be of considerable importance to us.

Theorem A.1.10 Let R be a commutative ring of *prime* characteristic p. Then if $q = p^n$, we have

$$(a+b)^q = a^q + b^q \quad \text{and} \quad (a-b)^q = a^q - b^q$$

Proof. Since the binomial formula holds in any commutative ring, we have

$$(a+b)^p = \sum_{k=0}^{p} \binom{p}{k} a^k b^{p-k}$$

where

$$\binom{p}{k} = \frac{p(p-1)\cdots(p-k+1)}{k!}$$

But, for $0 < k < p$, the denominator $k!$ does not have the prime number p as a factor, and so the expression

$$\frac{(p-1)\cdots(p-k+1)}{k!}$$

must be an integer. Hence $\binom{p}{k}$ is divisible by the characteristic p and so is equal to 0 in R, and the binomial formula therefore reduces to

$$(a+b)^p = a^p + b^p$$

The second part is proved similarly. ∎

EXTENSION FIELDS

If E and F are fields under the same operations, and if $F \subseteq E$, then we say that F is a **subfield** of E and that E is an **extension field** of F. We write $F < E$ if E is an extension field of F.

If E is an extension field of F, then E can be viewed as a

vector space over F. If the vector space dimension of E over F is finite, we say that E is a **finite extension** of F and denote the dimension by [E:F].

Theorem A.1.11 Let $F < E < K$. Then K is finite over F if and only if K is finite over E and E is finite over F. Furthermore, in this case

$$[K:F] = [K:E][E:F]$$ □

THE PRIME FIELD

For any field F, the smallest subfield contained in F is called the **prime field** of F. It is the intersection of all subfields of F.

Theorem A.1.12 If $\text{char}(F) = 0$, then the prime field of F is isomorphic to the rational numbers $\mathbb{Q}$. If char(F) is a prime p, then the prime field of F is isomorphic to $\mathbb{Z}_p$. □

SIMPLE EXTENSIONS

Definition Let $F < E$, $\alpha \in E$. Then $F(\alpha)$ is defined to be the smallest field containing F and α. Any field of the form $F(\alpha)$ is called a **simple extension** of F. □

The element α is often called a *primitive element* of the field $F(\alpha)$. However, this will conflict with the common usage of the term primitive in the theory of finite fields, and so we will reserve this term for later use.

Theorem A.1.13 Let $F < E$, $\alpha \in E$ algebraic over F. Then $F(\alpha)$ is isomorphic to

$$H = \frac{F[x]}{\langle \text{irr}(\alpha,F) \rangle}$$ □

Proof. Let $\phi: F[x] \to E$ be the evaluation homomorphism defined by $\phi(f(x)) = f(\alpha)$. The kernel of ϕ is the ideal $\langle \text{irr}(\alpha,F) \rangle$, and so H is isomorphic to $\phi(F[x])$. Thus, we need only show that $\phi(F[x]) = F(\alpha)$. But $\alpha = \phi(x) \in \phi(F[x])$ and $F \subseteq \phi(F[x])$ imply that $F(\alpha) \subseteq H$. Also, $\phi(F[x]) = \{f(\alpha) \mid f(x) \in F[x]\} \subseteq F(\alpha)$. ∎

Thus, $F(\alpha)$ is the set of all polynomials in α of degree less than $\deg(\text{irr}(\alpha,F))$, with addition and multiplication modulo $\text{irr}(\alpha,F)$.

THE ROOTS OF POLYNOMIALS

Our next result says that every nonconstant polynomial has a root in some field.

Theorem A.1.14 Let F be a field, and let $f(x) \in F[x]$ be a nonconstant polynomial. Then there exist an extension E of F and an $\alpha \in E$ such that $f(\alpha) = 0$. □

Theorem A.1.15 Let $f(x) \in F[x]$. There exists an extension K of F such that $f(x)$ factors into linear factors over K. Hence, all of the roots of $f(x)$ lie in K. □

SPLITTING FIELDS

If a polynomial $f(x) \in F[x]$ factors into linear factors

$$f(x) = a(x - \alpha_1)(x - \alpha_2)\cdots(x - \alpha_n)$$

over an extension field K (that is, $\alpha_1, \ldots, \alpha_n \in K$), we say that $f(x)$ **splits** over K.

Definition Let $f(x) \in F[x]$. A **splitting field** for $f(x)$ is an extension field K of F with the property that $f(x)$ splits over K,

$$f(x) = \beta(x - \alpha_1)(x - \alpha_2)\cdots(x - \alpha_n)$$

and that $K = F(\alpha_1, \ldots, \alpha_n)$. □

Theorem A.1.16 Every polynomial $f(x) \in F[x]$ has a splitting field, and any two splitting fields for $f(x)$ are isomorphic. □

POLYNOMIALS

If F is a field, then $F[x]$ denotes the ring of all polynomials with coefficients in F. If $f(x) \in F[x]$, we say that $f(x)$ is a polynomial *over* F. If $f(x) = a_n x^n + a_{n-1} x^{n-1} + \cdots + a_0$ where $a_n \neq 0$, then the **degree** of $f(x)$ is n. The coefficient a_n is the **leading coefficient.** If $a_n = 1$, we say that $f(x)$ is **monic.**

THE DIVISION ALGORITHM AND ITS CONSEQUENCES

Let us begin by discussing the consequences of simple division.

Theorem A.1.17 (Division algorithm) Let $f(x) \in F[x]$ and $g(x) \in F[x]$, where $\deg g(x) > 0$. Then there exist unique $q(x)$ and $r(x)$ in $F[x]$ such that

$$f(x) = q(x)g(x) + r(x)$$

where $r(x) = 0$ or $0 \le \deg r(x) < \deg g(x)$. ▯

Corollary A.1.18 Let $f(x) \in F[x]$. Then α is a root of $f(x)$ if and only if $x - \alpha$ is a factor of $f(x)$ over F. ▯

Corollary A.1.19 A nonzero polynomial $f(x) \in F[x]$ can have at most $\deg f(x)$ roots in F or in any extension of F. ▯

Corollary A.1.20 The ring $F[x]$ is a principal ideal domain. That is, every ideal in $F[x]$ is generated by a single polynomial. ▯

Corollary A.1.21 Let $f(x)$ and $g(x)$ be polynomials over F. The **greatest common divisor** of $f(x)$ and $g(x)$ over F, denoted by $(f(x),g(x))$ or $\gcd(f(x),g(x))$, is the unique monic polynomial $p(x)$ over F for which

1) $p(x) \mid f(x)$ and $p(x) \mid g(x)$.
2) If $r(x) \mid f(x)$ and $r(x) \mid g(x)$, then $r(x) \mid p(x)$.

Furthermore, there exist polynomials $a(x)$ and $b(x)$ over F for which $\gcd(f(x),g(x)) = a(x)f(x) + b(x)g(x)$. ▯

Definition Let $f(x)$ and $g(x)$ be polynomials over F. If $\gcd(f(x),g(x)) = 1$, we say that $f(x)$ and $g(x)$ are **relatively prime** (over F). In particular, $f(x)$ and $g(x)$ are relatively prime if and only if there exist polynomials $a(x)$ and $b(x)$ over F for which $a(x)f(x) + b(x)g(x) = 1$. ▯

THE EUCLIDEAN ALGORITHM

For a discussion of the Euclidean algorithm, see Section 8.3.

IRREDUCIBLE POLYNOMIALS

Definition A nonconstant polynomial $f(x) \in F[x]$ is **irreducible** if whenever $f(x) = p(x)q(x)$, then one of $p(x)$ or $q(x)$ must be constant. ▯

Theorem A.1.22 If $f(x)$ is irreducible and $f(x) \mid p(x)q(x)$, then either $f(x) \mid p(x)$ or $f(x) \mid q(x)$. ▯

Theorem A.1.23 Every nonconstant polynomial in $F[x]$ can be written uniquely (up to order) as a product of irreducible polynomials. ▯

Theorem A.1.24 An ideal $I = \langle p(x) \rangle$ in $F[x]$ is maximal if and only if $p(x)$ is irreducible. □

Theorem A.1.25 The ring $E = F[x]/\langle p(x) \rangle$ is a field if and only if $p(x)$ is irreducible. Furthermore, if E is a field, then as a vector space over F, we have $\dim E = \deg p(x)$. □

COMMON ROOTS

Theorem A.1.26 Let $f(x)$ and $g(x)$ be polynomials over F.

1) Then $f(x)$ and $g(x)$ are relatively prime (over F) if and only if they have no common roots in *any* extension field of F.
2) Equivalently, $f(x)$ and $g(x)$ have a nontrivial common factor over F if and only if they have a common root in some extension K of F.

Proof. If $(f(x),g(x)) = 1$ then there exists polynomials $a(x)$ and $b(x)$ in $F[x]$ for which

$$a(x)f(x) + b(x)g(x) = 1$$

Since this holds in any extension field as well, we see that $f(x)$ and $g(x)$ can have no common roots. Conversely, if $(f(x),g(x)) = d(x) \neq 1$, then any root of $d(x)$ in an extension K of F is a common root of $f(x)$ and $g(x)$. ∎

Theorem A.1.27 Let $f(x)$ and $g(x)$ be polynomials over F. Then if $f(x)$ and $g(x)$ have a nontrivial common factor in some extension K of F, they have a nontrivial common factor in any extension of F.

Proof. Suppose $f(x)$ and $g(x)$ have a nontrivial common factor in an extension K of F. Then they have a common root over some extension of K, which is also an extension of F. Hence by Theorem A.1.26, they have a nontrivial common factor in F. ∎

Theorem A.1.28 Let $f(x)$ and $g(x)$ be polynomials over F. Suppose that $f(x)$ has no multiple roots in any extension of F. Then $f(x) \mid g(x)$ over F if and only if all of the roots of $f(x)$ in its splitting field are also roots of $g(x)$.

Proof. If $f(x) \mid g(x)$ over F, then clearly all roots of $f(x)$ in any extension are roots of $g(x)$. Conversely, let $p(x)$ be an irreducible factor of $f(x)$ over F. Then $p(x)$ and $g(x)$ have a common root in the splitting field of $f(x)$, and so they have a nontrivial common factor over F. Therefore, $p(x) \mid g(x)$. Since $f(x)$ has no multiple roots, all of its irreducible factors over F are distinct, and so their product, that is, $f(x)$, must also divide $g(x)$. ∎

THE MINIMAL POLYNOMIAL

Let E be an extension of F, and let $\alpha \in E$. Suppose that there is some polynomial $f(x) \in F[x]$ for which $f(\alpha) = 0$. Then the set

$$I = \{g(x) \in F[x] \mid g(\alpha) = 0\}$$

is nonempty, since it contains f(x). It is not hard to see that I is an ideal in F[x], and so it is generated by a single *monic* polynomial p(x). This polynomial p(x) is called the **minimal polynomial** of α over F and is denoted by $irr(\alpha,F)$.

Theorem A.1.30 Let $F < F$ be fields, and let $\alpha \in E$ have minimal polynomial m(x) over F.

1) The polynomial m(x) is the unique monic irreducible polynomial over F for which $m(\alpha) = 0$.
2) The polynomial m(x) is the unique monic polynomial of smallest degree over F for which $m(\alpha) = 0$.
3) The polynomial m(x) is the unique monic polynomial over F with the property that, for all $f(x) \in F[x]$, we have $f(\alpha) = 0$ if and only if $m(x) \mid f(x)$. □

MULTIPLE ROOTS

Theorem A.1.31 Let $f(x) \in F[x]$, and let E be a field containing all the roots of f(x). Then f(x) has a multiple root in E if and only if f(x) and its derivative f′(x) have a nontrivial common factor in F[x].
Proof. Suppose that α is a multiple root of f(x) in E. Then over E, we have

$$f(x) = (x-\alpha)^n g(x)$$

for $n > 1$ and $g(\alpha) \neq 0$. Hence,

$$f'(x) = n(x-\alpha)^{n-1}g(x) + (x-\alpha)^n g'(x)$$

Thus, $f(\alpha) = 0$ and $f'(\alpha) = 0$, and so f(x) and f′(x) are both divisible by $irr(\alpha,F)$ over F. Conversely, if f(x) has only simple roots in E, then

$$f(x) = (x-\alpha_1)\cdots(x-\alpha_n)$$

and it is easy to see that f′(x) is not divisible by any $x-\alpha_i$. But if f(x) and f′(x) were to have a nontrivial common factor in F[x], then in some extension of F, they would have a common factor of the form $x-\alpha_i$. Hence, f(x) and f′(x) have no nontrivial common factor. ∎

Corollary A.1.32 A nonconstant *irreducible* polynomial p(x) has multiple roots in some extension if and only if $p'(x) = 0$.

Proof. If $p'(x) = 0$, then $p(x)$ and $p'(x)$ have a nontrivial common factor $p(x)$. On the other hand, if $p'(x) \neq 0$, then since $p(x)$ is irreducible and does not divide $p'(x)$, they have no nontrivial common factors in any extension. □

Corollary A.1.33 If F has characteristic zero, no irreducible polynomial can have multiple roots in any extension. □

Corollary A.1.34 If $\mathrm{char}(F) = p \neq 0$, then an irreducible polynomial $p(x)$ has multiple roots in some extension if and only if $p(x) = q(x^p)$, for some polynomial $q(x)$. □

Corollary A.1.35 If F is a *finite* field, then no irreducible polynomial in $F[x]$ can have multiple roots in any extension.
Proof. Let $\mathrm{char}(F) = p$. Since by Fermat's Theorem, $a^p = a \bmod p$, we see that any polynomial of the form $q(x^p)$ satisfies

$$\begin{aligned} q(x^p) &= a_0 + a_1 x^p + \cdots + a_n x^{pn} \\ &= a_0^p + a_1^p x^p + \cdots + a_n^p x^{np} \\ &= (a_0 + a_1 x + \cdots + a_n x^n)^p \end{aligned}$$

and so is not irreducible. ∎

We should note that in infinite fields of nonzero characteristic, there are irreducible polynomials with multiple roots.

Definition An irreducible polynomial $p(x) \in F[x]$ is **separable** if it has no multiple roots in any extension field. □

Corollary A.1.36 If F is a *finite* field, then all irreducible polynomials in $F[x]$ are separable. □

A.2 Möbius Inversion

Möbius inversion is a method for inverting certain types of sums. The classical form of Möbius inversion was originally developed, in 1935, independently by P. Hall and L. Weisner. However, in 1964, Gian-Carlo Rota generalized the classical form to apply to a much wider range of situations.

To describe the concept in its fullest generality, we require some facts about partially ordered sets.

PARTIALLY ORDERED SETS

Definition A **partial order** on a nonempty set P is a binary relation, denoted by $\leq$ and read "less than or equal to," with the following properties.

1) **(reflexivity)** For all $a \in P$,
$$a \leq a$$

2) **(antisymmetry)** For all $a,b \in P$,
$$a \leq b \text{ and } b \leq a \text{ implies } a = b$$

3) **(transitivity)** For all $a,b,c \in P$,
$$a \leq b \text{ and } b \leq c \text{ implies } a \leq c$$
□

Definition A **partially ordered set** is a nonempty set P, together with a partial order $\leq$ defined on P. The expression $a \leq b$ is read "a is less than or equal to b." If $a,b \in P$, we denote the fact that a is *not* less than or equal to b by $a \nleq b$. Also, we denote the fact that $a \leq b$, but $a \neq b$, by $a < b$.

If there exists an element $z \in P$ for which $z \leq x$ for all $x \in P$, we call z a **zero** element and denote it by 0. Similarly, if there exists an element $y \in P$ for which $x \leq y$ for all $x \in P$, then we call y a **one** and denote it by 1. □

As is customary, when the partial order $\leq$ is understood, we will use the phrase "let P be a partially ordered set."

Note that, in a partially ordered set, it is possible that not all elements are comparable. In other words, it is possible to have $x,y \in P$ with the property that $x \nleq y$ and $y \nleq x$. Thus, in general, $x \nleq y$ is *not* equivalent to $y \leq x$. A partially ordered set in which every pair of elements is comparable is called a **totally ordered set** or a **linearly ordered set**.

Example A.2.1

1) The set $\mathbb{R}$ of real numbers, with the usual binary relation $\leq$, is a partially ordered set. It is also a totally ordered set.
2) The set $\mathbb{N}$ of natural numbers, together with the binary relation of divides, is a partially ordered set. It is customary to write $n \mid m$ (rather than $n \leq m$) to indicate that n divides m.
3) Let S be any set, and let $\mathcal{P}(S)$ be the power set of S, that is, the set of all subsets of S. Then $\mathcal{P}(S)$, together with the subset relation $\subseteq$, is a partially ordered set. ☐

Definition Let P be a partially ordered set. For $a,b \in P$, the **(closed) interval** $[a,b]$ is the set

$$[a,b] = \{x \in P \mid a \leq x \leq b\}$$

We say that the partially ordered set P is **locally finite** if every closed interval is a finite set. ☐

Notice that, if P is locally finite and contains a zero element 0, then the set $\{x \in P \mid x \leq a\}$ is finite for all $a \in P$, for it is the same as the interval $[0,a]$.

THE INCIDENCE ALGEBRA OF A PARTIALLY ORDERED SET

Now let P be a locally finite partially ordered set, and let F be a field. We set

$$\mathcal{A}(P) = \{f: P \times P \to F \mid f(x,y) = 0 \text{ if } x \nleq y\}$$

Addition and scalar multiplication are defined on $\mathcal{A}(P)$ by

$$(f+g)(x,y) = f(x,y) + g(x,y)$$

and

$$(kf)(x,y) = k[f(x,y)]$$

We also define multiplication by

$$(f*g)(x,y) = \sum_{x \leq z \leq y} f(x,z)g(z,y)$$

the sum being finite, since P is assumed to be locally finite. Using these definitions, it is not hard to show that $\mathcal{A}(P)$ is an algebra, called the **incidence algebra** of P. The identity in this algebra is

$$\delta(x,y) = \begin{cases} 1 & \text{if } x = y \\ 0 & \text{if } x \neq y \end{cases}$$

The next theorem characterizes those elements of $\mathcal{A}(P)$ that have multiplicative inverses.

Theorem A.2.1 An element $f \in \mathcal{A}(P)$ is invertible if and only if $f(x,x) \neq 0$ for all $x \in P$.
Proof. An inverse g of f must satisfy

$$\text{(A.2.1)} \qquad \sum_{x \leq z \leq y} f(x,z)g(z,y) = \delta(x,y)$$

In particular, for $x = y$, we get

$$f(x,x)g(x,x) = 1$$

This shows the necessity and also that $g(x,x)$ must satisfy

$$\text{(A.2.2)} \qquad g(x,x) = \frac{1}{f(x,x)}$$

Equation (A.2.2) defines $g(x,y)$ when the interval $[x,y]$ has cardinality 1, that is, when $x = y$. We can use (A.2.1) to define $g(x,y)$ for intervals $[x,y]$ of all cardinalities.

Suppose that $g(x,y)$ has been defined for all intervals with cardinality at most n, and let $[x,y]$ have cardinality $n+1$. Then, by (A.2.1), since $x \neq y$, we get

$$f(x,x)g(x,y) = -\sum_{x < z \leq y} f(x,z)g(z,y)$$

But $g(z,y)$ is defined for $z > x$ since $[z,y]$ has cardinality at most n, and so we can use this to define $g(x,y)$. ∎

Definition The function $\zeta \in \mathcal{A}(P)$, defined by

$$\zeta(x,y) = \begin{cases} 1 & \text{if } x \leq y \\ 0 & \text{if } x \not\leq y \end{cases}$$

is called the **zeta function.** Its inverse $\mu(x,y)$ is called the **Möbius function.** □

The next result follows from the appropriate definitions.

Theorem A.2.2 The Möbius function is uniquely determined by any of the following conditions.

1) $\mu(x,x) = 1$ and, for $x < y$,

$$\sum_{x \leq z \leq y} \mu(z,y) = 0$$

2) $\mu(x,x) = 1$ and, for $x < y$,

$$\sum_{x \le z \le y} \mu(x,z) = 0$$

3) $\mu(x,x) = 1$ and, for $x < y$,

$$\mu(x,y) = -\sum_{x < z \le y} \mu(z,y)$$

4) $\mu(x,x) = 1$ and, for $x < y$,

$$\mu(x,y) = -\sum_{x \le z < y} \mu(x,z)$$

□

Now we come to the main result.

Theorem A.2.3 (Möbius Inversion) Let P be a locally finite partially ordered set with zero element 0. If f and g are functions from P to the field F, then

(A.2.4) $$g(x) = \sum_{y \le x} f(y) \quad \Rightarrow \quad f(x) = \sum_{y \le x} g(y)\mu(y,x)$$

If P is a locally finite partially ordered set with 1, then

(A.2.5) $$g(x) = \sum_{x \le y} f(y) \quad \Rightarrow \quad f(x) = \sum_{x \le y} \mu(x,y)g(y)$$

Proof. Since all sums are finite, we have, for any x,

$$\begin{aligned}
\sum_{y \le x} g(y)\mu(y,x) &= \sum_{y \le x}\left[\sum_{z \le y} f(z)\right]\mu(y,x) \\
&= \sum_{z \le x} \sum_{z \le y \le x} f(z)\mu(y,x) \\
&= \sum_{z \le x} f(z) \sum_{z \le y \le x} \mu(y,x) \\
&= \sum_{z \le x} f(z)\delta(z,x) = f(x)
\end{aligned}$$

The rest of the theorem is proved similarly. ∎

The formulas (A.2.4) and (A.2.5) are called **Möbius inversion formulas.**

Example A.2.2 (Subsets) Let $P = \mathcal{P}(S)$ be the set of all subsets of a finite set S, partially ordered by set inclusion. We will use the

notation $\subseteq$ for subset and $\subset$ for *proper* subset. (In the text, we use $\subset$ for subset.) The zeta function is

$$\zeta(A,B) = \begin{cases} 1 & \text{if } A \subseteq B \\ 0 & \text{otherwise} \end{cases}$$

The Möbius function μ is computed as follows. From Theorem A.2.2, we have

$$\mu(A,A) = 1$$

and

$$\mu(A,B) = -\sum_{A \subseteq X \subset B} \mu(A,X)$$

So, for $x,y,z \notin A$, we have

$$\begin{aligned}
\mu(A,A \cup \{x\}) &= -\mu(A,A) = -1 \\
\mu(A,A \cup \{x,y\}) &= -\mu(A,A) - \mu(A,A \cup \{x\}) - \mu(A,A \cup \{y\}) \\
&= -1 + 1 + 1 = 1 \\
\mu(A,A \cup \{x,y,z\}) &= -\mu(A,A) - \mu(A,A \cup \{x\}) - \mu(A,A \cup \{y\}) \\
&= -\mu(A,A \cup \{x,y\}) - \mu(A,A \cup \{x,z\}) - \mu(A,A \cup \{y,z\}) \\
&= -1 + 1 + 1 + 1 - 1 - 1 - 1 = -1
\end{aligned}$$

It begins to appear that the values of μ alternate between +1 and −1 and that

$$\mu(A,B) = \begin{cases} (-1)^{|B-A|} & \text{if } A \subseteq B \\ 0 & \text{otherwise} \end{cases}$$

To verify this, we have $\mu(A,A) = 1$ and, for $A \subseteq B$,

$$\sum_{A \subseteq X \subseteq B} (-1)^{|B-A|} = \sum_{k=0}^{|B-A|} \binom{|B-A|}{k} (-1)^k = 0$$

Now let $P_1, \ldots, P_n$ be properties that the elements of a set S may or may not possess. For $K \subseteq \{1,\ldots,k\}$, let $E(K)$ be the number of elements of S that have properties P_i for $i \in K$, *and no others.* Let $F(K)$ be the number of elements of S that have *at least* properties P_i, for $i \in K$. Then

$$F(K) = \sum_{K \subseteq L} E(L)$$

Hence, by Möbius inversion,

$$E(K) = \sum_{K \subseteq L} (-1)^{|L-K|} F(L)$$

In particular, if $K = \emptyset$ is the empty set, then

$$E(\emptyset) = \sum_{L \subseteq S} (-1)^{|L|} F(L)$$

But $E(\emptyset)$ is the number of elements of S that have *none* of the properties, and so we get

$$\text{Number elements with no properties} = \sum_{k \geq 0} (-1)^k \sum_{i_1,\ldots,i_k} F(\{i_1,\ldots,i_k\})$$

This formula is the wellknown Principle of Inclusion-Exclusion, which we now see is just a special case of Möbius inversion. □

CLASSICAL MÖBIUS INVERSION

Consider the partially ordered set $\mathbb{N}$ of natural numbers, ordered by division. That is, x is less than or equal to y if and only if x divides y, which we will denote by $x \mid y$. Notice that the natural number 1 (and *not* 0) is the zero element in this partially ordered set, since $1 \mid n$ for any natural number n.

In this case, the Möbius function $\mu(x,y)$ depends only on the ratio y/x, and is given by

$$\mu(x,y) = \mu(\tfrac{y}{x}) = \begin{cases} 1 & \text{if } \frac{y}{x} = 1 \\ (-1)^k & \text{if } \frac{y}{x} = p_1 p_2 \cdots p_k \text{ for } \textit{distinct} \text{ primes } p_i \\ 0 & \text{otherwise} \end{cases}$$

Notice that the "otherwise" case can occur if either $x \nmid y$ (x does not divide y) or if $p^2 \mid (y/x)$ for some prime p. Thus, the value of $\mu(x,y)$ depends on the nature of the prime decomposition of the ratio y/x.

To verify that this is indeed the Möbius function, we first observe that $\mu(x,x) = \mu(1) = 1$. Now let $x \mid y$, $x \neq y$ and

$$\frac{y}{x} = p_1^{e_1} p_2^{e_2} \cdots p_n^{e_n}$$

where the p_i are *distinct* primes. Then

$$\sum_{x \mid z \mid y} \mu(\tfrac{z}{x}) = \sum_{1 \mid \frac{z}{x} \mid \frac{y}{x}} \mu(\tfrac{z}{x}) = \sum_{1 \mid k \mid \frac{y}{x}} \mu(k) = \sum_{1 \leq j \leq n} \binom{n}{j} (-1)^j = 0$$

Now, in the present context, the Möbius inversion formula becomes

$$g(n) = \sum_{k \mid n} f(k) \quad \Rightarrow \quad f(n) = \sum_{k \mid n} g(k)\mu(\tfrac{n}{k})$$

This is the important classical formula, which often goes by the name Möbius inversion formula. ▯

MULTIPLICATIVE VERSION OF MÖBIUS INVERSION

We now present a multiplicative version of the Möbius inversion formula.

Theorem A.2.4 Let P be a locally finite partially ordered set with zero element 0. If f and g are functions from P to F, then

$$g(x) = \prod_{y \le x} f(y) \quad \Rightarrow \quad f(x) = \prod_{y \le x} [g(y)]^{\mu(y,x)}$$

Proof. Since all products are finite, we have, for any x,

$$\begin{aligned}\prod_{y \le x} [g(y)]^{\mu(y,x)} &= \prod_{y \le x}\left[\prod_{z \le y} f(z)\right]^{\mu(y,x)} \\ &= \prod_{z \le x}\ \prod_{z \le y \le x} [f(z)]^{\mu(y,x)} \\ &= \prod_{z \le x} f(z)^{\sum_{z \le y \le x} \mu(y,x)} \\ &= \prod_{z \le x} f(z)^{\delta(z,x)} = f(x)\end{aligned}$$ ▮

Example A.2.3 Let $P = \mathbb{N}$, and let F be the field of rational functions in x. Consider the formula

$$x^n - 1 = \prod_{k \mid n} Q_k(x)$$

Then, if we let $f(k) = Q_k(x)$ and $g(n) = x^n - 1$, Theorem A.2.4 gives

$$Q_n(x) = \prod_{k \mid n} (x^k - 1)^{\mu(n/k)} = \prod_{k \mid n} (x^{n/k} - 1)^{\mu(k)}$$ ▯

A.3 Binomial Inequalities

Inequalities involving binomial coefficients are often very useful in information and coding theory. We discuss some of the more useful ones in this section.

For $0 \leq \lambda \leq 1$ and $\mu = 1 - \lambda$, we let

$$H_q(\lambda) = \lambda \log_q \frac{1}{\lambda} + \mu \log_q \frac{1}{\mu}$$

Note that $H_2(\lambda) = H(\lambda)$ is the entropy function of Chapter 1. Note also that, for any integer $q \geq 2$,

$$q^{H_q(\lambda)} = 2^{H(\lambda)}$$

If $f(n)$ is a function of n, the expression $o(f(n))$ (read "little oh of $f(n)$") is used to represent a quantity for which

$$\lim_{n \to \infty} \frac{o(f(n))}{f(n)} = 0$$

For example, we have $\log n = o(n)$, since

$$\lim_{n \to \infty} \frac{\log n}{n} = 0$$

This can also be written

$$\frac{\log n}{n} = o(1)$$

since $o(1)$ represents a quantity that tends to 0 as $n \to \infty$.

As another example, if $0 \leq \lambda \leq 1$, then

$$\frac{\lfloor \lambda n \rfloor}{n} = \lambda + \frac{\lfloor \lambda n \rfloor - \lambda n}{n} = \lambda + o(1) \tag{A.3.1}$$

Also, the Mean Value Theorem can be used to show that

$$\log \lfloor \lambda n \rfloor = \log(\lambda n) + o(1)$$

We will have use for Stirling's formula, which can take many different forms, two of which are given below.

Theorem A.3.1 (Stirling's formula)

a) $\log n! = n \log n - n + o(n)$

b) $n! = \sqrt{2\pi n}\, n^n e^{-n} \exp\left(\frac{1}{12n} - \frac{1}{360n^3} + \cdots\right)$

Furthermore, formula b) underestimates n! when an even number (including zero) of terms of the series in parentheses are taken and overestimates n! when an odd number of terms are taken. □

INEQUALITIES INVOLVING A SINGLE BINOMIAL COEFFICIENT

Theorem A.3.2 We have

$$\binom{n}{k} \leq \frac{n^n}{k^k(n-k)^{n-k}}$$

Proof. For $0 \leq k \leq n$, we have

$$n^n = [k + (n-k)]^n = \sum_{i=0}^{n} \binom{n}{i} k^i (n-k)^{n-i} \geq \binom{n}{k} k^k (n-k)^{n-k}$$ ∎

The first version of Stirling's formula gives the following.

Theorem A.3.3 We have

$$\log\binom{n}{k} = n \log n - k \log k \ - (n-k) \log(n-k) + o(n)$$

Proof. Stirling's formula (a) gives

$$\begin{aligned} \log\binom{n}{k} &= \log n! - \log k! - \log(n-k)! \\ &= [n \log n - n + o(n)] - [k \log k - k + o(k)] \\ &\quad - [(n-k)\log(n-k) - (n-k) + o(n-k)] \end{aligned}$$

Collecting all terms that are $o(n)$, the result follows. ∎

If $0 \leq \lambda \leq 1$ and $\mu = 1 - \lambda$, Theorem A.3.3 gives

$$\log_q\binom{n}{\lfloor \lambda n \rfloor} = n \log n - \lfloor \lambda n \rfloor \log \lfloor \lambda n \rfloor - \lceil \mu n \rceil \log \lceil \mu n \rceil + o(n)$$

Hence, using (A.3.1), we have

$$\begin{aligned} n^{-1} \log\binom{n}{\lfloor \lambda n \rfloor} &= \log n - \frac{\lfloor \lambda n \rfloor}{n} \log \lfloor \lambda n \rfloor - \frac{\lceil \mu n \rceil}{n} \log \lceil \mu n \rceil + o(1) \\ &= \log n - \lambda \log(\lambda n) - \mu \log(\mu n) + o(1) \\ &= -\lambda \log(\lambda) - \mu \log(\mu) + o(1) \\ &= H_q(\lambda) + o(1) \end{aligned}$$

We have proved the following.

Corollary A.3.4 For $0 \le \lambda \le 1$,

$$n^{-1} \log_q \binom{n}{\lfloor \lambda n \rfloor} = H_q(\lambda) + o(1) \qquad \square$$

Additional estimates for the binomial coefficients can be obtained as follows. Let $0 < \lambda < 1$ and $\mu = 1 - \lambda$. Assuming that λn is an integer, we can use Stirling's formula (b) to obtain a lower bound for

$$\binom{n}{\lambda n} = \frac{n!}{(\lambda n)!(\mu n)!}$$

In particular, we underestimate $n!$ by taking no terms of the series and overestimate $(\lambda n)!$ and $(\mu n)!$ by taking one term of the series, to get

$$\text{(A.3.2)} \qquad \binom{n}{\lambda n} > \frac{1}{\sqrt{2\pi n \lambda \mu}} \frac{1}{\lambda^{\lambda n} \mu^{\mu n}} \exp\left(-\frac{1}{12\lambda n} - \frac{1}{12\mu n}\right)$$

Now, if at least one of λn or μn is at least 3, we have

$$\frac{1}{12\lambda n} + \frac{1}{12\mu n} < \frac{1}{12} + \frac{1}{36} = \frac{1}{9}$$

and since $\exp(-\frac{1}{9}) > \sqrt{\pi}/2$, (A.3.2) gives

$$\text{(A.3.3)} \qquad \binom{n}{\lambda n} > \frac{1}{\sqrt{8n\lambda\mu}} \frac{1}{\lambda^{\lambda n} \mu^{\mu n}}$$

It can also be shown directly that (A.3.3) holds when both λn and μn are less than 3.

We can obtain an upper bound by taking one term in the series in Stirling's formula to overestimate $n!$ and two terms to underestimate $(\lambda n)!$ and $(\mu n)!$. This gives

(A.3.4)

$$\binom{n}{\lambda n} < \frac{1}{\sqrt{2\pi n \lambda \mu}} \frac{1}{\lambda^{\lambda n} \mu^{\mu n}} \exp\left(\frac{1}{12n} - \frac{1}{12\lambda n} - \frac{1}{12\mu n} + \frac{1}{360(\lambda n)^3} + \frac{1}{360(\mu n)^3}\right)$$

Since the expression on the right is symmetric in λ and μ, we may assume that $\mu \le \lambda$. Hence,

$$\frac{1}{360(\lambda n)^3} \le \frac{1}{360(\mu n)^3} \le \frac{1}{360\mu n}$$

and since $\frac{1}{12n} < \frac{1}{12\lambda n}$, we have

$$\frac{1}{12n} - \frac{1}{12\lambda n} - \frac{1}{12\mu n} + \frac{1}{360(\lambda n)^3} + \frac{1}{360(\mu n)^3}$$

$$\leq \frac{1}{12n} - \frac{1}{12\lambda n} - \frac{1}{12\mu n} + \frac{1}{360(\mu n)} + \frac{1}{360(\mu n)}$$

$$= \frac{1}{12n} - \frac{1}{12\lambda n} - \frac{14}{180(\mu n)} < 0$$

Hence (A.3.4) gives

$$\binom{n}{\lambda n} < \frac{1}{\sqrt{2\pi n\lambda\mu}} \frac{1}{\lambda^{\lambda n}\mu^{\mu n}}$$

Putting these bounds together, we have

Theorem A.3.5 For $0 < \lambda < 1$, $\mu = 1 - \lambda$, $\lambda n \in \mathbb{N}$, we have

$$\frac{1}{\sqrt{8n\lambda\mu}} \lambda^{-\lambda n}\mu^{-\mu n} < \binom{n}{\lambda n} < \frac{1}{\sqrt{2\pi n\lambda\mu}} \lambda^{-\lambda n}\mu^{-\mu n} \qquad \square$$

Since

$$\lambda^{-\lambda n}\mu^{-\mu n} = q^{-\lambda n \log_q \lambda - \mu n \log_q \mu} = 2^{nH(\lambda)}$$

Theorem A.3.5 has the following corollary.

Corollary A.3.6 For $0 < \lambda < 1$, $\mu = 1 - \lambda$, $\lambda n \in \mathbb{N}$, we have

$$\frac{1}{\sqrt{8n\lambda\mu}} q^{nH_q(\lambda)} < \binom{n}{\lambda n} < \frac{1}{\sqrt{2\pi n\lambda\mu}} q^{nH_q(\lambda)} \qquad \square$$

INEQUALITIES INVOLVING SUMS OF BINOMIAL COEFFICIENTS

For $0 < p < 1$, let us write

$$B_p(n,m) = \sum_{k=0}^{m} \binom{n}{k} p^k (1-p)^{n-k}$$

Then we have the following bounds on $B_p(n,m)$.

Theorem A.3.7 For $0 \leq m \leq np$,

$$\binom{n}{m} p^m (1-p)^{n-m} \leq B_p(n,m) \leq (1+m)\binom{n}{m} p^m (1-p)^{n-m}$$

Proof. For $0 \leq k \leq m$, let t_k be the k-th term in the sum for $B_p(n,m)$. We claim that, for $m \leq np$, the largest term is t_m. For if $0 \leq k < j \leq m \leq np$, we have

$$t_k \le t_j \iff \frac{1}{k!(n-k)!}\, p^k(1-p)^{-k} \le \frac{1}{j!(n-j)!}\, p^j(1-p)^{-j}$$

$$\iff \frac{j!}{k!} \le \frac{(n-k)!}{(n-j)!}\left(\frac{p}{1-p}\right)^{j-k}$$

$$\iff j(j-1)\cdots(k+1) \le (n-k)(n-k-1)\cdots(n-j+1)\left(\frac{p}{1-p}\right)^{j-k}$$

But since $j \le np$, the left hand side is at most

$$j(j-1)\cdots(k+1) \le (np)^{j-k}$$

and since $k < j \le np$, the right hand side is at least

$$(n-np)^{j-k}\left(\frac{p}{1-p}\right)^{j-k} = (np)^{j-k}$$

This shows that $t_k \le t_j$, and so t_m is the largest term. Hence, this terms provides a lower bound on the sum, and since there are $(1+m)$ terms in the sum, the upper bound follows. ∎

In the proof of Theorem 1.2.8, we showed that, for $0 \le \lambda \le p$,

$$B_p(n,\lambda n) \le p^{\lambda n}(1-p)^{(1-\lambda)n}2^{nH(\lambda)}$$

We also have, using Corollary A.3.6,

$$B_p(n,\lambda n) = \sum_{k=0}^{\lambda n}\binom{n}{k}p^k(1-p)^{n-k} \ge \binom{n}{\lambda n}p^{\lambda n}(1-p)^{(1-\lambda)n}$$

$$> \frac{1}{\sqrt{8n\lambda\mu}}\, p^{\lambda n}(1-p)^{(1-\lambda)n}2^{nH(\lambda)}$$

Putting these bounds together gives

Theorem A.3.8 For $0 \le \lambda \le p < 1$, $\mu = 1-\lambda$, and $\lambda n \in \mathbb{N}$,

$$\frac{1}{\sqrt{8n\lambda\mu}}\, p^{\lambda n}(1-p)^{\mu n}2^{nH(\lambda)} < B_p(n,\lambda n) \le p^{\lambda n}(1-p)^{\mu n}2^{nH(\lambda)} \qquad \square$$

BOUNDS ON THE VOLUME OF A SPHERE

For an alphabet $\mathcal{A}$ of size q, we define the sphere about $\mathbf{x} \in A^n$ of radius r by

$$S_q(\mathbf{x},r) = \{\mathbf{y} \in \mathcal{A}^n \mid d(\mathbf{x},\mathbf{y}) \le r\}$$

where d is the Hamming distance function, defined in Section 4.2. The *volume* $V_q(n,r) = |S_q(\mathbf{x},r)|$ of this sphere is independent of the center $\mathbf{x}$, and since there are

$$\binom{n}{k}(q-1)^k$$

distinct words whose distance from the center is k, we have

$$V_q(n,r) = \sum_{k=0}^{r}\binom{n}{k}(q-1)^k$$

Now, we can write $V_q(n,r)$ in the form $B_p(n,r)$ by setting

$$p = \frac{q-1}{q}, \qquad 1-p = \frac{1}{q}$$

for then we have

$$p^k(1-p)^{n-k} = \frac{(q-1)^k}{q^n}$$

and so

(A.3.5) $$V_q(n,r) = q^n B_p(n,r)$$

Theorem A.3.7 then leads to the following result.

Theorem A.3.9 For $0 \le r \le \frac{q-1}{q}n$,

$$\binom{n}{r}(q-1)^r \le V_q(n,r) \le (1+r)\binom{n}{r}(q-1)^r \qquad \square$$

Taking $r = \lfloor \lambda n \rfloor$, where $0 \le \lambda \le \frac{q-1}{q}$, we get the following corollary.

Corollary A.3.10 For $0 \le \lambda \le \frac{q-1}{q}$,

$$\binom{n}{\lfloor \lambda n \rfloor}(q-1)^{\lfloor \lambda n \rfloor} \le V_q(n,\lfloor \lambda n \rfloor) \le (1+\lfloor \lambda n \rfloor)\binom{n}{\lfloor \lambda n \rfloor}(q-1)^{\lfloor \lambda n \rfloor} \qquad \square$$

Now consider the right-hand side of the inequality in Corollary A.3.10. Taking the logarithm gives

$$\begin{aligned}\log_q(\text{rhs}) &= \log_q\left[(1+\lfloor \lambda n \rfloor)\binom{n}{\lfloor \lambda n \rfloor}(q-1)^{\lfloor \lambda n \rfloor}\right] \\ &= \log_q(1+\lfloor \lambda n \rfloor) + \log_q\binom{n}{\lfloor \lambda n \rfloor} + \lfloor \lambda n \rfloor \log_q(q-1)\end{aligned}$$

Multiplying by n^{-1}, and using (A.3.1), we get

$$n^{-1}\log_q(\text{rhs}) = n^{-1}\log_q\binom{n}{\lfloor \lambda n \rfloor} + \lambda \log_q(q-1) + o(1)$$

Corollary A.3.4 then gives

$$n^{-1}\log_q(\text{rhs}) = H_q(\lambda) + \lambda \log_q(q-1) + o(1)$$

Since we get the same result from the left hand side, this proves the following.

Corollary A.3.11 For $0 \le \lambda \le \frac{q-1}{q}$,

$$\lim_{n\to\infty}\left[\frac{1}{n}\log_q V_q(n,\lfloor \lambda n \rfloor)\right] = H_q(\lambda) + \lambda \log_q(q-1) \qquad \square$$

Finally, using Theorem A.3.8, and (A.3.5), we get the following generalization of Theorem 1.2.8.

Theorem A.3.12 For $0 \le \lambda < 1 - \frac{1}{q}$, $\mu = 1-\lambda$, and $\lambda n \in \mathbb{N}$,

$$\frac{1}{\sqrt{8n\lambda\mu}}(q-1)^{\lambda n}2^{nH(\lambda)} < V_q(n,\lambda n) \le (q-1)^{\lambda n}2^{nH(\lambda)} \qquad \square$$

A.4 More on Finite Fields

In this section, we describe another method for computing the minimal polynomial of an element of a finite field, and we also present an algorithm for factoring polynomials over finite fields, which we will apply to the problem of finding primitive polynomials.

COMPUTING MINIMAL POLYNOMIALS

In Section 7.2, we discussed one method for computing minimal polynomials. Let us discuss another method here.

If β is an element of degree n over F_q, then the powers

$$1,\beta,\beta^2,\ldots,\beta^{n-1}$$

form a basis for F_{q^n} over F_q. Thus, for any element $\alpha \in F_{q^n}$, there exist constants $c_{i,j}$ in F_q for which

$$\begin{aligned} 1 &= c_{0,0} + c_{0,1}\beta + \cdots + c_{0,n-1}\beta^{n-1} \\ \alpha &= c_{1,0} + c_{1,1}\beta + \cdots + c_{1,n-1}\beta^{n-1} \\ \alpha^2 &= c_{2,0} + c_{2,1}\beta + \cdots + c_{2,n-1}\beta^{n-1} \\ &\vdots \\ \alpha^n &= c_{n,0} + c_{n,1}\beta + \cdots + c_{n,n-1}\beta^{n-1} \end{aligned} \tag{A.4.1}$$

Now let $p(x) = p_0 + p_1 x + \cdots + p_n x^n$ be a polynomial over F_q. Then $p(\alpha) = 0$ if and only if

$$\sum_{i=0}^{n} p_i \alpha^i = 0$$

that is, if and only if

$$0 = \sum_{i=0}^{n} p_i \sum_{j=0}^{n-1} c_{i,j}\beta^j = \sum_{j=0}^{n-1}\sum_{i=0}^{n} c_{i,j} p_i \beta^j$$

and this holds if and only if

$$\begin{aligned} 0 &= c_{0,0}p_0 + c_{1,0}p_1 + \cdots + c_{n,0}p_n \\ 0 &= c_{0,1}p_0 + c_{1,1}p_1 + \cdots + c_{n,1}p_n \\ 0 &= c_{0,2}p_0 + c_{1,2}p_1 + \cdots + c_{n,2}p_n \\ &\vdots \\ 0 &= c_{0,n-1}p_0 + c_{1,n-1}p_1 + \cdots + c_{n,n-1}p_n \end{aligned} \tag{A.4.2}$$

Our goal is thus to find a nontrivial solution to the system (A.4.2) with the property that the largest k for which $p_k \neq 0$ is as small as

possible, since this will give the minimal polynomial of α. Of course, this can be done by row reducing the coefficient matrix in the usual way and setting any independent variables appropriately.

More specifically, suppose that the minimal polynomial of α over F_q has degree s. Each power of α, being an element of the vector space F_{q^n} over F_q, can be thought of as a vector of length n over F_q, using (A.4.1). Thus, we think of α^k as the vector

$$(c_{k,0}, c_{k,1}, \ldots, c_{k,n-1})$$

Then each of the vectors $\alpha^s, \ldots, \alpha^{n-1}$ is a linear combination, over F_q, of the linearly independent vectors $1, \alpha, \ldots, \alpha^{s-1}$. In matrix terms, the coefficient matrix $(c_{i,j})$ of (A.4.1) has rank s and its first s rows are linearly independent.

But this matrix is the transpose of the coefficient matrix of (A.4.2), and so the latter matrix also has rank s, and its first s *columns* are linearly independent. This means that, when the coefficient matrix for (A.4.2) is brought to reduced row echelon form, its upper left corner will be the identity matrix of size s, and so we will be able to choose any values for $p_{s+1}, \ldots, p_n$. In particular, the choice

$$p_{s+1} = 1,\ p_{s+2} = \cdots p_n = 0$$

will give the minimal polynomial for α. Let us try an example.

Example A.4.1 Let $\beta \in F_{16}$ be a root of the irreducible polynomial $x^4 + x + 1$ over F_2. If $\alpha = \beta^5$, then the coefficient matrix of (A.4.1) is

$$M = \begin{bmatrix} 1 & 0 & 0 & 0 \\ 0 & 1 & 1 & 0 \\ 1 & 1 & 1 & 0 \\ 1 & 0 & 0 & 0 \\ 0 & 1 & 1 & 0 \end{bmatrix}$$

For example, the third row of this matrix is found by observing that

$$\alpha^2 = \beta^{10} = \beta^{4 \cdot 2 + 2} = (1+\beta)^2\beta^2 = (1+\beta^2)\beta^2 = \beta^2 + (1+\beta)$$

The transpose of M is the coefficient matrix of (A.4.2)

$$M^T = \begin{bmatrix} 1 & 0 & 1 & 1 & 0 \\ 0 & 1 & 1 & 0 & 1 \\ 0 & 1 & 1 & 0 & 1 \\ 0 & 0 & 0 & 0 & 0 \end{bmatrix}$$

Row reducing M^T gives

$$R = \begin{bmatrix} 1 & 0 & 1 & 1 & 0 \\ 0 & 1 & 1 & 0 & 1 \\ 0 & 0 & 0 & 0 & 0 \\ 0 & 0 & 0 & 0 & 0 \end{bmatrix}$$

and so we may set $p_4 = p_3 = 0$ and $p_2 = 1$, which implies that $p_0 = p_1 = 1$. Thus, the minimal polynomial for α is

$$p(x) = 1 + x + x^2$$ □

Example A.4.2 Let $\beta \in F_{64}$ be a root of the irreducible polynomial $x^6 + x + 1$ over F_2. If $\alpha = \beta^9$, then the coefficient matrix of (A.4.1) is found to be

$$M = \begin{bmatrix} 1 & 0 & 0 & 0 & 0 & 0 \\ 0 & 0 & 0 & 1 & 1 & 0 \\ 1 & 1 & 1 & 1 & 0 & 0 \\ 0 & 1 & 1 & 1 & 0 & 0 \\ 0 & 1 & 1 & 0 & 1 & 0 \\ 1 & 0 & 0 & 1 & 1 & 0 \\ 1 & 1 & 1 & 0 & 1 & 0 \end{bmatrix}$$

The transpose of M is

$$M^T = \begin{bmatrix} 1 & 0 & 1 & 0 & 0 & 1 & 1 \\ 0 & 0 & 1 & 1 & 1 & 0 & 1 \\ 0 & 0 & 1 & 1 & 1 & 0 & 1 \\ 0 & 1 & 1 & 1 & 0 & 1 & 0 \\ 0 & 1 & 0 & 0 & 1 & 1 & 1 \\ 0 & 0 & 0 & 0 & 0 & 0 & 0 \end{bmatrix}$$

Row reducing this matrix gives

$$R = \begin{bmatrix} 1 & 0 & 0 & 1 & 1 & 1 & 0 \\ 0 & 1 & 0 & 0 & 1 & 1 & 1 \\ 0 & 0 & 1 & 1 & 1 & 0 & 1 \\ 0 & 0 & 0 & 0 & 0 & 0 & 0 \\ 0 & 0 & 0 & 0 & 0 & 0 & 0 \\ 0 & 0 & 0 & 0 & 0 & 0 & 0 \end{bmatrix}$$

Hence, we may set $p_6 = p_5 = p_4 = 0$ and $p_3 = 1$, which implies that $p_0 = 1$, $p_1 = 0$, and $p_2 = 1$. The minimal polynomial for α is therefore $p(x) = 1 + x^2 + x^3$. □

AN ALGORITHM FOR FACTORING POLYNOMIALS

Let us now describe an algorithm for factoring polynomials, known as **Berlekamp's algorithm.** As we will see, this algorithm is most suited to the case where the base field F_q is small, since it requires the taking of q greatest common divisors.

We begin with a preliminary result, known as the *Chinese remainder theorem.*

Theorem A.4.1 (Chinese Remainder Theorem) Let $f_1(x),\ldots,f_n(x)$ be nonzero polynomials over F_q that are pairwise relatively prime, and let $p_1(x),\ldots,p_n(x)$ be arbitrary polynomials over F_q. Then the simultaneous congruences

$$\begin{aligned} a(x) &\equiv p_1(x) \bmod f_1(x) \\ &\vdots \\ a(x) &\equiv p_n(x) \bmod f_n(x) \end{aligned} \tag{A.4.3}$$

Have a solution $a(x)$, which is unique modulo the product $f(x) = f_1(x)\cdots f_n(x)$.

Proof. As for the uniqueness, suppose that $a(x)$ and $b(x)$ are solutions to (A.4.3). Then

$$a(x) \equiv b(x) \bmod(f_i(x))$$

for all i, and so

$$f_i(x) \mid a(x) - b(x)$$

which implies, since the polynomials $f_i(x)$ are pairwise relatively prime, that

$$f(x) \mid a(x) - b(x)$$

Thus, $a(x) \equiv b(x) \bmod(f(x))$.

Now let

$$\widehat{f}_i(x) = \frac{f(x)}{f_i(x)}$$

Then since $\widehat{f}_i(x)$ and $f_i(x)$ are relatively prime, there exist polynomials $u_i(x)$ and $v_i(x)$ for which

$$u_i(x)\widehat{f}_i(x) + v_i(x)f_i(x) = 1$$

In particular, $u_i(x)$ is the inverse of $\widehat{f}_i(x)$ modulo $f_i(x)$.

Finally, if

$$a(x) = u_1(x)\widehat{f}_1(x)p_1(x) + \cdots + u_n(x)\widehat{f}_n(x)p_n(x)$$

then since $f_i(x) \mid \widehat{f}_j(x)$ for $j \neq i$, we have

$$a(x) \equiv u_i(x)\hat{f}_i(x)p_i(x) \equiv p_i(x) \bmod(f_i(x))$$

which shows that $a(x)$ is a solution to (A.4.3). ∎

Now we are ready to describe Berlekamp's algorithm. Let $f(x)$ be a monic polynomial of degree n over F_q. The first step is to compute

$$g(x) = \gcd(f(x),f'(x))$$

where $f'(x)$ is the derivative of $f(x)$. If $g(x)$ has positive degree, then it is a factor of $f(x)$, and we can apply the algorithm to $g(x)$ and $f(x)/g(x)$ separately. On the other hand, if $g(x)$ is a constant, then $f(x)$ has no multiple roots in any extension, and so it must factor into the product

$$f(x) = f_1(x)\cdots f_k(x)$$

of k *distinct* irreducible polynomials $f_i(x)$.

Thus, we may reduce the factorization problem to one of factoring polynomials with distinct irreducible factors, and so we assume from now on that $f(x)$ has only distinct irreducible factors.

The basis for Berlekamp's algorithm is the following result.

Theorem A.4.2 Let $f(x)$ be a monic polynomial over F_q. If $h(x) \in F_q[x]$ has the property that

(A.4.4) $$h^q(x) \equiv h(x) \bmod(f(x))$$

then

(A.4.5) $$f(x) = \prod_{c \in F_q} \gcd(f(x),h(x)-c)$$

Proof. Certainly, each polynomial $\gcd(f(x),h(x)-c)$ divides $f(x)$, and since the polynomials $h(x)-c$ are pairwise relatively prime, we deduce that the product on the right side of (A.4.5) divides $f(x)$.

On the other hand, since

$$x^q - x = \prod_{c \in F_q} (x-c)$$

we have

$$h^q(x) - h(x) = \prod_{c \in F_q} (h(x)-c)$$

and so

(A.4.6) $$f(x) \,\Big|\, \prod_{c \in F_q} (h(x)-c)$$

which implies that $f(x)$ divides the right-hand side of (A.4.5). Hence, the two sides of (A.4.5) divide each other, and since they are both monic, they must be equal. ▮

Unfortunately, (A.4.5) does not always give a complete factorization of $f(x)$, since the polynomial $\gcd(f(x),h(x)-c)$ may be reducible. Furthermore, if $h(x) \equiv c \bmod(f(x))$ for some $c \in F_q$, then

$$\gcd(f(x),h(x)-c) = f(x)$$

and so (A.4.5) is the trivial factorization of $f(x)$.

On the other hand, if $h(x)$ has the property that (A.4.5) gives a nontrivial factorization of $f(x)$, we refer to $h(x)$ as an **f-reducing polynomial.** Certainly, any nonconstant $h(x)$ satisfying (A.4.4) and $\deg(h(x)) < n$ is f-reducing.

So, let us consider the congruence

$$\text{(A.4.7)} \qquad h^q(x) \equiv h(x) \bmod(f(x)), \quad \deg(h(x)) < n$$

Before attempting to solve this congruence, we determine the number of solutions.

If $h(x)$ is a solution to (A.4.7), then according to (A.4.6), every irreducible factor $f_i(x)$ of $f(x)$ is a factor of $h(x)-c_i$, for some $c_i \in F_q$. Hence, we have

$$\text{(A.4.8)} \qquad h(x) \equiv c_i \bmod(f_i(x)), \quad \deg(h(x)) < n$$

where $(c_1,\ldots,c_k)$ is some k-tuple over F_q.

On the other hand, since the polynomials $f_i(x)$ are pairwise relatively prime, the Chinese remainder theorem tells us that, for any k-tuple $(c_1,\ldots,c_k)$ over F_q, there is a unique solution $h(x)$ of the system (A.4.8). Furthermore, this solution has the property that

$$h^q(x) \equiv c_i^q \equiv c_i \equiv h(x) \bmod(f_i(x))$$

and so it satisfies (A.4.7).

Thus, we see that there is a one-to-one correspondence between solutions to (A.4.7) and (A.4.8), and since there is a unique solution to (A.4.8) for every ordered k-tuple $(c_1,\ldots,c_k)$ over F_q, we conclude that there are precisely q^k solutions of (A.4.7).

Furthermore, if we denote this correspondence by

$$\sigma(h(x)) = (c_1,\ldots,c_k)$$

then σ is a vector space isomorphism from the vector space of all solutions to (A.4.7) to the vector space $V(k,q)$.

To solve (A.4.7), we begin by constructing the $n \times n$ matrix $A = (a_{i,j})$ defined by the congruences

(A.4.9) $$x^{iq} \equiv \sum_{j=0}^{n-1} a_{i,j}x^j \bmod(f(x)), \quad 0 \le i \le n-1$$

If $h(x) = h_0 + h_1 x + \cdots + h_{n-1}x^{n-1}$ is a solution to (A.4.7), then since $h_i^q = h_i$, we have

$$h(x) \equiv h^q(x) \equiv \left(\sum_{i=0}^{n-1} h_i x^i\right)^q = \sum_{i=0}^{n-1} h_i x^{iq}$$

$$= \sum_{i=0}^{n-1} h_i \sum_{j=0}^{n-1} a_{i,j}x^j = \sum_{j=0}^{n-1}\left(\sum_{i=0}^{n-1} a_{i,j}h_i\right)x^j$$

Comparing coefficients of the last expression with those of $h(x)$ gives

$$h_j = \sum_{i=0}^{n-1} a_{i,j}h_i$$

for all $j = 0,\ldots,n-1$. In matrix terms, this is equivalent to

$$(h_0,\ldots,h_{n-1})A = (h_0,\ldots,h_{n-1})$$

or

(A.4.10) $$(h_0,\ldots,h_{n-1})(A - \mathbf{I}_n) = \mathbf{0}_n$$

We have seen that the solution space of this matrix equation has dimension k over F_q, and so the matrix $A - \mathbf{I}_n$ has rank $r = n - k$.

Thus, once the matrix A is constructed, we determine its rank r. This tells us that the polynomial $f(x)$ factors into $k = n - r$ distinct irreducible factors. Hence, if $k = 1$, we may conclude that $f(x)$ is irreducible and stop the algorithm. If $k > 1$, then we find a basis

$$h_1(x),\ldots,h_k(x)$$

for the solution space of (A.4.10), by column reducing the matrix $A - \mathbf{I}_n$. The first basis polynomial may be taken to be $h_1(x) = 1$, which is always a solution to (A.4.10). The polynomial $h_2(x)$ is a nontrivial solution $h(x)$ of degree at most $n-1$, which is therefore an f-reducing polynomial. Then we may apply Theorem A.4.2.

If this does not give all k factors of $f(x)$, we can then apply Theorem A.4.2 to the factors that we do get, this time using the f-reducing polynomial $h_3(x)$.

To see that this will eventually produce a complete factorization of $f(x)$, we show that for any pair of distinct irreducible factors $f_u(x)$ and $f_v(x)$, there is a j for which

$$f_i(x) \mid h_j(x) - c \quad \text{but} \quad f_j(x) \nmid h_j(x) - c$$

for some $c \in F_q$. This will show that, by the time we have used $h_j(x)$ in Theorem A.4.2, we will have separated $f_u(x)$ and $f_v(x)$ into distinct factors of $f(x)$.

Recall that the map

$$\sigma(h(x)) = (c_1,\ldots,c_k)$$

defined earlier, is a vector space isomorphism. This implies that the vectors

$$\sigma(h_j(x)) = \mathbf{c}_j = (c_{j,1},\ldots,c_{j,k})$$

form a basis for $V(k,q)$. This implies that, for any $1 \le u,v \le k$, we cannot have $c_{j,u} = c_{j,v}$ for all j, for then the $\mathbf{c}_j$ would not span F_q^k.

Let us assume that $c_{j,u} \neq c_{j,v}$. Then since

$$h_j(x) \equiv c_{j,u} \bmod(f_u(x))$$

and

$$h_j(x) \equiv c_{j,v} \bmod(f_v(x))$$

we cannot have

$$h_j(x) \equiv c_{j,u} \bmod(f_v(x))$$

for that would imply that

$$c_{j,u} \equiv c_{j,v} \bmod(f_v(x))$$

which is equivalent to $c_{j,u} = c_{j,v}$. Therefore,

$$f_u(x) \mid h_j(x) - c_{j,u} \quad \text{but} \quad f_v(x) \nmid h_j(x) - c_{j,u}$$

Thus, we conclude that sufficient applications of Theorem A.4.2, using the basis polynomials $h_2(x),\ldots,h_k(x)$, will eventually produce a complete factorization of $f(x)$.

FINDING PRIMITIVE POLYNOMIALS

We can use Berlekamp's algorithm to determine all primitive polynomials of a certain order, by factoring an appropriate cyclotomic polynomial. Recall that an irreducible polynomial $p(x)$ is a primitive polynomial for F_{q^n} over F_q if the order of $p(x)$ is $q^n - 1$, and so any root of $p(x)$ is a primitive (q^n-1)-st root of unity. Furthermore, since $p(x)$ is irreducible, its degree is $[F_{q^n}:F_q] = n$.

Since an irreducible polynomial cannot be primitive for two different extensions of F_q, the phrases *primitive polynomial of order* $q^n - 1$ *over* F_q and *primitive polynomial of degree* n *over* F_q have the same meaning.

Recall from Section 7.3 that the e-th cyclotomic polynomial $Q_e(x)$ over F_q, where $e = q^n - 1$, is the monic polynomial over F_q whose roots are the e-th roots of unity over F_q, and so its irreducible

factors are the primitive polynomials of order e, or equivalently, of degree n, over F_q. Finally, recall that $Q_e(x)$ is given by

$$\text{(A.4.11)} \qquad Q_e(x) = \prod_{d \mid n} (x^{n/d} - 1)^{\mu(d)}$$

where $\mu(m)$ is the Möbius function, defined in Section A.2 by

$$\mu(m) = \begin{cases} 1 & \text{if } m = 1 \\ (-1)^k & \text{if } m = p_1 p_2 \cdots p_k \text{ for \textit{distinct} primes } p_i \\ 0 & \text{otherwise} \end{cases}$$

Example A.4.3 Let us find all primitive polynomials of degree $n = 4$ over F_2. In this case, $e = 2^4 - 1 = 15$, and so any such polynomial has order 15, and is a primitive polynomial for F_{16} over F_2.

According to (A.4.11), the cyclotomic polynomial $Q_{15}(x)$ is

$$\begin{aligned} Q_{15}(x) &= (x^{15} - 1)(x^5 - 1)^{-1}(x^3 - 1)^{-1}(x - 1) \\ &= x^8 + x^7 + x^5 + x^4 + x^3 + x + 1 \end{aligned}$$

The system of congruences (A.4.9) is

$$\begin{array}{llllllllll} x^0 & \equiv & 1 & & & & & & & \\ x^2 & \equiv & & & x^2 & & & & & \\ x^4 & \equiv & & & & & x^4 & & & \\ x^6 & \equiv & & & & & & & x^6 & \\ x^8 & \equiv & 1+ & x+ & & x^3+ & x^4+ & x^5+ & & x^7 \\ x^{10} & \equiv & 1+ & & & & & x^5 & & \\ x^{12} & \equiv & & & x^2+ & & & & & x^7 \\ x^{14} & \equiv & 1+ & & x^2+ & x^3+ & x^4+ & & x^6+ & x^7 \end{array}$$

Hence, the 8×8 matrix A is

$$A = \begin{bmatrix} 1 & 0 & 0 & 0 & 0 & 0 & 0 & 0 \\ 0 & 0 & 1 & 0 & 0 & 0 & 0 & 0 \\ 0 & 0 & 0 & 0 & 1 & 0 & 0 & 0 \\ 0 & 0 & 0 & 0 & 0 & 0 & 1 & 0 \\ 1 & 1 & 0 & 1 & 1 & 1 & 0 & 1 \\ 1 & 0 & 0 & 0 & 0 & 1 & 0 & 0 \\ 0 & 0 & 1 & 0 & 0 & 0 & 0 & 1 \\ 1 & 0 & 1 & 1 & 1 & 0 & 1 & 1 \end{bmatrix}$$

and so

$$A - I_8 = \begin{bmatrix} 0&0&0&0&0&0&0&0 \\ 0&1&1&0&0&0&0&0 \\ 0&0&1&0&1&0&0&0 \\ 0&0&0&1&0&0&1&0 \\ 1&1&0&1&0&1&0&1 \\ 1&0&0&0&0&0&0&0 \\ 0&0&1&0&0&0&1&1 \\ 1&0&1&1&1&0&1&0 \end{bmatrix}$$

Column reducing this matrix, we obtain

$$R = \begin{bmatrix} 0&0&0&0&0&0&0&0 \\ 1&0&0&0&0&0&0&0 \\ 0&1&0&0&0&0&0&0 \\ 0&0&1&0&0&0&0&0 \\ 0&0&0&1&0&0&0&0 \\ 0&0&0&0&1&0&0&0 \\ 0&0&0&0&0&1&0&0 \\ 0&1&1&0&1&0&0&0 \end{bmatrix}$$

Since this matrix has rank 6, the polynomial $Q_{15}(x)$ factors into $k = 8-6 = 2$ irreducible factors over F_2.

From the matrix R, we also see that $x_1 = \alpha$ and $x_8 = \beta$ may be chosen arbitrarily, and so the solution space to (A.4.7) is

$$\{(\alpha,0,\beta,\beta,0,\beta,0,\beta) \mid \alpha,\beta \in F_2\}$$

Hence, a basis for this space consists of the two polynomials

$$h_1(x) = (1,0,0,0,0,0,0,0) = 1$$

and

$$h_2(x) = (0,0,1,1,0,1,0,1) = x^2 + x^3 + x^5 + x^7$$

where $h_2(x)$ is Q_{15}-reducing.

Next, we employ the Euclidean algorithm to deduce that

$$\gcd(Q_{15}(x),h_2(x)) = 1 + x^3 + x^4$$

and

$$\gcd(Q_{15}(x),h_2(x) - 1) = 1 + x + x^4$$

and so the complete factorization of $Q_{15}(x)$ is

$$Q_{15}(x) = (1 + x^3 + x^4)(1 + x + x^4)$$

This tells us that there are exactly two primitive polynomials of degree 4 over F_2, namely $1 + x^3 + x^4$ and $1 + x + x^4$. □

Table of Monic Irreducible Polynomials

$a_n a_{n-1} \ldots a_0;e$ is the polynomial $a_n x^n + a_{n-1} x^{n-1} + \ldots + a_0$ of order e

Polynomials over F_2			
Degree 1	100101101;255	1100010101;511	10101000011;1023
10;1	100111001; 17	1100011111;511	10101010111;1023
11;1	100111111; 85	1100100011;511	10101100001; 93
Degree 2	101001101;255	1100110001;511	10101100111; 341
111;3	101011111;255	1100111011;511	10101101011;1023
Degree 3	101100011;255	1101001001; 73	10110000101;1023
1011;7	101100101;255	1101001111;511	10110001111;1023
1101;7	101101001;255	1101011011;511	10110010111;1023
Degree 4	101110001;255	1101100001;511	10110011011; 341
10011;15	101110111; 85	1101101011;511	10110100001;1023
11001;15	101111011; 85	1101101101;511	10110101011; 341
11111; 5	110000111;255	1101110011;511	10110111001; 341
Degree 5	110001011; 85	1101111111;511	10111000001; 341
100101;31	110001101;255	1110000101;511	10111000111;1023
101001;31	110011111; 51	1110001111;511	10111100101;1023
101111;31	110100011; 85	1110100001; 73	10111110111;1023
110111;31	110101001;255	1110110101;511	10111111011;1023
111011;31	110110001; 51	1110111001;511	11000010011;1023
111101;31	110111101; 85	1111000111;511	11000010101;1023
Degree 6	111000011;255	1111001011;511	11000100011; 33
1000011;63	111001111;255	1111001101;511	11000100101;1023
1001001; 9	111010111; 17	1111010101;511	11000110001; 341
1010111;21	111011101; 85	1111011001;511	11000110111;1023
1011011;63	111100111;255	1111100011;511	11001000011;1023
1100001;63	111110011; 51	1111101001;511	11001001111;1023
1100111;63	111110101;255	1111111011;511	11001010001; 341
1101101;63	111111001; 85	Degree 10	11001011011;1023
1110011;63	Degree 9	10000001001;1023	11001111001;1023
1110101;21	1000000011; 73	10000001111; 341	11001111111;1023
Degree 7	1000010001;511	10000011011;1023	11010000101; 93
10000011;127	1000010111; 73	10000011101; 341	11010001001;1023
10001001;127	1000011011;511	10000100111;1023	11010100111; 93
10001111;127	1000100001;511	10000101101;1023	11010101101; 341
10010001;127	1000101101;511	10000110101; 93	11010110101;1023
10011101;127	1000110011;511	10001000111; 341	11010111111; 341
10100111;127	1001001011; 73	10001010011; 341	11011000001;1023
10101011;127	1001011001;511	10001100011; 341	11011001101; 341
10111001;127	1001011111;511	10001100101;1023	11011010011;1023
10111111;127	1001100101; 73	10001101111;1023	11011011111;1023
11000001;127	1001101001;511	10010000001;1023	11011110111; 341
11001011;127	1001101111;511	10010001011;1023	11011111101;1023
11010011;127	1001110111;511	10010011001; 341	11100001111; 341
11010101;127	1001111101;511	10010101001; 33	11100010001; 341
11100101;127	1010000111;511	10010101111; 341	11100010111;1023
11101111;127	1010010101;511	10011000101;1023	11100011101;1023
11110001;127	1010011001; 73	10011001001; 341	11100100001;1023
11110111;127	1010100011;511	10011010111;1023	11100101011; 93
11111101;127	1010100101;511	10011100111;1023	11100110101; 341
Degree 8	1010101111;511	10011101101; 341	11100111001;1023
100011011; 51	1010110111;511	10011110011;1023	11101000111;1023
100011101;255	1010111101;511	10011111111;1023	11101001101;1023
100101011;255	1011001111;511	10100001011; 93	11101010101;1023
	1011010001;511	10100001101;1023	11101011001;1023
	1011011011;511	10100011001;1023	11101100011;1023
	1011110101;511	10100011111; 341	11101111011; 341
	1011111001;511	10100100011;1023	11101111101;1023
	1100000001; 73	10100110001;1023	11110000001; 341
	1100010011;511	10100111101;1023	11110000111; 341
			11110001101;1023

11110010011;1023
11110101001; 341
11110110001;1023
11111000101; 341
11111011011;1023
11111101011; 341
11111110011;1023
11111111001;1023
11111111111; 11

Degree 11
100000000101;2047
100000010111;2047
100000101011;2047
100000101101;2047
100001000111;2047
100001100011;2047
100001100101;2047
100001110001;2047
100001111011;2047
100010001101;2047
100010010101;2047
100010011111;2047
100010101001;2047
100010110001;2047
100011000011; 89
100011001111;2047
100011010001;2047
100011100001;2047
100011100111;2047
100011101011;2047
100011110101;2047
100100001101;2047
100100010011;2047
100100100101;2047
100100101001;2047
100100110111; 89
100100111011;2047
100100111101;2047
100101000101;2047
100101001001;2047
100101010001;2047
100101011011;2047
100101110011;2047
100101110101;2047
100101111111;2047
100110000011;2047
100110001111;2047
100110101011;2047
100110101101;2047
100110111001;2047
100111000111;2047
100111011001;2047
100111100101;2047
100111101111; 89
100111110111;2047
101000000001;2047
101000000111;2047
101000010011;2047
101000010101;2047
101000101001;2047
101001001001;2047
101001100001;2047
101001101101;2047
101001111001;2047
101001111111;2047
101010000101;2047
101010010001;2047
101010011101;2047
101010100111;2047
101010101011;2047
101010110011;2047
101010110101;2047
101011010101;2047
101011011111;2047
101011100011; 23
101011101001;2047
101011101111;2047
101011110001;2047
101011111011;2047
101100000011;2047
101100001001;2047
101100010001;2047
101100110011;2047
101100111111;2047
101101000001;2047
101101001011;2047
101101011001;2047
101101011111;2047
101101100101;2047
101101101111;2047
101101111101;2047
101110000111;2047
101110001011;2047
101110010011;2047
101110010101;2047
101110101111;2047
101110110111;2047
101110111101;2047
101111001001;2047
101111011011;2047
101111011101;2047
101111100111;2047
101111101101;2047
110000001011;2047
110000001101;2047
110000011001;2047
110000011111;2047
110000110001; 89
110001010111;2047
110001100001;2047
110001101011;2047
110001110011;2047
110001110101; 23
110010000101;2047
110010001001;2047
110010010111;2047
110010011011;2047
110010011101;2047
110010110011;2047
110010111111;2047
110011000111;2047
110011001101;2047
110011010011;2047
110011010101;2047
110011100011;2047
110011101001;2047
110011110111;2047
110100000011;2047
110100001111;2047
110100011101;2047
110100100111;2047
110100101101;2047
110101000001;2047
110101000111;2047
110101010101;2047
110101011001;2047
110101100011;2047
110101101111;2047
110101110001;2047
110110010011;2047
110110011111;2047
110110101001;2047
110110111011;2047
110110111101;2047
110111001001;2047
110111010111;2047
110111011011;2047
110111100001;2047
110111100111;2047
110111110101;2047
110111111111; 89
111000000101;2047
111000011101;2047
111000100001;2047
111000100111;2047
111000101011;2047
111000110011;2047
111000111001;2047
111001000111;2047
111001001011;2047
111001010101;2047
111001011111;2047
111001110001;2047
111001111011;2047
111001111101;2047
111010000001;2047
111010010011;2047
111010011111;2047
111010100011;2047
111010111011;2047
111011001001; 89
111011001111;2047
111011011101;2047
111011110011;2047
111011111001;2047
111100001011;2047
111100011001;2047
111100110001;2047
111100110111;2047
111101011101;2047
111101101011;2047
111101101101;2047
111101110101;2047
111101111001; 89
111110000011;2047
111110010001;2047
111110010111;2047
111110011011;2047
111110100111;2047
111110101101;2047
111110110101;2047
111111001101;2047
111111010011;2047
111111100101;2047
111111101001;2047
111111111011; 89

Polynomials over F_3

Degree 1
10;1
11;1
12;1

Degree 2
101;4
112;8
122;8

Degree 3
1021;26
1022;13
1102;13
1112;13
1121;26
1201;26
1211;26
1222;13

Degree 4
10012;80
10022;80
10102;16
10111;40
10121;40
10202;16
11002;80
11021;20
11101;40
11111; 5
11122;80
11222;80
12002;80
12011;20
12101;40
12112;80
12121;10
12212;80

Degree 5
100021;242
100022;121
100112;121
100211;242
101011;242
101012;121
101102;121
101122;121
101201;242
101221;242
102101;242
102112;121
102122;11
102202;121
102211;242
102221; 22
110002;121
110012;121
110021;242
110101;242
110111;242
110122;121
111011;242
111121;242
111211;242
111212;121
112001;242
112022;121
112102; 11
112111;242
112201;242
112202;121
120001;242
120011;242
120022;121
120202;121
120212;121
120221;242
121012;121
121111;242
121112;121
121222;121
122002;121
122021;242
122101;242
122102;121
122201; 22
122212;121

Degree 6
1000012;728
1000022;728
1000111;364
1000121;364
1000201; 52
1001012;728
1001021;364

1001101; 91	1122202;728	10021001;2186	10220101;2186	11120111;2186
1001122;104	1122221;364	10021112;1093	10220222;1093	11120122;1093
1001221;182	1200002;728	10021202;1093	10221122;1093	11120212;1093
1002011;364	1200022; 56	10022002;1093	10221202;1093	11120221;2186
1002022;728	1200121;364	10022021;2186	10221212;1093	11121001;2186
1002101;182	1201001;364	10022101;2186	10221221;2186	11121101;2186
1002112;104	1201111;182	10022212;1093	10222012;1093	11121202;1093
1002211; 91	1201121;182	10100011;2186	10222021;2186	11122021;2186
1010201; 52	1201201;364	10100012;1093	10222111;2186	11122112;1093
1010212;728	1201202;728	10100102;1093	10222202;1093	11122201;2186
1010222;728	1202002;728	10100122;1093	10222211;2186	11122222;1093
1011001; 91	1202021; 28	10100201;2186	11000101;2186	11200201;2186
1011011;364	1202101;364	10100221;2186	11000222;1093	11200202;1093
1011022;728	1202122;728	10101101;2186	11001022;1093	11201012;1093
1011122;728	1202222;728	10101112;1093	11001112;1093	11201021;2186
1012001;182	1210001;364	10101202;1093	11001211;2186	11201101;2186
1012012;728	1210021;364	10101211;2186	11002012;1093	11201111;2186
1012021;364	1210112;728	10102102;1093	11002022;1093	11201221;2186
1012112;728	1210202;728	10102201;2186	11002121;2186	11201222;1093
1020001; 52	1210211; 91	10110022;1093	11002202;1093	11202002;1093
1020101; 52	1211021;182	10110101;2186	11010001;2186	11202121;2186
1020112;728	1211201; 91	10110211;2186	11010022;1093	11202211;2186
1020122;728	1211212;728	10111001;2186	11010121;2186	11202212;1093
1021021;364	1212011; 91	10111102;1093	11010221;2186	11210002;1093
1021102; 56	1212022;728	10111121;2186	11011111;2186	11210011;2186
1021112;728	1212121; 14	10111201;2186	11011202;1093	11210021;2186
1021121; 91	1212122;728	10112002;1093	11012002;1093	11210101;2186
1022011;364	1212212;728	10112012;1093	11012102;1093	11211001;2186
1022102; 56	1220102;728	10112021;2186	11012212;1093	11211022;1093
1022111;182	1220111;182	10112111;2186	11020021;2186	11211122;1093
1022122;728	1220212;728	10112122;1093	11020022;1093	11211212;1093
1100002;728	1221001;182	10120021;2186	11020102;1093	11211221;2186
1100012; 56	1221002;104	10120112;1093	11020112;1093	11212012;1093
1100111;364	1221112;104	10120202;1093	11020201;2186	11212112;1093
1101002;728	1221202;728	10121002;1093	11020222;1093	11212202;1093
1101011; 28	1221211;364	10121102;1093	11021111;2186	11220001;2186
1101101;364	1222022;728	10121201;2186	11021122;1093	11220112;1093
1101112;728	1222102;728	10121222;1093	11021201;2186	11220211;2186
1101212;728	1222112;104	10122001;2186	11021212;1093	11221022;1093
1102001;364	1222211;364	10122011;2186	11022101;2186	11221102;1093
1102111; 91	1222222;728	10122022;1093	11022122;1093	11221112;1093
1102121; 91		10122212;1093	11022211;2186	11221121;2186
1102201;364	Degree 7	10122221;2186	11022221;2186	11222011;2186
1102202;728	10000102;1093	10200001;2186	11100002;1093	11222102;1093
1110001;364	10000121;2186	10200002;1093	11100022;1093	11222122;1093
1110011;364	10000201;2186	10200101;2186	11100121;2186	11222201;2186
1110122;728	10000222;1093	10200112;1093	11100212;1093	11222221;2186
1110202;728	10001011;2186	10200202;1093	11101012;1093	12000121;2186
1110221;182	10001012;1093	10200211;2186	11101022;1093	12000202;1093
1111012;728	10001102;1093	10201021;2186	11101102;1093	12001021;2186
1111021;182	10001111;2186	10201022;1093	11101111;2186	12001112;1093
1111111; 7	10001201;2186	10201121;2186	11101121;2186	12001211;2186
1111112;728	10001212;1093	10201222;1093	11102002;1093	12002011;2186
1111222;728	10002112;1093	10202011;2186	11102111;2186	12002021;2186
1112011; 91	10002122;1093	10202012;1093	11102222;1093	12002101;2186
1112201;182	10002211;2186	10210001;2186	11110001;2186	12002222;1093
1112222;728	10002221;2186	10210121;2186	11110012;1093	12010021;2186
1120102;728	10010122;1093	10210202;1093	11110111;2186	12010022;1093
1120121; 91	10010222;1093	10211101;2186	11110112;1093	12010102;1093
1120222;728	10011002;1093	10211111;2186	11110211;2186	12010121;2186
1121012;728	10011101;2186	10211122;1093	11110222;1093	12010201;2186
1121102;728	10011211;2186	10211221;2186	11111011;2186	12010211;2186
1121122;104	10012001;2186	10212011;2186	11111021;2186	12011102;1093
1121212;728	10012022;1093	10212022;1093	11111201;2186	12011111;2186
1121221;364	10012111;2186	10212101;2186	11111222;1093	12011212;1093
1122001; 91	10012202;1093	10212112;1093	11112011;2186	12011221;2186
1122002;104	10020121;2186	10212212;1093	11112221;2186	12012112;1093
1122122;104	10020221;2186	10220002;1093	11120102;1093	12012122;1093

12012202;1093	12101212;1093	12120011;2186	12201212;1093	12220012;1093
12012221;2186	12101222;1093	12120112;1093	12202001;2186	12220022;1093
12020002;1093	12102001;2186	12120121;2186	12202111;2186	12220202;1093
12020021;2186	12102121;2186	12120211;2186	12202112;1093	12221002;1093
12020122;1093	12102212;1093	12120212;1093	12202222;1093	12221021;2186
12020222;1093	12110111;2186	12121012;1093	12210002;1093	12221111;2186
12021101;2186	12110122;1093	12121022;1093	12210112;1093	12221122;1093
12021212;1093	12110201;2186	12121102;1093	12210211;2186	12221221;2186
12022001;2186	12110212;1093	12121121;2186	12211021;2186	12222011;2186
12022111;2186	12110221;2186	12122012;1093	12211201;2186	12222101;2186
12022201;2186	12111002;1093	12122122;1093	12211211;2186	12222211;2186
12100001;2186	12111101;2186	12200101;2186	12211222;1093	
12100021;2186	12111202;1093	12200102;1093	12212012;1093	
12100111;2186	12112022;1093	12201011;2186	12212102;1093	
12100222;1093	12112102;1093	12201022;1093	12212122;1093	
12101011;2186	12112121;2186	12201121;2186	12212201;2186	
12101021;2186	12112211;2186	12201122;1093	12212221;2186	
12101201;2186	12120002;1093	12201202;1093	12220001;2186	

Table of Primitive Polynomials over F_2

If $deg(f(x)) = d$ *then* $o(f(x)) = 2^d - 1$ *and* $f(x)$ *is primitive for* F_{2^d} *over* F_2.

$x+1$	$x^{11}+x^2+1$	$x^{21}+x^2+1$
x^2+x+1	$x^{12}+x^6+x^4+x+1$	$x^{22}+x^1+1$
x^3+x+1	$x^{13}+x^4+x^3+x+1$	$x^{23}+x^5+1$
x^4+x+1	$x^{14}+x^5+x^3+x+1$	$x^{24}+x^4+x^3+x+1$
x^5+x^2+1	$x^{15}+x+1$	$x^{25}+x^3+1$
x^6+x+1	$x^{16}+x^5+x^3+x^2+1$	$x^{26}+x^6+x^2+x+1$
x^7+x+1	$x^{17}+x^3+1$	$x^{27}+x^5+x^2+x+1$
$x^8+x^4+x^3+x^2+1$	$x^{18}+x^5+x^2+x+1$	$x^{28}+x^3+x^3+1$
x^9+x^4+1	$x^{19}+x^5+x^2+x+1$	$x^{29}+x^2+x^2+1$
$x^{10}+x^3+1$	$x^{20}+x^3+1$	$x^{30}+x^6+x^4+x+1$

Table of Primitive Polynomials over F_3

If $deg(f(x)) = d$ *then* $o(f(x)) = 3^d - 1$ *and* $f(x)$ *is primitive for* F_{3^d} *over* F_3.

$x+1$	x^6+x^5+2	$x^{11}+x^{10}+x^4+1$
x^2+x+2	$x^7+x^6+x^4+1$	$x^{12}+x^{11}+x^7+2$
x^3+2x^2+1	x^8+x^5+2	$x^{13}+x^{12}+x^6+1$
x^4+x^3+2	$x^9+x^7+x^5+1$	$x^{14}+x^{13}+2$
$x^5+x^4+x^2+1$	$x^{10}+x^9+x^7+2$	$x^{15}+x^{14}+x^4+1$

Additional Primitive Polynomials

If $deg(f(x)) = d$ *then* $o(f(x)) = q^d - 1$ *and* $f(x)$ *is primitive for* F_{q^d} *over* F_q.

x^2+x+2 (q=5)	x^2+x+3 (q=7)	x^3+x^2+5 (q=11)
x^3+x^2+2 (q=5)	x^3+x^2+x+2 (q=7)	x^4+x+2 (q =11)
x^4+x^3+x+3 (q=5)	$x^4+x^3+x^2+3$ (q=7)	x^2+x+2 (q=13)
x^5+x^3+2 (q=5)	x^5+x^4+4 (q=7)	$x^3+10x+7$ (q=13)
x^6+x^5+2 (q=5)	x^2+x+7 (q=11)	$x^4+101x+2$ (q=13)

Field Tables

GF(2^2)

0	01
1	10
2	11

GF(2^3)

0	001
1	010
2	100
3	011
4	110
5	111
6	101

GF(2^4)

0	0001
1	0010
2	0100
3	1000
4	0011
5	0110
6	1100
7	1011
8	0101
9	1010
10	0111
11	1110
12	1111
13	1101
14	1001

GF(2^5)

0	00001
1	00010
2	00100
3	01000
4	10000
5	00101
6	01010
7	10100
8	01101
9	11010
10	10001
11	00111
12	01110
13	11100
14	11101
15	11111
16	11011
17	10011
18	00011
19	00110
20	01100
21	11000
22	10101
23	01111
24	11110
25	11001
26	10111
27	01011
28	10110
29	01001
30	10010

GF(2^6)

0	000001
1	000010
2	000100
3	001000
4	010000
5	100000
6	000011
7	000110
8	001100
9	011000
10	110000
11	100011
12	000101
13	001010
14	010100
15	101000
16	010011
17	100110
18	001111
19	011110
20	111100
21	111011
22	110101
23	101001
24	010001
25	100010
26	000111
27	001110
28	011100
29	111000
30	110011
31	100101
32	001001
33	010010
34	100100
35	001011
36	010110
37	101100
38	011011
39	110110
40	101111
41	011101
42	111010
43	110111
44	101101
45	011001
46	110010
47	100111
48	001101
49	011010
50	110100
51	101011
52	010101
53	101010
54	010111
55	101110
56	011111
57	111110
58	111111
59	111101
60	111001
61	110001
62	100001

GF(2^7)

0	0000001
1	0000010
2	0000100
3	0001000
4	0010000
5	0100000
6	1000000
7	0000011
8	0000110
9	0001100
10	0011000
11	0110000
12	1100000
13	1000011
14	0000101
15	0001010
16	0010100
17	0101000
18	1010000
19	0100011
20	1000110
21	0001111
22	0011110
23	0111100
24	1111000
25	1110011
26	1100101
27	1001001
28	0010001
29	0100010
30	1000100
31	0001011
32	0010110
33	0101100
34	1011000
35	0110011
36	1100110
37	1001111
38	0011101
39	0111010
40	1110100
41	1101011
42	1010101
43	0101001
44	1010010
45	0100111
46	1001110
47	0011111
48	0111110
49	1111100
50	1111011
51	1110101
52	1101001
53	1010001
54	0100001
55	1000010
56	0000111
57	0001110
58	0011100
59	0111000
60	1110000
61	1100011
62	1000101
63	0001001
64	0010010
65	0100100
66	1001000
67	0010011
68	0100110
69	1001100
70	0011011
71	0110110
72	1101100
73	1011011
74	0110101
75	1101010
76	1010111
77	0101101
78	1011010
79	0110111
80	1101110
81	1011111
82	0111101
83	1111010
84	1110111
85	1101101
86	1011001
87	0110001
88	1100010
89	1000111
90	0001101
91	0011010
92	0110100
93	1101000
94	1010011
95	0100101
96	1001010
97	0010111
98	0101110
99	1011100
100	0111011
101	1110110
102	1101111
103	1011101
104	0111001
105	1110010
106	1100111
107	1001101
108	0011001
109	0110010
110	1100100
111	1001011
112	0010101
113	0101010
114	1010100
115	0101011
116	1010110
117	0101111
118	1011110
119	0111111
120	1111110
121	1111111
122	1111101
123	1111001
124	1110001
125	1100001
126	1000001

GF(2^8)

0	00000001
1	00000010
2	00000100
3	00001000
4	00010000
5	00100000
6	01000000
7	10000000
8	00011101
9	00111010
10	01110100
11	11101000
12	11001101
13	10000111
14	00010011
15	00100110
16	01001100
17	10011000
18	00101101
19	01011010
20	10110100
21	01110101
22	11101010
23	11001001
24	10001111
25	00000011
26	00000110
27	00001100
28	00011000
29	00110000
30	01100000
31	11000000
32	10011101
33	00100111
34	01001110
35	10011100
36	00100101
37	01001010
38	10010100
39	00110101
40	01101010

41 11010100	106 00110100	171 10110011	236 11001011
42 10110101	107 01101000	172 01111011	237 10001011
43 01110111	108 11010000	173 11110110	238 00001011
44 11101110	109 10111101	174 11110001	239 00010110
45 11000001	110 01100111	175 11111111	240 00101100
46 10011111	111 11001110	176 11100011	241 01011000
47 00100011	112 10000001	177 11011011	242 10110000
48 01000110	113 00011111	178 10101011	243 01111101
49 10001100	114 00111110	179 01001011	244 11111010
50 00000101	115 01111100	180 10010110	245 11101001
51 00001010	116 11111000	181 00110001	246 11001111
52 00010100	117 11101101	182 01100010	247 10000011
53 00101000	118 11000111	183 11000100	248 00011011
54 01010000	119 10010011	184 10010101	249 00110110
55 10100000	120 00111011	185 00110111	250 01101100
56 01011101	121 01110110	186 01101110	251 11011000
57 10111010	122 11101100	187 11011100	252 10101101
58 01101001	123 11000101	188 10100101	253 01000111
59 11010010	124 10010111	189 01010111	254 10001110
60 10111001	125 00110011	190 10101110	
61 01101111	126 01100110	191 01000001	GF(3^2)
62 11011110	127 11001100	192 10000010	0 01
63 10100001	128 10000101	193 00011001	1 10
64 01011111	129 00010111	194 00110010	2 21
65 10111110	130 00101110	195 01100100	3 22
66 01100001	131 01011100	196 11001000	4 02
67 11000010	132 10111000	197 10001101	5 20
68 10011001	133 01101101	198 00000111	6 12
69 00101111	134 11011010	199 00001110	7 11
70 01011110	135 10101001	200 00011100	
71 10111100	136 01001111	201 00111000	GF(3^3)
72 01100101	137 10011110	202 01110000	0 001
73 11001010	138 00100001	203 11100000	1 010
74 10001001	139 01000010	204 11011101	2 100
75 00001111	140 10000100	205 10100111	3 102
76 00011110	141 00010101	206 01010011	4 122
77 00111100	142 00101010	207 10100110	5 022
78 01111000	143 01010100	208 01010001	6 220
79 11110000	144 10101000	209 10100010	7 101
80 11111101	145 01001101	210 01011001	8 112
81 11100111	146 10011010	211 10110010	9 222
82 11010011	147 00101001	212 01111001	10 121
83 10111011	148 01010010	213 11110010	11 012
84 01101011	149 10100100	214 11111001	12 120
85 11010110	150 01010101	215 11101111	13 002
86 10110001	151 10101010	216 11000011	14 020
87 01111111	152 01001001	217 10011011	15 200
88 11111110	153 10010010	218 00101011	16 201
89 11100001	154 00111001	219 01010110	17 211
90 11011111	155 01110010	220 10101100	18 011
91 10100011	156 11100100	221 01000101	19 110
92 01011011	157 11010101	222 10001010	20 202
93 10110110	158 10110111	223 00001001	21 221
94 01110001	159 01110011	224 00010010	22 111
95 11100010	160 11100110	225 00100100	23 212
96 11011001	161 11010001	226 01001000	24 021
97 10101111	162 10111111	227 10010000	25 210
98 01000011	163 01100011	228 00111101	
99 10000110	164 11000110	229 01111010	GF(3^4)
100 00010001	165 10010001	230 11110100	0 0001
101 00100010	166 00111111	231 11110101	1 0010
102 01000100	167 01111110	232 11110111	2 0100
103 10001000	168 11111100	233 11110011	3 1000
104 00001101	169 11100101	234 11111011	4 2001
105 00011010	170 11010111	235 11101011	5 1012

6	2121
7	2212
8	0122
9	1220
10	1201
11	1011
12	2111
13	2112
14	2122
15	2222
16	0222
17	2220
18	0202
19	2020
20	1202
21	1021
22	2211
23	0112
24	1120
25	0201
26	2010
27	1102
28	0021
29	0210
30	2100
31	2002
32	1022
33	2221
34	0212
35	2120
36	2202
37	0022
38	0220
39	2200
40	0002
41	0020
42	0200
43	2000
44	1002
45	2021
46	1212
47	1121
48	0211
49	2110
50	2102
51	2022
52	1222
53	1221
54	1211
55	1111
56	0111
57	1110
58	0101
59	1010
60	2101
61	2012
62	1122
63	0221
64	2210
65	0102
66	1020
67	2201
68	0012
69	0120
70	1200
71	1001
72	2011
73	1112
74	0121
75	1210
76	1101
77	0011
78	0110
79	1100

$GF(3^5)$

0	00001
1	00010
2	00100
3	01000
4	10000
5	20202
6	12121
7	11112
8	01022
9	10220
10	22102
11	01121
12	11210
13	02002
14	20020
15	10001
16	20212
17	12221
18	12112
19	11022
20	00122
21	01220
22	12200
23	12202
24	12222
25	12122
26	11122
27	01122
28	11220
29	02102
30	21020
31	20001
32	10111
33	21012
34	20221
35	12011
36	10012
37	20022
38	10021
39	20112
40	11221
41	02112
42	21120
43	21001
44	20111
45	11211
46	02012
47	20120
48	11001
49	00212
50	02120
51	21200
52	22101
53	01111
54	11110
55	01002
56	10020
57	20102
58	11121
59	01112
60	11120
61	01102
62	11020
63	00102
64	01020
65	10200
66	22202
67	02121
68	21210
69	22201
70	02111
71	21110
72	21201
73	22111
74	01211
75	12110
76	11002
77	00222
78	02220
79	22200
80	02101
81	21010
82	20201
83	12111
84	11012
85	00022
86	00220
87	02200
88	22000
89	00101
90	01010
91	10100
92	21202
93	22121
94	01011
95	10110
96	21002
97	20121
98	11011
99	00012
100	00120
101	01200
102	12000
103	10202
104	22222
105	02021
106	20210
107	12201
108	12212
109	12022
110	10122
111	21122
112	21021
113	20011
114	10211
115	22012
116	00221
117	02210
118	22100
119	01101
120	11010
121	00002
122	00020
123	00200
124	02000
125	20000
126	10101
127	21212
128	22221
129	02011
130	20110
131	11201
132	02212
133	22120
134	01001
135	10010
136	20002
137	10121
138	21112
139	21221
140	22011
141	00211
142	02110
143	21100
144	21101
145	21111
146	21211
147	22211
148	02211
149	22110
150	01201
151	12010
152	10002
153	20222
154	12021
155	10112
156	21022
157	20021
158	10011
159	20012
160	10221
161	22112
162	01221
163	12210
164	12002
165	10222
166	22122
167	01021
168	10210
169	22002
170	00121
171	01210
172	12100
173	11202
174	02222
175	22220
176	02001
177	20010
178	10201
179	22212
180	02221
181	22210
182	02201
183	22010
184	00201
185	02010
186	20100
187	11101
188	01212
189	12120
190	11102
191	01222
192	12220
193	12102
194	11222
195	02122
196	21220
197	22001
198	00111
199	01110
200	11100
201	01202
202	12020
203	10102
204	21222
205	22021
206	00011
207	00110
208	01100
209	11000
210	00202
211	02020
212	20200
213	12101
214	11212
215	02022
216	20220
217	12001
218	10212
219	22022
220	00021
221	00210
222	02100
223	21000
224	20101
225	11111
226	01012
227	10120
228	21102
229	21121
230	21011
231	20211
232	12211
233	12012
234	10022
235	20122
236	11021
237	00112
238	01120
239	11200
240	02202
241	22020

GF(5^2)

0	01
1	10
2	43
3	42
4	32
5	44
6	02
7	20
8	31
9	34
10	14
11	33
12	04
13	40
14	12
15	13
16	23
17	11
18	03
19	30
20	24
21	21
22	41
23	22

GF(7^2)

0	01
1	10
2	64
3	53
4	56
5	16
6	54
7	66
8	03
9	30
10	45
11	12
12	14
13	34
14	15
15	44
16	02
17	20
18	51
19	36
20	35
21	25
22	31
23	55
24	06
25	60
26	13
27	24
28	21
29	61
30	23
31	11
32	04
33	40
34	32
35	65
36	63
37	43
38	62
39	33
40	05
41	50
42	26
43	41
44	42
45	52
46	46
47	22

GF(11^2)

0	01
1	10
2	A4
3	57
4	29
5	78
6	16
7	54
8	A9
9	A7
10	87
11	AA
12	07
13	70
14	46
15	25
16	38
17	51
18	79
19	26
20	48
21	45
22	15
23	44
24	05
25	50
26	69
27	32
28	A1
29	27
30	58
31	39
32	61
33	62
34	72
35	66
36	02
37	20
38	98
39	A3
40	47
41	35
42	21
43	A8
44	97
45	93
46	53
47	99
48	03
49	30
50	81
51	4A
52	65
53	A2
54	37
55	41
56	85
57	8A
58	2A
59	88
60	0A
61	A0
62	17
63	64
64	92
65	43
66	A5
67	67
68	12
69	14
70	34
71	11
72	04
73	40
74	75
75	96
76	83
77	6A
78	42
79	95
80	73
81	76
82	A6
83	77
84	06
85	60
86	52
87	89
88	1A
89	94
90	63
91	82
92	5A
93	59
94	49
95	55
96	09
97	90
98	23
99	18
100	74
101	86
102	9A
103	13
104	24
105	28
106	68
107	22
108	08
109	80
110	3A
111	71
112	56
113	19
114	84
115	7A
116	36
117	31
118	91
119	33

GF(13^2)

0	01
1	10
2	CB
3	C2
4	32
5	C7
6	82
7	7A
8	3C
9	97
10	B8
11	A4
12	76
13	CC
14	02
15	20
16	B9
17	B4
18	64
19	B1
20	34
21	17
22	6B
23	51
24	93
25	78
26	1C
27	BB
28	04
29	40
30	95
31	98
32	C8
33	92
34	68
35	21
36	C9
37	A2
38	56
39	13
40	2B
41	99
42	08
43	80
44	5A
45	53
46	B3
47	54
48	C3
49	42
50	B5
51	74
52	AC
53	26
54	49
55	55
56	03
57	30
58	A7
59	A6
60	96
61	A8
62	B6
63	84
64	9A
65	18
66	7B
67	4C
68	85
69	AA
70	06
71	60
72	71
73	7C
74	5C
75	73
76	9C
77	38
78	57
79	23
80	19
81	8B
82	3A
83	77
84	0C
85	C0
86	12
87	1B
88	AB
89	16
90	5B
91	63
92	A1
93	46
94	25
95	39
96	67
97	11
98	0B
99	B0
100	24
101	29
102	79
103	2C
104	A9
105	C6
106	72
107	8C
108	4A
109	65
110	C1
111	22
112	09
113	90
114	48
115	45
116	15
117	4B
118	75
119	BC
120	14
121	3B
122	87
123	CA
124	B2
125	44
126	05
127	50
128	83
129	8A
130	2A
131	89
132	1A
133	9B
134	28
135	69
136	31
137	B7
138	94
139	88
140	0A
141	A0
142	36
143	37
144	47
145	35
146	27
147	59
148	43
149	C5
150	62
151	91
152	58
153	33
154	07
155	70
156	6C
157	61
158	81
159	6A
160	41
161	A5
162	86
163	BA
164	C4
165	52
166	A3
167	66

Factorization of $x^n - 1$ over F_2

n	Factorization
1	$x+1$
3	$(x+1)(x^2+x+1)$
5	$(x+1)(x^4+x^3+x^2+x+1)$
7	$(x+1)(x^3+x+1)(x^3+x^2+1)$
9	$(x+1)(x^2+x+1)(x^6+x^3+1)$
11	$(x+1)(x^{10}+x^9+\cdots+1)$
13	$(x+1)(x^{12}+x^{11}+\cdots+1)$
15	$(x+1)(x^2+x+1)(x^4+x+1)(x^4+x^3+1)(x^4+x^3+x^2+x+1)$
17	$(x+1)(x^8+x^5+x^4+x^3+1)(x^8+x^7+x^6+x^4+x^2+x+1)$
19	$(x+1)(x^{18}+x^{17}+\cdots+1)$
21	$(x+1)(x^2+x+1)(x^3+x^2+1)(x^3+x+1)(x^6+x^4+x^2+x+1)\cdot$ $(x^6+x^5+x^4+x^2+1)$
23	$(x+1)(x^{11}+x^9+x^7+x^6+x^5+x+1)(x^{11}+x^{10}+x^6+x^5+x^4+x^2+1)$
25	$(x+1)(x^4+x^3+x^2+x+1)(x^{20}+x^{15}+x^{10}+x^5+1)$

Note Further factorizations can be obtained by observing that, over F_2,

$$x^{2^k n} - 1 = (x^n - 1)^{2^k}$$

The Krawtchouk Polynomials $K_k(i;n,2)$

n=4

k \ i	0	1	2	3	4
0	1	1	1	1	1
1	4	2	0	-2	-4
2	6	0	-2	0	6
3	4	-2	0	2	-4
4	1	-1	1	-1	1

n=5

k \ i	0	1	2	3	4	5
0	1	1	1	1	1	1
1	5	3	1	-1	-3	-5
2	10	2	-2	-2	2	10
3	10	-2	-2	2	2	-10
4	5	-3	1	1	-3	5
5	1	-1	1	-1	1	-1

n=6

k \ i	0	1	2	3	4	5	6
0	1	1	1	1	1	1	1
1	6	4	2	0	-2	-4	-6
2	15	5	-1	-3	-1	5	15
3	20	0	-4	0	4	0	-20
4	15	-5	-1	3	-1	-5	15
5	6	-4	2	0	-2	4	-6
6	1	-1	1	-1	1	-1	1

n=7

k \ i	0	1	2	3	4	5	6	7
0	1	1	1	1	1	1	1	1
1	7	5	3	1	-1	-3	-5	-7
2	21	9	1	-3	-3	1	9	21
3	35	5	-5	-3	3	5	-5	-35
4	35	-5	-5	3	3	-5	-5	35
5	21	-9	1	3	-3	-1	9	-21
6	7	-5	3	-1	-1	3	-5	7
7	1	-1	1	-1	1	-1	1	-1

n=8

k \ i	0	1	2	3	4	5	6	7	8
0	1	1	1	1	1	1	1	1	1
1	8	6	4	2	0	-2	-4	-6	-8
2	28	14	4	-2	-4	-2	4	14	28
3	56	14	-4	-6	0	6	4	-14	-56
4	70	0	-10	0	6	0	-10	0	70
5	56	-14	-4	6	0	-6	4	14	-56
6	28	-14	4	2	-4	2	4	-14	28
7	8	-6	4	-2	0	2	-4	6	-8
8	1	-1	1	-1	1	-1	1	-1	1

n=9

k \ i	0	1	2	3	4	5	6	7	8	9
0	1	1	1	1	1	1	1	1	1	1
1	9	7	5	3	1	-1	-3	-5	-7	-9
2	36	20	8	0	-4	-4	0	8	20	36
3	84	28	0	-8	-4	4	8	0	-28	-84
4	126	14	-14	-6	6	6	-6	-14	14	126
5	126	-14	-14	6	6	-6	-6	14	14	-126
6	84	-28	0	8	-4	-4	8	0	-28	84
7	36	-20	8	0	-4	4	0	-8	20	-36
8	9	-7	5	-3	1	1	-3	5	-7	9
9	1	-1	1	-1	1	-1	1	-1	1	-1

n=10

k \ i	0	1	2	3	4	5	6	7	8	9	10
0	1	1	1	1	1	1	1	1	1	1	1
1	10	8	6	4	2	0	-2	-4	-6	-8	-10
2	45	27	13	3	-3	-5	-3	3	13	27	45
3	120	48	8	-8	-8	0	8	8	-8	-48	-120
4	210	42	-14	-14	2	10	2	-14	-14	42	210
5	252	0	-28	0	12	0	-12	0	28	0	-252
6	210	-42	-14	14	2	-10	2	14	-14	-42	210
7	120	-48	8	8	-8	0	8	-8	-8	48	-120
8	45	-27	13	-3	-3	5	-3	-3	13	-27	45
9	10	-8	6	-4	2	0	-2	4	-6	8	-10
10	1	-1	1	-1	1	-1	1	-1	1	-1	1

n=11

k \ i	0	1	2	3	4	5	6	7	8	9	10	11
0	1	1	1	1	1	1	1	1	1	1	1	1
1	11	9	7	5	3	1	-1	-3	-5	-7	-9	-11
2	55	35	19	7	-1	-5	-5	-1	7	19	35	55
3	165	75	21	-5	-11	-5	5	11	5	-21	-75	-165
4	330	90	-6	-22	-6	10	10	-6	-22	-6	90	330
5	462	42	-42	-14	14	10	-10	-14	14	42	-42	-462
6	462	-42	-42	14	14	-10	-10	14	14	-42	-42	462
7	330	-90	-6	22	-6	-10	10	6	-22	6	90	-330
8	165	-75	21	5	-11	5	5	-11	5	21	-75	165
9	55	-35	19	-7	-1	5	-5	1	7	-19	35	-55
10	11	-9	7	-5	3	-1	-1	3	-5	7	-9	11
11	1	-1	1	-1	1	-1	1	-1	1	-1	1	-1

n=12

k \ i	0	1	2	3	4	5	6	7	8	9	10	11	12
0	1	1	1	1	1	1	1	1	1	1	1	1	1
1	12	10	8	6	4	2	0	-2	-4	-6	-8	-10	-12
2	66	44	26	12	2	-4	-6	-4	2	12	26	44	66
3	220	110	40	2	-12	-10	0	10	12	-2	-40	-110	-220
4	495	165	15	-27	-17	5	15	5	-17	-27	15	165	495
5	792	132	-48	-36	8	20	0	-20	-8	36	48	-132	-792
6	924	0	-84	0	28	0	-20	0	28	0	-84	0	924
7	792	-132	-48	36	8	-20	0	20	-8	-36	48	132	-792
8	495	-165	15	27	-17	-5	15	-5	-17	27	15	-165	495
9	220	-110	40	-2	-12	10	0	-10	12	2	-40	110	-220
10	66	-44	26	-12	2	4	-6	4	2	-12	26	-44	66
11	12	-10	8	-6	4	-2	0	2	-4	6	-8	10	-12
12	1	-1	1	-1	1	-1	1	-1	1	-1	1	-1	1

n=13

k \ i	0	1	2	3	4	5	6	7	8	9	10	11	12	13
0	1	1	1	1	1	1	1	1	1	1	1	1	1	1
1	13	11	9	7	5	3	1	-1	-3	-5	-7	-9	-11	-13
2	78	54	34	18	6	-2	-6	-6	-2	6	18	34	54	78
3	286	154	66	14	-10	-14	-6	6	14	10	-14	-66	-154	-286
4	715	275	55	-25	-29	-5	15	15	-5	-29	-25	55	275	715
5	1287	297	-33	-63	-9	25	15	-15	-25	9	63	33	-297	-1287
6	1716	132	-132	-36	36	20	-20	-20	20	36	-36	-132	132	1716
7	1716	-132	-132	36	36	-20	-20	20	20	-36	-36	132	132	-1716
8	1287	-297	-33	63	-9	-25	15	15	-25	-9	63	-33	-297	1287
9	715	-275	55	25	-29	5	15	-15	-5	29	-25	-55	275	-715
10	286	-154	66	-14	-10	14	-6	-6	14	-10	-14	66	-154	286
11	78	-54	34	-18	6	2	-6	6	-2	-6	18	-34	54	-78
12	13	-11	9	-7	5	-3	1	1	-3	5	-7	9	-11	13
13	1	-1	1	-1	1	-1	1	-1	1	-1	1	-1	1	-1

n=14

k \ i	0	1	2	3	4	5	6	7	8	9	10	11	12	13	14
0	1	1	1	1	1	1	1	1	1	1	1	1	1	1	1
1	14	12	10	8	6	4	2	0	-2	-4	-6	-8	-10	-12	-14
2	91	65	43	25	11	1	-5	-7	-5	1	11	25	43	65	91
3	364	208	100	32	-4	-16	-12	0	12	16	4	-32	-100	-208	-364
4	1001	429	121	-11	-39	-19	9	21	9	-19	-39	-11	121	429	1001
5	2002	572	22	-88	-38	20	30	0	-30	-20	38	88	-22	-572	-2002
6	3003	429	-165	-99	27	45	-5	-35	-5	45	27	-99	-165	429	3003
7	3432	0	-264	0	72	0	-40	0	40	0	-72	0	264	0	-3432
8	3003	-429	-165	99	27	-45	-5	35	-5	-45	27	99	-165	-429	3003
9	2002	-572	22	88	-38	-20	30	0	-30	20	38	-88	-22	572	-2002
10	1001	-429	121	11	-39	19	9	-21	9	19	-39	11	121	-429	1001
11	364	-208	100	-32	-4	16	-12	0	12	-16	4	32	-100	208	-364
12	91	-65	43	-25	11	-1	-5	7	-5	-1	11	-25	43	-65	91
13	14	-12	10	-8	6	-4	2	0	-2	4	-6	8	-10	12	-14
14	1	-1	1	-1	1	-1	1	-1	1	-1	1	-1	1	-1	1

n=15

k \ i	0	1	2	3	4	5	6	7	8	9	10	11	12	13	14	15
0	1	1	1	1	1	1	1	1	1	1	1	1	1	1	1	1
1	15	13	11	9	7	5	3	1	-1	-3	-5	-7	-9	-11	-13	-15
2	105	77	53	33	17	5	-3	-7	-7	-3	5	17	33	53	77	105
3	455	273	143	57	7	-15	-17	-7	7	17	15	-7	-57	-143	-273	-455
4	1365	637	221	21	-43	-35	-3	21	21	-3	-35	-43	21	221	637	1365
5	3003	1001	143	-99	-77	1	39	21	-21	-39	-1	77	99	-143	-1001	-3003
6	5005	1001	-143	-187	-11	65	25	-35	-35	25	65	-11	-187	-143	1001	5005
7	6435	429	-429	-99	99	45	-45	-35	35	45	-45	-99	99	429	-429	-6435
8	6435	-429	-429	99	99	-45	-45	35	35	-45	-45	99	99	-429	-429	6435
9	5005	-1001	-143	187	-11	-65	25	35	-35	-25	65	11	-187	143	1001	-5005
10	3003	-1001	143	99	-77	-1	39	-21	-21	39	-1	-77	99	143	-1001	3003
11	1365	-637	221	-21	-43	35	-3	-21	21	3	-35	43	21	-221	637	-1365
12	455	-273	143	-57	7	15	-17	7	7	-17	15	7	-57	143	-273	455
13	105	-77	53	-33	17	-5	-3	7	-7	3	5	-17	33	-53	77	-105
14	15	-13	11	-9	7	-5	3	-1	-1	3	-5	7	-9	11	-13	15
15	1	-1	1	-1	1	-1	1	-1	1	-1	1	-1	1	-1	1	-1

References

Books on Information and Coding Theory

Abramson, Norman, *Information Theory and Coding*, McGraw-Hill, 1963.

Adámek, Jiří, *Foundations of Coding*, John-Wiley and Sons, 1991.

Ash, Robert, *Information Theory*, Dover Publications, 1965.

Berlekamp, Elwyn, *Algebraic Coding Theory*, Agean Park Press, 1984.

Berlekamp, Elwyn, ed., *Key Papers in the Development of Coding Theory*, IEEE Press, 1974.

Blahut, Richard, *Principles and Practice of Information Theory*, Addison-Wesley, 1987.

Blahut, Richard, *Theory and Practice of Error Control Codes*, Addison-Wesley, 1983.

Blake, Ian and Mullin, Ronald, *An Introduction to Algebraic and Combinatorial Coding Theory*, Academic Press, 1976.

Feinstein, Amiel, *Foundations of Information Theory*, McGraw-Hill, 1958.

Gallagher, Robert, *Information Theory and Reliable Communication*, John Wiley and Sons, 1968.

Hamming, Richard, *Coding and Information Theory*, Second Edition, Prentice-Hall, 1986.

Hill, Raymond, *A First Course in Coding Theory*, Clarendon Press, Oxford, 1986.

Jones, D.S., *Elementary Information Theory*, Clarendon Press, Oxford, 1979.

Lin, Shu and Costello, Daniel, *Error Control Coding*, Prentice-Hall, 1983.

MacWilliams, F.J. and Sloane, N.J.A., *The Theory of Error-Correcting Codes*, North-Holland, 1977.

Mansuripur, Masud, *Introduction to Information Theory*, Prentice-Hall, 1987.

McEliece, Robert, *The Theory of Information and Coding Theory*, Cambridge University Press, 1977.

Peterson, W. and Weldon, E.J., *Error-Correcting Codes*, MIT Press, 1972.

Pless, Vera, *Introduction to the Theory of Error-Correcting Codes*, Second Edition, John-Wiley and Sons, 1989.

Slepian, David, ed., *Key Papers in the Development of Information Theory*, IEEE Press, 1974.

Thompson, Thomas, *From Error-Correcting Codes Through Sphere-Packing to Simple Groups*, Carus Mathematical Monographs, Mathematical Association of America, 1983.

van Lint, J.H., *Introduction to Coding Theory*, Springer-Verlag, 1982.

Vanstone, Scott and van Oorschot, Paul, *An Introduction to Error Correcting Codes with Applications*, Kluwer Academic Publishers, 1989.

Additional References

Anderson, Ian, *Combinatorial Designs*, Ellis Horwood Limited, John Wiley, 1990.

Bose, R.C. and Ray-Chaudhuri, D.K., On a class of error correcting group codes, *Info. and Control*, 3 (1960) 68-79.

Bose, R.C. and Ray-Chaudhuri, D.K., Further results on error correcting binary group codes, *Info. and Control*, 3 (1960) 279-290.

Delsarte, P., Bounds for unrestricted codes, by linear programming, *Philips Res. Reports*, 27 (1972) 272-289.

Delsarte, P., Four fundamental parameters of a code and their combinatorial significance, *Info. and Control*, 23 (1973) 407-438.

Delsarte, P., An algebraic approach to the association schemes of coding theory, *Philips Res. Reports Supplements No. 10*, (1973).

Delsarte, P. and Goethals, J.M., Unrestricted codes with the Golay parameters are unique, *Disctere Math.*, 12 (1975) 211-224.

Gleason, A.M., Weight polynomials of self-dual codes and the MacWilliams identity, in *Actes Congres Internl. de Mathematiques*, 3 (1970) (Gauthiers-Villars, Paris, 1971) 211-215.

Hocquenghem, A., Codes correcteurs d'erreurs, *Chiffres* (Paris) 2 (1959) 147-156.

Justesen, J., A class of constructive asymptotically good algebraic

codes, *IEEE Trans. Info. Theory*, 18 (1972) 652-656.

Kalbfleisch, J.G. and Stanton, R.G., Maximal and minimal coverings of (k−1)-tuples by k-tuples, *Pacific J. Math.*, 26 (1968) 131-140.

Lidl, Rudolf and Niederreiter, Harald, *Introduction to Finite Fields and Their Applications*, Cambridge University Press, 1986.

Lindström, K., The nonexistence of unknown nearly perfect binary codes, *Ann. Univ. Turku., Ser. A., No. 169*, (1975) 3-28.

MacWilliams, F.J., Sloane, N.J.A. and Thompson, J.G., Good self-dual codes exist, *Discrete Math.*, 3 (1972) 153-162.

Meggitt, J.E., Error correcting codes for correcting bursts of errors, *IBM Res. Devel.*, 4 (1960) 329-334.

Pless, Vera, On the uniqueness of the Golay codes, *J.Comb. Theory*, 5 (1968) 215-228.

Pless, V. and Pierce, N.J., Self-dual codes over GF(q) satisfy a modified Varshamov bound, *Info. and Control*, 23 (1973) 35-40.

Roman, Steven, *The Umbral Calculus*, Academic Press, 1984.

Rota, G.-C., On the foundations of combinatorial theory I: Theory of Möbius functions, *Z. Wahrscheinlichkeitstheorie* 2 (1964) 340-368.

Schönheim, J., On maximal systems of k-tuples, *Stud. Sci. Math. Hungar.*, 1 (1966) 363-368.

Shannon, C.E., A mathematical theory of communication, *Bell Syst. Tech. J.*, 27 (1948) 379-423 and 623-656. (See also Slepian, ed. 1974.)

Sloane, N.J.A., Error-correcting codes and invariant theory: new applications of a nineteenth-century technique, *Amer. Math. Monthly*, 84 (1977) 82-107.

Tietäväinen, A., On the nonexistence of perfect codes over finite fields, *SIAM J. Appl. Math.*, 24 (1973) 88-96.

Vasil'ev, J.L., On nongroup close-packed codes (in Russian), *Probl. Kibernet.*, 8 (1962) 337-339, translated in *Probleme der Kybernitek* 8 (1965) 375-378.

Symbol Index

Index

Graduate Texts in Mathematics

continued from page ii

119 ROTMAN. An Introduction to Algebraic Topology.
120 ZIEMER. Weakly Differentiable Functions: Sobolev Spaces and Functions of Bounded Variation.
121 LANG. Cyclotomic Fields I and II. Combined 2nd ed.
122 REMMERT. Theory of Complex Functions. *Readings in Mathematics*
123 EBBINGHAUS/HERMES et al. Numbers. *Readings in Mathematics*
124 DUBROVIN/FOMENKO/NOVIKOV. Modern Geometry—Methods and Applications. Part III.
125 BERENSTEIN/GAY. Complex Variables: An Introduction.
126 BOREL. Linear Algebraic Groups. 2nd ed.
127 MASSEY. A Basic Course in Algebraic Topology.
128 RAUCH. Partial Differential Equations.
129 FULTON/HARRIS. Representation Theory: A First Course. *Readings in Mathematics*
130 DODSON/POSTON. Tensor Geometry.
131 LAM. A First Course in Noncommutative Rings.
132 BEARDON. Iteration of Rational Functions.
133 HARRIS. Algebraic Geometry: A First Course.
134 ROMAN. Coding and Information Theory.
135 ROMAN. Advanced Linear Algebra.
136 ADKINS/WEINTRAUB. Algebra: An Approach via Module Theory.
137 AXLER/BOURDON/RAMEY. Harmonic Function Theory.
138 COHEN. A Course in Computational Algebraic Number Theory.
139 BREDON. Topology and Geometry.
140 AUBIN. Optima and Equilibria. An Introduction to Nonlinear Analysis.
141 BECKER/WEISPFENNING/KREDEL. Gröbner Bases. A Computational Approach to Commutative Algebra.
142 LANG. Real and Functional Analysis. 3rd ed.
143 DOOB. Measure Theory.
144 DENNIS/FARB. Noncommutative Algebra.
145 VICK. Homology Theory. An Introduction to Algebraic Topology. 2nd ed.
146 BRIDGES. Computability: A Mathematical Sketchbook.
147 ROSENBERG. Algebraic K-Theory and Its Applications.
148 ROTMAN. An Introduction to the Theory of Groups. 4th ed.
149 RATCLIFFE. Foundations of Hyperbolic Manifolds.
150 EISENBUD. Commutative Algebra with a View Toward Algebraic Geometry.
151 SILVERMAN. Advanced Topics in the Arithmetic of Elliptic Curves.
152 ZIEGLER. Lectures on Polytopes.
153 FULTON. Algebraic Topology: A First Course.
154 BROWN/PEARCY. An Introduction to Analysis.
155 KASSEL. Quantum Groups.
156 KECHRIS. Classical Descriptive Set Theory.
157 MALLIAVIN. Integration and Probability.
158 ROMAN. Field Theory.
159 CONWAY. Functions of One Complex Variable II.
160 LANG. Differential and Riemannian Manifolds.
161 BORWEIN/ERDÉLYI. Polynomials and Polynomial Inequalities.
162 ALPERIN/BELL. Groups and Representations.
163 DIXON/MORTIMER. Permutation Groups.
164 NATHANSON. Additive Number Theory: The Classical Bases.
165 NATHANSON. Additive Number Theory: Inverse Problems and the Geometry of Sumsets.
166 SHARPE. Differential Geometry: Cartan's Generalization of Klein's Erlangen Program.
167 MORANDI. Field and Galois Theory.
168 EWALD. Combinatorial Convexity and Algebraic Geometry.
169 BHATIA. Matrix Analysis.
170 BREDON. Sheaf Theory. 2nd ed.
171 PETERSEN. Riemannian Geometry.
172 REMMERT. Classical Topics in Complex Function Theory.
173 DIESTEL. Graph Theory.
174 BRIDGES. Foundations of Real and Abstract Analysis.
175 LICKORISH. An Introduction to Knot Theory.
176 LEE. Riemannian Manifolds.
177 NEWMAN. Analytic Number Theory.
178 CLARKE/LEDYAEV/STERN/WOLENSKI. Nonsmooth Analysis and Control Theory.

9 780387 978123